T0210998

Lecture Notes in Artificial Intelligence   10412

Subseries of Lecture Notes in Computer Science

More information about this series at http://www.springer.com/series/1244

Gang Li · Yong Ge · Zili Zhang
Zhi Jin · Michael Blumenstein (Eds.)

Knowledge Science, Engineering and Management

10th International Conference, KSEM 2017
Melbourne, VIC, Australia, August 19–20, 2017
Proceedings

Editors
Gang Li (ID)
Deakin University
Burwood, VIC
Australia

Yong Ge
University of Arizona
Tucson, AZ
USA

Zili Zhang
Southwest University
Chongqing
China

Zhi Jin
Peking University
Beijing
China

Michael Blumenstein
University of Technology Sydney
Sydney, NSW
Australia

ISSN 0302-9743   ISSN 1611-3349   (electronic)
Lecture Notes in Artificial Intelligence
ISBN 978-3-319-63557-6   ISBN 978-3-319-63558-3   (eBook)
DOI 10.1007/978-3-319-63558-3

Library of Congress Control Number: 2017946695

LNCS Sublibrary: SL7 – Artificial Intelligence

Printed on acid-free paper

This Springer imprint is published by Springer Nature
The registered company is Springer International Publishing AG
The registered company address is: Gewerbestrasse 11, 6330 Cham, Switzerland

Preface

The International Conference on Knowledge Science, Engineering and Management (KSEM) provides a forum for researchers in the broad areas of knowledge science, knowledge engineering, and knowledge management to exchange ideas and to report state-of-the-art research results. KSEM 2017 was the tenth in this series, building on the success of nine events held in: Guilin, China (KSEM 2006); Melbourne, Australia (KSEM 2007); Vienna, Austria (KSEM 2009); Belfast, UK (KSEM 2010); Irvine, USA (KSEM 2011); Dalian, China (KSEM 2013); Sibiu, Romania (KSEM 2014); Chongqing, China (KSEM 2015); and Passau, Germany (KSEM 2016).

The selection process this year was competitive. KSEM 2017 received 134 submissions, and each submitted paper was reviewed by at least three members of the Program Committee (PC). Following this independent review, there were discussions among reviewers and PC chairs. A total of 35 papers were selected as full papers (26.1%), and another 12 papers were selected as short papers (8.9%), yielding a combined acceptance rate of 35%. Moreover, we were honored to have two prestigious scholars giving keynote speeches at the conference: Prof. Zhi-Hua Zhou (Nanjing University, China) and Prof. Geoff Webb (Monash University, Australia). The abstract of Prof. Webb's talk is included in this volume.

We would like to thank everyone who participated in the development of the KSEM 2017 program. In particular, we would give special thanks to the PC, for their diligence and concern for the quality of the program, and also for their detailed feedback to the authors. The general organization of the conference also relied on the efforts of the KSEM 2017 Organizing Committee. We especially thank Dr. Shaowu Liu (University of Technology Sydney, Australia) and Dr. Huy Quan Vu (Victoria University, Australia) for the general administrative issues and for maintaining the conference website.

Moreover, we would like to express our gratitude to the KSEM Steering Committee chair, Prof. Hui Xiong (Rutgers University, USA), as well as KSEM 2017 general co-chairs, Prof. Michael Blumenstein (University of Technology Sydney, Australia) Prof. Zili Zhang (Southwest University, China) and Prof. Zhi Jin (Peking University, China). We are also grateful to the team at Springer led by Alfred Hofmann for the publication of this volume.

Finally and most importantly, we thank all the authors, who are the primary reason why KSEM 2017 is so exciting and why it is the premier forum for presentation and discussion of innovative ideas, research results, and experience from around the world.

June 2017
                                                                    Gang Li
                                                                    Yong Ge

Organization

KSEM 2017 was co-located with the flagship international AI conference, IJCAI 2017, in Melbourne, Australia, during August 19–20, 2017.

KSEM Steering Committee

Steering Committee

David Bell	Queen's University Belfast, UK
Yaxin Bi	Ulster University, Belfast, UK
Cungen Cao	Chinese Academy of Sciences, China
Zhi Jin	Peking University, China
Dimitris Karagiannis	Deputy Chair, University of Vienna, Austria
Claudiu Kifor	Lucian Blaga University of Sibiu, Romania
Jérome Lang	University Paul Sabatier, France
Ruqian Lu (Honorary Chair)	Chinese Academy of Sciences, China
Yoshiteru Nakamori	JAIST, Japan
Jorg Siekmann	German Research Centre of Artificial Intelligence, Germany
Eric Tsui	The Hong Kong Polytechnic University, Hong Kong, SAR China
Zongtuo Wang	Dalian Science and Technology University, China
Kwok Kee Wei	City University of Hong Kong, Hong Kong, SAR China
Martin Wirsing	Ludwig-Maximilians-Universität München, Germany
Hui Xiong (Chair)	The State University of New Jersey, Rutgers, USA
Mingsheng Ying	Tsinghua University, China
Chengqi Zhang	Past Chair, University of Technology, Sydney, Australia
Zili Zhang	Southwest University, China

KSEM 2017 Organizing Committee

General Co-chairs

Michael Blumenstein	University of Technology, Sydney, Australia
Zili Zhang	Southwest University, China
Zhi Jin	Peking University, China

Program Co-chairs

Gang Li	Deakin University, Australia
Yong Ge	University of Arizona, USA

Publicity Co-chairs

Elena-Teodora Miron   University of Vienna, Austria
Ge Li   Peking University, China

Organization Co-chairs

Shaowu Liu   University of Technology Sydney, Australia
Huy Quan Vu   Victoria University, Australia

KSEM 2017 Program Committee

Andreas Albrecht   Middlesex University, UK
Klaus-Dieter Althoff   DFKI/University of Hildesheim, Germany
Serge Autexier   DFKI, Germany
Costin Badica   University of Craiova, Romania
Salem Benferhat   Université d'Artois, France
Philippe Besnard   CNRS/IRIT, France
Remus Brad   Lucian Blaga University of Sibiu, Romania
Krysia Broda   Imperial College, UK
Robert Andrei   Babeş-Bolyai University, Romania
  Buchmann
Enhong Chen   University of Science and Technology of China, China
Paolo Ciancarini   University of Bologna, Italy
Ireneusz Czarnowski   Gdynia Maritime University, Poland
Richard Dapoigny   LISTIC/Polytech Savoie, France
Yong Deng   Southwest University, China
Juan Manuel Dodero   Universidad de Cádiz, Spain
Josep Domenech   Universitat Politècnica de València, Spain
Josep Domingo-Ferrer   Universitat Rovira i Virgili, Spain
Susan Elias   VIT University Chennai Campus, India
Dieter Fensel   University of Innsbruck, Austria
Hans-Georg Fill   University of Bamberg, Germany
Yanjie Fu   Missouri University of Science and Technology, USA
Fausto Giunchiglia   DISI, University of Trento, Italy
Vijayabharadwaj Gsr   VIT University Chennai Campus, India
Jiaxin Han   Xi'an Shiyou University, China
Ming He   University of Science and Technology of China, China
Knut Hinkelmann   FHNW University of Applied Sciences and Arts
                               Northwestern Switzerland, Switzerland
Zhisheng Huang   Vrije University Amsterdam, The Netherlands
Van Nam Huynh   JAIST, Japan
Tan Jianlong   Institute of Information Engineering, Chinese Academy
                               of Sciences, China
Fang Jin   Texas Tech University, USA
Mouna Kamel   IRIT, Université Paul Sabatier, Toulouse, France
Claudiu Kifor   Lucian Blaga University of Sibiu, Romania

Tong Xu	University of Science and Technology of China, China
Yuemei Xu	Beijing Foreign Studies University, China
Yating Yang	Xinjiang Technical Institute of Physics and Chemistry, Chinese Academy of Sciences, China
Jingyuan Yang	Rutgers University, USA
Feng Yi	Institute of Information Engineering, Chinese Academy of Sciences, China
Qingtian Zeng	Shandong University of Science and Technology, China
Chunxia Zhang	Beijing Institute of Technology, China
Le Zhang	Southwest University, China
Songmao Zhang	Chinese Academy of Sciences, China
Hongke Zhao	University of Science and Technology of China, China
Hao Zhong	Rutgers University, USA
Shuigeng Zhou	Fudan University, China
Yan Ziqi	Beijing Jiaotong University, China
Jiali Zuo	Jiangxi Normal University, China

KSEM 2017 External Reviewers

Zaenal Akbar	STI Innsbruck, Institute of Computer Science, University of Innsbruck, Austria
Nikolina Bader	University of Bamberg, Germany
Boran Taylan Balci	STI Innsbruck, Institute of Computer Science, University of Innsbruck, Austria
Kien Do	Deakin University, Australia
Elias Kärle	STI Innsbruck, Institute of Computer Science, University of Innsbruck, Austria
Jesús Manjón	Universitat Rovira i Virgili, Spain
Sergio Martínez	Universitat Rovira i Virgili, Spain
Thanh-Dat Nguyen	Lucian Blaga University of Sibiu, Romania and Quy Nhon University, Vietnam
Sergiu Nicolaescu	Lucian Blaga University of Sibiu, Romania
Oleksandra Panasiuk	STI Innsbruck, Institute of Computer Science, University of Innsbruck, Austria
Javier Parra	Universitat Rovira i Virgili, Spain
Umutcan Simsek	STI Innsbruck, Institute of Computer Science, University of Innsbruck, Austria
Andreas Steffan	University of Bamberg, Germany
Luis Miguel del Vasto	Universitat Rovira i Virgili, Spain
Shuai Zheng	Centers for Disease Control and Prevention, USA

Learning from Non-stationary Distributions
(Invited Speech)

Geoff Webb

Monash University, Clayton, Australia
geoff.webb@monash.edu

Abstract. The world is dynamic - in a constant state of flux - but most learned models are static. Models learned from historical data are likely to decline in accuracy over time. This talk presents theoretical tools for analyzing non-stationary distributions and some insights that they provide. Shortcomings of standard approaches to learning from non-stationary distributions are discussed together with strategies for developing more effective techniques.

Keywords: Non-stationary distributions Learning

Contents

Knowledge Management

Knowledge Integration

Knowledge Retrieval

Recommendation Algorithms and Systems

Text Mining and Document Analysis

Learning Sparse Overcomplete Word Vectors Without Intermediate Dense Representations

Yunchuan Chen[1,2], Ge Li[1,3(✉)], and Zhi Jin[1,3(✉)]

[1] Key Laboratory of High Confidence Software Technologies,
Peking University, Beijing, China
chenyunchuan11@mails.ucas.ac.cn, {lige,zhijin}@pku.edu.cn
[2] University of Chinese Academy of Sciences, Beijing, China
[3] Institute of Software, Peking University, Beijing, China

Abstract. Dense word representation models have attracted a lot of interest for their promising performances in various natural language processing (NLP) tasks. However, dense word vectors are uninterpretable, inseparable, and time and space consuming. We propose a model to learn sparse word representations directly from the plain text, rather than most existing methods that learn sparse vectors from intermediate dense word embeddings. Additionally, we design an efficient algorithm based on noise-contrastive estimation (NCE) to train the model. Moreover, a clustering-based adaptive updating scheme for noise distributions is introduced for effective learning when NCE is applied. Experimental results show that the resulting sparse word vectors are comparable to dense vectors on the word analogy tasks. Our models outperform dense word vectors on the word similarity tasks. The sparse word vectors are much more interpretable, according to the sparse vector visualization and the word intruder identification experiments.

1 Introduction

Word representation learning is aimed at associating each word with a syntactically and semantically rich feature vector. The learned word vectors could serve as input features for higher-level algorithms in NLP applications. Based on the distributional hypothesis [11], a variety of methods have been proposed in the NLP community, such as clustering-based methods [1], matrix-based methods [13,15], and neural network-based methods [3,17]. Recently, neural network-based methods have dominated word representation learning because of their effectiveness in a variety of natural language processing (NLP) tasks, such as part-of-speech tagging, semantic role labeling [3], parsing [26], sentiment analysis [27], language modeling [28], paraphrase detection [5] and dialogue analysis [12]. Neural network-based methods often represent words with dense, real-valued vectors.

However, dense representations are often criticized on their interpretability, separability, and complexity aspects. For neural network induced word vectors,

© Springer International Publishing AG 2017
G. Li et al. (Eds.): KSEM 2017, LNAI 10412, pp. 3–15, 2017.
DOI: 10.1007/978-3-319-63558-3_1

we do not know what feature is represented by each dimension, which corresponds to an implicit feature. There are complex dependencies between these underlying implicit features and human interpretable features such as *noun*, *adjective*, *plural* or *singular*, etc. It is therefore difficult to understand how a word vector-based computational model works. In addition, dense word vectors go against some widely believed properties a good high-level representation should have. For example, it is thought features should be represented in a separable way, or related to each other through simple dependencies [8], but dense word vectors are generally not separable. In addition, it usually results in high time or space complexity models when applied in downstream NLP tasks.

Sparse representation is considered as a potential choice for interpretable word representations and can be used to design time and space efficient algorithms for downstream NLP tasks. In the image, speech, or signal processing field, sparse overcomplete representations have been widely used as a way to improve separability and interpretability [4,21,25], and to increase stability in the presence of noise [4]. In NLP, sparsity constraints are useful in various applications, such as POS-tagging [30], dependency parsing [16] and document classification [32]. It has been shown that imposing sparse Dirichlet priors in Latent Dirichlet Allocation (LDA) is useful for downstream NLP tasks like POS-tagging [30], and improves interpretability [23]. Experiments show that the gathered descriptions for a given word are typically limited to approximately 20–30 features in norming studies [31].

In this paper, we propose a new, principled sparse representation method that learns sparse overcomplete word representations directly from the raw unlabeled text. We also design an easy-to-parallelize algorithm, which is based on noise-contrastive estimation (NCE) to train our proposed model. Additionally, we propose a clustering-based adaptive updating scheme for noise distributions used by NCE for effective learning. This updating scheme makes the noise distributions approximate the data distribution, and thus pushes the learning improves with fewer noise samples.

We evaluate our model on the word analogy, word similarity, and word intrusion tasks. The first two tasks are used to examine the expressive power of the learned representations and the last task is for interpretability. On the word analogy task, the results show that our proposed sparse model can achieve competitive performance with the state-of-the-art models under the same settings. On the word similarity task, the proposed model outperforms the competitors. For the interpretability, experimental results demonstrate that the sparse word vectors are much more interpretable.

2   Related Work

Learning sparse word vectors is booming along with dense vector representations. We put the existing sparse word vector learning methods into two categories: matrix-based methods and neural network-based methods.

The matrix-based methods can be divided into two steps. The first step is to construct a co-occurrence matrix M of size $V \times C$, where V is the vocabulary size and C is the context size. The w-th row $M_{w,:}$ is the initial representation of the w-th word. The second step is to apply a dimension reduction method to map M to a sparse matrix M' of size $V \times d$, where $d \ll C$. For example, Murphy et al. (2012) improved the interpretability of word vectors by introducing sparse and non-negative constrains into matrix factorization [20]. Levy and Goldberg (2014) showed that the sparse word vectors of word-context co-occurrence PPMI statistics also possess linguistic regularities that present in dense neural embeddings [14].

The neural network-based methods usually transforms dense neural embeddings into sparse ones. These methods generally require two steps of learning procedures. The first step is to train an embedding model, such as CBoW, SKIPGRAM [17] and GLOVE [24] to obtain dense feature vectors of the words in the vocabulary. The second step is to learn the sparse representation of each word by fixing the dense word embeddings. For example, Faruqui et al. (2015) transformed dense word vectors into sparse representations using dictionary learning method and showed the resulting sparse vectors are more similar to the interpretable features typically used in NLP [6]. Chen et al. (2016) proposed to use sparse linear combination of common words to represent uncommon ones, which results in sparse representation of words that is effective of compressing neural language models.

In summary, it is tricky to construct the co-occurrence matrix and design the dimension reduction algorithm to obtain good sparse word vectors. For neural network-based methods, the two-step pipelines may lose the sparse structure of words. This is because the prior that each word has a sparse structure is not imposed to learn dense embeddings, which could potentially lose the information for the sparse vector learning. Hence, it is preferred to learn sparse vectors without intermediate dense word embeddings like the method Sun et al. (2016) proposed [29]. The method Sun et al. proposed does not learn overcomplete sparse word vectors.

3   Our Model

In this section, we will talk about a model that try to discover the fundamental elements that constitute each word. We call these fundamental elements *word atoms*. Word atoms and words are analogs to atoms and molecules in Chemistry, respectively. The types of atoms are very small, but they can make up a huge number of different molecules. Likewise, we expect the limited word atoms could represent a large number of words in the vocabulary. A word atom can be regarded as an indivisible semantic or syntactic object.

The design philosophy of our model is similar to that of SKIPGRAM, i.e., "good representations result in good performances to predict context words". In addition, we assume that each word is composed of a few *word atoms*. In detail, each word is assumed to be represented by a sparse, linear combination of word atoms' vectors. This assumption is similar to but different from Chen et al.'s [2], which assume that each uncommon word is represented by a sparse, linear combination of the common ones. A word should not have too many semantic or syntactic components, so the sparseness assumption is reasonable.

Before introducing the mathematical model of representing words, we describe briefly the denotations. Let the vocabulary $\mathcal{V} = \{w_1, w_2, \ldots, w_n\}$, contexts $\mathcal{C} = \{c_1, c_2, \ldots, c_{n'}\}$. Each column of $\boldsymbol{B} \in \mathbb{R}^{d \times n_b}$, and $\boldsymbol{C} \in \mathbb{R}^{d \times n_c}$ are word and context atoms, respectively.[1] n_b and n_c are the number of word and context atoms, respectively; d is the dimension of atom vectors. For any given word $w \in \mathcal{V}$, its vector representation is $\boldsymbol{w} = \boldsymbol{B}\boldsymbol{\alpha}$, where $\boldsymbol{\alpha}$, called the sparse representation of w, is the coefficients that are used to combine word atoms to make up the word. Similarly, for a context $c \in \mathcal{C}$, its vector representation is $\boldsymbol{c} = \boldsymbol{C}\boldsymbol{\beta}$, where $\boldsymbol{\beta}$ is the sparse representation of c.

We think good word representations are helpful to predict a word's surrounding context words. The *softmax* model is used to model the distribution of a word's surrounding contexts.

$$p(c \mid w) = \frac{\exp(\boldsymbol{w}^\top \boldsymbol{c})}{\sum_{c' \in \mathcal{C}} \exp(\boldsymbol{w}^\top \boldsymbol{c}')} = \frac{\exp(\boldsymbol{\alpha}^\top \boldsymbol{B}^\top \boldsymbol{C}\boldsymbol{\beta})}{\sum_j \exp(\boldsymbol{\alpha}^\top \boldsymbol{B}^\top \boldsymbol{C}\boldsymbol{\beta}_j)}. \tag{1}$$

The word-context pair (w, c) is drawing from plain text. Concretely, for input plain text that consists of N words $w_1, w_2, \ldots, w_N$, the word-context pair (w_i, w_j) is drawn such that $|j - i| < \ell/2$, where ℓ is the window size.

It is difficult to train model (1) with the maximum likelihood estimation because of the difficulty of computing the normalization constant (a.k.a. partition function) for each word. In the literature, there are several methods to confront the partition function of a *single distribution*, such as MCMC-based algorithms, pseudo-likelihood, (denoising) score matching, Noise-Contrastive Estimation (NCE) [10]. But not all of these methods can be applied to discrete-input models like (1).

3.1 Parameter Estimation

We will adopt NCE to train model (1). The basic idea of NCE is to train a logistic regression classifier to discriminate between samples from the data distribution and samples from some "noise" distributions. It is a parameter estimation technique that is asymptotically unbiased and is suitable to estimate the parameters of a model with few number of random variables [8]. And it is also applicable to discrete-input models. One issue to apply NCE to train model (1) is that our model is a series of distributions that share the same parameters, which does not accommodate to NCE's setting. Following the work using NCE to train neural language models [2,19] and word embeddings [18], we define the training objective as the expectation of all distributions' NCE objective functions.

---

[1] The same as a word is composed of word atoms, we also assume that a context is composed of context atoms. In this paper, we will use surrounding words as contexts and thus $\mathcal{V} = \mathcal{C}$. The number of word and context atoms are also set to be equal, i.e., $n_b = n_c$.

In our situation, where the number of estimated conditional distributions is fairly small, we could learn a parameter corresponds to the partition function of each conditional distribution following the standard procedures NCE suggested to handle unnormalized probabilities [10]. Denote these parameters as a vector $z = (z_1, z_2, \ldots, z_V)$. Suppose to draw k negative instances per positive instance. Taking the sparseness requirement on the parameter α and β into consideration, the resulting parameter estimation model for (1) is

$$\arg \max_{\theta} J(\theta) - \lambda h(S_\alpha, S_\beta), \tag{2}$$

where $\theta = \{B, C, S_\alpha, S_\beta, z\}$, $S_\alpha = (\alpha_1, \alpha_2, \ldots, \alpha_V)$, $S_\beta = (\beta_1, \beta_2, \ldots, \beta_V)$ and

$$J(\theta) = \sum_{(w_i, c_i, s_i) \in \mathcal{D}} \ln \left(\frac{1 - s_i}{2} + s_i \sigma_k(\alpha_i^\top B^\top C \beta_i + z_{w_i} - \ln p_n(w_i)) \right),$$

$$h(S_\alpha, S_\beta) = \sum_{j=1}^{|V|} (\|\alpha_j\|_1 + \|\beta_j\|_1),$$

where $\sigma_k(x) = 1/(1 + k \cdot \exp(-x))$ is the logistic function parameterized by k; z_{w_i} is a parameter corresponds to the partition function of distribution $p(\cdot \mid w_i)$; $s_i = +1$ and -1 is a variable indicates whether the corresponding instance is extracted from the corpus (namely, positive instances) or drawn from a noise distribution (namely, negative instances); $\mathcal{D}$ is the training dataset, including positive and negative instances; λ is a hyperparameter used to control the degree of sparseness of S_α and S_β.

The first term of the optimization problem (2), i.e., $J(\theta)$, is derived from applying NCE to model (1). The second term—which is a ℓ_1 regularization—encourages sparse solutions for α and β.

Because our goal is not to obtain an accurate prediction model but rather the vector representations of the word atoms and the sparse codes, following Mikolov et al.'s suggestion in [17], we adopt a simplified version of NCE, which called "negative sampling (NS)" to learn word representations. This is done by redefine the first term of model (2) as

$$J(\theta) = \sum_{(w_i, c_i, s_i) \in \mathcal{D}} \log \sigma(s_i \alpha_i^\top B^\top C \beta_i),$$

where $\sigma(x) = 1/(1 + \exp(-x))$ is the *sigmoid* function.

Denote model (2) as SPVEC ignoring the concrete definition of $J(\theta)$. We use a suffix to indicate which training algorithm is applied: when NCE (NS) is used, we call the model SPVEC-NCE (SPVEC-NS). Note that the SPVEC model has two sets of word representations: one for target words and one for context words, which is the same as SKIPGRAM. For SKIPGRAM, the word analogy test experiments show that target word embeddings and context word embeddings have similar structures: both embeddings encode the relationship between words by the difference of corresponding words. Additionally, the sparse representation of words is a description of the structure of a word: it determines how a word is composed from word atoms. Therefore, a natural question is that: is it possible to enforce the words and contexts to have the same sparse representations? This can be done by setting S_α and S_β to share an identical parameter set, which introduces another variant of SPVEC. We denote it with a prefix: S-SPVEC, which means the sparse vectors are shared.

Both NS and NCE require some noise distributions to draw negative instances. When NS is used, an identical distribution $p_n(w) \propto \#(w)^{0.75}$ is used. When NCE is used, dynamic distributions are used to draw negative instances, which is inspired by *self-contrastive estimation* (SCE) [9]. According to the theory of NCE, this estimation method can learn effectively when the negative samples are drawing from a distribution similar to the data distribution. The model is approaching the data distribution along with the training. The SCE therefore suggest to copy the trained model as new noise distributions during training. But it is intractable in our scenario, where there are V multinomial distributions, each of which has V parameters to describe. To make it tractable, we apply a clustering method to group the distributions and compute a delegate noise distribution for each group. Concretely, we use M distributions that are updated after every 5% of progress using the following three steps.

1. Compute the dense words and contexts embeddings: $U = BS_\alpha$, $V = CS_\beta$.
2. Apply k-means to cluster word embeddings into M classes.
3. Specify a distribution for every cluster $\mathcal{X}$ using the following formula.

$$P_\mathcal{X}(c = i) = \left[\sum_{w \in \mathcal{X}} \frac{\hat{p}_d(w)}{\sum_{w' \in \mathcal{X}} \hat{p}_d(w')} \text{softmax}(V^\top w) \right]_i,$$

where $\hat{p}_d$ is the word frequency distribution; $\text{softmax}(y) = \exp(y)/\sum_i \exp(y_i)$.

The cluster dependent noise distributions have the property that for any $w \in \mathcal{X}$, $p_\mathcal{X}(c) \approx p(c \mid w)$. In principle, other clustering methods could also be applied instead of k-means. The k-means clustering method is satisfactory in our experiments.

3.2   Optimization Algorithm

In this subsection, we introduce an easy-to-parallelize algorithm to train SPVEC. This algorithm is based on Stochastic Proximal Gradient Descent (SPGD) [22]. Take SPVEC-NS as an example.[2] Define the per-instance loss function as $f = -\log \sigma(s\alpha^\top B^\top C\beta)$. Suppose the gradients of per-instance loss w.r.t all parameters are obtained, the parameters are updated by the following formulas.

---

[2] SPVEC-NCE's learning algorithm could be derived similarly.

$$\alpha^{(t+1)} = \text{prox}_{\eta_t h}\left(\alpha^{(t)} - \eta_t \frac{\partial f}{\partial \alpha}\right), \qquad \beta^{(t+1)} = \text{prox}_{\eta_t h}\left(\beta^{(t)} - \eta_t \frac{\partial f}{\partial \beta}\right), \quad (3)$$

$$B^{(t+1)} = B^{(t)} - \eta_t \frac{\partial f}{\partial B}, \qquad\qquad C^{(t+1)} = C^{(t)} - \eta_t \frac{\partial f}{\partial C}, \qquad (4)$$

where $h(x) = \lambda\|x\|_1$; η_t is the learning rate at the t-th step; $\text{prox}_g(x) = \arg\min_u \left(g(u) + \frac{1}{2}\|u - x\|_2^2\right)$ is the proximal mapping. In detail,

$$\left(\text{prox}_{\eta h}(x)\right)_i = \begin{cases} x_i - \lambda\eta, & x_i > \lambda\eta, \\ 0, & -\lambda\eta \le x_i \le \lambda\eta, \\ x_i + \lambda\eta, & x_i < -\lambda\eta. \end{cases}$$

However, it is inefficient to update the model for every training instance. Therefore, we design an algorithm that updates parameters on mini-batches. The core idea is to update the parameters based on the loss function defined on mini-batches, which is carefully arranged so that the gradients can be expressed by simple matrix-matrix products.

We represent a mini-batch using a vector $w \in \mathbb{N}^m$ and a matrix $c \in \mathbb{N}^{(k+1)\times m}$, where m is the size of mini-batches. The index vector w and the first row of c (denoted by[3] $c_{1,:}$) form a set of positive instances, i.e., a pair $(w_i, c_{1,i})$ is a positive instance. Similarly, w and $c_{i,:}, 1 < i \le k+1$ form negative instances. Denote[4]

$$s_{1,:} = +1 \qquad\qquad s_{2:k+1,:} = -1$$
$$\bar{\alpha} = (S_\alpha)_{:,w} \in \mathbb{R}^{n_b \times m}, \qquad \bar{\beta} = (S_\beta)_{:,c} \in \mathbb{R}^{n_c \times (k+1) \times m},$$
$$a = B\bar{\alpha} \in \mathbb{R}^{d \times m}, \qquad b = C\bar{\beta} \in \mathbb{R}^{d \times (k+1) \times m},$$
$$z_{:,j} = b_{:,:,j}^\top a_{:,j}, \qquad \Sigma_{ij} = \sigma(s_{ij} z_{ij}),$$
$$\gamma_{ij} = -s_{ij}(1 - \Sigma_{ij}), \qquad Q = (\bar{\beta}_{:,:,j}\gamma_{:,j})_{j=1,2,\dots,m},$$
$$M = C^\top a, \qquad N = (b_{:,:,j}\gamma_{:,j})_{j=1,2,\dots,m},$$

Redefine f as the loss on mini-batches $f(\bar{\alpha}, \bar{\beta}, B, C) = -\sum_{i=1}^{k+1}\sum_{j=1}^{m} \log \Sigma_{ij}$. We can prove

$$\frac{\partial f}{\partial B} = N\bar{\alpha}^\top, \qquad \frac{\partial f}{\partial C} = aQ^\top, \qquad\qquad (5)$$

$$\frac{\partial f}{\partial a} = N, \qquad \frac{\partial f}{\partial \bar{\alpha}} = \frac{\partial a}{\partial \bar{\alpha}}\frac{\partial f}{\partial a} = B^\top N, \qquad (6)$$

$$\frac{\partial f}{\partial \bar{\beta}_{:,:,j}} = M_{:,j}\gamma_{:,j}^\top. \qquad\qquad (7)$$

In summery, the mini-batch SPGD is repeatedly applying Eqs. (5)–(7) to calculate the gradients of parameters and using (3)–(4) to update the parameters.

---

[3] We use index convention from Python except that indexes start with 1.

[4] We define the product of a matrix $A \in \mathbb{R}^{m \times n}$ and a 3-way tensor $B \in \mathbb{R}^{n \times p \times q}$ to be a 3-way tensor C such that $C_{:,i,j} = AB_{:,i,j}$.

Note that there could be duplicated word or context index in w or c and duplicated gradients w.r.t α and β should be combined before updating when the program is running in parallel. All gradients are calculated by matrix products, which means it is easy to leverage existing high performance algebra libraries to parallelize the computations.

4   Evaluation

In this section, we will evaluate the resulting sparse representations on two similarity-based tasks and investigate the interpretability of our SPVEC model.

4.1   Experimental Settings

We use the English Wikipedia dump (July, 2014) as the corpus to train all the models. After some preprocessing such as document extraction, markup removing, sentence splitting, tokenization, lowercasing and text normalization, the plain text corpus contains about 1.6 billion running words.

The hyper parameters for SPVEC are given as follows. The number of word or context atoms is set to be 1024 and the dimension of these atom embeddings is set to be 200. The ℓ_1 regularization penalty λ was set empirically such that the overall sparsity of words exceeds 95% for all variants of SPVEC. The size of mini-batch is set to be 1024. The learning rate is dynamically updated using formula $\alpha = \alpha_0 - (\alpha_0 - \alpha_{\text{end}})g$, where $g \in [0, 1]$ is the training progress, the initial learning rate α_0 and the minimum learning rate α_{end} are set to be 5×10^{-5} and 1×10^{-6}, respectively. Following Mikolov et al.'s work [17], we use windows with random sizes to draw positive instances and the largest distance between a target word and a context word is 8. During training, we draw 8 negative instances for each positive instance.

After training, we perform an extra operation to further increase the sparsity of the sparse representations. This is done by setting the values that is less than a small fraction of the largest element of the vector (in absolute sense) be zero, i.e., setting α_i be 0 if $|\alpha_i| < \xi \cdot \max\{|\alpha_j| : 1 \le j \le n_b\}$. In practice, ξ is set to be 0.05.

For SKIPGRAM, CBOW and SC[5], we train them using the released tools on the same corpus with the same settings as our models if possible for fair comparison. The first two models, which are implemented in the `word2vec` tool[6] are both trained with negative sampling since NCE is not implemented in the tool. The PPMI matrix is built based on the word-context co-occurrence counts with window size as 8.

4.2   Word Analogy

The word analogy task can be used to evaluate models' ability to encode linguistic regularities between words, which is introduced by Mikolov et al. [17]. We

---

[5] https://github.com/mfaruqui/sparse-coding.
[6] https://github.com/dav/word2vec.

Table 1. Results of word analogy and word intrusion tasks. We report accuracy (%) for word analogy task.

Model	Dim.	Sparsity	DistRatio	Google			MSR			
				Sem.	Syn.	Total	Adj.	Nouns	Verbs	Total
SkipGram	200	0.0%	1.11	73.9	70.8	71.9	66.7	63.3	67.5	66.5
CBoW	200	0.0%	1.08	72.2	69.8	70.7	65.8	61.9	66.1	65.2
Sparse CBoW[a]	300	90.1%	1.39	73.2	67.5	70.1	-	-	-	-
SC (SG)[b]	1024	94.3%	1.27	67.4	60.1	62.7	59.6	57.4	58.9	58.8
SC (CBoW)[c]	1024	93.6%	1.21	68.9	63.4	65.4	60.3	57.8	59.2	59.3
PPMI	60000	90.8%	1.31	74.0	40.3	52.3	38.2	35.5	37.7	37.4
SpVec-ns	1024	96.1%	1.44	70.8	68.1	69.1	63.4	60.0	64.7	63.3
SpVec-nce	1024	95.1%	1.46	69.5	68.5	68.9	64.1	62.3	64.4	63.9
s-SpVec-ns	1024	96.3%	1.47	70.5	68.7	69.3	65.2	63.1	64.8	64.6
s-SpVec-ncs	1024	96.5%	1.51	70.1	69.4	69.6	65.0	63.9	65.3	65.0

[a]This line is adopted from [29].
[b]The input matrix of SC is the 200d vectors of SkipGram in the first row.
[c] The input matrix of SC is the 200d vectors of CBoW in the second row.

use two word analogy test sets, namely, Google and MSR, both containing test case like "*run* is to *running* as *walk* is to *walking*". The Google dataset[7] contains 19,544 analogy questions, which can be categorized into semantic and morpho-syntactic related subsets [17]. The MSR dataset[8] contains 8,000 analogy questions, categorized according to part-of-speech; all of them are morpho-syntactic.

This task is to predict the last word of the analogy questions, pretending it is missing. Following Mikolov et al.'s work [17], for question "a is to b as c is to __", we apply $d = \arg\max_{d \in \mathcal{V} \setminus \{a,b,c\}} \cos(c - a + b, d)$ to fill the blank. Table 1 shows the result on word analogy tasks. It shows that word analogy is more challenging for sparse models. None of sparse models outperforms SkipGram or CBoW. Nevertheless, SpVec models can achieve similar performance comparing with SkipGram or CBoW. We also find that all variants of SpVec have similar performance on this task.

4.3 Word Similarity

One important indicator to assess the quality of word representations is the clustering property—similar words should have similar vectors. We use WordSim353 dataset [7] to investigate the similarity aspect of the resulting word vectors. This dataset contains 353 word pairs along with their similarity/relatedness scores. We use the sparse word vectors to retrieve and rank the most similar words. For every word w in WordSim353, we rank its similar words by cosine similarities.

---

[7] https://github.com/dav/word2vec/blob/master/data/questions-words.txt.
[8] http://research.microsoft.com/en-us/projects/rnn/.

The ground truth of w's similar words is a set $\mathcal{U}(w)$, which is a collection of all the words in WordSim353 that the similarity score with w is higher than 0.6.

The recall-precision curve is depicted by Fig. 1. We expect SPVEC's curve to be comparable to SC and higher than SKIPGRAM's, which in turn is expected to be higher than PPMI. This means that the similarities induced by SPVEC models are more consistent with human cognitions.

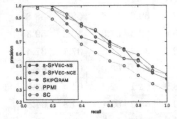 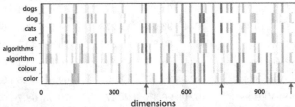

Fig. 1. Recall-precision curve when attempting to rank similar words above unsimilar ones

Fig. 2. Visualization of several selected words' sparse representation from S-SPVEC-NCE. Zeroes are white; negative (positive) values are blue (red). (Color figure online)

4.4 Interpretability

In this subsection, we talk about the interpretability of the learned sparse vectors. We visualize 8 selected words' sparse vectors from S-SPVEC-NCE in Fig. 2. We find that similar words have similar sparse patterns and dissimilar words possess different sparse codes. We also observe some interpretable patterns from this figure. For example, the dimensions marked by arrows clearly relate to the plural and singular aspects of words.

Following Sun et al.'s work [29], we evaluate the interpretability of our learned sparse word vectors quantitatively by word intrusion task. The details of construction test data for this task are described in [6,29]. Roughly, it sorts words dimensionally and chooses the top 5 and an intruder word to form an instance. An intruder word is a word from the bottom half of the sorted list that is in top 10% of a sorted list corresponds to another dimension.

We use DistRatio to measure the interpretability of word representations. DistRatio is defined to be the average ratio of the distance a_i to distance b_i, where a_i is the average distance between the intruder word and top words for the i-th instance; and b_i is the average distance between the top words for the i-th instance. This measure is first introduced by Sun et al. [29]. The higher the ratio is, the stronger interpretability the representation possesses.

Table 1 presents the DistRatio of our models and their competing ones. It shows that the interpretability of dense models is weak while sparse representations illustrate much stronger interpretability. This confirms that the sparse representations are more interpretable than the dense ones. Moreover, the SPVEC variants outperform other sparse representation models significantly on the interpretability aspect. Compared to SC, the reason might be that our method directly learns the sparse word vectors from the data instead of transforming pre-trained dense vectors to sparse codes that SC does, and thus can avoid the information loss caused by the pipeline of learning sparse representations.

5   Conclusion

In this paper, we propose a method to learn sparse word vectors directly from the plain text, which is based on two assumptions: (1) each word is composed of a few fundamental elements and (2) good representations result in good performances to predict context words. We also give an efficient and easy-to-parallelize algorithm that based on NCE to train the proposed model. Additionally, a clustering-based adaptive updating scheme for noise distributions is proposed for effective learning when NCE is applied.

The experimental results on word analogy tasks show that the performance loss due to imposing sparse structure on word representations is limited. On the word similarity task, our models outperform dense representations like SKIP-GRAM, which is considered to be a strong competitor. On the interpretability aspect, the sparse representations are more interpretable than dense ones. The experiments demonstrate the effectiveness of our learned sparse vectors in interpretability.

Acknowledgement. This research is supported by the National Basic Research Program of China (the 973 Program) under Grant No. 2015CB352201 and the National Natural Science Foundation of China under Grant Nos. 61421091, 61232015 and 61502014.

References

1. Brown, P.F., Della Pietra, V.J., de Souza, P.V., Lai, J.C., Mercer, R.L.: Class-based n-gram models of natural language. Comput. Linguist. **18**(4), 467–479 (1992)
2. Chen, Y., Mou, L., Xu, Y., Li, G., Jin, Z.: Compressing neural language models by sparse word representations. In: Proceedings of ACL (2016)
3. Collobert, R., Weston, J., Bottou, L., Karlen, M., Kavukcuoglu, K., Kuksa, P.: Natural language processing (almost) from scratch. J. Mach. Learn. Res. **12**, 2493–2537 (2011)
4. Donoho, D.L., Elad, M., Temlyakov, V.N.: Stable recovery of sparse overcomplete representations in the presence of noise. IEEE Trans. Inf. Theory **52**, 6–18 (2006)
5. Erk, K., Padó, S.: A structured vector space model for word meaning in context. In: Proceedings of EMNLP, pp. 897–906, Morristown, NJ, USA (2008)
6. Faruqui, M., Tsvetkov, Y., Yogatama, D., Dyer, C., Smith, N.A.: Sparse overcomplete word vector representations. In: Proceedings of ACL, pp. 1491–1500 (2015)

7. Finkelstein, L., Gabrilovich, E., Matias, Y., Rivlin, E., Solan, Z., Wolfman, G., Ruppin, E.: Placing search in context: the concept revisited. ACM Trans. Inf. Syst. **20**(1), 116–131 (2002)
8. Goodfellow, I., Bengio, Y., Courville, A.: Deep Learning. MIT Press, Cambridge (2016)
9. Goodfellow, I.J.: On distinguishability criteria for estimating generative models. arXiv, December 2014
10. Gutmann, M., Hyvärinen, A.: Noise-contrastive estimation of unnormalized statistical models, with applications to natural image statistics. J. Mach. Learn. Res. **13**(1), 307–361 (2012)
11. Harris, Z.S.: Distributional structure. Word **10**, 146–162 (1954)
12. Kalchbrenner, N., Blunsom, P.: Recurrent convolutional neural networks for discourse compositionality. arXiv.org, June 2013
13. Landauer, T.K., Foltz, P.W., Laham, D.: An introduction to latent semantic analysis. Discourse Process. **25**(2–3), 259–284 (1998)
14. Levy, O., Goldberg, Y.: Linguistic regularities in sparse and explicit word representations. In: Conference on Natural Language Learning, pp. 171–180 (2014)
15. Lund, K., Burgess, C., Atchley, R.A.: Semantic and associative priming in high-dimensional semantic space. In: Annual Conference of the Cognitive Science Society, vol. 17, pp. 660–665 (1995)
16. Martins, A.F.T., Smith, N.A., Figueiredo, M.A.T., Aguiar, P.M.Q.: Structured sparsity in structured prediction. In: Proceedings of EMNLP (2011)
17. Mikolov, T., Chen, K., Corrado, G., Dean, J.: Efficient estimation of word representations in vector space. In: Computing Research Repository (2013)
18. Mnih, A., Kavukcuoglu, K.: Learning word embeddings efficiently with noise-contrastive estimation. In: NIPS, pp. 2265–2273 (2013)
19. Mnih, A., Teh, Y.W.: A fast and simple algorithm for training neural probabilistic language models. arXiv preprint arXiv:1206.6426 (2012)
20. Murphy, B., Talukdar, P.P., Mitchell, T.M.: Learning effective and interpretable semantic models using non-negative sparse embedding. In: Proceedings of COLING. ACL (2012)
21. Olshausen, B.A., Field, D.J.: Sparse coding with an overcomplete basis set: a strategy employed by v1? Vis. Res. **37**(23), 3311–3325 (1997)
22. Parikh, N., Boyd, S.: Proximal algorithms. Found. Trends® Optim. **1**(3), 127–239 (2014)
23. Paul, M., Dredze, M.: Factorial LDA: sparse multi-dimensional text models. In: Advances in Neural Information Processing (2012)
24. Pennington, J., Socher, R., Manning, C.D.: GloVe: global vectors for word representation. In: Proceedings of ACL (2014)
25. Sivaram, G.S.V.S., Nemala, S.K., Elhilali, M., Tran, T.D., Hermansky, H.: Sparse coding for speech recognition. In: ICASSP, pp. 4346–4349 (2010)
26. Socher, R., Bauer, J., Manning, C.D., Ng, A.Y.: Parsing with compositional vector grammars. In: Proceedings of ACL (2013)
27. Socher, R., Chen, D., Manning, C.D.: Reasoning with neural tensor networks for knowledge base completion. In: NIPS (2013)
28. Soutner, D., Müller, L.: Continuous distributed representations of words as input of LSTM network language model. In: Sojka, P., Horák, A., Kopeček, I., Pala, K. (eds.) TSD 2014. LNCS, vol. 8655, pp. 150–157. Springer, Cham (2014). doi:10.1007/978-3-319-10816-2_19

29. Sun, F., Guo, J., Lan, Y., Xu, J., Cheng, X.: Sparse word embeddings using ℓ_1 regularized online learning. In: Proceedings of the 25th International Joint Conference on Artificial Intelligence, New York, USA, pp. 959–966 (2016)
30. Toutanova, K., Johnson, M.: A Bayesian LDA-based model for semi-supervised part-of-speech tagging. In: NIPS (2007)
31. Vinson, D.P., Vigliocco, G.: Semantic feature production norms for a large set of objects and events. Behav. Res. Methods **40**(1), 183–190 (2008)
32. Yogatama, D., Smith, N.A.: Linguistic structured sparsity in text categorization. In: Proceedings of ACL (2014)

A Study of Distributed Semantic Representations for Automated Essay Scoring

Cancan Jin, Ben He$^{(\boxtimes)}$, and Jungang Xu$^{(\boxtimes)}$

School of Computer and Control Engineering,
University of Chinese Academy of Sciences,
Geese-resting Lake Campus, Beijing 101408, China
jincancan15@mails.ucas.ac.cn, {benhe,xujg}@ucas.ac.cn

Abstract. Automated essay scoring (AES) applies machine learning and NLP techniques to automatically rate essays written in an educational setting, by which the workload of human raters is considerably reduced. Current AES systems utilize common text features such as essay length, *tf-idf* weight, and the number of grammar errors to learn a scoring function. Despite the effectiveness brought by those common features, the semantics within the essay text is not well considered. To this end, this paper presents a study of the usefulness of the distributed semantic representations to AES. Novel features based on word or paragraph embeddings are combined with the common text features in order to improve the effectiveness of the AES systems. Evaluation results show that the use of the distributed semantic representations are beneficial for the task of AES.

Keywords: Automated essay scoring · Distributed semantic representations · Embeddings

1 Introduction

Automated essay scoring (AES) is usually considered as a machine learning problem [1–3] where learning algorithms such as K-nearest neighbor and support vector machines for ranking are applied to learn a rating model for a given essay prompt with a set of training essays rated by human assessors [4]. Currently, the AES systems have been widely used in large-scale English writing tests, e.g. Graduate Record Examination (GRE), to reduce the human efforts in the writing assessments.

In general, existing AES systems are based on a number of common text features that are not linked to intuitive dimensions of semantics or writing quality, such as lexical complexity, grammar errors, syntactic complexity, organization and development, coherence, etc. However, these shallow text features are not able to represent the semantic content of essays, resulting in limited robustness and effectiveness [5].

© Springer International Publishing AG 2017
G. Li et al. (Eds.): KSEM 2017, LNAI 10412, pp. 16–28, 2017.
DOI: 10.1007/978-3-319-63558-3_2

Recently, the word level, phrase level, and sentence level semantic representations of documents are successfully applied to compute the syntactic and semantic similarity in quite a few natural language processing (NLP) tasks. For example, Tomas et al. propose a method based on representations of words and sentences that achieves promising results for movie rating prediction on a crawl of IMDB [8]. Richard et al. propose a method based on the continuous representations of sentences that obtains good performance on the Stanford background dataset [9]. There are also efforts in developing methods for extracting the semantic representations of documents [9–11]. For instance, a simple approach is to use a weighted average of all word vectors in the document [12]. A more sophisticated approach is to learn continuous distributed vector representations for pieces of texts [13]. For the task AES, there has been little success of the application of the semantic features as far as we are aware of.

To this end, this paper presents an investigation in the usefulness of various novel features in indicating the writing quality of essays. The new features are derived based on different approaches to generating distributed representations of words, paragraphs, and documents, including latent Dirichlet allocation (LDA) [17], Word2Vec [18], and PV-DBOW [13]. Experimental results on the publicly available dataset ASAP indicate that the new features based on the semantic similarity features and distributed semantic representations of essays achieve higher agreement with human raters than the use of only the common text features. In our evaluation, the use of the new features can achieve up to 12.33% improvement in Kappa, and 18.61% improvement in nRMSE against the baseline.

2 Common Text Features

This section introduces common text features widely used in the previous AES methods [1,7,14,15], which are listed in Table 1. The detailed description of the features is given below.

- *Statistics of word length*: The *mean* and *variance* of word length in characters. These can be indicators for the degree of complexity a writer can master since the unusual words tend to be longer. The number of unique words appeared in an essay, normalized by the essay length in words.
- *Statistics of sentence length*: The *mean* and *variance* of sentence length in words. The variety of the length of sentences potentially reflects the complexity of syntactics.
- *Statistics of essay length*: The essay length is measured by the number of words and the number of characters in an essay. Essays are usually written under a time limit, so the essay length can be a useful predictor of the productivity of the writer. The fourth root of essay length in words is proved to be highly correlated with the essay score [15].
- *Clauses*: The *mean* number of clauses in each sentence, normalized by the number of sentence in an essay. The *maximum* number of clauses of a sentence in an essay. The *mean length* of sentences that contain at least one clause.

Table 1. Common text features.

No.	Feature
1	Mean and variance word length in characters
2	Mean length of clauses
3	Essay length in characters and words
4	Number of spelling errors
5	The number of prepositions and commas
6	Mean number of clauses per sentence
7	Mean and variance of sentence length in words
8	Maximum number of clauses of a sentence
9	Semantic vector similarity based on LSA
10	Mean cosine similarity of word vectors by $tf\text{-}idf$
11	The average height of the parser tree of each sentence in an essay
12	Word bigram/trigram frequency tf divided by collection frequency TF
13	POS bigram/trigram frequency tf divided by collection frequency TF

- *Sentence structure*: The number of prepositions and commas in each sentence, normalized by words in sentences. The average height of the parser tree of each sentence in an essay. The average of the sum of the depth of all nodes in a parser tree of each sentence in an essay. The more complicated the sentences are, the higher complexity the parser trees exhibit. It is therefore necessary to utilize the sentence structure to indicate the essay quality.
- *Spelling errors*: Grammatical or spelling errors are one of the most obvious indicators of bad essays, which are detected by the spelling check API provided by LanguageTool[1].
- *Word bigram and trigram*: The level of grammar and fluency of an essay can be measured by the mean tf/TF of word bigrams and trigrams [16] (tf is the frequency of bigram/trigram in a single essay and TF is the frequency of bigram/trigram in the whole essay collection). We assume a bigram or trigram with high tf/TF as a grammar error because high tf/TF means that this kind of bigram or trigram is not commonly used in the whole essay collection but appears in the specific essay.
- *POS bigram and trigram*: Mean tf/TF of POS bigrams and trigrams. The Part-of-Speech tagging of each word is done by the Stanford Parser[2].
- *Word vector similarity*: Mean cosine similarity of word vectors, in which the element is the term frequency multiplied by inverse document frequency (tf-idf) of each word. It is calculated as the weighted mean of cosine similarities and the weight is set as the corresponding essay score.

---

[1] https://www.languagetool.org.
[2] http://nlp.stanford.edu/software/corenlp.shtml.

– *Semantic vector similarity*: Semantic vectors are generated by Latent Semantic Analysis [6]. The calculation of mean cosine similarity of semantic vectors is the same with word vector similarity.

Each feature is normalized to be within [0, 1]. The features introduced in this section include most of the common text features used in recent studies on AES, which lead to state-of-the-art results [1,4,7,16,20]. Therefore, the AES system trained by those common text features is used as the baseline in this paper.

3 Semantic Representations for AES

This section introduces the semantic features involved in this study. Section 3.1 introduces the methods used for learning the semantic representations of essays, from which the semantic features are generated, as in Sect. 3.2.

3.1 Methods for Vector Representations

Other than the previously applied LSA approach in [7], we propose to generate semantic features based on the following recent methods for the vector representations. A brief introduction of how to obtain semantic embeddings of essays through these learning algorithms is given below.

Latent Dirichlet Allocation (LDA) is a generative probabilistic model of a corpus [17]. The basic idea is that documents are represented as random mixtures over latent topics, where each topic is characterized by a distribution over words. The probabilistic distribution on all topics of a document is considered as a kind of semantic representation of a document. Using LDA, the ith dimension of the semantic representation of the essay is given by the probability that the essay belongs to topic i. To analyze the effectiveness of the different number of topics, the number of topics is set as 5, 6, 7, 8, 9, 10, respectively. Those settings are found to be the most effective in the preliminary experiments. Results obtained using the different settings above are also presented in Sect. 5.2.

Continuous Skip-gram Model is used to learn high-quality distributed vector representations that capture syntactic and semantic relationships between words [18]. In continuous skip-gram model, every word is mapped to a unique vector, and all the word vectors are stacked in a word embedding matrix W generated by the model. The weighted mean of word embeddings of words appeared in an essay is used as a semantic representation of the essay, which the weight is set as the *tf/TF*. The ith dimension of the semantic representation of the essay is given by:

$$word\_Vec_i = \frac{\sum_{j=1}^{n} weight_j * W_{ji}}{n} \tag{1}$$

where n is the number of unique words in an essay, $weight_j$ is the *tf/TF* of the jth word in the essay, and W_{ji} is the ith dimension of the word vector of the jth word in the essay.

In this paper, we use two different datasets to learn the word embeddings, one is the publicly available ASAP dataset (see Sect. 4.1 for details), and the other is the GoogleNews dataset[3]. The latter is a large-scale news corpus which may lead to better training of word embeddings. The word embeddings obtained on ASAP is trained by Word2Vec, and the dimension of word embedding is set as 50, 100, 200 and 300, respectively. Those settings are found to be the most effective in the preliminary experiments. Results obtained using the different settings above are also presented in Sect. 5.2. Using GoogleNews, the word vectors are pre-trained on 100 billion words of Google news dataset and are of length 300.

Distributed Bag of Words Version of Paragraph Vectors Model: Tomas et al. propose the distributed bag of words version of paragraph vector model (PV-DBOW) [13]. The PV-DBOW model learns the paragraph vector based on the continuous skip-gram model. A notable difference between the outcome of the PV-DBOW model and the continuous skip-gram model is that the PV-DBOW model generates the vector representations of paragraphs, in addition to the word vectors. The paragraph vector representations are obtained by the PV-DBOW model trained on the ASAP dataset. GoogleNews is not used as it only comes with word embeddings. The same as the word embeddings, the dimension of paragraph embedding is set to 50, 100, 200, and 300, respectively, for effectiveness reason.

3.2   Semantic Features

In this paper, we present two ways to using the semantic representations of essays for generating the semantic features for AES, namely Vector Similarity and Dimension Extension.

Vector Similarity: It is calculated as the mean of all weighted cosine similarities between the given essay and the other essays for a given prompt. Assuming $w_1, w_2, ..., w_m$ are the semantic representations of essays in the specific essay set, Sim_i is the Vector Similarity of the ith essay:

$$Sim_i = \frac{\sum_{j=1, j!=i}^{m} r_j \cdot \frac{\vec{w_i} \cdot \vec{w_j}}{||\vec{w_i}|| \times ||\vec{w_j}||}}{(m-1) \cdot \sum_{j=1, j!=i}^{m} r_j} \tag{2}$$

where m is the number of essays associated to the given prompt, and r_j is the actual rating of the jth essay. Using Vector Similarity, only a single semantic feature is generated from the essay embeddings.

Dimension Extension: The feature vector of a given essay is extended by the entire semantic representations of the essay. Each dimension of the essay embedding is regarded as a semantic feature of the essay. In other words, the size of the feature vector of the given essay is extended by the number of dimensions of the entire semantic representations of the essay.

---

[3] https://drive.google.com/file/d/0B7XkCwpI5KDYNlNUTTlSS21pQmM/edit?usp=
sharing.

Table 2. Semantic features.

Semantic features	Feature description
lda_sim_k	Vector Similarity feature learned by LDA The number of topics k is set as 5, 6, 7, 8, 9, 10, respectively
lda_vec_k	Dimension Extension feature learned by LDA The number of topics k is set as 5, 6, 7, 8, 9, 10, respectively
word_sim_k	Vector Similarity feature learned from ASAP by skip-gram The number of dimensions k is set as 50, 100, 200, 300, respectively
word_vec_k	Dimension Extension feature learned from ASAP by skip-gram The number of dimensions k is set as 50, 100, 200, 300, respectively
google_sim_300	Vector Similarity feature learned from GoogleNews The dimension of the word embeddings is 300
google_vec_300	Dimension Extension feature learned from GoogleNews The dimension of the word embeddings is 300
para_sim_k	Vector Similarity feature learned from ASAP by PV-DBOW The number of dimensions k is set as 50, 100, 200, 300, respectively
para_vec_k	Dimension Extension feature learned from ASAP by PV-DBOW The number of dimensions k is set as 50, 100, 200, 300, respectively

We can generate a list of semantic features through the above methods on the basis of semantic representations of essays introduced in Sect. 3.1.

A list of the semantic features used in this paper is summarized in Table 2. For example, when the *para_vec_100* feature is used, the essay feature vector is extended by the semantic paragraph vector with 100 dimensions. Instead, if *para_sim_100* is used, the learned paragraph embeddings have 100 dimensions, and the essay feature vector is extended by a single dimension, which is the similarity between the semantic paragraph vector of essay and other essays in the same essay set.

4 Experimental Settings

This section presents our experimental design, including the dataset used, the evaluation metrics of the AES system, and the learning algorithms.

4.1 Dataset

The dataset used in our experiments comes from the Automated Student Assessment Prize (ASAP)[4]. Dataset in this competition consists of eight essay sets. Each essay set was generated from a specific prompt. All essays received a resolved score, namely the actual rating, from professional human raters. As the official test data is no longer available, the evaluation is done by 10-fold cross-validation on the training data, split by random partitioning.

---

[4] https://www.kaggle.com/c/asap-aes/data.

4.2 Evaluation Metrics and Learning Algorithms

In this paper, we use Kappa, Pearson correlation coefficient, Spearman correlation coefficient and normalized root-mean-squared error to evaluate the agreement between the ratings given by the AES system and the actual ratings. They are widely accepted as reasonable evaluation measures for the AES systems [14, 19, 20].

Quadratic Weighted Kappa is a statistical metric which is used to measure inter-rater agreement. Quadratic weighted Kappa takes the degree of disagreement between raters into account. The kappa metric is computed by the mean of the kappa values across all essay sets after applying the Fisher Transformation[5], instead of the average of the raw kappa values over all essay sets.

Pearson Correlation Coefficient [21] is used to measure the strength of a linear association between two variables.

Spearman Correlation Coefficient [22] assesses how well the relationship between two variables can be described using a monotonic function.

In addition to the above three human-machine agreement metrics, the **Normalized root-mean-squared error** (nRMSE) [23] measures the prediction error of the essay ratings. The ratings of a given essay topic are normalized to be within $[0, 1]$ such that the errors among different prompts are comparable. *Different from the other three metrics, a lower nRMSE value indicates better effectiveness.* The nRMSE reported in the results is averaged over all test essays in the whole dataset. All statistical tests are based on Analysis of Variance (ANOVA).

In this paper, we use K-nearest neighbor (KNN) and support vector machines for ranking (SVM-rank) to predict ratings of essays. These two classical algorithms are widely used in recent studies on the AES systems [14, 15].

K-nearest Neighbors (KNN) [24] is a classical classification algorithm commonly used in automated essay scoring. Using KNN, we select the K essays in the training collection that are most similar to the test essay. Then the predicted score of the test essay is the average of the scores of the K essays. The parameter K is set by grid search on the ASAP validation set.

For **SVM-rank**, the linear kernel function is used in the experiments. The parameter C, which controls the trade-off between empirical loss and regularizer, is set by grid search on the ASAP validation set. To determine the final rating of a given essay, we take the average rating of k essays whose scores are closest to the given essay. The parameter k is also set by grid search on the ASAP validation set. We use the implementation of SVM-rank in SVMrank package[6].

5 Evaluation Design and Results

5.1 Evaluation Design

In order to examine the effectiveness of the semantic features when applied to AES, the experiments conducted in this paper are organized as follows.

---

[5] https://www.kaggle.com/c/asap-aes/details/evaluation.

[6] http://svmlight.joachims.org/.

Table 3. Performance of the semantic features generated by Word2Vec with different numbers of dimensions.

Methods	Vector similarity						
	Metrics	Base	word _sim_50	word _sim_100	word _sim_200	word _sim_300	google _sim_300
SVM_rank	Kappa	.7423	.7656	**.7716**	.7600	.7695	.7592
	Pearson	.7793	.8115	**.8189**	.8082	.8109	.8079
	Spearman	.7355	.7702	**.7797**	.7749	.7722	.7678
	nRMSE	.1709	.1532	**.1486**	.1533	.1530	.1571
KNN	Kappa	.7103	.7728	.7746	.7727	.7734	**.7754**
	Pearson	.7429	.7890	.7881	.7890	.7845	**.7928**
	Spearman	.7225	.7768	.7766	.7746	.7765	**.7781**
	nRMSE	.1811	.1650	.1648	.1649	.1655	**.1638**
Methods	Dimension extension						
	Metrics	Base	word _vec_50	word _vec_100	word _vec_200	word _vec_300	google _vec_300
SVM_rank	Kappa	.7423	**.7895**	.7784	.7863	.7766	.7817
	Pearson	.7793	**.8292**	.8215	.8242	.8018	.8169
	Spearman	.7355	**.7912**	.7812	.7862	.7850	.7817
	nRMSE	.1709	**.1454**	.1459	.1464	.1569	.1463
KNN	Kappa	.7103	.7190	.7028	.7182	.7114	**.7365**
	Pearson	.7429	.7492	.7428	.7421	.7501	**.7678**
	Spearman	.7225	.6950	.6905	.6995	.6978	**.7396**
	nRMSE	.1811	.1716	.1718	.1703	.1706	**.1672**

Effectiveness of the Individual Semantic Features with Different Numbers of Dimensions: To investigate the effectiveness of the semantic features with different numbers of dimensions, six sets of experiments are conducted. Each set of experiments corresponds to one specific semantic feature, in addition to the baseline that uses the common text features in Table 2, as presented in Tables 3, 4 and 5, respectively.

Comparison to the Baseline Using a Combination of the Best Individual Semantic Features Based on Vector Similarity and Dimension Extension: In these experiments, we compare the semantic features to the baseline that uses the common text features. The semantic features are *word_sim*, *word_vec*, *lda_sim*, *lda_vec*, *para_sim*, *para_vec* and *sim_best+vec_best*. Out of these semantic features with different embedding dimensions presented in Tables 3, 4 and 5, we choose the best individual semantic feature in each table to be evaluated against the baseline. *sim_best+vec_best* denotes the concatenation of the best Vector Similarity feature and the best Dimension Extension feature out of Tables 3, 4 and 5, which correspond to Word2Vec, PV-DBOW, and LDA, respectively.

Table 4. Performance of the semantic features generated by PV-DBOW with different numbers of dimensions.

Methods	Vector similarity					
	Metrics	Base	para_sim_50	para_sim_100	para_sim_200	para_sim_300
SVM_rank	Kappa	.7423	.7631	.7624	**.7707**	.7641
	Pearson	.7793	.8085	.8056	**.8198**	.8083
	Spearman	.7355	.7684	.7633	**.7733**	.7696
	nRMSE	.1709	.1549	.1543	**.1523**	.1535
KNN	Kappa	.7103	.7663	**.7669**	.7627	.7626
	Pearson	.7429	.7821	**.7831**	.7818	.7803
	Spearman	.7225	.7610	**.7620**	.7592	.7903
	nRMSE	.1811	.1648	**.1646**	.1649	.1650
Methods	Dimension extension					
	Metrics	Base	para_vec_50	para_vec_100	para_vec_200	para_vec_300
SVM_rank	Kappa	.7423	**.7731**	.7691	.7624	.7671
	Pearson	.7793	**.8205**	.8121	.8079	.8099
	Spearman	.7355	**.7838**	.7702	.7706	.7656
	nRMSE	.1709	**.1498**	.1545	.1544	.1545
KNN	Kappa	.7103	.7892	**.7908**	.7866	.7907
	Pearson	.7429	.8116	**.8128**	.8126	.8104
	Spearman	.7225	.7943	**.7969**	.7946	.7920
	nRMSE	.1811	.1625	**.1617**	.1620	.1621

Baseline uses all the common text features in Table 1 to learn a rating model for AES. The results are listed in Table 6. The last column in Table 6 presents the results of *sim_best+vec_best*.

Take KNN for example, *lda_sim/vec_6*, *google_sim/vec_300*, and *para_sim /vec_100* are compared with the baseline as they are the best out of the different settings of parameter k. *sim_best+vec_best* means that the feature set used is the concatenation of *Baseline*, *para_vec_100* and *google_sim_300*.

5.2   Evaluation Results

Firstly, the performance of the individual semantic features are evaluated. Tables 3, 4 and 5 present the evaluation results brought by the use of individual semantic features in addition to the common text features, with respect to different numbers of dimensions. Each of the tables corresponds to the results of the semantic features generated by a single learning method, i.e. Word2Vec, PV-DBOW, or LDA. In Tables 3, 4 and 5, the best result of each semantic feature is in **bold**.

According to Tables 3, 4 and 5, the effectiveness of the semantic features is in general stable with different numbers of embedding dimensions in different evaluation metrics. Therefore, changing this parameter setting does not have

Table 5. Performance of the semantic features generated by LDA with different numbers of dimensions.

Methods	Vector similarity							
	Metrics	Base	lda_sim_5	lda_sim_6	lda_sim_7	lda_sim_8	lda_sim_9	lda_sim_10
SVM_rank	Kappa	.7423	.7657	.7555	.7579	**.7692**	.7616	.7495
	Pearson	.7793	.8131	.8040	.8062	**.8139**	.8129	.7998
	Spearman	.7355	.7716	.7674	.7656	**.7742**	.7740	.7595
	nRMSE	.1709	.1535	.1561	.1551	**.1525**	.1531	.1561
KNN	Kappa	.7103	.7252	**.7553**	.7304	.7355	.7204	.7378
	Pearson	.7429	.7518	**.7718**	.7643	.7483	.7356	.7595
	Spearman	.7225	.7292	**.7478**	.7321	.7293	.7288	.7311
	nRMSE	.1811	.1713	**.1650**	.1675	.1696	.1695	.1685
Methods	Dimension extension							
	Metrics	Base	lda_vec_5	lda_vec_6	lda_vec_7	lda_vec_8	lda_vec_9	lda_vec_10
SVM_rank	Kappa	.7423	.7679	**.7713**	.7647	.7671	.7566	.7603
	Pearson	.7793	.8124	**.8163**	.8134	.8121	.8026	.8094
	Spearman	.7355	.7674	**.7752**	.7682	.7652	.7669	.7745
	nRMSE	.1709	.1525	**.1523**	.1533	.1528	.1541	.1540
KNN	Kappa	.7103	.6811	**.7239**	.7098	.6931	.6831	.6640
	Pearson	.7429	.7183	**.7532**	.7430	.7288	.7312	.7292
	Spearman	.7225	.7006	**.7298**	.7170	.7072	.7112	.7097
	nRMSE	.1811	.1851	**.1708**	.1781	.1845	.1801	.1809

a significant impact on the performance of the individual features. Moreover, according to Table 3, the word embeddings learned from ASAP appears to have slight better performance than those learned from GoogleNews when SVM-rank is used, and the other way around when KNN is used. In addition, comparing the evaluation results of using Vector Similarity and Dimension Extension of the same embeddings, we find no conclusive results. When SVM-rank is used, the Vector Similarity features have overall slightly better performance than the Dimension Extension features. However, when KNN is used, the Vector Similarity features have better performance when generated by Word2Vec (Table 3) and LDA (Table 5), while the Dimension Extension features gave better performance when generated by PV-DBOW. Such diverse results suggest the potential usefulness to combine the best individual semantic features based on Vector Similarity and Dimension Extension, respectively.

Next, the best Vector Similarity and Dimension Extension features are combined in order to make the best use of those semantic features. Table 6 compares the use of the combination of the best semantic features against the baseline. The last column in Table 6 presents the results of sim_best+vec_best, the concatenation of the baseline features, and the best features generated by Vector Similarity and Dimension Extension, respectively. A * indicates a statistically significant improvement over the baseline according to the ANOVA test. According to Table 6, all semantic features we present in this study have improvements over the baseline, and sim_best+vec_best has the best performance in all cases.

Table 6. Main evaluation result: best individual features (columns 3–8) against the baseline, and the combination (the last column) of the best Vector Similarity (sim) and Dimension Extension (vec) features against the baseline.

SVM_rank	Base	lda _sim_8	lda _vec_6	word _sim_100	word _vec_50	para _sim_200	para _vec_50	word _sim_100 +word _vec_50
Kappa	.7423	.7692 +3.62%*	.7713 +3.91%*	.7716 +3.95%*	.7895 +6.36%*	.7707 +3.83%*	.7731 +4.15%*	**.8016** **+7.99%∗**
Pearson	.7793	.8139 +4.44%*	.8163 +4.75%*	.8189 +5.08%*	.8292 +6.40%*	.8198 +5.20%*	.8205 +5.29%*	**.8374** **+7.46%∗**
Spearman	.7355	.7742 +5.26%*	.7752 +5.40%*	.7797 +6.01%*	.7912 +7.57%*	.7733 +5.14%*	.7838 +6.57%*	**.8031** **+9.19%∗**
nRMSE	.1709	.1525 −10.77%*	.1523 −10.88%*	.1486 −13.05%*	.1454 −14.92%*	.1523 −10.88%*	.1498 −12.34%*	**.1391** **−18.61%∗**
KNN	Base	lda _sim_6	lda _vec_6	google _sim_300	google _vec_300	para _sim_100	para _vec_100	google _sim_300 +para _vec_100
Kappa	.7103	.7553 +6.34%*	.7239 +1.91%*	.7754 +9.16%*	.7365 +3.69%*	.76697.97%*	.7908 +11.33%*	**.7979** **+12.33%∗**
Pearson	.7429	.7718 +3.89%*	.7532 +1.39%	.7928 +6.72%*	.7678 +3.35%*	.7831 +5.41%*	.8128 +9.41%*	**.8161** **+9.85%∗**
Spearman	.7225	.7478 +3.50%*	.7298 +1.01%	.7781 +7.69%*	.7396 +2.37%*	.7620 +5.47%*	.7969 +10.30%*	**.8001** **+10.74%∗**
nRMSE	.1811	.1650 −8.89%*	.1708 −5.69%*	.1638 −9.55%*	.1672 −7.68%*	.1646 −9.11%*	.1617 −10.71%*	**.1612** **−10.99%∗**

This shows that it is beneficial to combine the semantic features generated by both methods. When using SVM-rank, the features generated by Dimension Extension have overall better performance than those generated by Vector Similarity and the effectiveness of features generated by word embeddings outperform the features generated by PV-DBOW and LDA.

Using SVM-rank, the improvements brought by all semantic features generated by Vector Similarity and Dimension Extension are statistically significant when the effectiveness is measured by all four evaluation metrics. Using KNN, google_sim_300 outperforms para_sim_100 and lda_sim_6, and para_vec_100 has better performance than google_vec_300 and lda_vec_6. According to the ANOVA significance test, the improvements brought by google_sim_300, google_vec_300, para_sim_100, para_vec_100, lda_sim_6 and sim_best+vec_best are statistically significant when the effectiveness is measured by Kappa, Pearson and Spearman. All improvements are statistically significant when the effectiveness is measured by nRMSE.

Overall, the results show that the use of the semantic features can indeed improve the effectiveness of AES on top of the common text features. As shown in Table 6, it is particularly encouraging that a combination of the best features can achieve up to 12.33% improvement in Kappa, and 18.61% improvement in nRMSE. Therefore, it is also recommended to combine the best features generated by Vector Similarity and Dimension Extension, in order to achieve the maximized performance of AES. It is widely accepted that the agreement between professional human raters ranges from 0.70 to 0.80, measured by quadratic weighted Kappa or Pearson's correlation [3]. In Table 6, the semantic features achieve a Kappa of 0.8016 and a Pearson's correlation of 0.8374, suggesting their potential usefulness in automated essay scoring.

6   Conclusions and Future Work

In summary, this paper presents an investigation on the effectiveness of using semantic vector representations for the task of automated essay scoring (AES). According to the evaluation results on the standard ASAP English dataset, the effectiveness brought by our proposed semantic representations of essays depends on the learning algorithms and the evaluation metrics used. On the other hand, the effectiveness of individual semantic features is stable with respect to different numbers of dimensions. Results show that statistically significant improvement over the baseline can be achieved by applying our proposed semantic features listed in Table 2. Results also show that the concatenation of the best features generated by Vector Similarity and Dimension Extension, namely feature *sim_best+vec_best* has the best effectiveness among all features involved in this investigation. Moreover, the semantic features based on word embeddings lead to better effectiveness than those based on LDA embeddings and paragraph embeddings.

In the future, we plan to continue the research by mining effective features based on different sources of information, e.g. the structure of a given essay. We also plan to further improve this work by using the embeddings as input to a deep neural network, in order to learn an AES model.

Acknowledgments. This work is supported by the National Natural Science Foundation of China (61472391).

References

1. Attali, Y., Burstein, J.: Automated essay scoring with e-rater. J. Technol. Learn. Assess. **4**(3), 7–15 (2006)
2. Dikli, S.: An overview of automated scoring of essays. J. Technol. Learn. Assess. **5**(1), 5–21 (2006)
3. Williamson, D.M.: A framework for implementing automated scoring. In: Annual Meeting of the American Educational Research Association and the National Council on Measurement in Education, San Diego, CA (2009)
4. Chen, H., He, B., Luo, T., Li, B.: A ranked-based learning approach to automated essay scoring. In: 2012 Second International Conference on Cloud and Green Computing (CGC), pp. 448–455 (2012)
5. Yongwei, Y., Buckendahl, C.W., Juszkiewicz, P.J., et al.: A review of strategies for validating computer-automated scoring. Appl. Measur. Educ. **15**(4), 391–412 (2002)
6. Dumais, S.T.: Latent semantic analysis. Annu. Rev. Inf. Sci. Technol. **38**(1), 188–230 (2004)

7. Foltz, P.W., Laham, D., Landauer, T.K.: Automated essay scoring: applications to educational technology. In: World Conference on Educational Multimedia, Hypermedia and Telecommunications, vol. 1999(1), pp. 939–944 (1999)

8. Mesnil, G., Mikolov, T., Ranzato, M., Bengio, Y.: Ensemble of generative and discriminative techniques for sentiment analysis of movie reviews. CoRR, abs/1412.5335 (2006)

9. Socher, R., Lin, C.C., Ng, A.Y., Manning, C.: Parsing natural scenes and natural language with recursive neural networks. In: ICML, pp. 129–136 (2011)

10. Zanzotto, F.M., Korkontzelos, I., Fallucchi, F., Manandhar, S.: Estimating linear models for compositional distributional semantics. In: COLING, pp. 1263–1271 (2010)

11. Wang, S., Manning, C.D.: Baselines and bigrams: simple, good sentiment and topic classification. In: ACL (2), pp. 90–94 (2012)

12. Lai, S., Xu, L., Liu, K., Zhao, J.: Recurrent convolutional neural networks for text classification. In: AAAI, pp. 2267–2273 (2015)

13. Le, Q.V., Mikolov, T.: Distributed representations of sentences and documents. ICML, vol. 32, pp. 1188–1196 (2014)

14. Chen, H., He, B.: Automated essay scoring by maximizing human-machine agreement. In: EMNLP, pp. 1741–1752 (2013)

15. Shermis, M.D., Burstein, J.C.: Automated Essay Scoring: A Cross-Disciplinary Perspective. Routledge, Abingdon (2003)

16. Briscoe, T., Medlock, B., Andersen, Ø.: Automated assessment of ESOL free text examinations. University of Cambridge Computer Laboratory Technical reports, vol. 790 (2010)

17. Blei, D.M., Ng, A.Y., Jordan, M.I.: Latent dirichlet allocation. J. Mach. Learn. Res. **3**, 993–1022 (2003)

18. Mikolov, T., Chen, K., Corrado, G., Dean, J.: Efficient estimation of word representations in vector space In: Proceedings of Workshop at ICLR (2013)

19. Shermis, M.D., Burstein, J.: Handbook of Automated Essay Evaluation: Current Applications and New Directions. Routledge, Abingdon (2013)

20. Yannakoudakis, H., Briscoe, T., Medlock, B.: A new dataset and method for automatically grading ESOL texts. In: ACL, pp. 180–189 (2011)

21. Lawrence, I., Lin, K.: A concordance correlation coefficient to evaluate reproducibility. Biometrics **45**, 255–268 (1989)

22. Croux, C., Dehon, C.: Influence functions of the Spearman and Kendall correlation measures. Stat. Methods Appl. **19**(4), 497–515 (2010)

23. Hyndman, R.J., Koehler, A.B.: Another look at measures of forecast accuracy. Int. J. Forecast. **22**(4), 679–688 (2006)

24. Altman, N.S.: An introduction to kernel and nearest-neighbor nonparametric regression. Am. Stat. **46**(3), 175–185 (1992)

Weakly Supervised Feature Compression Based Topic Model for Sentiment Classification

Yan Hu, Xiaofei Xu, and Li Li(✉)

School of Computer and Information Science,
Southwest University, Chongqing 400715, China
guyuemuzhi@sina.cn, lily@swu.edu.cn

Abstract. Sentiment classification aims to use automatic tools to explore the subjective information like opinions and attitudes from user comments. Most of existing methods are centered on the semantic relationships and the extraction of syntactic feature, while the document topic feature is ignored. In this paper, a weakly supervised hierarchical model called external knowledge-based Latent Dirichlet Allocation (ELDA) is proposed to extract document topic feature. First of all, we take advantage of ELDA to compress document feature and increase the polarity weight of document topic feature. And then, we train a classifier based on the topic feature using SVM. Experiment results on one English dataset and one Chinese dataset show that our method can outperform the state-of-the-art models by at least 4% in terms of accuracy.

Keywords: Sentiment classification · Topic feature · External knowledge · SVM

1 Introduction

With the rapid development of Web 2.0, various types of social media such as product reviews, microblogs and forums have provided a wealth of information that can be very helpful to access the general public's sentiment and opinions. Exploring these sentiment and opinions towards to different topics are useful in many applications [12]. Besides, the sentiment mining about user generated content has played an important role in user interest mining, election poll and crisis management [5,15]. Therefore, sentiment classification has been a hot research topic in many fields.

There are many machine learning-based methods to explore sentiment by considering sentiment classification as text classification [24]. Mainly these methods are applying supervised machine learning techniques to train a sentiment classifier from a set of labeled data and predict the unlabeled data sentiment [3,13]. It is a key issue how to extract more complex and valuable feature [25]. A variety of feature extraction methods have been proposed, including single words [16], and other novel models [22]. Nevertheless, all these methods don't take the semantic relationships between words and the document topic feature

© Springer International Publishing AG 2017
G. Li et al. (Eds.): KSEM 2017, LNAI 10412, pp. 29–41, 2017.
DOI: 10.1007/978-3-319-63558-3_3

into consideration. Many studies take advantage of external knowledge to build a classification model on both labeled training data and hidden topics discovered from external knowledge [7,14,20,27]. Here we extract the topic of document as feature to reduce feature dimension.

Although much classification work has been done in detecting topic [1,4], these work mainly focus on discovering the topic of document, without any consideration about sentiment. Inspired by the idea of using external knowledge mentioned above, we present a general framework for building a classifier with compressed topic feature instead of word feature discovered from document integrated with external knowledge. In this paper, we focus on document-level sentiment classification in conjunction with topic analysis model LDA [1], and powerful machine learning method SVM.

The rest of this paper is organized as follows. In Sect. 2, we briefly review several representative works related to our method. Section 3 mainly introduces the Latent Dirichlet Allocation (LDA). Section 4 mainly presents our weakly supervised feature compression approach based on topic model in detail and the experiment results based on Chinese reviews and English reviews are illustrated and discussed in Sect. 5. Finally, Sect. 6 is the conclusion.

2   Related Work

A great deal of work has been focused on the problem of sentiment classification at various levels, from document level, to sentence and word level. At the early stage, Turney [17] used an unsupervised learning algorithm with mutual information to predict the semantic orientation at the word level. Pang et al. proposed a semi-supervised machine learning algorithm by employing a subjectivity detector at the sentence level [11]. At the document level, Wu [21] trained sentiment classifiers for multiple domains in a collaborative way based on multi-task learning. Zhang et al. mainly proposed a method for sentiment classification based on word2vec and SVM^{perf} [26].

By contrast, the supervised classification methods aim to train a sentiment classifier using labeled corpus. Pang et al. used SVM as the classifier to classify sentiment for the first time by using the n-grams model [13]. Wang et al. used document frequency (DF), information gain (IG), chi-squared statistic (CHI) and mutual information (MI) to choose features and then used boolean weighting method to set feature weight and constructed a vector space model [18,19]. Besides, a pre-training method applied to deep neural networks based on restricted Boltzmann machine was proposed to gain stable classification performance over short text [10].

The above work mainly focused on sentiment classification without considering about the topic feature of the document which can reduce the document feature dimension. Some other work has been done regarding to this direction. Earlier topic models such as LDA and PLSA [6] have mainly focused on extracting topics. Those models have been extended to explore text sentiment as shown in Sentiment-LDA and Dependency-Sentiment-LDA [8]. Yang et al. regard the visual feature as a mixture of Gaussian and treat the corpus of comments as a mixture of topic model [23]. Multilingual Supervised Latent Dirichlet Allocation (MLSLDA) is proposed for sentiment analysis on a multilingual corpus [2].

Motivated by these observations, we mainly take advantage of topic model to extract document topic feature and reduce the feature dimension. Here we use topic model to extract document topic feature and supervised method to classify the document instead of sentiment lexicon to represent the polarity of word. At the same time we can add the polarity feature weight by integrating with the external knowledge. More details will be discussed subsequently.

3   Hidden Topic Analysis Model

Topic model [1] represents each document as mixtures of (latent) topics, where each topic is a probability distribution over words. Consider a collection of M documents containing words from the vocabulary of terms $\{1, \ldots, V\}$, and let $\{1, \ldots, K\}$ be a set of topics. LDA was developed based on an assumption of document generation process depicted in both Algorithm 1.

Algorithm 1. Generation process for LDA

Require: α, β, Document m
for all topics $k \in [1, K]$ do
   sample word distribution $\phi_k \sim \mathrm{Dir}(\beta)$
end for
for all documents $m \in [1, M]$
   sample topic distribution $\theta_m \sim \mathrm{Dir}(\alpha)$
   for all words $n \in [1, N_m]$
      sample topic index $z_{m,n} \sim \mathrm{Mult}(\theta_m)$
      sample term for word $w_{m,n} \sim \mathrm{Mult}(\phi_{z_{m,n}})$
   end for
end for

Where α and β are Dirichlet parameters, θ_m is the topic distribution for document m, ϕ_k is the word distribution for topic k and N_m is the length of document m. $z_{m,n}$ is the topic index of n_{th} word in document m and $w_{m,n}$ is a particular word for word placeholder [m, n]. We can write the joint distribution of all known and hidden variables given the Dirichlet parameters as follows.

$$p(w_m, z_m, \theta_m, \phi | \alpha, \beta) = p(\phi|\beta) \prod_{n=1}^{N_m} p(w_{m,n}|\phi_{z_{m,n}}) p(z_{m,n}|\theta_m) p(\theta_m|\alpha) \quad (1)$$

The likelihood of a document w_m is obtained by integrating over θ_m, ϕ and summing over z_m as follows.

$$p(w_m|\alpha, \beta) = \iint p(\theta_m|\alpha) p(\phi|\beta) \prod_{n=1}^{N_m} p(w_{m,n}|\theta_m, \phi) d\phi d\theta_m \quad (2)$$

Finally, the likelihood of the whole data collection $W = \{w_m\}_{m=1}^M$ is as follows:

$$p(W|\alpha, \beta) = \prod_{m=1}^{M} p(w_m|\alpha, \beta) \tag{3}$$

The solution to estimate the posterior distribution is to use Gibbs sampling which is a popular Markov chain Monte Carlo algorithm to sample from complex high dimensional distributions. In each step, the algorithm randomly samples a topic assignment for a word w_i conditioned on the topic assignments of all other words. More formally, in each Gibbs sampling step, the algorithm replaces z_i by a topic drawn from the distribution $p(z_i|z_{\neg i}, w)$ which is given by:

$$p(z_i = k|z_{\neg i}, w, \alpha, \beta) = \frac{n_{k,\neg i}^{(t)} + \beta_t}{\left[\sum_{v=1}^{V} n_k^{(v)} + \beta_v\right] - 1} \frac{n_{m,\neg i}^{(k)} + \alpha_k}{\left[\sum_{j=1}^{K} n_m^{(j)} + \alpha_j\right] - 1} \tag{4}$$

Above, $n_{k,\neg i}^{(t)}$ is the number of times of word t which is assigned to topic k excluding the current assignment and $n_{m,\neg i}^{(k)}$ is the number of words in document m assigned to topic k except the current assignment. Samples obtained from the Markov chain are then used to estimate the distributions ϕ and θ as follows.

$$\phi_{k,t} = \frac{n_k^{(t)} + \beta_t}{\sum_{v=1}^{V} n_k^{(v)} + \beta_v} \tag{5}$$

$$\theta_{m,k} = \frac{n_m^{(k)} + \alpha_k}{\sum_{j=1}^{K} n_m^{(j)} + \alpha_j} \tag{6}$$

4 Weakly Supervised Feature Compression Based ELDA

The standard LDA topic model is completely unsupervised, and determines topics for words entirely based on the co-occurrence patterns of words across documents. Here we mainly take advantage of weakly supervised topic model to extract document topic feature.

We leverage an external knowledge to provide a form of weak supervision for the traditional LDA model, aiming at improving document polarity feature weight on English dataset and Chinese dataset. The English and Chinese external knowledge A which all contain 5000 positive reviews and 5000 negative reviews are obtained from the NLPCC 2014 sentiment analysis task[1]. Here the external knowledge should be high-quality for the classified document, so the data type should be same. We mainly use the external knowledge to bias the weight of

---

[1] http://tcci.ccf.org.cn/conference/2014/.

document polarity, so the topic feature which can represent polarity is most important. Phan et al. integrates external knowledge topic distribution with the original document and builds a classifier on both labeled training data and hidden topics discovered from the external knowledge [14]. By contrast, the purpose of our approach is to bias the topic-word distribution $\phi_{k,t}$ in favor of the words that are frequently annotated with topic k in A, while the document-topic distribution $\theta_{m,k}$ is in favor of topics that frequently occur in document m in A.

The problem of statistical inference involves estimating the probability distribution $\phi_{k,t}$ over words associated with each topic k, the distribution $\theta_{m,k}$ over topics for each document m.

Given the hyper-parameters of dirichlet α and β, posterior distribution of $\phi_{k,t}$ on A is defined as $P(\phi_{k,t}|\beta, A)$ which is equivalent to $Dir(\phi_{k,t}|\beta_t + \delta_k^t)$. Here, δ_k^t is the number of times of term t which is assigned topic k in A. Similarly, we can show that $\theta_{m,k}$ has a Dirichlet posterior with hyper-parameters $\alpha_k + \delta_m^k$ where δ_m^k is the number of annotated terms in document m that are assigned topic k in A. Finally, we get the conditional probability that $z_i = k$ (for word $w_i = t$ in document m) given the observed external knowledge A as:

$$p(z_i = k|z_{\neg i}, w, \alpha, \beta) = \frac{n_{k,\neg i}^{(t)} + \beta_t + \delta_k^t}{\left[\sum_{v=1}^{V} n_k^{(v)} + \beta_v + \delta_k^v\right] - 1} \frac{n_{m,\neg i}^{(k)} + \alpha_k + \delta_m^k}{\left[\sum_{j=1}^{K} n_m^{(j)} + \alpha_j + \delta_m^j\right] - 1} \quad (7)$$

Above, $n_{k,\neg i}^{(t)}$ is the number of times of word t which is assigned to topic k excluding the current assignment and $n_{m,\neg i}^{(k)}$ is the number of words in document m assigned to topic k except the current assignment.

In this paper, we incorporate external knowledge A into the new Gibbs sampling equation above by adding the prior counts δ_k^t and δ_m^k observed in A to the counts $n_{k,\neg i}^{(t)}$ and $n_{m,\neg i}^{(k)}$, respectively. This has the effect of biasing the topic assignment for a word (in each Gibbs sampling step) in favor of topics that frequently appear in A for either the word or the document containing it. Furthermore, this topic bias spreads to other occurrences of the word in the document corpus as well as to co-occurring words, and recursively through them to more words. Samples obtained from the Markov chain are then used to estimate the distributions ϕ and θ as follows.

$$\phi_{k,t} = \frac{n_k^{(t)} + \beta_t + \delta_k^t}{\sum_{v=1}^{V} n_k^{(v)} + \beta_v + \delta_k^v} \quad (8)$$

$$\theta_{m,k} = \frac{n_m^{(k)} + \alpha_k + \delta_m^k}{\sum_{j=1}^{K} n_m^{(j)} + \alpha_j + \delta_m^j} \quad (9)$$

Firstly, we run Gibbs sampling on the external knowledge A using Algorithm 1 (z_a denotes the topic assignment for words in A). Regarding the combination of the external knowledge A and documents D as $A \cup D$, we only sample topic for words in D while keeping the topics for words in A fixed at z_a and save the result using Algorithm 2 where M represents the document number of $A \cup D$.

Algorithm 2. Generation process for ELDA

Require: α, β, Document $A \cup D$
for all topics $k \in [1, K]$ do
    sample word distribution $\phi_k \sim \text{Dir}(\beta)$
end for
for all documents $m \in [1, M]$
  if $m <$ the document number of A
    for all words $n \in [1, N_m]$
      keep topic index as z_a saved from Algorithm 1
      sample term for word $w_{m,n} \sim \text{Mult}(\phi_{z_{m,n}})$
    end for
  else
    sample topic distribution $\theta_m \sim \text{Dir}(\alpha)$
    for all words $n \in [1, N_m]$
      sample topic index $z_{m,n} \sim \text{Mult}(\theta_m)$
      sample term for word $w_{m,n} \sim \text{Mult}(\phi_{z_{m,n}})$
    end for
  end if
end for

It is proved that the above method can bias the document-topic distribution to improve the corresponding document polarity feature weight. Here we describe how we present the document feature to build a classifier. Document m contains N_m words as $\{w_1, w_2, \ldots, w_{N_m}\}$. At last, we use ELDA to get the document-topic distribution $\theta_{m,k}$ to represent feature of document m as $\{p_{m1}, p_{m2}, \ldots, p_{mk}\}$ which can reduce document dimension to get the compressed feature at the same time. p_{mk} represents the probability that topic k is associated with document m.

5 Experiment Results

In this section, we mainly present the experiment results. Firstly we introduce two real-word sentiment datasets in the experiment. We evaluate the performance of our approach by comparing it with other methods and explore the influence of parameter setting on the performance of our approach.

5.1 Data Preparation

Two publicly available datasets are used in our experiment which contain the famous Amazon product review dataset [21] first used and the hotel review dataset in Chinese[2]. The Amazon dataset contains four different categories of product reviews crawled from Amazon.com including Book, DVD, Electronics and Kitchen. The detailed statistics are summarized in Table 1.

---

[2] http://www.datatang.com.

Table 1. Statistic of datasets

Dataset	Book	DVD	Electronics	Kitchen	Total	Hotel
Positive	1000	1000	1000	1000	4000	3000
Negative	1000	1000	1000	1000	4000	3000

Several preprocessing steps are taken before experiments. All English words are converted to lower cases. We mainly use JieBa Tokenization[3] to process hotel review dataset and all stopwords are removed. We take advantage of 5-fold cross-validation to select the most appropriate parameters of the classifier. We randomly allocate 80% of the data as training dataset, and the rest is used as testing dataset. We conduct each experiment ten times independently and select the average result as the final result.

5.2   Performance Evaluation

Firstly, we conduct some comparison experiments to demonstrate the effectiveness of our approach. Here we compare the proposed method with other baselines in terms of accuracy, precision, recall and F1-score. The detailed baselines are shown as follows.

- **NB-SVM:** We use the Naive Bayes with Support Vector Machine (NB-SVM) that can compute a log-ratio vector between the average word counts extracted from the positive reviews and the average word counts extracted from the negative reviews to consider a supervised weight of the counts.
- **Word2Vec-RNN:** It is used for extracting the feature of document word. This method maps each word to n-dimensional real vector. We regard these vectors as the feature of each document and the input of recurrent neural network(RNN).
- **LDA-SVM:** It mainly extracts the hidden topics from user reviews [9]. It then takes the topic distributions of reviews as the document features to train SVM as a classifier. We use this method to compare the effectiveness of joint modeling part of the proposed model and this alternative method.
- **sLDA-SVM:** It is a supervised latent dirichlet allocation, a statistical model of labelled documents [18]. A response variable (sentiment label) associated with each document is added to the LDA model.
- **ELDA-SVM:** It employs the proposed ELDA method which leverages external knowledge to bias the document-topic distribution. It can improve document sentiment feature weight and compress document feature. Here we empirically set parameters $\alpha = 50/k$, $\beta = 0.1$.

The experimental results are presented in Table 2. Overall, ELDA-SVM outperforms the baselines in both datasets. NB-SVM only considers word-of-bag feature and Word2Vec-RNN just translates the document word to the

---

[3] https://github.com/fxsjy/jieba.

Table 2. Performs of sentiment classification

Data	Method	Accuracy	Precision	Recall	F1-score
Book	NB-SVM	79%	85.5%	75.66%	80.28%
	Word2Vec-RNN	77.75%	75%	78.95%	76.92%
	LDA-SVM	83.75%	83.98%	84.39%	84.18%
	sLDA-SVM	87.5%	89.95%	84.58%	87.18%
	ELDA-SVM	**91.25%**	**91.79%**	**90.4%**	**91.09%**
DVD	NB-SVM	81%	88.5%	76.96%	82.33%
	Word2Vec-RNN	77.75%	79%	77.03%	78.02%
	LDA-SVM	85%	86.67%	83.25%	84.92%
	sLDA-SVM	89%	**94.12%**	84.21%	88.89%
	ELDA-SVM	**90.25%**	93.3%	**87.44%**	**90.27%**
Electronics	NB-SVM	86.5%	86%	86.87%	86.43%
	Word2Vec-RNN	81.5%	80%	82.47%	81.22%
	LDA-SVM	87.5%	87.89%	86.08%	86.98%
	sLDA-SVM	87%	88.4%	86.73%	87.56%
	ELDA-SVM	**94.5%**	**93.24%**	**96.02%**	**94.06%**
Kitchen	NB-SVM	83.25%	83.5%	83.08%	83.29%
	Word2Vec-RNN	80.5%	80%	80.81%	80.4%
	LDA-SVM	86.75%	86.41%	87.68%	87.04%
	sLDA-SVM	86.25%	86.21%	86.63%	86.42%
	ELDA-SVM	**95%**	**95.94%**	**94.03%**	**94.97%**
Total	NB-SVM	83.5%	85.75%	82.06%	83.86%
	Word2Vec-RNN	79.63%	79.13%	77.92%	79.52%
	LDA-SVM	84.75%	84.52%	84.94%	84.73%
	sLDA-SVM	87.5%	87.08%	87.95%	87.52%
	ELDA-SVM	**92.5%**	**92.8%**	**91.86%**	**92.33%**
Hotel	NB-SVM	88.5%	86%	90.53%	88.21%
	Word2Vec-RNN	83.25%	82.41%	83.67%	83.04%
	LDA-SVM	85.63%	85.62%	84.74%	85.18%
	sLDA-SVM	90.18%	91.1%	89.56%	90.31%
	ELDA-SVM	**92.25%**	**91.17%**	**93.19%**	**92.17%**

n-dimensional vector. LDA-SVM incorporates topic models to represent the feature of review information. However it doesn't present the sentiment feature specially. Although sLDA-SVM introduces supervised dirichlet allocation, it is more focusing on document label without considering the document word sentiment weight. By contrast, the proposed method compresses the document feature and increases the weight of sentiment feature with the introduction of the external knowledge.

5.3 Results with Different Topics

We also conduct experiment to see how classification accuracy changes if we change the number of hidden topics. We estimate LDA model, sLDA model and ELDA model for the Amazon review dataset and the hotel review dataset with different topic number $T \in \{20, 40, 60, 80, 100, 120, 140, 160\}$. After doing topic inference, SVM classifiers are built on the training dataset according to different topic number. Then we use the remaining dataset to test and measure classification accuracy. Here we choose to show the change of classification accuracy in the Book, Electronic, all Amazon dataset and hotel dataset. As can be seen from Fig. 1(a, b, c), LDA-SVM, sLDA-SVM and ELDA-SVM have a significant improvement over the baselines in all of dataset. Especially in Book, there is around 12% improvement at $T = 140$ over the NB-SVM baseline and around 14% improvement at $T = 140$ over the Word2Vec-RNN baseline. From Fig. 1(d) we also can find that ELDA-SVM can perform significantly better than all the state-of-the-art baseline methods when $T = 100$ and $T = 120$ in terms of accuracy on Chinese dataset.

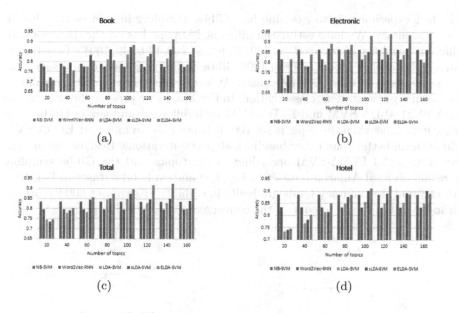

Fig. 1. (a)–(d) Accuracy with varying number of topics

When comparing LDA-SVM with ELDA-SVM, we can find they keep the same tendency and they all perform much better (actually the best results in the experiment) when $T = 120$ on electronic dataset. We also can find that ELDA-SVM can perform better than sLDA-SVM along with the rising topic number. The ELDA-SVM appears to be a more appropriate model design for document sentiment classification.

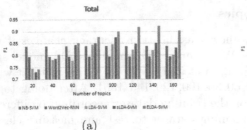
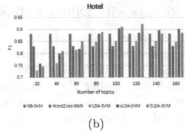

(a)   (b)

Fig. 2. (a)–(b) F1 score on Total dataset and Hotel dataset

The evaluation of F1 score with different topic number for total dataset and hotel dataset is shown in Fig. 2(a, b). It also clearly shows that ELDA-SVM can achieve higher performance in document sentiment classification.

5.4 Results with Different Gibbs Sampling Iterations

The last experiment is to examine how Gibbs Sampling influences the classification accuracy. We have estimated different LDA models on the datasets with different numbers of topics $T \in \{20, 40, 60, 80, 100, 120, 140, 160\}$. To estimate parameters of each model, we ran 1000 Gibbs Sampling iteration, and saved the sampling result at every 200 iterations. At last, we use these results to validate the accuracy of sentiment classification. In Fig. 3(a, b) we present the accuracy of LDA-SVM, sLDA-SVM and ELDA-SVM with different numbers of Gibbs sampling iterations when the topic is set 100. It is not hard to find that ELDA-SVM can perform better than other baselines after 600 iterations. Besides, we present the accuracy of ELDA-SVM according to the topics and the Gibbs sampling iteration on total Amazon dataset in Fig. 4(a) and on hotel dataset in Fig. 4(b). We can find the accuracy tends gradually in stability after about 600 iterations, although it is difficult to control the convergence of Gibbs Sampling.

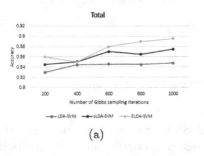
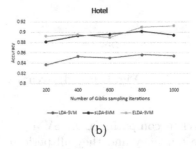

(a)   (b)

Fig. 3. (a)–(b) Accuracy changes with Gibbs sampling iterations

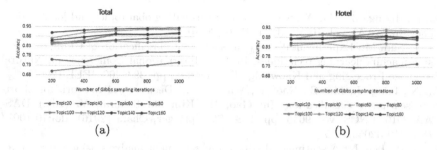

Fig. 4. (a)–(b) Accuracy changes with topic and Gibbs sampling iterations

6 Conclusion

In this paper, we introduce ELDA, a LDA-based new method for extracting document feature, to reach the goal of feature compression. Here we regard the hidden topic as the feature of the document and leverage external knowledge to bias the document-topic distribution. Then we use SVM as the classifier and the experimental results clearly show that our method can outperform the baselines significantly on both datasets in terms of classification accuracy.

Acknowledgement. This work was supported by Natural Science Foundation of China (No. 61170192). Li L. is the corresponding author for the paper.

References

1. Blei, D.M., Ng, A.Y., Jordan, M.I.: Latent dirichlet allocation. J. Mach. Learn. Res. 3(Jan), 993–1022 (2003)
2. Boyd-Graber, J., Resnik, P.: Holistic sentiment analysis across languages: multilingual supervised latent dirichlet allocation. In: Proceedings of the 2010 Conference on Empirical Methods in Natural Language Processing, pp. 45–55. Association for Computational Linguistics (2010)
3. Go, A., Bhayani, R., Huang, L.: Twitter sentiment classification using distant supervision. CS224N Project Report, Stanford, vol. 1, p. 12 (2009)
4. Griffiths, T.L., Steyvers, M.: Finding scientific topics. Proc. Nat. Acad. Sci. 101(Suppl. 1), 5228–5235 (2004)
5. Hirschberg, J., Manning, C.D.: Advances in natural language processing. Science 349(6245), 261–266 (2015)
6. Hofmann, T.: Probabilistic latent semantic indexing. In: Proceedings of the 22nd Annual International ACM SIGIR Conference on Research and Development in Information Retrieval, pp. 50–57. ACM (1999)
7. Kataria, S.S., Kumar, K.S., Rastogi, R.R., Sen, P., Sengamedu, S.H.: Entity disambiguation with hierarchical topic models. In: Proceedings of the 17th ACM SIGKDD International Conference on Knowledge Discovery and Data Mining, pp. 1037–1045. ACM (2011)

8. Li, F., Huang, M., Zhu, X.: Sentiment analysis with global topics and local dependency. In: AAAI, vol. 10, pp. 1371–1376 (2010)
9. Li, J., Sun, M.: Experimental study on sentiment classification of Chinese review using machine learning techniques. In: 2007 International Conference on Natural Language Processing and Knowledge Engineering, pp. 393–400. IEEE (2007)
10. Li, X., Pang, J., Mo, B., Rao, Y., Wang, F.L.: Deep neural network for short-text sentiment classification. In: Gao, H., Kim, J., Sakurai, Y. (eds.) DAS-FAA 2016. LNCS, vol. 9645, pp. 168–175. Springer, Cham (2016). doi:10.1007/978-3-319-32055-7_15
11. Pang, B., Lee, L.: A sentimental education: sentiment analysis using subjectivity summarization based on minimum cuts. In: Proceedings of the 42nd Annual Meeting on Association for Computational Linguistics, p. 271. Association for Computational Linguistics (2004)
12. Pang, B., Lee, L.: Opinion mining and sentiment analysis. Found. Trends Inf. Retrieval **2**(1–2), 1–135 (2008)
13. Pang, B., Lee, L., Vaithyanathan, S.: Thumbs up? Sentiment classification using machine learning techniques. In: Proceedings of the ACL 2002 Conference on Empirical Methods in Natural Language Processing-Volume 10, pp. 79–86. Association for Computational Linguistics (2002)
14. Phan, X.-H., Nguyen, L.-M., Horiguchi, S.: Learning to classify short and sparse text & web with hidden topics from large-scale data collections. In: Proceedings of the 17th International Conference on World Wide Web, pp. 91–100. ACM (2008)
15. Ren, F., Ye, W.: Predicting user-topic opinions in twitter with social and topical context. IEEE Trans. Affect. Comput. **4**(4), 412–424 (2013)
16. Tan, S., Zhang, J.: An empirical study of sentiment analysis for Chinese documents. Expert Syst. Appl. **34**(4), 2622–2629 (2008)
17. Turney, P., Littman, M.L.: Unsupervised learning of semantic orientation from a hundred-billion-word corpus (2002)
18. Wang, H., Yin, P., Yao, J., Liu, J.N.K.: Text feature selection for sentiment classification of Chinese online reviews. J. Exp. Theoret. Artif. Intell. **25**(4), 425–439 (2013)
19. Wang, H., Yin, P., Zheng, L., Liu, J.N.K.: Sentiment classification of online reviews: using sentence-based language model. J. Exp. Theoret. Artif. Intell. **26**(1), 13–31 (2014)
20. Wang, J., Li, L., Tan, F., Zhu, Y., Feng, W.: Detecting hotspot information using multi-attribute based topic model. PLoS ONE **10**(10), e0140539 (2015)
21. Wu, F., Huang, Y.: Collaborative multi-domain sentiment classification. In: 2015 IEEE International Conference on Data Mining (ICDM), pp. 459–468. IEEE (2015)
22. Xia, R., Zong, C.: Exploring the use of word relation features for sentiment classification. In: Proceedings of the 23rd International Conference on Computational Linguistics: Posters, pp. 1336–1344. Association for Computational Linguistics (2010)
23. Yang, Y., Jia, J., Zhang, S., Boya, W., Chen, Q., Li, J., Xing, C., Tang, J.: How do your friends on social media disclose your emotions? In: AAAI, vol. 14, pp. 1–7 (2014)
24. Yin, P., Wang, H., Zheng, L.: Sentiment classification of Chinese online reviews: analysing and improving supervised machine learning. Int. J. Web Eng. Technol. **7**(4), 381–398 (2012)
25. Zhai, Z., Hua, X., Kang, B., Jia, P.: Exploiting effective features for Chinese sentiment classification. Expert Syst. Appl. **38**(8), 9139–9146 (2011)

26. Zhang, D., Hua, X., Zengcai, S., Yunfeng, X.: Chinese comments sentiment classification based on word2vec and SVM perf. Expert Syst. Appl. **42**(4), 1857–1863 (2015)
27. Zhu, Y., Li, L., Luo, L.: Learning to classify short text with topic model and external knowledge. In: Wang, M. (ed.) KSEM 2013. LNCS, vol. 8041, pp. 493–503. Springer, Heidelberg (2013). doi:10.1007/978-3-642-39787-5_41

An Effective Gated and Attention-Based Neural Network Model for Fine-Grained Financial Target-Dependent Sentiment Analysis

Mengxiao Jiang[1], Jianxiang Wang[1], Man Lan[1,2(✉)], and Yuanbin Wu[1,2]

[1] Department of Computer Science and Software Engineering,
East China Normal University, Shanghai 200062, People's Republic of China
{51151201080,51141201062}@stu.ecnu.edu.cn, {mlan,ybwu}@cs.ecnu.edu.cn
[2] Shanghai Key Laboratory of Multidimensional Information Processing,
Shanghai, China

Abstract. In this work, we propose an effective neural network architecture GABi-LSTM to address fine-grained financial target-dependent sentiment analysis from financial microblogs and news. We first adopt a gated mechanism to adaptively integrate character level and word level embeddings for word representation, then present an attention-based Bi-LSTM component to embed target-dependent information into sentence representation, and finally use a linear regression layer to predict sentiment score with respect to target company. Comparative experiments on financial benchmark datasets show that our proposed GABi-LSTM model outperforms baselines and previous top systems by a large margin and achieves the state-of-the-art performance.

Keywords: Target-dependent sentiment analysis · Financial domain · Attention neural network · Gate mechanism · Stock market prediction

1 Introduction

Sentiment analysis deals with the treatment of opinion, sentiment, and subjectivity in product reviews and microblogs [1–3], which has gained significant popularity over the past few years in natural language processing (NLP). In financial domain, sentiment analysis has been proved to be useful in many economic and financial applications, such as market dynamic prediction [4], movie box office forecasting [5], stock prices forecasting [6], etc. In order to predict stock market, many studies have been carried out on different data sources, e.g., microblogs [7,8], reviews [9], Tweets [10–12] and financial news [9,12,13].

Clearly, the *sentiment* in financial domain indicates the stock market movement of target company, i.e., bullish (optimistic; believing that the stock price will increase) or bearish (pessimistic; believing that the stock price will decline), which is not exactly the same as the positive or negative emotions in product reviews. For example, in product reviews, *"increasing price"* usually indicates

© Springer International Publishing AG 2017
G. Li et al. (Eds.): KSEM 2017, LNAI 10412, pp. 42–54, 2017.
DOI: 10.1007/978-3-319-63558-3_4

bad news (i.e., negative emotion), but in financial domain *"increasing price"* of stock is welcome news. To make it easier to look at stock news, the social media services Twitter and StockTwits have been curating a \$-tagged stock information "cashtag" (a "\$" followed by a ticker symbol, e.g., \$GOOG). For example, a message from StockTwits with ID 7744550 *"Following this morning's EPS, I blew out \$ZMH and added to \$DNKN (longs)."*, expressed negative and positive sentiments for two cashtags \$ZMH and \$DNKN, respectively.

In this work, we focus on fine-grained sentiment analysis on financial microblogs and news. Given above example, we not only identify sentiment orientation but also predict sentiment scores for target cashtags (in this case, the sentiment scores for two cashtags are marked as -0.462 and 0.528 by human annotators). To address it, we propose a gated and attention-based neural network model GABi-LSTM. Specifically, the first convolutional layer extends widely-used word embedding into fine-grained character-level embedding in order to capture morphological information and alleviate rare word problem (such as symbols and numbers appearing frequently in financial domain) as well. Moreover, to effectively integrate character embedding into word embedding for word representation, we propose a novel gate mechanism to adaptively choose most appropriate way to represent each word rather than simple concatenation operation. Most important, we propose an attention mechanism to select out the most relevant part from text in terms of target company (i.e., cashtag in microblogs or company name in news) and to assign them higher weight for sentence representation. In this way, our proposed model not only takes financial domain into consideration but also embeds target-dependent information into representation.

The main contributions of this work are summarized as follows:

- Our proposed neural network model GABi-LSTM is capable of automatically conducting fine-grained target-dependent sentiment analysis without complicated human-made features or pre-defined sentiment lexicons.
- Comparative experiments on two benchmark datasets in financial microblogs and news showed that our GABi-LSTM model achieves the state-of-the-art performance and outperforms the baseline systems by a large margin.

2   Related Work

A vast amount of research has been dedicated to detect the correlations between media data and stock market. For example, [14] proved that investor sentiment is positively associated with future stock price crash risk and [7] revealed the effectiveness of microblogs for stock market prediction. Previous studies performed financial sentiment analysis on different data sources, e.g., microblog data [7,8], film reviews [9], Tweets [10–12] and financial news [9,12,13], with different methds, e.g., sentiment-based trading strategy [9] and time series models [8,10].

In recent years, most previous work adopted elaborately designed NLP features with machine learning algorithms [11–13,15–17]. Besides, a hybrid method with machine learning and lexicon-based approach [12,16] has been widely used and [12] achieved the state-of-the-art performance recently. Meanwhile, with the development of deep learning methods, various neural network models have been proposed. For example, [18] combines the Convolution Neural Network (CNN) and Long Short-Term Memory (LSTM) and [19] adopts a Bidirectional Gated Recurrent Unit (Bi-GRU). However, these neural network models are still in infancy to deal with fine-grained sentiment analysis on financial microblogs and news and their performances are much lower than traditional methods.

3   The Proposed Neural Network Model GABi-LSTM

3.1   Motivation

The proposed architecture of GABi-LSTM model is motivated by the observations and analysis we made on as follows.

First, fine-grained financial sentiment analysis on microblogs is a domain-dependent task. The words *"put"*, *"call"* and *"long"* regarded as neutral words in traditional sentiment analysis but express a strong sentiment tendency in stocks prediction, which results in many pre-defined sentiment lexicons in traditional NLP tasks not suitable for financial sentiment analysis.

Second, unlike product reviews or news domain, financial microblogs have several characteristics: (1) the sentence length is relatively short; (2) they usually contain a lot of numbers and non-English characters, e.g., "\$, %, +, −, #, @, -"; (3) many sentences are incomplete, such as *"Worst performers today"*. These differences inspire us to encode words in both character-level and word-level embeddings and then integrate them using a gate mechanism.

Third, the fine-grained target-dependent sentiment analysis task aims to predict the sentiment score associated with a target company. In order to capture the relationship between the target company and contexts, an attention mechanism is proposed to embed target company information into sentence representation.

3.2   The Overview of Model Architecture

Given a sentence associated with a target company as input, the model is designed to return a sentiment score reflecting the bullish or bearish sentiment of the target company. Figure 1 shows the architecture of the proposed model GABi-LSTM, which consists of three modules. First, a **word representation** module is to map each word in the given input sentence into a vector through:

- **Word-level Embedding Layer** maps each word into a vector using a pre-trained word lookup table.
- **Charcter-level Embedding Layer** maps each word into a vector at the character level using a CNN.
- **Gate Layer** adaptively integrates word-level and character-level representations using a gate mechanism to get the final word representation.

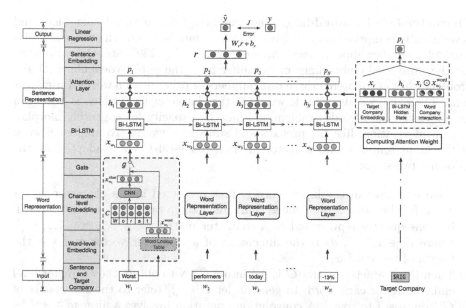

Fig. 1. Architecture of the proposed model GABi-LSTM.

Then, a **sentence representation** module is to learn sentence vector through:

- **Bi-LSTM Layer** models word interactions based on word representations.
- **Attention Layer** produces an attention (i.e., weight) vector over hidden states of Bi-LSTM using an attention mechanism, which is to select the most relevant words in terms of given target company in the sentence.
- **Sentence Embedding Layer** obtains sentence vector representation.

Finally, a **Linear Regression** Layer is to perform sentiment score prediction. The detailed description of these three modules in GABi-LSTM is as follows.

3.3   Word Representation with Gated Char- and Word- Embedding

Word representation module aims to map each word of the input sentence to a low-dimensional vector. For each word, (1) a pre-trained word lookup table is used to get the word-level embedding; (2) a character-level convolutional neural network (Char-CNN) is adopted to get the character-level embedding; (3) then, a gate mechanism is followed to adaptively integrate word-level and char-level embeddings to represent the word.

Word-Level Embedding. In this module, each word w_i in the input sentence is projected to a low-dimensional vector $x_{w_i}^{word} \in \mathbb{R}^{d_w}$ by a pre-trained word lookup table $E \in \mathbb{R}^{|V| \times d_w}$: $x_{w_i}^{word} = E v_{w_i}$, where $|V|$ is the vocabulary size, d_w is the dimension of a word vector and $v_{w_i} \in \mathbb{R}^{|V|}$ is the one-hot representation of w_i. The word look up table is obtained from publicly available *Google word2vec*[1] trained on 100 billion words from Google News with Skip-gram model [20].

---

[1] https://code.google.com/archive/p/word2vec.

Character-Level Embedding. The sentences from financial microblogs and news usually comprise many different forms of numbers, but the frequency of a particular number appears very low, such as "+5%", "$30", etc. Only considering the word-level embeddings might suffer from the out-of-vocabulary (OOV) problem when encountered these rare numbers or other unseen words during test time. Therefore, we also encode words at the character level which is beneficial for alleviating rare words problem and has the capacity for capturing morphological information, like the prefixes and suffixes of words. To map each word of the input sentence into a vector, a character-level CNN is adopted by the following two steps:

1. Each character in the input word is first projected into a character vector $x_{c_j} \in \mathbb{R}^{d_c}$ by using a pre-defined character lookup table. Thus each word will be consequently represented as a character matrix $C = [x_{c_1}, x_{c_2}, \ldots, x_{c_s}]$, where $C \in \mathbb{R}^{d_c \times s}$, d_c is the dimension of a character vector and s is the length of the word w_i.

2. Then the character matrix C is fed into a CNN to obtain the character level embedding for each word. In general, let $C[i : j]$ refer to the sub-matrix of C from row i to row j. A convolution operation involves a filter $w \in \mathbb{R}^{h \times d_c}$, which is applied to a window of h characters to produce a new *feature map* $c \in \mathbb{R}^{s-h+1}$ for this filter:

$$c_i = f(w \cdot C[i : i + h - 1] + b) \tag{1}$$

where $i = 1, \ldots, s - h + 1$, $\cdot$ is a sum over element-wise multiplications, f is a non-linear function such as the hyperbolic tangent and $b \in \mathbb{R}$ is a bias term. The max pooling operation is then applied over each feature map to take the maximum value $\hat{c} = max\{c\}$. We use n filters and then all n outputs generated from each feature map by the max pooling operation are concatenated to get the character-level vector representation $x_{w_i}^{char} \in \mathbb{R}^n$ of the input word w_i, defined as: $x_{w_i}^{char} = \hat{c}_1 \oplus \hat{c}_2 \ldots \hat{c}_n$.

Gate Mechanism. The gate is a way to optionally let information through, which is composed of a sigmoid neural net layer and a pointwise multiplication operation. The sigmoid layer describes how much of each component should be let through. Note that word-level embedding can capture the semantic and syntactic information and character-level embedding can capture the morphological information well. Considering these two embeddings have their own advantages and can not be replaced with each other, we propose a gate mechanism to adaptively choose the most appropriate way to represent each word based on these two types of word vectors, defined as:

$$x'^{char}_{w_i} = W_d x_{w_i}^{char} + b_d \tag{2}$$

$$g = \sigma(W_g[x_{w_i}^{word}, x'^{char}_{w_i}] + b_g) \tag{3}$$

$$x_{w_i} = g x_{w_i}^{word} + (1 - g) x'^{char}_{w_i} \tag{4}$$

where $\boldsymbol{W}_d \in \mathbb{R}^{d_w \times n}$, $\boldsymbol{b}_d \in \mathbb{R}^{d_w}$ are the transform parameters to project $\boldsymbol{x}_{w_i}^{char}$ into the same dimension space as $\boldsymbol{x}_{w_i}^{word}$ to get $\boldsymbol{x'}_{w_i}^{char} \in \mathbb{R}^{d_w}$. [ ] means the concatenation operation of vectors, $\sigma(\cdot)$ is a sigmoid function, $\boldsymbol{w}_g \in \mathbb{R}^{2d_w}$ and $b_g \in \mathbb{R}$. And $g \in \mathbb{R}$ is the gate expressing the weight between $\boldsymbol{x}_{w_i}^{word}$ and $\boldsymbol{x'}$. Finally, we obtain $\boldsymbol{x}_{w_i}$ as the final representation for the input word w_i.

3.4 Sentence Representation with Attention-Based Bi-LSTM

Sentence Representation Module is used to represent the input sentence as a fixed size vector. LSTM is proved to be particularly useful for modeling sequential data and can specifically address the long-term dependency problem compared with the conventional recurrent neural network (RNN). However, LSTM can only capture the interaction between the current word and preceding words, but ignore successive words which are significant for sentence representation. Therefore, we utilize bidirectional LSTM (Bi-LSTM) to get the sentence representation, which has a forward LSTM and a backward LSTM used to capture the information from both past and future words. Moreover, in order to capture the most relevant part from the sentence according to the target company, we proposed to introduce the attention mechanism to embed target company into sentence representation.

Bi-LSTM. Given the input sentence $S = [\boldsymbol{x}_{w_1}, \boldsymbol{x}_{w_2}, \ldots, \boldsymbol{x}_{w_N}]$, for each time step i, the step of LSTM computation corresponds to:

$$i_i = \sigma(\boldsymbol{W}_i[\boldsymbol{x}_{w_i}, \boldsymbol{h}_{i-1}] + \boldsymbol{b}_i) \tag{5}$$

$$\boldsymbol{f}_i = \sigma(\boldsymbol{W}_f[\boldsymbol{x}_{w_i}, \boldsymbol{h}_{i-1}] + \boldsymbol{b}_f) \tag{6}$$

$$\boldsymbol{o}_i = \sigma(\boldsymbol{W}_o[\boldsymbol{x}_{w_i}, \boldsymbol{h}_{i-1}] + \boldsymbol{b}_o) \tag{7}$$

$$\tilde{\boldsymbol{c}}_i = tanh(\boldsymbol{W}_c[\boldsymbol{x}_{w_i}, \boldsymbol{h}_{i-1}] + \boldsymbol{b}_c) \tag{8}$$

$$\boldsymbol{c}_i = \boldsymbol{i}_i \odot \tilde{\boldsymbol{c}}_i + \boldsymbol{f}_i \odot \boldsymbol{c}_{i-1} \tag{9}$$

$$\boldsymbol{h}_i = \boldsymbol{o}_i \odot tanh(\boldsymbol{c}_i) \tag{10}$$

where $\odot$ denotes element-wise multiplication. $\boldsymbol{i}_i, \boldsymbol{f}_i, \boldsymbol{o}_i, \boldsymbol{c}_i$ denote the input gate, forget gate, output gate, memory cell respectively. Moreover, since a Bi-LSTM is used in our model, at each position i of the sequence, we obtain a forward hidden state $\overrightarrow{\boldsymbol{h}}_i$ and a backward hidden state $\overleftarrow{\boldsymbol{h}}_i$, where $\overrightarrow{\boldsymbol{h}}_i, \overleftarrow{\boldsymbol{h}}_i \in \mathbb{R}^{d_h}$. We concatenate them to produce the intermediate state, i.e., $\boldsymbol{h}_i = [\overrightarrow{\boldsymbol{h}}_i, \overleftarrow{\boldsymbol{h}}_i] \in \mathbb{R}^{2d_h}$.

Attention Mechanism. As we mentioned before, the sentiment score is associated with a target company in the input sentence, and it varies widely with respect to different companies. Therefore, we design an attention mechanism which is capable of capturing the most relevant part of sentence in response to a given company and gives them a higher weight.

The target company is represented as a vector $x_t \in \mathbb{R}^{d_w}$ through the word lookup table E. The attention mechanism will produce an attention weight vector $p = [p_1, p_2, \cdots, p_N]$ used to weight the hidden states $H = [h_1, h_2, \cdots, h_N]$ output by Bi-LSTM. Each attention weight p_i for h_i is computed by following,

$$p_i = softmax(W_a[x_t, h_i, x_t \odot x_{w_i}^{word}] + b_a) \tag{11}$$

where $W_a \in \mathbb{R}^{2(d_w + d_h)}$, $b_a \in \mathbb{R}$, $softmax(z_i) = e^{z_i} / \sum_j e^{z_i}$. p is a attention (probability) vector over the inputs and can be viewed as the weights of the words measuring to what degree our model should pay attention to. Next, we sum over the state h_i weighted by the attention vector p to compute the representation $r \in \mathbb{R}^{2d_h}$ for the input sentence: $r = \sum_{i=1}^{N} p_i h_i$.

3.5   Linear Regression

Since the sentiment scores for the given sentences are continuous, this prediction task requires a regression. Instead of using a softmax classifier, we use a linear function in the output layer, defined as,

$$\hat{y} = W_r r + b_r \tag{12}$$

where $W_r \in \mathbb{R}^{2d_h}$ and $b_r \in \mathbb{R}$ are the trainable weight and bias parameters respectively, $\hat{y}$ is the sentiment score predicted by the model. We define the training loss (to be minimized) as the mean square error (MSE) between the predicted $\hat{y}$ and ground-truth y:

$$J(\theta) = \frac{1}{2m} \sum_{k=1}^{m} ||\hat{y}^{(k)} - y^{(k)}||^2 \tag{13}$$

where $y^{(k)}$ and $\hat{y}^{(k)}$ are gold and predicted labels, m is instances count.

3.6   Parameter Learning

We fix the lengths of sentences (number of words) to be 50 and the lengths of words (number of characters) to be 30, and apply truncating or zero-padding when necessary. The dimensions for character embeddings, word embeddings and target embeddings are 30, 300 and 300 respectively. The other parameters are initialized by sampling from a uniform distribution $U(-0.5, 0.5)$. In the Char-CNN module, we choose three groups of 50 filters, with filter window sizes of (2, 3, 4). The dimensions of the hidden states d_h in Bi-LSTM is set to 200 and 100 for microblog and news data respectively. We use AdaGrad [21] with a learning rate of 0.001 and a minibatch size 64 to train the model.

4   Experiments

4.1   Datasets

We use the two datasets of SemEval 2017 Task 5[2], consisting of financial microblogs messages (from Twitter or StockTwits) and news statements or headlines collected by [22]. Each instance is associated with a target company (i.e., cashtag in microblogs or company name in news). Specially, the data derived from microblog messages is sentence fragments (namely *"Spans"*) which are extracted from the raw messege according to the target company, whereas the data derived from news statements or headlines is complete sentences (namely *"Title"*). Besides, for microblog messages, we rebuild the raw messege (namely *"Text"*) of training and test set respectively with the official API of Twitter and StockTwits. In our experiments, for both data sources we randomly selected 80% as training dataset and the remaining 20% as development dataset. Table 1 shows the statistics of the datasets.

Table 1. Statistics of two datasets in SemEval 2017 Task 5.

Source		Dataset	Instance	Positive	Negative	Neutral
Microblog	Twitter	Train	765	246	510	9
		Test	365	116	243	6
	StockTwits	Train	934	330	586	18
		Test	429	141	280	8
News		Train	1156	658	460	38
		Test	491	276	203	12

4.2   Evaluation Measure

The evaluation measure is *weighted cosine similarity* (*WCS*). The scores are conceptualised as vectors, where each dimension represents a company within a given microblog message or news text. Note that the both vectors will have the same number of dimensions as the companies for which sentiment needs to be assigned will be given in the input data. Cosine similarity and cosine_weight are calculated as in Eq. (14),

$$cosine(G, P) = \frac{\sum_{i=1}^{m} G_i \times P_i}{\sqrt{\sum_{i=1}^{m} G_i^2} \times \sqrt{\sum_{i=1}^{m} P_i^2}}; \quad cosine\_weight = \frac{m}{M} \quad (14)$$

where m is the number of instances predicted by system, M is the total number of instances in test data, P and G are the vectors of predicted scores and gold standard scores for m predicted instances, respectively. Since not all instances would be predicted by system, the *cosine_weight* is a penalty weight and the final score is the product of the cosine similarity and the cosine_weight (i.e., $WCS = cosine\_weight * cosine(G, P)$).

---

[2] http://alt.qcri.org/semeval2017/task5/index.php?id=data-and-tools.

4.3　Experimental Results

Table 2 shows the experiments of our proposed models and several baselines on microblog and news test datasets. Standard CNN, LSTM and Bi-LSTM with word-only embeddings serve as three baselines. The CNN utilizes the vector after the max pooling layer, the LSTM and Bi-LSTM adopt the average of hidden state vectors to represent the input sentence. We see that these models have consistent results on both microblog and news datasets and observe the following findings.

Table 2. Comparison of different methods on microblog and news datasets in terms of *WCS* (%), ⊕ denotes concatenation operator.

Sentence representation	Word representation	*WCS* (%)		
		Microblog		News
		Spans	Text	Title
CNN	Word only	72.56	66.48	70.65
LSTM	Word only	74.11	67.37	72.35
Bi-LSTM	Word only	75.32	68.40	73.30
Bi-LSTM	Char only	73.53	68.57	69.42
	Word ⊕ Char	76.45	70.15	75.31
	Gated word & Char	77.66	71.37	76.16
Attention-based Bi-LSTM	Word only	77.52	70.41	75.05
	Word ⊕ Char	78.81	72.82	76.52
	Gated word & Char	**79.21**	**73.47**	**77.43**

First, from the first group of results (first three rows) in the Table 2, Bi-LSTM outperforms other two standard CNN and LSTM with only word level embeddings. This demonstrates the importance of taking into account both past and future information of the context in encoding the sentence, which is consistent with previous studies. Meanwhile, we also see that LSTM performs better than CNN, which proves the effectiveness of LSTM in modeling sequential sentence.

Second, in the second group of experiment, we choose Bi-LSTM as the sentence representation model and investigate the performance of different word representations. We find that the integration of character level and word level embeddings performs better than any of the individual embedding. This is not surprising since the integration can take into account both the advantages of character and word level embeddings for word representation. That is, the word-level embedding is able to capture semantic and syntactic information and the character-level embedding takes the word morphological information and rare symbols into consideration, which is crucial for sentiment analysis in financial domain. Furthermore, regarding integration strategy, the gate mechanism outperforms the simple and straightforward concatenation of these two types of embeddings, which indicates that the gate mechanism does help to choose a better way for word representation.

Third, from the third group of results, we find that no matter what kind of word representation is taken, attention-based Bi-LSTM achieves substantially higher performance over the Bi-LSTM. This is because the attention mechanism captures the most relevant part of the sentence in terms of the target company and embeds the target information into the final sentence representation. This demonstrates the effectiveness of the attention mechanism for capturing interaction between target company and context. Overall, GABi-LSTM model achieves the best performance using the proposed attention-based Bi-LSTM model incorporating with *Gated Word & Char* embeddings, which indicates the effectiveness of the proposed method.

Moreover, comparing the experimental results on *Spans* and the *Text* in microblog dataset, we find that *Spans* achieves results highly superior to the *Text*. The reasons are from: (1) The *Text* contains more noises and non-standard words, such as the hashtag, slangs and elongated words; (2) The *Text* may contain more than one companies and their sentiments are mixed in one message.

4.4   Comparison with the State-of-the-Art Systems

Table 3 shows the comparison between our proposed GABi-LSTM model and top systems reported in SemEval 2017 Task 5, where *ML, DL* and *Lex* denote *Machine Learning, Deep Learning* and *Lexicon* respectively. Most studies have adopted a combination of deep learning and pre-defined sentiment lexicons [16,18,19,23], or a method of traditional NLP features with pre-defined sentiment lexicons [12,25]. Several studies adopt a single deep learning [24] or a pure machine learning method [15,17]. From Table 3, we find that the proposed GABi-LSTM outperforms the state-of-the-art systems on both datasets. Specificly, the GABi-LSTM system improves approximately 2% on microblog dataset, and 3% on news dataset compared with the reported best systems. Overall, the experimental results show the superiority of the proposed model GABi-LSTM in dealing with the fine-grained financial target-dependent sentiment analysis. Moreover, another advantage of the proposed model is that it does not require expensive hand-crafted features or external resources (pre-defined lexicons).

Table 3. Comparison with the top systems reported in SemEval 2017.

Rank	Microblog			News		
	Technique	System	WCS (%)	Technique	System	WCS (%)
1	Hybrid (ML, Lex)	ECNU [12]	77.79	Hybrid (DL, Lex)	Fortia-FBK [23]	74.52
2	Hybrid (DL, Lex)	IITP [18]	75.13	Hybrid (DL,Lex)	RiTUAL-UH [19]	74.37
3	ML	SSN_LMRG1 [15]	73.47	ML	TakeLab [17]	73.27
4	Hybrid (DL, Lex)	HHU [16]	72.96	DL	Lancaster A [24]	73.20
5	Hybrid (ML, Lex)	IITPB [25]	72.56	Hybrid (ML,Lex)	ECNU [12]	71.01
	Our model	GABi-LSTM	**79.21**	Our model	GABi-LSTM	**77.43**

4.5 Qualitative Visualization Analysis

To further understand the attention mechanism within our model, we performed visualized analysis on a randomly selected example from the microblog dataset: *"Worst performers today: $RIG −13%, $EK −10%, $MGM $IO −6%, $CAR −5.5%/best stock: $WTS +15%"*. Figure 2 shows the relative magnitudes of the attention weights with respect to target companies for the individual words in the example sentence. We see that the words like "worst", "$MGM", "−6" and "%" are assigned higher attention values when the target company is "$MGM", while words such as "best", "$WTS" and "+15" contribute greatly to the target company "$WTS". Additionally, we observe that words such as "today", which are rather irrelevant with respect to the target company, indeed have lower attention scores. The results show the effectiveness of our attention mechanism in capturing the most relevant fragments to the target company in the sentence.

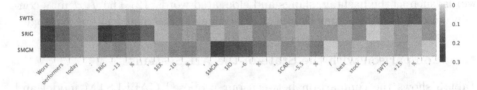

Fig. 2. Visualized analysis for the example from microblog dataset. $RIG, $MGM and $WTS are target companys. Darker shades of blue indicate stronger attention values.

5 Concluding Remarks

To address fine-grained target-dependent sentiment analysis from financial microblogs and news, we proposed an effective neural network model GABi-LSTM, consisting of a CNN layer to achieve character-level embedding in order to capture morphological information and alleviate rare word problem, a gated component to integrate character embedding into word embedding for word representation and an attention-based Bi-LSTM component to embed target-dependent information into sentence representation. Extensive comparative experiments on financial benchmark datasets show the effectiveness of GABi-LSTM.

Acknowledgments. This work is supported by grants from Science and Technology Commission of Shanghai Municipality (14DZ2260800 and 15ZR1410700), Shanghai Collaborative Innovation Center of Trustworthy Software for Internet of Things (ZF1213) and NSFC (61402175).

References

1. Pang, B., Lee, L., et al.: Opinion mining and sentiment analysis. Found. Trends® Inf. Retrieval, **2**(1–2), 1–135 (2008)
2. Pak, A., Paroubek, P.: Twitter as a corpus for sentiment analysis and opinion mining. In: LREc, vol. 10 (2010)
3. Liu, B.: Sentiment analysis and opinion mining. Synth. Lect. Hum. Lang. Technol. **5**(1), 1–167 (2012)
4. Nassirtoussi, A., Aghabozorgi, S., Wah, T., Ngo, D.: Text mining of news-headlines for forex market prediction: a multi-layer dimension reduction algorithm with semantics and sentiment. Expert Syst. Appl. **42**(1), 306–324 (2015)
5. Asur, S., Huberman, B.: Predicting the future with social media. In: 2010 IEEE/WIC/ACM International Conference on WI-IAT, vol. 1, pp. 492–499 (2010)
6. Oh, C., Sheng, O.: Investigating predictive power of stock micro blog sentiment in forecasting future stock price directional movement (2011)
7. Oliveira, N., Cortez, P., Areal, N.: The impact of microblogging data for stock market prediction: using Twitter to predict returns, volatility, trading volume and survey sentiment indices. Expert Syst. Appl. **73**, 125–144 (2017)
8. Wang, Y.: Stock market forecasting with financial micro-blog based on sentiment and time series analysis. J. Shanghai Jiaotong Univ. (Sci.) **22**(2), 173–179 (2017)
9. Kazemian, S., Zhao, S., Penn, G.: Evaluating sentiment analysis in the context of securities trading. In: ACL (2016)
10. Si, J., Mukherjee, A., Liu, B., Li, Q., Li, H., Deng, X.: Exploiting topic based Twitter sentiment for stock prediction. In: ACL (2) (2013)
11. Mittal, A., Goel, A.: Stock prediction using twitter sentiment analysis. Standford University, CS229, vol. 15 (2012)
12. Jiang, M., Lan, M.: An ensemble of regression algorithms with effective features for fine-grained sentiment analysis in financial domain. In: SemEval-2017
13. Van de Kauter, M., Breesch, D., Hoste, V.: Fine-grained analysis of explicit and implicit sentiment in financial news articles. Expert Syst. Appl. **42**(11), 4999–5010 (2015)
14. Yin, Y., Tian, R.: Investor sentiment, financial report quality and stock price crash risk: role of short-sales constraints. Emerg. Markets Finan. Trade **53**(3), 493–510 (2017)
15. Milton Rajendram, S., Angel Deborah, S., Mirnalinee, T.T.: SSN MLRG1 at SemEval-2017 task 5: finegrained sentiment analysis using multiple kernel Gaussian process regression model. In: SemEval-2017
16. Romberg, J., Cabanski, T., Conrad, S.: HHU at SemEval-2017 task 5: fine-grained sentiment analysis on financial data using machine learning methods. In: SemEval-2017
17. Šnajder, J., Rotim, L., Tutek, M.: Takelab at semeval-2017 task 5: linear aggregation of word embeddings for fine-grained sentiment analysis of financial news. In: SemEval-2017
18. Akhtar, M.S., Ekbal, A., Ghosal, D., Bhatnagar, S., Bhattacharyya, P.: IITP at SemEval-2017 task 5: an ensemble of deep learning and feature based models for financial sentiment analysis. In: SemEval-2017
19. Maharjan, S., Kar, S., Solorio, T.: RiTUAL-UH at SemEval-2017 task 5: sentiment analysis on financial data using neural networks. In: SemEval-2017
20. Mikolov, T., Sutskever, I., Chen, K., Corrado, G., Dean, J.: Distributed representations of words and phrases and their compositionality. In: Advances in Neural Information Processing Systems, pp. 3111–3119 (2013)

21. Duchi, J., Hazan, E., Singer, Y.: Adaptive subgradient methods for online learning and stochastic optimization. J. Mach. Learn. Res. **12**, 2121–2159 (2011)
22. Davis, B., Cortis, K., Vasiliu, L., Koumpis, A., McDermott, R., Handschuh, S.: Social sentiment indices powered by x-scores. In: ALLDATA 2016, p. 21 (2016)
23. Ferradans, S., Guerini, M., Mansar, Y., Gatti, L., Staiano, J.: Fortia-FBK at SemEval-2017 task 5: bullish or bearish? Inferring sentiment towards brands from financial news headlines. In: SemEval-2017
24. Andrew, M., Paul, R.: Lancaster a at SemEval-2017 task 5: evaluation metrics matter: predicting sentiment from financial news headlines. In: SemEval-2017
25. Sethi, A., Ekbal, A., Biemann, C., Akhtar, M.S., Kumar, A., Bhattacharyya, P.: IITPB at SemEval-2017 task 5: sentiment prediction in financial text. In: SemEval-2017

A Hidden Astroturfing Detection Approach Base on Emotion Analysis

Tong Chen[1], Noora Hashim Alallaq[2], Wenjia Niu[1(✉)], Yingdi Wang[1],
Xiaoxuan Bai[1], Jingjing Liu[1], Yingxiao Xiang[1], Tong Wu[3(✉)],
and Jiqiang Liu[1]

[1] Beijing Key Laboratory of Security and Privacy in Intelligent Transportation,
Beijing Jiaotong University, 3 Shangyuan Village, Haidian District,
Beijing 100044, China
niuwj@bjtu.edu.cn
[2] School of Information Technology, Deakin University,
Burwood, VIC 3125, Australia
[3] First Research Institute of The Ministry of Public Security of PRC,
Tsinghua University, Beijing, China
tongwu@mail.tsinghua.edu.cn

Abstract. This paper aims to take detection of hidden astroturfing based on emotion analysis. We propose a hidden astroturfing detection method which combines emotion analysis and unfair rating detection together. This approach contains five functional modules as: a data crawling module, pre-processing module, bag-of-word establishment module, emotion mining and analysis module and matching module. We give ROC curve (AUC) to evaluate the approach proposed in this paper. The results show that this method can realize the detection of implicit astroturfing under the prerequisite of improving the emotion classification accuracy. Our work discovers and studies a new hidden astroturfing characteristic, and construct a corpus manually for text emotion classification that establish a basis for our future research.

Keywords: Astroturfing detection · Emotion analysis · Algorithm

1 Introduction

The development of internet technologies has seen the emergence of a *Astroturfing*, a phenomenon which has becomes a hot research topic recent years. In most cases of astroturfing, hired groups or individuals support arguments or claims favoured by employers while challenging critics and denying adverse claims [1]. In order to prevent these unfair or even illegal situations, studying the detection and identification of astroturfing is necessary.

W. Niu and T. Wu—This research is supported by the National Natural Science Foundation of China (No. 61672091).

© Springer International Publishing AG 2017
G. Li et al. (Eds.): KSEM 2017, LNAI 10412, pp. 55–66, 2017.
DOI: 10.1007/978-3-319-63558-3_5

Confrontation is very important in the context of astroturfing. *Confrontation* refers to the confrontation that astroturfing against existing variety of water military identification and detection technologies. Avoiding public disclosure of covert practice is important to the success of astroturfing if it is to succeed in influencing public opinions. If astroturfers are identified, their accounts will usually be cancelled by sites. It will be very costly to re-register accounts and such disclosure will reduce its credibility. Astroturfer are thus extremely focused on *concealment* when they take actions. Emotion has close relationship with concealment in the context of astroturfing. Astroturfing can add negative feedback to the review of highly praised goods, or in contrast add positive feedback to the review of lowly regarded goods. This is the mainstream of astroturfing, is relatively easy to discover. However, if astroturfing gives a good rating to highly praised product and a negative review in the text comment, can covertly influence public opinion. In its attempt at concealment, this practice differs is different from the main method way of astroturfing, and constitutes a separate trend in astroturfing. The contrary-to-emotional rating is an important clue in hidden astroturfing detection.

Serval relevant works [6,10,11] in this area have some shortcomings, we can make a summary as follows:

1. Ignoring the feature of astroturfing that emotion may be is inconsistent with rating. As many works in this field either focus on opinion mining or unfair ratings detection only, they do not pay attention to the relationship between both sides. This paper mainly focuses on this inconsistency to identify the abnormal astroturfing behaviour.
2. Not paying attention to the processing speed of astroturfing detection. With the rapid development of e-commerce, more and more users begin shopping online which increases the scale of user data set. For a data set like this, processing efficiency naturally becomes the focus of our attention. Therefore, how to deal with a big data set, and detect abnormal astroturfing behaviour quickly is the research topic of this paper.

This paper proposes a hidden emotion-based approach to detect this kind of abnormal user behaviour. In order to achieve this number of tasks must be accomplished, including: the processing of data according to several basic features of astroturfing to filter of original users' data recording, removing redundant data to reduce the amount of data processed later, using the machine learning method SVM to undertake emotion classification (which can reduce the processing time), and setting matching rules to do quickly match user comment emotion and rating.

The structure of this paper is as follows: Sect. 2 will discuss related work in this field; Sect. 3 gives schematic representation of the architecture in this paper; Sect. 4 will describe in detail the implementation of each functional module contained in the architecture proposed in this paper; Sect. 5 gives the experiment and evaluation.

2   Related Work

As review opinion is the content feature of astroturfing, while rating belongs to the category of behaviour, so the method proposed in this paper is to use multiple features to undertake the detection of astroturfing. This section briefly surveys some related works in the field, including works that discuss the use multiple features to detect abnormal users, the extraction of opinion from reviews, unfair ratings detection and review spam detection. Meanwhile, we introduce the unique contribution of this paper, and make comparison with others.

Astroturfing Detection Base on Multiple Features. In the field of using multiple features to detect astroturfing, there exist large number of works. Jiang et al. take advantage of behaviour-based method and content-based method to accomplish astroturfing detection [4]. Ghosh and Vismanat combine content-based method and structure-based method together to undertake the detection [3].

Review Opinion Extracted and Unfair Rating Detection. In the context of opinion mining, there are serval works that deal with the problem of extracting positive or negative opinions from product reviews [7,9]. These works achieve the extraction of opinion from reviews, but they cannot detect make a detection of astroturfing through the emotion extracted.

Wu et al. propose a distortion criterion to identify suspicious reviews [13], addressing the trustworthiness of reviews. Meanwhile, Dellarocas propose and evaluate a set of mechanisms which can eliminate or significantly reduce the effects of unfair ratings [2].

All these works, in the context of unfair rating detection as discussed above, consider the serious impact of dishonest ratings, and propose detection approaches to deal with this problem. However, they do not directly detect or review spam activities.

Review Spam Detection. This is a relatively new filed that has not been well studied. Jindal and Liu give a first attempt in studying opinion spam in reviews, identifying three types of spam [5]. However, this work does not pay attention to the relationship between review opinion and ratings.

Emotion-Based Astroturfing Detection and Comparison with Others. This paper focuses on emotion-based astroturfing detection, and proposes a hidden astroturfing detection method can the inconsistency of review opinion and rating. In contrast to works discussed above, our method can not only detect review opinion and consider unfairness of users' rating, but can also connect emotion with rating as an indicative factor in the detection of astroturfing.

3   Hidden Astroturfing Detection Method

As shown in Fig. 1 is the schematic representation of the proposed architecture. The proposed emotion-based astroturfing detection method architecture

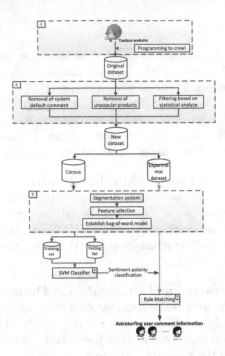

Fig. 1. Schematic representation of the emotion-based astroturfing detection method framework

is divided into the following functional modules: data crawling module, pre-processing module, bag-of-word establishment module, emotion mining and analysis module and matching module.

In this section, we will describe in detail the implementation of each functional module contained in the architecture proposed in this paper, and show how to achieve hidden detection of astroturfing while improving the accuracy of emotion classification.

3.1   Data Preparation

Taobao is the largest e-commerce platform in China. In order to maintain a strong position under the severe business competition on the platform, companies use of astroturfing to further their interest has become an open secret. Hence we choose this site as the data source for our experiment.

For each product on this site, users can give a text review of their user experience, and give a rating in the range of 1 to 5 stars. For research purposes, we programmed a crawl of data set from this site, which contains user and product attributes as follows: (a) users' text comments; (b) star rating; (c) user ID; (d) product shop id; (e) comment time.

3.2 Pre-processing Operations

We take several pre-processing steps to the original data set before we use it.

1. *Removal of unpopular products:* We only focused on products that attracted sufficient attention, in other words these products must possess sufficient users' reviews and ratings. This is because the basic features of astroturfing are large numbers and significant impact on public opinion. If a product contains few comments, it cannot fulfil these requirements. Therefore, in this paper we restrict the selection of products to those above a threshold number of reviews.
2. *Removal of system default comment:* The evaluation system of Taobao is designed such that, if the user does not provide a comment for a set of time after receive the product, then it will be automatically given a comment like system default praise and a 5 stars rating. Obviously, comments and ratings like this do not make sense for our research, therefore in our experiment we omit system default comments.
3. *Filtering based on statistical analysis:* We carry out a statistical analysis of each different product in the original data set based on the inherent character- istics of astroturfing. For each product, we generate a line chart which takes time as its horizontal axis and the number of reviews as vertical axis, setting a minimum number of reviews as restricting threshold. For all line charts generated, we find out the area in which the number of reviews exceeds the threshold we set, furthermore under unit period. Finally, we extract users' review which fall in this area of the line charts to create as a new data set as data base for the following operations.

3.3 Bag-of-Word Module Construction

For the bag-of-words construction, we first need to extract user comments to create a data set $D_{comment}$ which is used as database in this step. As there is no separation between words in Chinese naturally, the first step of this phase is to undertake sentences segmentation in $D_{comment}$. The Chinese word segmentation system used in this paper is *ICTCLAS* (Institute of Computing Technology, Chi- nese Lexical Analysis System), which possesses a high performance and accuracy.

The next step, is the selection of classification features. In this paper, we choose CHI [14] to do the selection. This method is used to measure the sta- tistical correlation strength between the feature t_j and the document category C_j, and assumes that the χ^2 distribution with first-order degrees of freedom is satisfied between t_j and C_j. The higher χ^2 a feature possesses the more infor- mation it contains, and the greater the feature correlation with this class. CHI is calculated as follow:

$$\chi^2(t_j, C_j) = \frac{N \times (AD - CB)^2}{(A + C) \times (B + D) \times (A + B) \times (C + D)}$$

where A indicates the number of documents that contain features t_j and belongs to class C_i; B indicates the number of documents that contain the features t_j

but not belongs to category C_j; C indicates the number of documents belong to class C_j but not including feature t_j; D indicates the number of documents that do not belong to category C_j and do not contain the feature t_j. With this method, we can achieve a better classification result by reducing more than 85% feature dimension.

Finally, we get a classification feature set $F = (f_1, f_2, \ldots, f_m)$ to establish the bag-of-words model, which transforms each sentence in $D_{comment}$ into a vector for the following operation. Let $\{f_1, \ldots, f_m\}$ generated above be the m feature vectors that appear in $D_{comment}$, let $n_i(s)$ be the number of times that f_i appears in the sentence s contained in $D_{comment}$. Then each sentence s can be represented as a vector

$$s := (n_1(s), n_2(s), \ldots, n_m(s))$$

3.4   Emotion Mining and Analysis

Emotion mining and analysis can be transformed in to a classification problem, so we can use machine learning methods to solve it. The prerequisite for using this method is the preparation for training and testing set.

Training and Testing Set Preparation. In this paper, we divide the corpus generated manually into two parts: one for training, the other used for testing.

For all parts, we tag their emotional categories manually dividing them into two classes: Positive(P), Negative(N). We choose to split it into two categories because in [10] the author establish that this division achieves the highest precision. Each class of emotion is described as follows:

1. Positive- the sentence expresses a positive emotion in general. For example, "This product is really good, I like it very much!" we will tag sentence like this to be Positive(P).
2. Negative- the sentence express a negative emotion in general. For example, "The phone is unfriendly, and battery of it discharge so fast." we will tag sentence like this to be Negative(N).

Classifier Selection. The three most frequently used machine learning methods in classification tasks are Naive Bayes, Support Vector Machine and Maximum Entropy. Pang et al.'s comparison of these three approaches [7], Support Vector Machine(SVM) obtained the highest accuracy. Therefore, we choose SVM to implement the opinion mining.

In the traditional classification problem, the basic idea of SVM is not to divide samples into two categories simply, but to search for the hyperplane that maximizes the margin. The solution can written as

$$\omega := \sum_j \alpha_j c_j s_j, \alpha_j \geq 0$$

where α_j can be obtained by solving a optimization problem, $c_j \in -1, +1$ represented as the positive and negative samples, and s_j is the support vector.

We use the prepared training set to contribute this SVM classifier. Finally, we classify the experimental data with the trained classifier to predict their emotional categories.

3.5   Matching

In order to identify astroturfing, we set rules for matching after opinion mining of users' comments. matching user review and rating, rules proposed in this paper are as follows:

$$IF(emotion = negative)AND(rating = highrating)THEN(outputtheuser'sID)$$

$$IF(emotion = positive)AND(rating = lowrating)THEN(outputtheuser'sID)$$

where *high rating* and *low rating* are distinguished by setting the threshold. In this paper, we use this method to do the mapping between user comment emotion and corresponding scores, subsequently outputting the hidden astorturfing user.

3.6   Summary

Using functional modules described above, the algorithm **HDM** (hidden astroturfing detection method) proposed in this paper can be summarized as follow.

In this algorithm, input $D_{original}$ is the data set that was crawled from Taobao site, output L_{ID} is the collection whose owner is astroturfing. In the first part, $D_{remunpopular}$ is the data set after the removal of unpopular products, and $D_{remdefault}$ represents the data set after the removal all system default comments. In the second part of this algorithm, using segmentation system on D_{new} gives the word list L_{word}, then using it transform sentences in D_{new} into a vector that gives D which used later. In the third part, we construct the corpus used for training of SVM to get a classifier which can take classification of data set to get the emotion for each sentence. In the last part, we set rules to do the matching between emotion and rating, and output incompatible user IDs.

The specific process steps of the algorithm are as follows:

Algorithm HDM

$Input: D_{original}$
$Output: L_{ID}$

---

1. *Pre − processing* :
      i. $D_{remunpopular} \longrightarrow D_{remdefault}$
      ii. $D_{remdefault} \longrightarrow D_{new}$
2. *New dataset* :
      i. $D_{new} \xrightarrow{segmentation\ system} L_{word}$
      ii. $D_{new}(s, emotion) \xrightarrow{L_{word}} D(s, emotion)$
3. *Emotion mining and anaylsis* :
      i.   *construct corpus C*
      ii. *training with C get the classifier*
      iii. *use this classifier to do classification*
4. *Matching* :
      IF (*emotion* = *negative*) *AND* (*rating* = *high rating*)*THEN* (*output user's ID*)
      IF (*emotion* = *positive*) *AND* (*rating* = *low rating*)*THEN* (*output user's ID*)
5. L_{ID}

In this way, we achieve the purpose that undertakes hidden detection of astroturfing while improving the accuracy of emotion classification.

4   Experiment and Evaluation

In this section, we introduce the experimental preparation phase first. In the second part we identify hidden astroturfing in accordance with the new clue proposed in this paper: that user comment emotion is inconsistent with rating.

In the following two sections these phases described in detail.

4.1   Experiment Setup

The experiment setup consists of the following steps:

1. To crawl the data set, we use multiple severs which are set up on the cloud to do the data crawling on Taobao. For study purposes, we chose ten types of hot commodities under the clothing category as the target of crawling.

   We choose products belonging to the same domain because the author in paper [10] evaluates the affect of domain on precision, and shows that in the context of text opinion mining choosing tweets in same domain achieves the highest accuracy. Applying the same principle and extending its findings to this paper, we explore users comment emotion mining under a single commodity domain to obtain the highest classification precision. On the other hand, the reason for choosing hot commodities is that they possess a large amount of valuable uses information, including comments, ratings, comment time and users ID, and the probability that astroturfing exist in such product is higher. These conditions favourable to the detection of hidden astroturfing. In the crawled data set, we sampled 10,000 users evaluation uniformly to create a corpus used for model construction, and used the remaining part as the experimental data. The author in [10] shows that this size of corpus is sufficient for text opinion mining and can achieve ideal experimental accuracy.

2. In this paper, we chose traditional SVM as baseline methods to evaluate the emotion classification accuracy of HDM proposed in this paper.

3. The corpus is generated randomly according to the proportion of the number for each product to ensure the randomness of the corpus, thus ensuring the accuracy of our experiment.

   In addition, to generate the corpus used later, five manual taggers are employed to annotate the emotion of user comments. They are students majoring in computer science who are familiar with searching and reading product comments on Taobao. Each tagger reads 2,000 user comments, and gives a judgment to determine the polarity of emotion, classifying each comment as into positive (P) or negative (N).

4. In this paper, we use the ROC curve to evaluate the classification accuracy of our experiment. In our case, the distribution of favourable and negative ratings is not uniform in the crawled data set. Therefore, we implement an ROC curve to evaluate the accuracy in this paper.

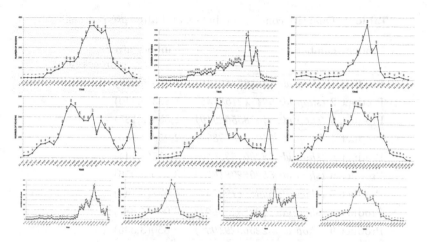

Fig. 2. Statistical analysis resulting

ROC curve plots TPR against FPR, and points on this curve represent the performance of each classifier.

$$TPR = TP/(TP + FN), FPR = FP/(TN + FP)$$

4.2 Evaluation Result

In this part, we introduce the results of the experiment in detail.

Resulting of Pre-processing Filter. We first give the results of statistical analysis of pre-processing filter in the pre-processing module.

As shown in Fig. 2 are the results for each product after statistical processing. In this paper, we generate ten analysis charts in total as follows. For each graph, we can thus filter the user comments according to a characteristic of astroturfing, namely that within a short period of time the number of comments increase rapidly. We only select reviews that meet the condition that the number of reviews exceeds the threshold we set a given period of time.

The original crawling data set can be filtered to reduce the amount of data to be processed later. Table 1 shows quantity comparison before and after pre-processing to the original data set, we can found that reduced by nearly 20% of the original data set.

Evaluation Result of Processing Speed. In order to measure and make comparison of the processing speed for our method purposed in this paper, we recording the *processing time* of our method and baseline method separately as a evaluation index.

To the same data set (43,337 user reviews in total) crawled from Web in the data preparation module, recording the *processing time* required for taking abnormal user behavior detection based on hidden astroturfing detection

Table 1. Quantity comparison before and after pre-processing

Product number	Number after (before pre-processing)
Product 1 *(shop id: 538867385545)*	4,953(5,200)
Product 2 *(shop id: 534142168735)*	5,823(5,940)
Product 3 *(shop id: 540785552910)*	1,354(2,185)
Product 4 *(shop id: 540355817870)*	2,958(4,155)
Product 5 *(shop id: 539967533600)*	4,915(5,144)
Product 6 *(shop id: 539062784072)*	3,370(5,056)
Product 7 *(shop id: 533824123534)*	4,632(5,144)
Product 8 *(shop id: 540853480604)*	3,009(4,882)
Product 9 *(shop id: 522566294932)*	2,650(3,336)
Product 10 *(shop id: 540093321953)*	1,590(2,405)

mechanism (HDM) purposed in this paper, and traditional machine learning mechanism (TDM) to do the detection which act as the baseline method for comparison respectively.

We can see that HDM method purposed in this paper only need 47,450 ms to complete the detection, while 267,345 ms are required when using the TMD method. As can be seen that, HDM proposed in this paper can improves the processing speed greatly.

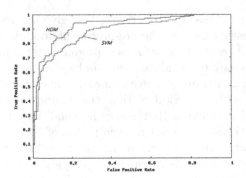

Fig. 3. ROC comparison between HDM and SVM.

The Evaluation Result for Accuracy of Emotion Classification. In this paper, we examine the accuracy of text emotion classification, and make comparison of AUC between HDM proposed in this paper and traditional SVM method.

The ROC curves are shown in Fig. 3. We can see that the AUC value for HDM is 0.9248, and the AUC value for traditional SVM is 0.8820. This result shows

Fig. 4. Interface of experimental output system.

that HDM proposed in this paper achieves better accuracy than traditional SVM. In other words, the pre-processing module can delete redundant data effectively, and improve the accuracy of emotion classification.

Experimental Output Result. Finally, we give the output result of our experiment.

In total 43,447 user reviews we crawled for study, 154 comment records have been discovered in which user review sentiment is inconsistent with their rating. Shown in Fig. 4 is the interface of experimental output system, we can see that the interface is divided into three parts. *Processing time*, is the processing time required for this experiment in total. The *ROC curve* gives an accuracy evaluation of the text emotion classification. In *detection result*, we give the astroturfing user records discovered by our method. Achieving the purpose that take detection of hidden astroturfing whose comment emotion is inconsistent with rating based on emotion analysis.

5 Conclusion

In this paper, we have focused on the detection of hidden astroturfing based on emotion analysis. The aspects included in this research can be summarized as follows: first, to combine the mining of review emotion and ratings, focusing on a new astroturfing characteristic that user comment emotion is inconsistent with ratings, and to use this characteristic to uncover of hidden astroturfing; second, the study of how to carry out this detection rapidly.

We derive a hidden astroturfing detection method (HDM) to solve this problem that improves the accuracy of emotion classification while ensuring rapid detection. This is in part achieved by the removal of system default comments and unpopular products in the pre-processing stage and meanwhile filtering a according to the basic features of astroturfing on the original users' data recording. In this way, we eliminate the redundant data, and make the training set used for machine learning later more effective. Furthermore, we construct the corpus manually for emotion classification, thus improve the accuracy of classification compared with the baseline method TDM used in this paper. For our future work, we will explore and study more hidden astroturfing behaviour characteristics and propose mechanisms to detect them.

References

1. Cho, C.H., Martens, M.L., Kim, H., et al.: Astroturfing global warming: it isn't always greener on the other side of the fence. J. Bus. Eth. **104**(4), 571–587 (2011)
2. Dellarocas, C.: Immunizing online reputation reporting systems against unfair ratings and discriminatory behavior. In: Proceedings of the 2nd ACM Conference on Electronic Commerce, pp. 150–157. ACM (2000)
3. Ghosh, S., Viswanath, B., Kooti, F., et al.: Understanding and combating link farming in the twitter social network. In: Proceedings of the 21st International Conference on World Wide Web, pp. 61–70. ACM (2012)
4. Jiang, M., Beutel, A., Cui, P., et al.: A general suspiciousness metric for dense blocks in multimodal data. In: 2015 IEEE International Conference on Data Mining (ICDM), pp. 781–786. IEEE (2015)
5. Jindal, N., Liu, B.: Opinion spam and analysis. In: Proceedings of the 2008 International Conference on Web Search and Data Mining, pp. 219–230. ACM (2008)
6. Pang, B., Lee, L.: A sentimental education: sentiment analysis using subjectivity summarization based on minimum cuts. In: Proceedings of the 42nd Annual Meeting on Association for Computational Linguistics, p. 271. Association for Computational Linguistics (2004)
7. Pang, B., Lee, L.: Opinion mining and sentiment analysis. Found. Trends? Inf. Retr. **2**(1C2), 1–135 (2008)
8. Pang, B., Lee, L., Vaithyanathan, S.: Thumbs up? Sentiment classification using machine learning techniques. In: Proceedings of the ACL-02 Conference on Empirical Methods in Natural Language Processing-Volume 10, pp. 79–86. Association for Computational Linguistics (2002)
9. Popescu, A.M., Etzioni, O.: Extracting product features and opinions from reviews. In: Kao, A., Poteet, S.R. (eds.) Natural Language Processing and Text Mining, pp. 9–28. Springer, London (2007). doi:10.1007/978-1-84628-754-1_2
10. Sidorov, G., et al.: Empirical study of machine learning based approach for opinion mining in tweets. In: Batyrshin, I., González Mendoza, M. (eds.) MICAI 2012. LNCS, vol. 7629, pp. 1–14. Springer, Heidelberg (2013). doi:10.1007/978-3-642-37807-2_1
11. Thelwall, M., Wilkinson, D., Uppal, S.: Data mining emotion in social network communication: gender differences in MySpace. J. Assoc. Inf. Sci. Technol. **61**(1), 190–199 (2010)
12. Wilson, T., Hoffmann, P., Somasundaran, S., et al.: OpinionFinder: a system for subjectivity analysis. In: Proceedings of HLT/EMNLP on Interactive Demonstrations, pp. 34–35. Association for Computational Linguistics (2005)
13. Wu, G., Greene, D., Smyth, B., et al.: Distortion as a validation criterion in the identification of suspicious reviews. In: Proceedings of the First Workshop on Social Media Analytics, pp. 10–13. ACM (2010)
14. Zheng, Z., Srihari, S.N.: Text categorization using modified-CHI feature selection and document/term frequencies. In: ICMLA, pp. 143–146 (2002)

Leveraging Term Co-occurrence Distance and Strong Classification Features for Short Text Feature Selection

Huifang Ma[✉], Yuying Xing, Shuang Wang, and Miao Li

College of Computer Science and Engineering, Northwest Normal University,
Lanzhou, China
mahuifang@yeah.net

Abstract. In this paper, a short text feature selection method based on term co-occurrence distance and strong classification features is presented. On the one hand, co-occurrence distance between terms in each document is considered to determine the co-occurrence distance correlation, based on which the correlation weight for each term can be defined. On the other hand, the improved expected cross entropy is defined to obtain the weight of a term in a particular class with strong class indication. All terms of each class is sorted in a descending order based on their weights and top-k terms are selected as feature terms. Experiments show that our method can improve the effectiveness of short text feature selection.

Keywords: Short text · Co-occurrence distance · Strong classification feature · Expected cross entropy · Feature selection

1 Introduction

In recent years, with the rapid growing of Web and social media, more and more information exist in the form of short texts and tend to grow explosively. Different kinds of feature selection approaches have been put forward to reduce dimensionality in the past years. To be more specific, there are two main methods of feature extraction [1]. One is feature selection, which refers to choosing a subset of features from the original features and the feature space is optimally reduced by a certain criterion. The other is feature extraction, which means that a set of new features is constructed from the original features. They are used either in isolation or in combination.

Term weighting has been proved to be an effective way to improve the expressiveness of short text classification. There are two kinds of traditional weighting methods: unsupervised methods, such as term frequency (TF), term frequency-inverse document frequency (TF*IDF) [2] and supervised methods, such as information gain (IG), expected cross entropy (ECE), $tf*\chi^2$ and so on. From the point of view of co-occurrence between terms, the two terms are considered to be related if they frequently co-occur with each other in the entire corpus. Due to the fact that short text contains few words, the co-occurrence distance between two terms can also cause a certain influence on their semantic relation. Standing from the angle of different classes, if one term distributes more evenly between each class, it hardly makes any contribution

© Springer International Publishing AG 2017
G. Li et al. (Eds.): KSEM 2017, LNAI 10412, pp. 67–75, 2017.
DOI: 10.1007/978-3-319-63558-3_6

to classification. If one term is mainly focused on a particular class, while at the same time does not exist in other class, such word is able to represent the characteristics of this class. Similarly, if a feature-word is distributed in one class, it will provide a good representation of feature of this class.

Based on the above consideration, a new feature selection method for short text is proposed, which is called Leveraging Term Co-occurrence Distance and Strong Classification Features for Short Text Feature Selection, CDCFS, for short. First of all, co-occurrence distance between two terms are considered to calculate the correlation degree between them, which can reveal the hidden semantic correlation between terms. And then, we present an improved ECE method, which can fully take category information of short texts into consideration. Finally, the weight of each term is calculated combining the above information. Experimental results on both Chinese data sets and English data sets show our method significantly enhance the effectiveness of short text feature selection compared with existing methods.

The remainder of this paper is organized as follows. The relevant theoretical knowledge are presented in Sect. 2. Section 3 introduces the proposed leveraging term co-occurrence distance and strong classification features for short text feature selection. Experimental designs and findings are presented in Sect. 4. Section 5 concludes the proposed work and points out our future work.

2   Problem Preliminaries

In this paper, we focus on paper titles of computer science both in Chinese and English. Let $D = \{d_1, d_2, \ldots, d_m\}$ be the short text corpus, where m is the number of short texts in D. $W = \{t_1, t_2, \ldots, t_n\}$ denotes the terms of D, where n is the number of unique words in D. $C = \{c_1, c_2, \ldots, c_l\}$ is the collection of class labels, where l is the number of class labels.

2.1   Correlation of Two Terms in a Text

The correlation between term t_i and term t_j in a certain short text d_s is defined as follows [3]:

$$cor_{d_s}(t_i, t_j) = \begin{cases} e^{-dist_{d_s}(t_i,t_j)} & t_i \in d_s \ and \ t_j \in d_s \\ 0 & t_i \notin d_s \ or \ t_j \notin d_s \end{cases} \quad (1)$$

where $dist_{d_s}(t_i, t_j)$ is co-occurrence distance of term t_i and term t_j in short text d_s [4], i.e. the number of terms between t_i and t_j in d_s, $|j-i|$. Our method adopts co-occurrence distance computational method, which takes term context into account and makes the calculation become more reasonable.

2.2   Expected Cross Entropy

Expected Cross Entropy, or ECE, is a kind of feature selection measure which is based on information theory and takes category knowledge into consideration. It also considers both word frequency and relations between term and class. The bigger the ECE value, the stronger indication the feature-term has for the purpose of classification, which is defined as follows:

$$ECE(t_i) = P(t_i) \sum_r p(C_r|t_i) \log_2 \frac{P(C_r|t_i)}{P(C_r)} \tag{2}$$

where $P(t_i)$ shows the probability of term t_i appeared in the corpus. $P(C_r)$ indicates the possibility of short texts belonging to class C_r in the corpus. $P(C_r|t_i)$ is conditional probability of class C_r given a specific term t_i.

3 The Proposed Approach

3.1 Terming Weighting Method Based on Co-occurrence Distance

Short text feature selection is a crucial step of short text mining, which will eventually be helpful to capture key information from massive data quickly and effectively [5]. The purpose of feature selection is removing words which can rarely affect decision to improve correctness and efficient of short text classification. Traditional term weighting methods [6, 7] do not consider correlation semantic information and category distribution information of terms. To address this problem, we proposed CDCFS and it mainly contains the following steps:

(1) Preprocessing the initial short text set D′ of k classes to get the new short text set D and term set T.
(2) Taking advantage of co-occurrence distance between terms to calculate the correlation weight of each term in a certain short text.
(3) Calculating ECE″ value of each term based on the ECE′ value.
(4) Combining (2) and (3) to get the final weight of each term for all classes, then ranking terms in descending order and taking the top-k terms as feature terms to form the new feature term set.

Explanations of notations used in this stage are shown as Table 1:
We define correlation of t_j given t_i as follows:

$$Cor(t_j|t_i) = \frac{\sum\limits_{d_s \in D} cor_{d_s}(t_i, t_j)}{\sum\limits_{d_s \in D} \sum\limits_{t_l \in T} cor_{d_s}(t_i, t_l)} \cdot \log_2 \frac{n}{Nei(t_j)} \tag{3}$$

where $Cor(t_j|t_i)$ indicates the co-occurrence condition of term t_i and term t_j, $\dfrac{\sum\limits_{d_s \in D} cor_{d_s}(t_i,t_j)}{\sum\limits_{d_s \in D}\sum\limits_{t_l \in T} cor_{d_s}(t_i,t_l)}$ shows probability that people think about term t_j when they see term t_i. Generally speaking, $Cor(t_j|t_i) \neq Cor(t_i|t_j)$, we then define correlation of term t_i and term t_j as,

$$Cor(t_i, t_j) = \frac{Cor(t_j|t_i) + Cor(t_i|t_j)}{2} \tag{4}$$

Table 1. Notation definition

Notation	Definition	Notation	Definition
D	Short text set	$cor_{d_s}(t_i, t_j)$	Correlation of term t_i and term t_j in short text d_s
T	Term set	$Nei(t_j)$	Term number of co-occurrence with term t_j in T
$Cor(t_j \| t_i)$	Co-occurrence of term t_j given term t_i	$Cor(t_i, t_j)$	Co-occurrence between term t_i and term t_j in D
m	Total number of short texts in D	$cow_{d_s}(t_i)$	Correlation weight of term t_i in short text d_s
N	Total number of terms in T	$Nei_{d_s}(t_i)$	Term number of co-occurrence with term t_j in d_s
w_{t_i}	Initial weight of term t_i	$cow_r(t_i)$	Correlation weight of term t_i in the C_r class
D_r	The C_r class short text set	$\|D_r(t_i)\|$	Short text number contains term t_i in the C_r class

The correlation weight of term t_i in the short text d_s is defined as:

$$cow_{d_s}(t_i) = w_{t_i} + \frac{\sum\limits_{t_j \in d_s} w_{t_j} \cdot Cor(t_i, t_j)}{Nei_{d_s}(t_i)} \tag{5}$$

where w_{t_i} is the word frequency of term t_i in short text d_s. $\sum\limits_{t_j \in d_s} w_{t_j} \cdot Cor(t_i, t_j)$ measures weights circumstance of all terms appeared together with term t_i in short text d_s. For all short texts of the C_r class where d_s exists, we can get

$$cow_r(t_i) = \frac{\sum\limits_{d_s \in D_r} cow_{d_s}(t_i)}{|D_r(t_i)|} \tag{6}$$

where $cow_r(t_i)$ measures overall condition of correlation weight of term t_i in the C_r class short text set.

3.2   Feature Dictionary Construction

The most important part of the proposed approach is to select the most representative words from a short text as features to construct the term dictionary. We use ECE'' (an improved ECE method) value to measure the topical-specificity of a word.

In most cases, a term which can be representative of Class A is likely to have little effect on distribution of Class B. Therefore, we should take the weights of this term in different classes into consideration, which can be calculated as follows:

$$ECE(t_i, C_r) = \begin{cases} P(t_i|C_r)P(C_r|t_i) \log_2 \frac{P(C_r|t_i)}{P(C_r)} & P(C_r) \le P(C_r|t_i) \\ P(t_i|C_r)P(C_r|t_i) \log_2 \frac{P(C_r)}{P(C_r|t_i)} & P(C_r) > P(C_r|t_i) \end{cases} \tag{7}$$

From Eq. (7), we know that the bigger the value of $P(C_r|t_i)$ is, the stronger the correlation of term t_i and Class C_r is. $\frac{P(C_r|t_i)}{P(C_r)}$ is larger, term t_i takes bigger effects on the C_r Class. When term t_i has a strong correlation with one class and weak correlations with other classes, the probability of this term to be selected will increase. The weight of term t_i in the C_r Class [8] is shown as Eq. (8):

$$ECE'(t_i, C_r) = \frac{\sum_{j \neq r} ECE(t_i, C_j)}{k - 1} \tag{8}$$

where $ECE'(t_i, C_r)$ is average weight of term t_i in total classes except the C_r class.

$$ECE''(t_i, C_r) = \frac{ECE(t_i, C_r)}{ECE'(t_i, C_r) + 0.01} \tag{9}$$

$ECE''(t_i, C_r)$ reflects the weight of term t_i in Class C_r. and the bigger the value is, the stronger indication of term t_i relative to Class C_r is. We can make use of Eq. (9) to gain ECE'' values of each term in Class C_r. The final weight calculation equation of term t_i in Class C_r is as follows:

$$W_{t_i} = cow_r(t_i) \times ECE''(t_i, C_r) \times idf(t_i) \tag{10}$$

where $cow_r(t_i)$ reveals significance of term t_i to short text set of Class C_r. $ECE''(t_i, C_r)$ shows that term t_i is indicative of categories of short texts. $idf(t_i)$ is inverse document frequency of term t_i.

We then sort terms of the C_r Class in a descending order of W_{t_i} value, and Top K terms are selected as feature terms. All classes in short text sets are treated similarly. Feature terms in each class are merged to reconstruct feature dictionary.

4 Experiments and Results Analysis

In this section, we present several experiments to prove the effectiveness of CDCFS in short text scenario. We will give experimental results and analysis compared with four baseline models.

4.1 Data Sets and Evaluation Metrics

In order to verify the effectiveness of our approach, we conducted several experiments on both Chinese data sets and English data sets, respectively. We adopted 15 classes with 150 paper titles obtained from CCF recommended list in Rank A and B as English data sets, and collected 6 classes with 12534 paper titles from CSCD as Chinese data sets. We then take 5-fold cross validation to show the effectiveness of our method.

We did a series of preprocessing work on both data sets including data denoising, stop words removal, and for Chinese data set, we utilized jieba, a Chinese text segmentation tool, as the tokenizer for the task of text segmentation to acquire the vocabulary [9]. We take advantage of F1-measure and accuracy as the evaluation of metrics [10].

4.2   Experimental Results and Analysis

We present experimental results on two tasks. Firstly, we visualize the selection results and evaluate our schemes for short text feature selection and compare the performances with four other selection methods. Secondly, classification performance is evaluated with different size of feature of our method and compare the performances with other methods.

The four feature selection strategies are CSFS (Using Strong Classification Features for Short Text Feature Selection), TCDS (Using Term Co-occurrence Distance for Short Text Feature Selection), TCS (Using Term Co-occurrence for Short Text Feature Selection) and TF*IDF method.

4.2.1   Comparison of Feature Dictionaries

We compare feature dictionaries obtained from the above 5 kinds of strategies to verify that our method can get a high accuracy for short text feature selection. Table 2 is the comparison of different feature selection methods.

We pick top 10 terms from 3598 terms in the class of Artificial Intelligence and Pattern Recognition, then we distinguish feature dictionaries using different methods. From Table 2 we can easily see that compared with TCS, TCDS which considers the co-occurrence distance between terms can better reveal terms. Our method has the strongest indication to representation of this category, which illustrates that both employing co-occurrence distance between terms to balance co-occurrence cases and considering strong classification feature is more effective.

4.2.2   Effect of Variation of Dictionary Size for Short Text Classification

We select different size of features ranging from 30 to 300 for our method and then apply SVM and k-NN for short text classification. Previous study on short text classification showed that most of the term-weighting schemes reach their peak values in the range of 20–45. Therefore, we parameterize k-NN by choosing different value k in this range and demonstrated the best performance using optimal k. As for the SVM algorithm, it is implemented based on the libsvm tool [11]. We take advantage of the

Table 2. Comparison of different feature selection methods

Method	TF*IDF	TCS	TCDS	CSFS	CDCFS
1	Algorithm	Algorithm	Space	Algorithm	**Learning**
2	Model	Network	Learning	Model	**Model**
3	Detection	Learning	Algorithm	Learning	**FR**
4	Learning	Graph	Cluster	Structure	**Cluster**
5	Feature	Study	Classify	Feature	**Classify**
6	Classify	Detection	Goal	Classify	**Sparse**
7	Optimization	Optimization	Local	Cluster	**Feature**
8	Cluster	Classify	Sparse	Sparse	**Supervised**
9	Strategy	Cluster	FR	FR	**Expression**
10	Graph	Recognition	Expression	Optimization	**NN**

(Note: NN is short for Neural Network and FR is short for Face Recognition.)

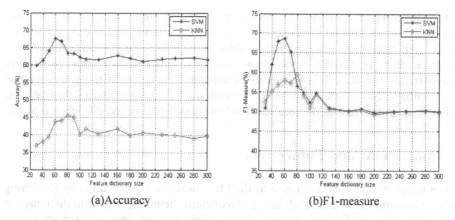

(a)Accuracy   (b)F1-measure

Fig. 1. Effect of feature dictionary size for short text classification

linear kernel functions since it has been reported to achieve competitive performance in the context of text categorization. The other parameters of SVM are set to their default values.

From Fig. 1 we can see that with setting of different feature size, the classification accuracy and F1-measure is different. More words do not lead to better performance. It shows a tendency of ascending firstly and then declining until reaching to be stable eventually for both SVM and k-NN classifier.

4.2.3   Classification Performance of Different Feature Selection Methods

In order to test and validate the impacts of employing diverse kinds of feature selection strategies for short text classification, different length of feature dictionary can be used and applied to libsvm to train classification model. The above results show that when the feature dictionary length is between 60 and 80, the classification accuracy rates reach their maximum, and the fluctuation range is more stable in SVM classifier than that of k-NN classifier. We then choose SVM classifier and apply it to both Chinese data sets and English data sets then observe the classification performance of feature dictionary of each strategy for short text classification. When the feature term number is 60, the accuracy and F1-Measure value on both datasets are as shown in Table 3, respectively.

From Table 3, we can make the following observations, in all cases with different feature selection methods, it is observed that our approach achieves the highest

Table 3. Classification performance of different feature selection methods

Method	(a) Chinese data sets		(b) English data sets	
	Accuracy	F1-measure	Accuracy	F1-measure
TF*IDF	51.83%	0.4374	46.49%	0.4633
CSFS	56.64%	0.4876	46.14%	0.4689
TCS	63.82%	0.5965	46.40%	0.4754
TCDS	64.96%	0.6241	49.73%	0.4829
CDCFS	67.53%	0.6802	51.53%	0.5129

performance on both data sets. This demonstrates the robustness of our method to noisy data. Moreover, compared to category information that short texts carried by their own, co-occurrence distance between terms in a document will cause a greater influence. At the same time, employing co-occurrence distance between two terms to balance co-occurrence condition between terms is apparently better than traditional method which utilizes. Nevertheless, adopting conventional TF*IDF method to classify short texts, its accuracy and F1-Measure value is least and the classified efficiency is worst.

5 Conclusions and Future Work

In this paper, we put forward a new method for short text classification by leveraging term co-occurrence distance and strong classification features. The term dictionary is built considering two aspects the correlation weight and the improved expected cross entropy. Thereafter the new feature for short text can be constructed. For the future work, we are interested to generalize this model for multi-label classification.

Acknowledgement. This work is supported by the National Natural Science Foundation of China (No. 61363058), Gansu province college students' innovation and entrepreneurship training program (201610736041), and the open fund of Key Laboratory of intelligent information processing Institute of computing technology of Chinese Academy of Sciences (IIP2014-4), the Natural Science Foundation of Gansu Province for Distinguished Young Scholars (1308RJDA007).

References

1. Ridder, D., Tax, D., Lei, B., et al.: Feature extraction and selection. In: Classification, Parameter Estimation and State Estimation, pp. 259–301. Wiley (2017)
2. Deng, Z.H., Luo, K.H., Yu, H.L.: A study of supervised term weighting scheme for sentiment analysis. Expert Syst. Appl. **41**(7), 3506–3513 (2014)
3. Song, S., Zhu, H., Chen, L.: Probabilistic correlation-based similarity measure on text records. Inf. Sci. **289**(1), 81–124 (2014)
4. Hua, W., Wang, Z., Wang, H., et al.: Short text understanding through lexical-semantic analysis. In: IEEE International Conference on Data Engineering, pp. 495–506. IEEE (2015)
5. Kotis, K., Papasalouros, A., Maragoudakis, M.: Mining query logs for learning useful ontologies: an incentive to SW content creation. Inf. J. Knowl. Eng. Data Min. **1**(4), 303–330 (2011)
6. Ma, H., Di, L., Zeng, X., Yan, L., Ma, Y.: Short text feature extension based on improved frequent term sets. In: Shi, Z., Vadera, S., Li, G. (eds.) IIP 2016. IAICT, vol. 486, pp. 169–178. Springer, Cham (2016). doi:10.1007/978-3-319-48390-0_18
7. Wang, L.: An improved method of short text feature extraction based on words co-occurrence. Appl. Mech. Mater. **519–520**, 840–843 (2014)
8. Ma, H., Zhou, R., Liu, F., Lu, X.: Effectively classifying short texts via improved lexical category and semantic features. In: Huang, D.-S., Bevilacqua, V., Premaratne, P. (eds.) ICIC 2016. LNCS, vol. 9771, pp. 163–174. Springer, Cham (2016). doi:10.1007/978-3-319-42291-6_16

9. Qin, W.L.: jiebaR: Chinese Text Segmentation (2016)
10. Gao, L., Zhou, S., Guan, J.: Effectively classifying short texts by structured sparse representation with dictionary filtering. Inf. Sci. **323**, 130–142 (2015)
11. Abdiansah, A., Wardoyo, R.: Time complexity analysis of support vector machines (SVM) in LibSVM. Int. J. Comput. Appl. **128**(3), 975–8887 (2015)

Formal Semantics and Fuzzy Logic

A Fuzzy Logic Based Policy Negotiation Model

Jieyu Zhan[1], Xudong Luo[2(✉)], Yuncheng Jiang[1(✉)], Wenjun Ma[1],
and Mukun Cao[3]

[1] School of Computer Science, South China Normal University, Guangzhou, China
ycjiang@scnu.edu.cn
[2] Department of Information and Management Science, Guangxi Normal University,
Guilin, China
luoxd@mailbox.gxnu.edu.cn
[3] School of Management, Xiamen University, Xiamen, China

Abstract. Few existing policy generation models can reflect the fact that self-interested stockholders often need to interact when generating a set of policies acceptable to them all. To this end, this paper proposes a negotiation model for policy generation. In this model, each negotiating agent employs a fuzzy logic system to evaluate each policy in a proposal made by others during their negotiation, and then uses a uninorm operator to aggregate the evaluations of all the single policies in the proposal to gain an overall evaluation of the proposal. Moreover, different negotiating agent can use different fuzzy reasoning systems. Finally, we do some experiments to reveal some insights into our model.

Keywords: Multi-issue negotiation · Proposal evaluation · Fuzzy logic · Uniform operator · Knowledge-based systems

1 Introduction

A policy is a statement of intent or a principle of action to guide decisions and achieve certain goals. For example, *disarmament* is a policy and *joining a military alliance* is another policy. Policy generation is a process of aggregating different stockholders' opinions or preferences to reach a set of policies acceptable to them all. An example of policy generation is that in a congress, the debate on the policy of *disarmament* can lead to an acceptable level, such as reducing five thousand soldiers. Therefore, policy generation is very important in political science, management science and economics [2]. However, most of the studies of policy generation have two limitations [2]: (i) For example, voting [4] (the most common used method for policy generation) cannot reflect that in real life sometimes policy generation is an interactive process among a group of people, for they need flexibly change their opinions in the light of more information revealed during policy generation. (ii) There are few effective methods for evaluating a set of policies when there are many alternative choices of policy set in fuzzy environments [18].

© Springer International Publishing AG 2017
G. Li et al. (Eds.): KSEM 2017, LNAI 10412, pp. 79–92, 2017.
DOI: 10.1007/978-3-319-63558-3_7

To address the above problems, researchers have tried to propose alternative models for policy generation, for example, negotiation models. In fact, negotiation is suitable for policy generation because it is a process in which a group of parties exchange information to obtain a mutually acceptable outcomes [3,10,11]. Thus, for instance, Zhang [19] develops a logic-based axiomatic model of bargaining to handle conflicting demands (e.g., policy). However, the model cannot handle the dynamic situation where players need to make some changes during a bargaining process. However, in real life policies often have to be generated in a dynamic environment. For example, in the Six-Party Talk on North Korea's Nuclear Program [19], different countries should dynamically change their proposals to result in a better outcome. Thus, Zhan et al. [18] propose a negotiation model in which agents can change their preferences over demands (e.g., policy) during a bargaining to increase the chance of reaching an agreement.

Nonetheless, the above negotiation models for policy generation still suffer from the second problem of lacking methods for evaluating a set of policies. This is mainly because such models are all ordinal models, where the utilities of policies are not measured by numerical values, but modelled by qualitative preference orderings over policies. However, when there are too many policies, it is hard to evaluate the combinations of different values of different policies. Actually, it is easy to use the ordinal model to depict the preference over policies intuitively, but it is hard to evaluate a set of policies due to the lack of quantitative assessments. So, an agreement in an ordinal model often has to be accord with some certain axioms [19], but its final result is not always optimal. Hence, a proposal evaluation method is required for negotiation-based policy generation models in order to search a set of policies agreeable to all the agents involved.

Our method takes two steps to evaluate a proposal during the course of negotiation: (i) to evaluate each policy in a proposal during negotiation; and (ii) to aggregate the evaluations of all the policies in the proposal to gain an overall evaluation of the proposal. For the first step, we suppose that each agent knows the importance degrees of their policies and the concession of their policies in a proposal compared with their original attitude toward to the policies. Then each negotiating agents employs a fuzzy logic system to calculate the utility of each policy in a proposal, because the fuzzy rules are the intuitive way to capture how the human user of each agent reasons the utility of each policy. In our model, different negotiating agents could have different fuzzy rules for they act on behalf of different human users. For the second step, each agent uses a uninorm operator [15] to generate the utility of a proposal of a policy set since uniform operators can hold some properties desired.

Our work in this paper advances the state of art in the field of negotiation-based policy generation models in the following aspects. (i) We design an agent architecture for a group of self-interested agents to negotiate a set of policies. (ii) With respect to the agent model, we propose a fuzzy logic based method to calculate agents' utilities, which negotiating agents use to evaluate their proposals of policies. (iii) Different negotiating agents can use different sets of fuzzy rules, different linguistic terms associated with fuzzy rules, and different membership functions even for the same linguistic terms.

The rest of this paper is organised as follows. Section 2 recaps basics about uninorm operator and fuzzy logic. Section 3 describes our negotiation-based model. Section 4 discusses our fuzzy logic method for evaluating a set of policies. Section 5 conducts some experiments to reveal some insights into our model. Section 6 discusses the related work. Finally, Sect. 7 concludes the paper with future work.

2 Preliminaries

This section recaps the concepts of uninorm operator [15] and fuzzy logic [16].

Firstly, we recall the concept of uninorm operator.

Definition 1. *A binary operator* $\oplus : [0,1] \times [0,1] \to [0,1]$ *is a uninorm operator that is increasing, associative and commutative, and there exists* $\tau \in [0,1]$ *such that:*

$$\forall a \in [0,1], a \oplus \tau = a, \tag{1}$$

Here τ *is said to be the unit element of a uninorm.*

The following formula is a specific uninorm operator [10]:

$$a_1 \oplus a_2 = \frac{(1-\tau)a_1 a_2}{(1-\tau)a_1 a_2 + \tau(1-a_1)(1-a_2)} \tag{2}$$

where $\tau \in (0,1)$ is the unit element.

The following lemma [10] reveals some properties of uninorm operator.

Lemma 1. *A uninorm operator* $\oplus$ *with unit element* τ *has the following properties:*

(i) $\forall a_1, a_2 \in (\tau, 1], a_1 \oplus a_2 \geqslant \max\{a_1, a_2\}$;
(ii) $\forall a_1, a_2 \in [0, \tau), a_1 \oplus a_2 \leqslant \min\{a_1, a_2\}$; *and*
(iii) $\forall a_1 \in [0, \tau), a_2 \in (\tau, 1], \min\{a_1, a_2\} \leqslant a_1 \oplus a_2 \leqslant \max\{a_1, a_2\}$.

The unit element of a uninorm operator actually is a threshold to distinguish whether an agent is satisfied with a policy in a proposal or not. If an evaluation is greater than the threshold, then the evaluation is regarded as being positive; otherwise, it is regarded as being negative. Thus, in Lemma 1, property (i) means that when the evaluations are both positive, the aggregated one is not less than either of the original evaluations; property (ii) means that when they are both negative, the aggregated one will enhance the negative effect, and is not greater than either of the original evaluations; and property (iii) means that when they are in conflict, the aggregated one is a compromise. Since the unit element of a uninorm operator actually is a threshold, it could vary from person to person to represent different personalities of different human users of the system developed in this paper.

Next, we recall the concept of fuzzy set [16] as follows:

Definition 2. *Let U be a set (called domain). A fuzzy set A on U is characterised by its membership function*

$$\mu_A : U \to [0,1]$$

and $\forall u \in U, \mu_A(u)$ is called the membership degree of u in fuzzy set A.

The following definition is about the fuzzy reasoning method of Mamdani [12].

Definition 3. *Let A_i be a Boolean combination of fuzzy sets $A_{i,1}, \cdots, A_{i,m}$, where $A_{i,j}$ is a fuzzy set defined on $U_{i,j}$ $(i = 1, \cdots, n; j = 1, \cdots, m)$, and B_i be a fuzzy set on U' $(i = 1, \cdots, n)$. Then when the inputs are $\mu_{A_{i,1}}(u_{i,1}), \cdots, \mu_{A_{i,m}}(u_{i,m})$, the output of such fuzzy rule $A_i \to B_i$ is fuzzy set B_i' defined as follows:*

$$\forall u' \in U', \mu_i(u') = \min\{f(\mu_{A_{i,1}}(u_{i,1}), \cdots, \mu_{A_{i,m}}(u_{i,m})), \mu_{B_i}(u')\}, \tag{3}$$

where f is obtained through replacing $A_{i,j}$ in A_i by $\mu_{i,j}(u_{i,j})$ and replacing "and", "or" and "not" in A_i by "min", "max" and "$1 - \mu$", respectively. And the output of all rules $A_1 \to B_1, \cdots, A_n \to B_n$, is fuzzy set M, which is defined as:

$$\forall u' \in U', \mu_M(u') = \max\{\mu_1(u'), \cdots, \mu_n(u')\}. \tag{4}$$

The result above is still a fuzzy set and we should turn the fuzzy output into a crisp one, which process is called defuzzification. We will employ the following well-known centroid method [12] for defuzzification in this paper:

Definition 4. *The centroid point u_{cen} of fuzzy set M given by formula (4) is:*

$$u_{cen} = \frac{\int_{U'} u' \mu_M(u') \mathrm{d}u'}{\int_{U'} \mu_M(u') \mathrm{d}u'}. \tag{5}$$

3   Model Definition

This section will define the main components of our policy negotiation model.

Firstly, we assume each agent has its own opinions on some policies (support or opposition). Policies can be represented by propositions. We describe a policy through two dimensions: one is about an agent's attitude to the policy (*e.g.*, support or reject *disarmament*), and the other is about how important it is to the agent. Formally, we have:

Definition 5. *For agent i, its policy structure $\mathscr{S}$ is a tuple of (X_i, Att_i, Imp_i), where:*

– *X_i is a finite set of propositions in a propositional language $\mathscr{L}$ and $\forall x \in X_i, x$ is a policy represented by a propositional variable;*

- $Att_i : X_i \rightarrow \{0, 1\}$ *is an attitude function, i.e.,* $\forall x \in X_i, Att_i(x)$ *is agent i's attitude to policy x; in particular,* $Att_i(x) = 1$ *means that the agent supports policy x and* $Att_i(x) = 0$ *means that the agent opposes policy x; and*
- $Imp_i : X_i \rightarrow (0, 1]$ *is an importance function, i.e.,* $\forall x \in X_i, Imp(x)$ *is the importance degree of x in set* X_i, *which represents how important policy x is to agent i.*

For a certain policy, an agent should have only one attitude in the beginning (either support or oppose the policy), and does not have an attitude to any irrelevant policy since those policies do not affect his utility of an agreement. For example, if agent i supports the policy of *disarmament*, then his attitude to this issue is $Att_i(disarmament) = 1$. In fact, opposing a policy can also be represented by a proposition [8, 18, 19], such as *oppose disarmament*, but we differentiate attitudes and policies in this paper because we aim to deal with both policies in continuous domains and discrete domains. In discrete domains, the policy, such as *joining a military alliance*, should be completely accepted or completely rejected by an agent. In the final agreement of policies, such a policy is either supported or opposed. Nevertheless, in continuous domains, a policy can be partially accepted in a final agreement, *i.e.*, the acceptance domain is divisible. For example, if the policy is *disarmament*, then supporting such a policy may mean to reduce ten-thousand soldiers and opposing it may mean to totally maintain the existing size of the army. Thus, in the agreement, *disarmament* can be partly accepted and opposed. A successful negotiation can lead to an acceptable level, such as reducing five thousand soldiers.

Our model uses an extension of alternating-offers protocol [13], *i.e.*, in every round one of the agents gives a proposal to the others. Formally, we have:

Definition 6. *A proposal function of agent i is a mapping* $O_i : X_i \bigcup X_{-i} \rightarrow [0, 1]$, *where* X_{-i} *denotes the policy set of all i's opponents. And* $\forall x \in X_i \bigcup X_{-i}, O_i(x)$ *is called agent i's acceptance degree of policy x. And* $o_i = \{O_i(x_1), \ldots, O_i(x_n)\}$ *is called a proposal of agent i, where n is the number of elements in* $X_i \bigcup X_{-i}$.

After formally defining the concept of proposal, we can present our method for proposal evaluation. Firstly, a negotiating agent uses a fuzzy reasoning system to evaluate a policy in a proposal. Clearly, if all of its acceptance degrees of policies in a proposal are the same as its attitudes to the policies, then its utility is the highest because it has no concession on any policy. On the contrary, when its opponent's policy proposal is completely against its attitudes, its utility is the lowest. Intuitively, the more concessions an agent makes on a policy, the less utility it gains from the policy; and the more important the policy is for the agent, the more the utility decreases. Formally, we have:

Definition 7. *For agent i, a local utility of policy x in proposal o is given by:*

$$u_i(x) = FR(Imp_i(x), Con_i(x)),$$ (6)

where $x \in X_i, FR$ *is the result of fuzzy reasoning based on a set of fuzzy rules (we will detail it in the next section);* Con_i *is a concession degree function, representing a degree to which agent i makes a concession on a policy, and defined as*

$$Con_i(x) = |Att_i(x) - O_i(x)|,$$ (7)

where $Att_i(x)$ and $O_i(x)$ are agent i's attitude and acceptance degree to policy x, respectively.

After calculating the local utility of each policy in a policy proposal for an agent, we should find a way to aggregate all the local utilities into a global utility to reflect the utility an agent can gain from a proposal. Here we apply the uninorm operator [15] to solve this problem, because the operator has the properties shown in Lemma 1, which satisfy the intuitions of local utility aggregation. Formally, we have:

Definition 8. *For agent i, a global utility of proposal o, denoted as $u_i(o)$, is given by:*

$$u_i(o) = \underset{j=1,\dots k}{\oplus} u_i(x_j),$$ (8)

where k is the number of elements in X_i.

In this paper, we employ formula (2) as the uninorm operator $\oplus$ in formula (8).

When calculating the utility of a policy in a proposal, we take the time cost into consideration in this model because the same proposal in different rounds has different utilities for an agent. Intuitively, the utility an agent gains from a proposal is lower than the same one in a previous round, and the utility becomes less and less as time goes on. Formally, we have:

Definition 9. *A global utility of proposal o with time constrains for agent i is given by:*

$$u_i^t(o) = u_i(o)\sigma^{\lambda-1},$$ (9)

where u_i is the global utility of agent $i, \sigma \in [0,1]$ is discount factor for decreasing the utility of the proposals as time passes, and λ refers to the λ-th round of the negotiation.

When a negotiating agent receives a proposal, it needs to decide whether to accept it or not, and thus it should have an acceptable threshold of utility in every round. That is, if one of the opponents' proposal results in a low utility for an agent, which is lower than its acceptable threshold in that round, the agent should reject it, and vice versa. If the agent rejects a proposal, it should generate another policy proposal, which utility is not lower than the threshold. The acceptable threshold of utility changes in every round because an agent has different expectations for the results in different rounds. For example, in real life it often happens that in the beginning of a negotiation, an agent expects the agreement that is the best to it, but as the time goes, it becomes increasingly willing to accept a suboptimal result in order to avoid breakdown of the negotiation. In this paper, we use $\hat{u}_i(\lambda)$ to represent the acceptable threshold of agent i in the λth round.

We use an extended protocol of bilateral negotiation [13] for our multilateral one of policy generation. More specifically, one of the agents initialises the negotiation randomly and then the others respond to it. The first proposal of every agent should indicate its attitude to the policies it concerns. Formally, the first proposal of agent i, denoted as $O_{i,1}$, should satisfy the following property: $\forall x_i \in X_i, O_{i,1}(x_i) = Att_i(x_i)$. This means that the agent should reveal its attitude at the beginning, because this kind of proposal can maximise its utility. That is, there is absolutely no concession on any policy. Of course, it is the best proposal, which the negotiating agent wants the most to be the agreement. Other agent can choose: (i) to accept the proposal, or (ii) to reject the proposal but make a new one, or (iii) to end the negotiation that results in the lowest utilities for all agents. If every negotiating agent has proposed a proposal in turn, then a negotiation round is completed and if there is no proposal supported by all negotiating agents, they will enter into a new round of negotiation. Some policies proposed by agent i may be irrelevant to agent j and different outcomes of this kind of policies do not influence j's utility. We call such policies as irrelevant to party j, but an agreement should also include such policies.

Obviously, an agreement appears if and only if a proposal proposed by one of the agents in negotiation is supported by all the other agents. Formally, we have:

Definition 10. *A proposal o proposed in the λ-th round in an n-agent negotiation is an agreement among the n agents if it satisfies: $\forall i \in \{1, 2, \ldots, n\}$,*

$$u_i^t(o) \geqslant \hat{u}_i(\lambda), \tag{10}$$

where $u_i^t(o)$ is the global utility of proposal o with time constrains for agent i and $\hat{u}_i(\lambda)$ is the acceptable threshold in λth round for agent i.

4 Fuzzy Rules

This section will discuss fuzzy rules that are used to evaluate local utilities of each policy in a policy proposal of an agent.

The local utility of a policy mainly depends on two factors: the importance degree and the concession degree. Here we use five terms (*i.e., very unimportant, unimportant, medium, important* and *very unimportant*) to depict different levels of importance of a policy. Similarly, we use four terms (*i.e., very high, high, low* and *very low*) to represent different levels of concession an agent makes on a policy and seven terms *i.e., very high, high, slightly high, medium, slight low, low* and *very low*) to indicate different levels of the output of fuzzy reasoning (*i.e.,* the local utility of a policy for an agent).

Moreover, we employ the trapezoidal-type fuzzy membership function:

$$\mu(x) = \begin{cases} 0 & \text{if } x \leqslant a, \\ \frac{x-a}{b-a} & \text{if } a \leqslant x \leqslant b, \\ 1 & \text{if } b \leqslant x \leqslant c, \\ \frac{d-x}{d-c} & \text{if } c \leqslant x \leqslant d, \\ 0 & \text{if } x \geqslant d. \end{cases} \tag{11}$$

Thus, we can draw the membership functions of these linguistic terms of importance degree, concession degree and local utility as shown in Figs. 1, 2 and 3. Moreover, if necessary, different users can take different kinds of membership functions to represent their linguistic terms of importance degree, concession degree and local utility according to their cognition. That is, even for the same linguistic term, for example, "important", if necessary, different users can have different kinds of membership functions.

A fuzzy reasoning is carried out according to fuzzy rules. Similarly to choosing linguistic terms and their membership functions, different human users can set different sets of fuzzy rules to calculate the local utility of a proposal. Here in order to illustrate how the fuzzy rules work, we give an example of fuzzy rules as displayed in Table 1. There Rule 1 means that if the importance degree of a policy is *very unimportant* and its concession degree is *not very high*, then its local utility is *very high*. Such a rule reflects the intuition that if a person does

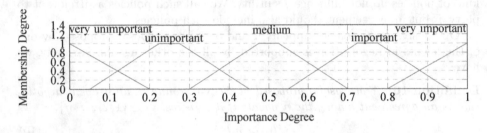

Fig. 1. Membership function of importance degree

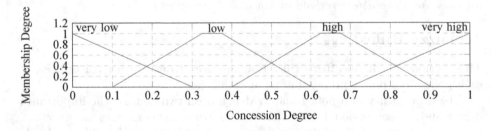

Fig. 2. Membership function of concession degree

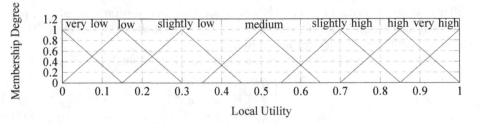

Fig. 3. Membership function of local utility

Table 1. Fuzzy rules

1	If importance degree is *very unimportant* and concession degree is *not very high* then local utility is *very high*
2	If importance degree is *not very important* and concession degree is *very low* then local utility is *very high*
3	If importance degree is *unimportant* and concession degree is *low* then local utility is *high*
4	If importance degree is *unimportant* and concession degree is *high* then local utility is *slightly high*
5	If importance degree is *unimportant* and concession degree is *very high* then local utility is *medium*
6	If importance degree is *medium* and concession degree is *low* then local utility is *slightly high*
7	If importance degree is *medium* and concession degree is *high* then local utility is *medium*
8	If importance degree is *medium* and concession degree is *very high* then local utility is *slightly low*
9	If importance degree is *important* and concession degree is *low* then local utility is *medium*
10	If importance degree is *important* and concession degree is *high* then local utility is *slight-low*
11	If importance degree is *important* and concession degree is *very high* then local utility is *low*
12	If importance degree is *very important* and concession degree is *low* then local utility is *slightly low*
13	If importance degree is *very important* and concession degree is *high* then local utility is *low*
14	If importance degree is *very important* and concession degree is *high* then local utility is *very low*

not make a very high concession on a very unimportant policy, then his local utility of the policy is very high. Similarly, we can understand other rules. Notice that these rules and the linguistic terms of inputs and output may vary from person to person. Those in this paper are just some of many possibilities.

5 Experiment

This section will do an experiment to reveal some insight into our model.

Assume Parties 1 and 2 are going to negotiate for four policies: *tax increase (TI)*, *disarmament (DA)*, *increase of educational investment (IEI)*, and *economical housings investment (EHI)*. These policies are in continuous domains, *i.e.*, the parties can give a proposal reflecting they can partially support or oppose some policies; the acceptance degree can take a value from $\{0, 0.2, 0.4, \cdots, 1\}$. The policy structures of both agents are shown in Table 2. That is, Party 1 supports TI and IEI, opposes EHI, and does not care about DA (we call it a policy irrelevant to Party 1 because it does not affect its global utility); while Party 2 supports DA, opposes TI and IEI, and EHI is irrelevant to it.

Table 2. Policy structure of agents

	Tax increase	Disarmament	Increase of educational investment	Economical housings investment
Attitude of party 1	1	N/A	1	0
Preference of party 1	0.8	N/A	0.3	0.2
Attitude of party 2	0	1	0	N/A
Preference of party 2	0.2	0.5	0.7	N/A

In this example, we assume that Agents 1 and 2 negotiate these policies on behalf of Parties 1 and 2 respectively, and have different kinds of linguistic terms of importance degree, concession degree and local utility. More specifically, what Agent 1 has are as shown in Figs. 1, 2 and 3, respectively, and it adopts fuzzy rules shown in Table 1; while Agent 2 has different ones as shown in Figs. 4, 5 and 6, and Agent 2's fuzzy rules are shown in Table 3.

In this example, an agent makes a proposal with the highest global utility first. If the opponent rejects this proposal and the agent also rejects its opponent's counter-proposal, the agent gives another proposal with the second highest global utility in a new round. In order to calculate the global utility of a proposal for an agent, we should firstly calculate the local utility of each relevant policy in a proposal for an agent.

In our experiment, we change the discount factors (*i.e.*, σ_1 and σ_2) of both agents randomly in $[0, 1]$ and see how different combinations of discount factors

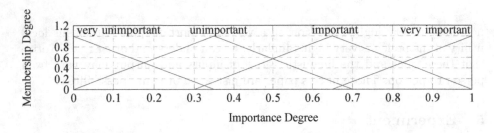

Fig. 4. Agent 2's membership function of importance degree

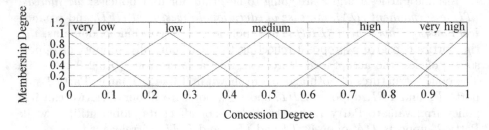

Fig. 5. Agent 2's membership function of concession degree

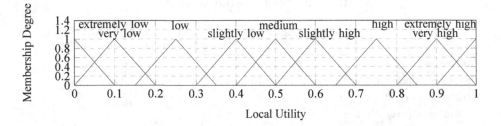

Fig. 6. Agent 2's membership function of local utility

Table 3. Fuzzy rules of Agent 2

1	If importance degree is *very unimportant* and concession degree is *not very high* then local utility is *high*
2	If importance degree is *very unimportant* and concession degree is *very high* then local utility is *extremely high*
3	If importance degree is *not very important* and concession degree is *very low* then local utility is *high*
4	If importance degree is *unimportant* and concession degree is *low* then local utility is *high*
5	If importance degree is *unimportant* and concession degree is *medium* then local utility is *slightly high*
6	If importance degree is *unimportant* and concession degree is *high* then local utility is *medium*
7	If importance degree is *unimportant* and concession degree is *very high* then local utility is *slight low*
8	If importance degree is *important* and concession degree is *low* then local utility is *medium*
9	If importance degree is *important* and concession degree is *medium* then local utility is *slightly low*
10	If importance degree is *important* and concession degree is *high* then local utility is *low*
11	If importance degree is *important* and concession degree is *very high* then local utility is *very low*
12	If importance degree is *very important* and concession degree is *low* then local utility is *slightly low*
12	If importance degree is *very important* and concession degree is *low* then local utility is *low*
13	If importance degree is *very important* and concession degree is *high* then local utility is *very low*
14	If importance degree is *very important* and concession degree is *high* then local utility is *extremely low*

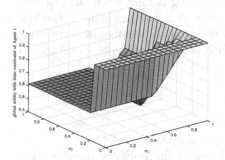

Fig. 7. Global utility with time constraint of Agent 1 with discount factor change

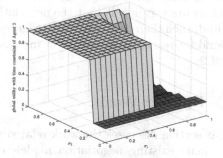

Fig. 8. Global utility with time constraint of Agent 2 with discount factor change

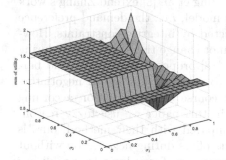

Fig. 9. Sum of utility of Agents 1 and 2 with discount factor change

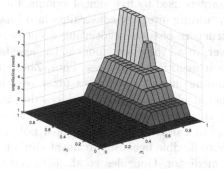

Fig. 10. Negotiation round with discount factor change

change: (i) the global utilities with time constraint of Agent 1 (see Fig. 7) and Agent 2 (see Fig. 8), (ii) the sum of utility with time constraint of the two Agents (see Fig. 9), and (iii) the number of negotiation round for reaching an agreement (see Fig. 10).

From Fig. 7, we can see that when the discount factor of Agent 1 is very high or the discount factor of Agent 2 is very low, the global utility with time constraint of Agent 1 is very high, meaning that if Agent 1 does not need to pay too much time cost during the course of the negotiation or Agent 2 has to pay a high time cost when negotiating with its opponent, Agent 1 has advantages for reaching an agreement that generates a higher utility for itself. On the contrary, Fig. 8 shows that if the discount factor of Agent 1 is low or the discount factor of Agent 2 is high, the Agent 2 will have advantages for reaching an agreement that generates a higher utility for itself. In Fig. 9, the sum of utility with time constraint of Agents 1 and 2 without obvious monotonicity. The maximum (*i.e.*, 1.8518) appears when $\sigma_1 = \sigma_2 = 1$ because no matter how many rounds for reaching an agreement is needed, there is no time cost in such situation; while the minimum (*i.e.*, 0.5821) appears when $\sigma_1 = 0.8$ and $\sigma_2 = 0.5$, because both the discount factors and the negotiation round are not low. The reason why the sum of utility with time constraint of Agents 1 and 2 is high when the discount factors are very low is that the negotiations ends with an agreement in the first round, then there is no time cost. Figure 10 shows that the number of negotiation round for reaching an agreement increases with the discount factors of Agents 1 and 2.

6   Related Work

This section discusses the work related to our negotiation model.

Some existing negotiation models can be used for policy generation. For example, Zhang and Zhang et al. [20] proposed a logic-based axiomatic model for political bargaining and other similar ones. There bargainers' demands (could be policies) are represented by logical statements and bargainers' preferences are total pre-orders. Zhang [19] later proved their concept of solution is uniquely characterised by five logical axioms. Further, Jing et al. [8] extend Zhang's work by adding integrity constraints into Zhang's model, *i.e.*, the demand preference structure of each negotiating agent is restricted by integrity constraints. However, in all the above models, the agents cannot change their preferences during their negotiation process. Thus, Zhan et al. [18] proposed a multi-demand bargaining model, which uses fuzzy reasoning to calculate how much negotiators should change their preferences during the course of a negotiation according to their personalities. However, differently in this paper we use fuzzy reasoning to calculate the utility of a policy. Compared with their method, ours is more flexible because our agent can adjust its offers during negotiation without a mediator. González et al. [6] based compensatory fuzzy logic to establish a negotiation model, which is also used to solve policy making problems. However, they study the problem from the view of cooperative game theory, different from our non-cooperative one. Moreover, unlike them, we focus on how to use fuzzy logic to reflect the relation between a policy's concession, importance and utility.

Also lots of researchers use fuzzy logic to deal with different aspects in automated negotiation. For example, Kolomvatsos et al. [9] propose an adaptive fuzzy logic system to help buyer agents to respond to sellers' offers, *i.e.*, the fuzzy logic system is used as a part of negotiation strategies. Similarly, Haberland et al. [7] also use a fuzzy logic system to adjust agent's negotiation strategy. Zhan and Luo [17] use type-2 fuzzy logic to determine the change of strategies according to the remaining time and opponents' cooperative degree. Adabi et al. [1] design a fuzzy grid market pressure determination system based negotiation protocol, which is enhanced by a fuzzy logic system in order to increase negotiators' utilities and their negotiation success rate and speed. Fu et al. [5] use fuzzy logic to make an offer, *i.e.*, to determine a price that balances the good customer relationship and profits for companies. Shojaiemehr and Rafsanjani [14] use fuzzy logic to match the right supplier with a buyers by modifying suppliers' offers automatically during negotiation. However, none of these models use fuzzy logic to evaluate offers during the course of a negotiation. Rather, firstly we use fuzzy logic to help agents to evaluate local utility of a proposal for one issue, and then we use a uniform operator to aggregate the local utility of individual issues to a global utility for the whole offer.

7 Conclusions

This paper proposes a negotiation model for policy generation. The evaluation method in the model can reason how concessions and preferences of policies determine an agent's utility of a policy. Actually, a negotiating agent uses a fuzzy reasoning system to calculate the local utility of a policy in a proposal for an agent according to intuitive fuzzy rules (and different agents could have different fuzzy reasoning systems). Then a uninorm operator is used to aggregate all the local utilities into a global utility that an agent can gain if the proposal is accepted. With respect to an example, we did experiments to reveal some insights into our model. In particular, the experimental results show that the discount factor can influence the global utility of a proposal with time constraint of negotiating agents, and the negotiation rounds for reaching an agreement. In the future, maybe the most interesting thing is to improve its negotiation protocol to reflect partnership among negotiators, because in policy generation, negotiators may form alliances to raise their interests.

Acknowledgements. This research was partially supported by the Natural Science Foundation of Guangdong Province, China (Nos. 2016A030313231 and S2012030006242), the National Natural Science Foundation of China (No. 61272066), the key project in universities in Guangdong Province of China (No. 2016KZDXM024), the Innovation project of postgraduate education in Guangdong Province of China (No. 2016SFKC_13), and the Project of Science and Technology in Guangzhou in China (Nos. 2014J4100031 and 201604010098).

References

1. Adabi, S., Movaghar, A., Masoud Rahmani, A., Beigy, H., Dastmalchy-Tabrizi, H.: A new fuzzy negotiation protocol for grid resource allocation. J. Netw. Comput. Appl. **37**(39), 89–126 (2014)

2. Arrow, K.J.: Social Choice and Individual Values, vol. 12. Yale University Press, New Haven (2012)
3. Davis, R., Smith, R.G.: Negotiation as a metaphor for distributed problem solving. Artif. Intell. **20**(1), 63–109 (1983)
4. Fishburn, P.C., Brams, S.J.: Paradoxes of preferential voting. Math. Mag. **56**(4), 207–214 (1983)
5. Fu, X., Zeng, X.J., Wang, D., Xu, D., Yang, L.: Fuzzy system approaches to negotiation pricing decision support. J. Intell. Fuzzy Syst. **29**(2), 685–699 (2015)
6. González, E., Espín, R.A., Fernández, E.: Negotiation based on fuzzy logic and knowledge engineering: some case studies. Group Decis. Negot. **25**(2), 373–397 (2016)
7. Haberland, V., Miles, S., Luck, M.: Adjustable fuzzy inference for adaptive grid resource negotiation. In: Fujita, K., Ito, T., Zhang, M., Robu, V. (eds.) Next Frontier in Agent-Based Complex Automated Negotiation. SCI, vol. 596, pp. 37–57. Springer, Tokyo (2015). doi:10.1007/978-4-431-55525-4_3
8. Jing, X., Zhang, D., Luo, X., Zhan, J.: A logical multidemand bargaining model with integrity constraints. Int. J. Intell. Syst. **31**(7), 673–697 (2016)
9. Kolomvatsos, K., Trivizakis, D., Hadjiefthymiades, S.: An adaptive fuzzy logic system for automated negotiations. Fuzzy Sets Syst. **269**, 135–152 (2015)
10. Luo, X., Jennings, N.R., Shadbolt, N., Leung, H.-F., Lee, J.H.-M.: A fuzzy constraint based model for bilateral, multi-issue negotiations in semi-competitive environments. Artif. Intell. **148**(1), 53–102 (2003)
11. Luo, X., Miao, C., Jennings, N.R., He, M., Shen, Z., Zhang, M.: KEMNAD: a knowledge engineering methodology for negotiating agent development. Comput. Intell. **28**(1), 51–105 (2012)
12. Mamdani, E.H., Assilian, S.: An experiment in linguistic synthesis with a fuzzy logic controller. Int. J. Man-Mach. Stud. **7**(1), 1–13 (1975)
13. Rubinstein, A.: Perfect equilibrium in a bargaining model. Econom.: J. Econom. Soc. **50**(1), 97–109 (1982)
14. Shojaiemehr, B., Rafsanjani, M.K.: A fuzzy system approach to multilateral automated negotiation in B2C e-commerce. Neural Comput. Appl. **25**(2), 367–377 (2014)
15. Yager, R.R., Rybalov, A.: Uninorm aggregation operators. Fuzzy Sets Syst. **80**(1), 111–120 (1996)
16. Zadeh, L.A.: Fuzzy sets. Inf. Control **8**(3), 338–353 (1965)
17. Zhan, J., Luo, X.: Adaptive conceding strategies for negotiating agents based on interval type-2 fuzzy logic. In: Lehner, F., Fteimi, N. (eds.) KSEM 2016. LNCS, vol. 9983, pp. 222–235. Springer, Cham (2016). doi:10.1007/978-3-319-47650-6_18
18. Zhan, J., Luo, X., Feng, C., Ma, W.: A fuzzy logic based bargaining model in discrete domains: axiom, elicitation and property. In: 2014 IEEE International Conference on Fuzzy Systems, pp. 424–431 (2014)
19. Zhang, D.: A logic-based axiomatic model of bargaining. Artif. Intell. **174**(16), 1307–1322 (2010)
20. Zhang, D., Zhang, Y.: An ordinal bargaining solution with fixed-point property. J. Artif. Intell. Res. **33**(1), 433–464 (2008)

f-$\mathcal{ALC}$(D)-LTL: A Fuzzy Spatio-Temporal Description Logic

Haitao Cheng[1(✉)] and Zongmin Ma[2]

[1] School of Computer Science,
Nanjing University of Posts and Telecommunications,
Nanjing 210023, China
chenghaitao@yahoo.com
[2] College of Computer Science and Technology,
Nanjing University of Aeronautics and Astronautics,
Nanjing 211106, China

Abstract. In order to achieve representation and reasoning of fuzzy spatio-temporal knowledge on the Semantic Web, in this paper, we propose a fuzzy spatio-temporal description logic f-$\mathcal{ALC}$(D)-LTL that extends spatial fuzzy description logic f-$\mathcal{ALC}$(D) with linear temporal logic (LTL). Firstly, we give a formal definition of syntax, semantics of the f-$\mathcal{ALC}$(D)-LTL. Then, we propose a tableau algorithm for reasoning fuzzy spatio-temporal knowledge, i.e., determining satisfiability problem of f-$\mathcal{ALC}$(D)-LTL formula. Finally, we show the termination, soundness, and completeness of the tableau algorithm.

Keywords: Fuzzy spatio-temporal knowledge · LTL · Description logic · Tableau algorithm · Reasoning

1 Introduction

Currently, many application fields are related to spatio-temporal information, for example, Geographic Information Systems (GIS) and spatio-temporal database [11]. Because of the increasing requirements of these spatio-temporal applications, the need for management of spatio-temporal knowledge has received much attention. The research issue related to the management technologies of spatio-temporal knowledge emerges, including representation and reasoning [6]. However, in many spatio-temporal applications domains, there are a lot of imprecise and uncertain temporal and spatial information as well as spatial topological relations [20]. With the emergence of fuzzy spatio-temporal knowledge, the representation and reasoning of fuzzy spatio-temporal knowledge have become one of the hot research issues in the fields of visual object tracking and GIS [16].

Description logics (DLs), as a formal language of knowledge representation, have been widely used in the fields of computer science and artificial intelligence. In particular, DLs are a logical basis of knowledge representation and reasoning

© Springer International Publishing AG 2017
G. Li et al. (Eds.): KSEM 2017, LNAI 10412, pp. 93–105, 2017.
DOI: 10.1007/978-3-319-63558-3_8

of the Semantic Web [8]. Hence, how to extend DLs to realize the representation and reasoning of fuzzy spatio-temporal knowledge needs to be solved.

Currently, in order to represent and reason spatial knowledge in the Semantic Web, many researchers have already proposed some extensions of DLs, such as $\mathcal{ALCRP}(D)$ [8], $\mathcal{ALCRP}^3(D)$ [10], $\mathcal{ALC}(C)$ [12], $\mathcal{ALC}(CDC)$ [4] and $\mathcal{ALC}_s$ [24]. However, the approaches described can only deal with crisp spatial information. For the reasoning of fuzzy spatial knowledge in the Semantic Web, Straccia [21] presents a spatial fuzzy description logic named $f\text{-}\mathcal{ALC}(D)$ which supports classic region and fuzzy RCC-8 spatial predicates. To support spatial reasoning in medical image domain, Hudelot et al. [9] propose a description logic named $\mathcal{ALC}(F)$ which defines concrete domain using fuzzy mathematical morphology. Although the tableau-based algorithm for reasoning fuzzy spatial knowledge is given, the $\mathcal{ALC}(F)$ does not prove the correctness of the tableau algorithm.

For temporal knowledge modeling, some temporal extensions of DLs have been investigated (see [13] for surveys). In these extensions, two common approaches for representing temporal knowledge have been widely used: time interval-based [2] and time point-based [23]. The time interval-based temporal description logics combine basic DLs with Allen's interval algebra [1], while the time point-based temporal description logics combine basic DLs with (linear) temporal logic (LTL) [14]. Although these temporal DLs cannot support the representation and reasoning of spatial knowledge, they give good hints and ideas for our works.

Although there have been some proposals for crisp/fuzzy spatial and temporal extensions to DLs, to the best of our knowledge, so far, there are no comprehensive reports on developing fuzzy spatio-temporal description logic. Developing this description logic is a natural way to extend spatial fuzzy description logic with temporal logic. Therefore, in this paper, we propose a fuzzy spatio-temporal description logic named $f\text{-}\mathcal{ALC}(D)\text{-}LTL$ that extends spatial fuzzy description logic $f\text{-}\mathcal{ALC}(D)$ [21] with linear temporal logic LTL. The $f\text{-}\mathcal{ALC}(D)\text{-}LTL$ can be interpreted over temporal sequences $\mathcal{I}(w)$ ($w \in \mathbb{N}$) of $f\text{-}\mathcal{ALC}(D)\text{-}LTL$ structure reflecting possible temporal evolutions of fuzzy spatial knowledge, especially fuzzy spatial relations. In brief, the paper makes the following contributions:

- We extend spatial fuzzy description logic $f\text{-}\mathcal{ALC}(D)$ with linear temporal logic LTL, and then develop a new fuzzy spatio-temporal description logic $f\text{-}\mathcal{ALC}(D)\text{-}LTL$. Furthermore, we define the syntax structure and semantic interpretation of $f\text{-}\mathcal{ALC}(D)\text{-}LTL$.
- We propose a tableau algorithm for deciding satisfiability of $f\text{-}\mathcal{ALC}(D)\text{-}LTL$ formulas. Then, we prove the termination, soundness, and completeness of the algorithm.

2   Spatial Fuzzy Description Logic

A spatial fuzzy description logic (called *fuzzy* $\mathcal{ALC}(D)$) proposed by Straccia [21] is regarded as a basic fuzzy DL extended with spatial fuzzy concrete domain. We recall the syntax and semantics of *fuzzy* $\mathcal{ALC}(D)$ [21, 22].

Definition 1 (Spatial fuzzy concrete domain $\mathcal{D}$). *A spatial fuzzy concrete domain is a pair $\mathcal{D} = (\triangle_\mathcal{D}, \Phi_\mathcal{D})$, where $\triangle_\mathcal{D}$ is a set of fuzzy regions and $\Phi_\mathcal{D}$ is a set of fuzzy RCC predicates d (d $\in$ {C, DC, P, PP, EQ, O, DR, PO, EC, NTP, TPP, NTTP}) over $\triangle_\mathcal{D}$ and an interpretation $d^\mathcal{D}: \triangle_\mathcal{D} \times \triangle_\mathcal{D} \rightarrow [0, 1]$, which is a fuzzy relation over $\triangle_\mathcal{D}$.*

Noted that the *fuzzy Region Connection Calculus* (*f*-RCC) is a fuzzy extension of original crisp RCC [15], which represents fuzzy topological relations such as {C, DC, P, PP, EQ, O, DR, PO, EC, NTP, TPP, NTTP}. For more details about fuzzy RCC relations, it can be found in [17,18].

Definition 2 (Syntax). *Let $\mathbf{C}, \mathbf{R}, \mathbf{T}, \mathbf{I}$ and $\mathbf{O}$ be a disjoint set of concept names, abstract roles names, concrete roles names, abstract individual names and fuzzy spatial regions names. Also, let $R \in \mathbf{R}$ be an abstract role and $T \in \mathbf{T}$ be a concrete role. Concepts (denoted by C or $D \in \mathbf{C}$) of fuzzy $\mathcal{ALC}(D)$ can be given by:*

$C, D = \top \mid \bot \mid A \mid \neg C \mid C \sqcap D \mid C \sqcup D \mid \forall R.C \mid \exists R.C \mid \forall (T_1, T_2).\mathbf{d} \mid \exists (T_1, T_2).\mathbf{d}$, where A is an atomic concept of C and $\mathbf{d}$ is a Boolean combination of fuzzy RCC predicate names.

The semantics of *fuzzy* $\mathcal{ALC}(D)$ is a pair $\mathcal{I} = (\triangle^\mathcal{I}, \bullet^{\mathcal{I}(w)})$ relative to the spatial fuzzy concrete domain $\mathcal{D} = (\triangle_\mathcal{D}, \Phi_\mathcal{D})$, where $\triangle^\mathcal{I}$ is a domain of interpretation with non-empty set and $\bullet^{\mathcal{I}(w)}$ is a fuzzy interpretation function that maps each abstract concept (role), individual, concrete role, and region into a membership function between 0 and 1.

A *fuzzy* $\mathcal{ALC}(D)$ knowledge base $\mathcal{K}$ consists of a fuzzy TBox $\mathcal{T}$ and a fuzzy ABox $\mathcal{A}$. A fuzzy TBox $\mathcal{T}$ is a finite set of fuzzy General Concept Inclusions (*fuzzy GCIs*) axioms of the form $\langle C \sqsubseteq D \bowtie k \rangle$ with $\bowtie \in \{\geq, >, <, \leq\}, k \in [0, 1]$. A fuzzy ABox $\mathcal{A}$ is a finite set of fuzzy assertions, which has four types of forms $\langle C(a) \bowtie k \rangle$, $\langle R(a, b) \bowtie k \rangle$, $\langle T(a, o) \bowtie k \rangle$ and $\langle \mathbf{d}(x, y) \bowtie k \rangle$. In knowledge base $\mathcal{K}$, we call *fuzzy GCIs* and fuzzy ABox *fuzzy* $\mathcal{ALC}(D)$ *axioms*. In the next section, we will replace propositional letters of LTL with *fuzzy* $\mathcal{ALC}(D)$ axioms to support fuzzy spatio-temporal reasoning.

3   Fuzzy Spatio-Temporal Description Logic *f*-$\mathcal{ALC}$(D)-LTL

In this section, we propose a fuzzy spatio-temporal description logic called *f*-$\mathcal{ALC}$(D)-LTL which is a temporal extension of *fuzzy* $\mathcal{ALC}$(D).

3.1   Syntax

By extending *f*-$\mathcal{ALC}(D)$ [21] with LTL, we obtain a fuzzy spatio-temporal description logic *f*-$\mathcal{ALC}(D)$-LTL. We first define *f*-$\mathcal{ALC}(D)$-LTL formulas.

Definition 3 (Atomic formulas). *The f-ALC(D)-LTL atomic formulas ϕ are given by:*

$\phi ::= \langle C \sqsubseteq D \bowtie n \rangle \mid \langle C(a) \bowtie n \rangle \mid \langle R(a,b) \bowtie n \rangle \mid \langle T(a,o) \bowtie n \rangle \mid \langle \mathbf{d}(o_1,o_2) \bowtie n \rangle$ *where* $C, D \in \mathbf{C}$, $R \in \mathbf{R}$, $T \in \mathbf{T}$, $a, b \in \mathbf{I}, o, o_1, o_2 \in \mathbf{O}$, $\bowtie \in \{\geq, >, <, \leq\}$, $n \in [0, 1]$.

Atomic formulas of form $\langle C \sqsubseteq D \bowtie n \rangle, \langle C(a) \bowtie n \rangle, \langle R(a,b) \bowtie n \rangle, \langle T(a,o) \bowtie n \rangle, \langle \mathbf{d}(o_1,o_2) \bowtie n \rangle$ and their negative are called *f-ALC(D)* literals.

Definition 4 (f-ALC(D)-LTL formulas). *The f-ALC(D)-LTL formulas φ is the smallest set containing the atomic formulas such that:*

– *if ϕ is an atomic formula, then ϕ is an f-ALC(D)-LTL formula;*
– *if φ_1, φ_2 are f-ALC(D)-LTL formulas, then so are $\neg\varphi_1, \circ\varphi_1, \varphi_1 \vee \varphi_2, \varphi_1 \wedge \varphi_2$ and $\varphi_1 U \varphi_2$, where temporal operators $\circ$ and U mean that next and until, respectively.*

We use abbreviations: $\Diamond\varphi = $ true $U\varphi, \Box\varphi = \neg\Diamond\neg\varphi, \varphi_1 \rightarrow \varphi_2 = \neg\varphi_1 \vee \varphi_2$. $\Diamond$ means "sometime in the future" and $\Box$ means "always in the future".

3.2   Models

Similar to [14], the semantics of *f-ALC(D)*-LTL is also based on temporal models, which are a natural combination of the semantic interpretation of *f-ALC(D)* and LTL. Now, we define a temporal model for *f-ALC(D)*-LTL.

Definition 5 (Temporal Model). *An f-ALC(D)-LTL temporal model is a pair $\mathfrak{M} = \langle \mathfrak{I}, \mathcal{I} \rangle$, where $\mathfrak{I} = \langle W, < \rangle$ denotes strict order over $(\langle \mathbb{N}, < \rangle)$ discrete time and $\mathcal{I}$ is a fuzzy assignment function associating with each time point $w \in W$ an f-ALC(D)-interpretation $\mathcal{I}(w) = (\triangle_\mathcal{D} \cup \triangle^\mathcal{I}, \bullet^{\mathcal{I}(w)})$, where $\triangle_\mathcal{D} \cup \triangle^\mathcal{I}$ is interpretation domain and $\bullet^{\mathcal{I}(w)}$ is interpretation function that maps:*

(1)   each abstract concept name $C_i \in \mathbf{C}$ to a function $C_i^{\mathcal{I}(w)} : \triangle^\mathcal{I} \rightarrow [0, 1]$;
(2)   each abstract role name $R_i \in \mathbf{R}$ to a function $R_i^{\mathcal{I}(w)} : \triangle^\mathcal{I} \times \triangle^\mathcal{I} \rightarrow [0, 1]$;
(3)   each concrete role name $T_i \in \mathbf{T}$ to a function $T_i^{\mathcal{I}(w)} : \triangle^\mathcal{I} \times \triangle_\mathcal{D} \rightarrow [0, 1]$;
(4)   each abstract individual $a_i \in \mathbf{I}$ to an element $a_i^{\mathcal{I}(w)} \in \triangle^\mathcal{I}$, for any $w, v \in W$, there is $a_i^{\mathcal{I}(w)} = a_i^{\mathcal{I}(v)}$;
(5)   each fuzzy RCC predicate name $\mathbf{d}_i$ to a function $\mathbf{d}_i^{\mathcal{I}(w)} : \triangle_\mathcal{D} \times \triangle_\mathcal{D} \rightarrow [0, 1]$;
(6)   each fuzzy region $o_i \in \mathbf{O}$ to an element $o_i^{\mathcal{I}(w)} \in \triangle^\mathcal{D}$, and for any $w, v \in W$, there is $O_i^{\mathcal{I}(w)} = O_i^{\mathcal{I}(v)}$ $(i = 1, 2, \cdots)$.

The semantic interpretation $\mathcal{I}$ of *f-ALC(D)*-LTL can be defined as an infinite sequence $\mathcal{I}(0), \mathcal{I}(1), \cdots, \mathcal{I}(w)$ of fuzzy interpretations that share the same domain $\triangle_\mathcal{D} \cup \triangle^\mathcal{I}$. For different time w, we have different interpretation $\mathcal{I}(w)$.

Definition 6 (Truth). *Given a temporal model* $\mathfrak{M} = \langle \mathfrak{S}, \mathcal{I} \rangle$ *and arbitrary time point (or state)* $w \in W$. *At time point* w, *the truth of f-ALC(D)-LTL formulas* φ *(denoted by* $(\mathfrak{M}, w) \vDash \varphi$*) is defined inductively as follows:*

$(\mathfrak{M}, w) \vDash \langle C \sqsubseteq D \bowtie n \rangle$ *iff* $\inf_{a \in \triangle^{\mathcal{I}}}(C^{\mathcal{I}}(a) \Rightarrow D^{\mathcal{I}}(a))$

$(\mathfrak{M}, w) \vDash \langle C(a) \bowtie n \rangle$ *iff* $C^{\mathcal{I}(w)}(a) \bowtie n$

$(\mathfrak{M}, w) \vDash \langle R(a, b) \bowtie n \rangle$ *iff* $R^{\mathcal{I}(w)}(a, b) \bowtie n$

$(\mathfrak{M}, w) \vDash \langle T(a, o) \bowtie n \rangle$ *iff* $T^{\mathcal{I}(w)}(a, o) \bowtie n$

$(\mathfrak{M}, w) \vDash \langle \mathbf{d}(o_1, o_2) \bowtie n \rangle$ *iff* $\mathbf{d}^{\mathcal{I}(w)}(o_1, o_2) \bowtie n$

$(\mathfrak{M}, w) \vDash \neg \varphi$ *iff* $(\mathfrak{M}, w) \nvDash \varphi$

$(\mathfrak{M}, w) \vDash \varphi_1 \vee \varphi_2$ *iff* $(\mathfrak{M}, w) \vDash \varphi_1 \vee (\mathfrak{M}, w) \vDash \varphi_2$

$(\mathfrak{M}, w) \vDash \varphi_1 \wedge \varphi_2$ *iff* $(\mathfrak{M}, w) \vDash \varphi_1 \wedge (\mathfrak{M}, w) \vDash \varphi_2$

$(\mathfrak{M}, w) \vDash \varphi_1 U \varphi_2$ *iff* $\exists v > w, \forall k \in [w, v]$ *s.t* $(\mathfrak{M}, v) \vDash \varphi_2$ *and* $(\mathfrak{M}, k) \vDash \varphi_1$

$(\mathfrak{M}, w) \vDash \Diamond \varphi$ *iff* $\exists v > w$, *such that* $(\mathfrak{M}, v) \vDash \varphi$

$(\mathfrak{M}, w) \vDash \Box \varphi$ *iff* $\forall v > w$, *such that* $(\mathfrak{M}, v) \vDash \varphi$

$(\mathfrak{M}, w) \vDash \circ \varphi$ *iff* $(\mathfrak{M}, w + 1) \vDash \varphi$.

Definition 7 (Satisfiability). *Let* $\mathfrak{M} = \langle \mathfrak{S}, \mathcal{I} \rangle$ *be a temporal model. An f-ALC(D)-LTL formulas* φ *is satisfiable iff there is a temporal model* $\mathfrak{M}$ *such that* $(\mathfrak{M}, w_0) \vDash \varphi$, *where* $w_0 \in W$ *is an initial state or an initial time point.*

The satisfiability problem of *f-ALC(D)-LTL* formulas φ is a basic reasoning problem. Other reasoning problems (e.g., concept satisfiability) can be reduced to this reasoning problem. The following section will give a tableau algorithm for deciding satisfiability of *f-ALC(D)-LTL* formula.

4   Hintikka Structures for *f-ALC*(D)-LTL

As mentioned in Definition 5, a temporal model is an infinite sequence of *f-ALC(D)*-interpretations and its interpretation domains are in a state of infinite expansions. Based on [19], we use Hintikka structure to solve the infinite expansions of interpretation domains. Before discussing Hintikka structure, we first give closure and Hintikka set of φ.

Definition 8 (Closure). *Given an f-ALC(D)-LTL formula* φ, *the closure of* φ, *which is denoted by* $cl(\varphi)$, *is inductively defined as follows:*

(1) $\varphi \in cl(\varphi)$;

(2) if $\neg \varphi \in cl(\varphi)$, *then* $\varphi \in cl(\varphi)$; *if* $\varphi \in cl(\varphi)$, *then* $sub(\varphi) \subseteq cl(\varphi)$; *if* $\varphi \in cl(\varphi)$, *then* $\neg \varphi \in cl(\varphi)$; *if* $\varphi_1 \wedge \varphi_2 \in cl(\varphi)$, *then* $\varphi_1, \varphi_2 \in cl(\varphi)$; *if* $\neg X\varphi \in cl(\varphi)$, *then* $X\neg \varphi \in cl(\varphi)(X$ *denotes* $\Diamond, \Box, \circ)$; *if* $\varphi_1 U \varphi_2 \in cl(\varphi)$, *then* $\circ(\varphi_1 U \varphi_2) \in cl(\varphi)$.

Hintikka set is the smallest saturated set of formulas. We define Hintikka set.

Definition 9 (Hintikka set). *Given an f-$\mathcal{ALC}(D)$-LTL formula φ. We say that $h \subseteq cl(\varphi)$ is a Hintikka set iff it satisfies the following nine conditions:*

(H1) if ϕ is f-$\mathcal{ALC}(D)$-LTL atomic formula, then $\phi \notin h$ or $\neg\phi \notin h$; (H2) if $\neg\neg\varphi \in h$ then $\varphi \in h$; (H3) if $\varphi_1 \vee \varphi_2 \in h$ then $\varphi_1 \in h$ or $\varphi_2 \in h$; (H4) if $\neg(\varphi_1 \vee \varphi_2) \in h$ then $\neg\varphi_1 \in h$ and $\neg\varphi_2 \in h$; (H5) if $\varphi_1 \wedge \varphi_2 \in h$ then $\varphi_1 \in h$ and $\varphi_2 \in h$; (H6) if $\neg(\varphi_1 \wedge \varphi_2) \in h$ then $\neg\varphi_1 \in h$ or $\neg\varphi_2 \in h$; (H7) if $\varphi_1 U \varphi_2 \in h$ then $\varphi_2 \in h$ or $\{\varphi_1, \circ(\varphi_1 U \varphi_2)\} \subseteq h$; (H8) if $\neg \circ \varphi \in h$ then $\circ\neg\varphi \in h$; (H9) if $\neg(\varphi_1 U \varphi_2) \in h$ then $\neg\varphi_2 \in h$, and $\neg\varphi_1 \in h$ or $\neg \circ (\varphi_1 U \varphi_2)\} \in h$.

Definition 10 (Basic Hintikka Structure). *Let φ be an arbitrary f-$\mathcal{ALC}(D)$-LTL formula. A basic Hintikka structure of φ is a pair $\mathcal{H} = (\Sigma, \mathcal{L})$, where:*

- *$\Sigma = \{\omega_0, \omega_1, \dots\}$ is nonempty set of states, where ω_0 denotes an initial state,*
- *$\mathcal{L} \colon \Sigma \to 2^{cl(\varphi)}$ is a function mapping each state into a subset of $cl(\varphi)$.*

Definition 11 (Hintikka structure). *Given a basic Hintikka structure $\mathcal{H} = (\Sigma, \mathcal{L})$. Let h be a Hintikka set, then $\mathcal{H}$ is called Hintikka structure iff it satisfies:*

(1) $\varphi \in \mathcal{L}(\omega_0)$ where ω_0 is an initial state;
(2) for any state $\omega_i \in \Sigma$ such that:

(a) $\mathcal{L}(\omega_i) \subseteq h$; (b) if $\square\varphi \in \mathcal{L}(\omega_i)$, then $\varphi \in \mathcal{L}(\omega_i)$ and $\circ\square\varphi \in \mathcal{L}(\omega_i)$; (c) if $\neg\Diamond\varphi \in \mathcal{L}(\omega_i)$, then $\neg\varphi \in \mathcal{L}(\omega_i)$ and $\neg \circ \square\varphi \in \mathcal{L}(\omega_i)$; (d) if $\Diamond\varphi \in \mathcal{L}(\omega_i)$, then $\varphi \in \mathcal{L}(\omega_i)$ and $\circ\square\varphi \in \mathcal{L}(\omega_i)$; (e) if $\neg\square\varphi \in \mathcal{L}(\omega_i)$, then $\neg\varphi \in \mathcal{L}(\omega_i)$ and $\neg \circ \square\varphi \in \mathcal{L}(\omega_i)$; (f) if $\circ\varphi \in \mathcal{L}(\omega_i)$, then $\varphi \in \mathcal{L}(\omega_i + 1)$; (g) if $\neg \circ \varphi \in \mathcal{L}(\omega_i)$, then $\neg\varphi \in \mathcal{L}(\omega_i + 1)$; (h) if $\varphi_1 U \varphi_2 \in \mathcal{L}(\omega_i)$, then $\exists v > \omega_i$, for all $k \in [\omega_i, v]$, such that $\varphi_2 \in \mathcal{L}(v)$ and $\varphi_1 \in \mathcal{L}(k)$.

Theorem 1. *Given an f-$\mathcal{ALC}(D)$-LTL formula φ. If there is a Hintikka structure for φ, then φ is satisfiable.*

Proof. By Definition 11, we have to show that if φ has a Hintikka structure, then there is a temporal model for φ. Let $\mathcal{H} = (\Sigma, \mathcal{L})$ be a Hintikka structure for φ. We define $\mathcal{H} = (\Sigma, \mathcal{L}')$ s.t if $\varphi \in \mathcal{L}'(\omega)$, then $(\Sigma, \omega) \vDash \varphi$; if $\neg\varphi \in \mathcal{L}'(\omega)$, then $(\Sigma, \omega) \vDash \neg\varphi$. Suppose that for each $\omega \in \Sigma$, if $\varphi \in \mathcal{L}(\omega)$, then $\varphi \in \mathcal{L}'(\omega)$, i.e., $\mathcal{H} \subseteq \mathcal{H}'$. Now, we have to show $\mathcal{H}'$ is a temporal model for φ. We prove by induction that $\mathcal{H}'$ is a temporal model for φ.

Basis: for any atomic formula ϕ, if $\phi \in \mathcal{L}(\omega)$, then $(\mathcal{H}', \omega) \vDash \varphi$; for any atomic formula $\neg\phi$, if $\neg\phi \in \mathcal{L}(\omega)$, then $(\mathcal{H}', \omega) \vDash \neg\varphi$.

Induction: (i) for $\varphi_1 \vee \varphi_2 \in \mathcal{L}(\omega)$, it follows from Definition 13 that $\varphi_1 \in \mathcal{L}(\omega)$ or $\varphi_2 \in \mathcal{L}(\omega)$. Since $\mathcal{H} \subseteq \mathcal{H}', \varphi_1 \in \mathcal{L}'(\omega)$ or $\varphi_2 \in \mathcal{L}'(\omega)$. Thus, $(\mathcal{H}', \omega) \vDash \varphi_1$ or $(\mathcal{H}', \omega) \vDash \varphi_2$. It follows from Definition 7 that $(\mathcal{H}', \omega) \vDash \varphi_1 \vee \varphi_2$. (ii) for $\circ\varphi \in \mathcal{L}(\omega)$, it follows that $\varphi \in \mathcal{L}(\omega + 1)$. Since $\mathcal{H} \subseteq \mathcal{H}', \varphi \in \mathcal{L}'(\omega + 1)$. Thus, $(\mathcal{H}', \omega + 1) \vDash \varphi$. The other formulas have a similar situation.

Thus, $\mathcal{H}'$ is a temporal model. Since $\mathcal{H} \subseteq \mathcal{H}', \mathcal{H}$ is also a temporal model. For the initial state $\omega_0 \in \Sigma$, $(\mathcal{H}, \omega_0) \vDash \varphi$. So, φ is satisfiable. $\square$

5 Reasoning in *f-ALC*(D)-LTL

In this section, we propose a tableau-based reasoning algorithm for determining satisfiability of *f-ALC*(D)-LTL formula. Our algorithm is based on the tableau algorithm for PLTL proposed by Wolper [25]. Similar to [7], our algorithm consists of two phases: (i) tableau construction, and (ii) tableau elimination. The first phase can obtain a complete tableau by applying a series of tableau rules for a given formula φ. The second phase can eliminate some unsatisfiable nodes of the complete tableau by repeatedly applying elimination rules.

5.1 Tableau Rules

The definition of tableau rules is based on the identities: $\Diamond\varphi \equiv \varphi \vee \Diamond\varphi, \Box\varphi \equiv \varphi \wedge \circ\Box\varphi, \varphi_1 U \varphi_2 \equiv \varphi_2 \vee (\varphi_1 \wedge \circ(\varphi_1 U \varphi_2))$.

Similar to [25], *f-ALC*(D)-LTL formulas of the form $\varphi_1 U \varphi_2$ or $\Diamond\varphi$ or $\Box\varphi$ are called *eventualities*. Those of form $\circ\varphi$ and $\neg \circ \varphi$ are called $\circ$-*formulas*.

Before we can give the tableau rules, we first introduce one notion: conjugated pairs of *f-ALC*(D) *literals*. If φ is an *f-ALC*(D) assertion, then φ_c is the conjugation of φ. In *f-ALC*(D), there are a total of four possible conjugated pairs: $\{\langle\varphi \geq n\rangle, \langle\varphi < m\rangle, n \geq m\}$, $\{\langle\varphi \geq n\rangle, \langle\varphi \leq m\rangle, n > m\}$, $\{\langle\varphi > n\rangle, \langle\varphi < m\rangle, n \geq m\}$, and $\{\langle\varphi > n\rangle, \langle\varphi \leq m\rangle, n > m\}$ with $m, n \in [0, 1]$. For instance, if $\varphi = (a, b)$: $R \geq 0.4$, then φ_c may be (a, b): $R < 0.3$.

Definition 12. *Given an f-ALC(D)-LTL formula φ. If φ is $\circ$-formula or f-ALC(D) literal, then we say φ is elementary.*

Table 1. α-formula tableau rules

Name	Formula α	α_1
$\neg\neg$-rule	$\neg\neg\varphi$	φ
$\wedge$-rule	$\varphi_1 \wedge \varphi_2$	φ_1, φ_2
$\Box$-rule	$\Box\varphi$	$\varphi, \circ\Box\varphi$
$\neg\Diamond$-rule	$\neg\Diamond\varphi$	$\neg\varphi, \neg \circ \Diamond\varphi$
$\neg\vee$-rule	$\neg(\varphi_1 \vee \varphi_2)$	$\neg\varphi_1, \neg\varphi_2$

Table 2. β-formula tableau rules

Name	Formula β	β_1	β_2
$\neg\wedge$-rule	$\neg(\varphi_1 \wedge \varphi_2)$	$\neg\varphi_1$	$\neg\varphi_2$
$\vee$-rule	$\varphi_1 \vee \varphi_2$	φ_1	φ_2
$\Diamond$-rule	$\Diamond\varphi$	φ	$\circ\Diamond\varphi$
$\neg\Box$-rule	$\neg\Box\varphi$	$\neg\varphi$	$\varphi, \neg \circ \Box\varphi$
U-rule	$\varphi_1 U \varphi_2$	φ_2	$\varphi_1, \circ(\varphi_1 U \varphi_2)$
$\neg U$-rule	$\neg(\varphi_1 U \varphi_2)$	$\neg\varphi_1, \neg\varphi_2$	$\neg\varphi_2, \neg \circ (\varphi_1 U \varphi_2)$

Table 3. $\circ$-formula tableau rules

Name	Formula γ	Formula γ_1
$\neg\circ$-rule	$\neg \circ \varphi$	$\neg\varphi$
$\circ$-rule	$\circ\varphi$	φ

By using tableau rules, the non-elementary formula φ can be extended into a finite number of elementary formulas. According to the method of rule classification in [5], we divide the tableau rules into three categories: α-formula tableau rules (see Table 1), β-formula tableau rules (see Table 2), and $\circ$-formula tableau rules (see Table 3). α, β, γ denote the formula to be decomposed and $\alpha_1, \beta_1, \beta_2, \gamma_1$ denote the decomposed elementary formula.

5.2 Tableau Construction

The purpose of the tableau construction phase is to construct tableau for a given $f\text{-}\mathcal{ALC}(D)\text{-}LTL$ formula φ. Before constructing tableau, we first define a tableau.

Definition 13 (Tableau). *Given a formula φ. A tableau for φ is a directed graph $T = (N, n_0, \rightarrow, L)$, where:*

- *N is a finite set of nodes (states)*
- *$n_0 \in N$ is a root node*
- *$\rightarrow$ denotes a binary relation over N. $\forall\, n \in N, \exists\, t \in N$ s.t $n \rightarrow t$*
- *L is a labeling function s.t $L(n_0) = \{\varphi\}$, for $n \in N \setminus \{n_0\}$.*

Each state $n \in N \setminus \{n_0\}$ corresponds to a time point $t \in \mathbb{N}$. For each state n, its label $L(n)$ corresponds to a Hintikka set of formula φ. An initial tableau graph is only a graph with a root node. By repeatedly applying the tableau rules given in the Tables 1, 2 and 3, the initial tableau graph is expanded.

Definition 14. *Let $T = (N, n_0, \rightarrow, L)$ be a tableau graph for a given $f\text{-}\mathcal{ALC}(D)\text{-}LTL$ formula φ. T is called complete iff none of the tableau rules shown in Tables 1, 2 and 3 is applicable to T. Otherwise, we call T is incomplete.*

In tableau graph, for some node $n \in N$, if $L(n)$ contains only elementary or marked formulas, then the node is called a *state*. If a node n is either an initial node or the immediate child of a *state*, then the node is called a *pre-state*. To avoid decomposing the same formula twice, we use φ^* to denote a formula that has been decomposed. The construction algorithm is shown in Algorithm 1.

5.3 Tableau Elimination

In order to decide satisfiability of $f\text{-}\mathcal{ALC}(D)\text{-}LTL$ formula φ, in this section, we have to eliminate the unsatisfiable nodes of the tableau graph. We first give the formal definition of unsatisfiable nodes and eventualities.

Definition 15 (Unsatisfiable nodes). *Let $T = (N, n_0, \rightarrow, L)$ be a complete tableau graph for a given $f\text{-}\mathcal{ALC}(D)\text{-}LTL$ formula φ. Aso let n be an arbitrary node and let $L(n)$ be a label of n. A node n is unsatisfiable iff one of the following situations holds at least:*

(1) $L(n)$ contains two fuzzy conjugated triples;
(2) $L(x)$ contains axiom $\langle C, <, 0 \rangle$;
(3) $L(x)$ contains axiom $\langle C, >, k \rangle$ with $k > 1$;
(4) There is a formula ϕ s.t $\circ\phi \in L(n)$ and $\neg \circ \neg\phi \notin L(n)$.

Algorithm 1. $ConTab(\varphi)$

Input: an *f-ALC*(S)-LTL formula φ
Output: a tableau graph T
 1: initialize tableau $T = (N, n_0 \rightarrow, L) = (\{n_0\}, n_0, \oslash, \{(n_0, \{\varphi\})\})$
 2: **if** $(L(n_0)$ is elementary **then**
 3:    return;
 4: **end if**
 5: **while** $(T$ is *incomplete*) **do**
 6:    **if** $(n \in N \&\& \alpha \in L(n) \&\& \alpha^* \notin L(n))$ **then**
 7:       create a new node n' as a child of node n and
 8:       label $L(n')$ with $L(n') = (L(n) - \{\alpha\}) \bigcup \{\alpha_1\} \bigcup \{\alpha^*\}$;
 9:    **end if**
10:    **if** $(n \in N \&\& \beta \in L(n) \&\& \beta^* \notin L(n))$ **then**
11:       create two new node n' and n'' as children of node n and
12:       the corresponding label $L(n')$ and $L(n'')$ with
13:       $L(n') = (L(n) - \{\beta\}) \bigcup \{\beta_1\} \bigcup \{\beta^*\}, L(n'') = (L(n) - \{\beta\}) \bigcup \{\beta_2\} \bigcup \{\beta^*\}$;
14:    **end if**
15:    **if** $(n$ is a *state* $\&\& \gamma \in L(n) \&\& \gamma^* \notin L(n))$ **then**
16:       create a new node n' as a child of n and label
17:       $L(n')$ with $L(n') = (L(n) - \{\gamma\}) \bigcup \{\gamma_1\} \bigcup \{\gamma^*\}$;
18:    **end if**
19: **end while**

Definition 16 (Unsatisfiable eventualities). *Given a pre-state n_i of a complete tableau graph and its label $L(n_i)$ containing eventualities formulas. If there is a path $n_i, \ldots, n_j$ ($i \geq 1$, $j \geq i + 1$) in the tableau such that $n_i, n_j \in N$ and $\{\varphi_2\} \in L(n_j)$, then the eventualities formulas in $L(n_i)$ are called satisfiable. Otherwise, they are unsatisfiable.*

In order to decide the satisfiability of *f-ALC*(D)-LTL formula φ, we need to eliminate the unsatisfiable nodes in the complete tableau graph. Algorithm 2 gives the tableau graph elimination algorithm *Eliminate(T)*. If the root node is not marked ($\times$), then the tableau graph is open. In this case, we say that the *f-ALC*(D)-LTL formula φ is satisfiable. Otherwise, we say φ is unsatisfiable.

5.4   Correctness

The correctness includes termination, soundness, and completeness. Our proof of the correctness is based on the proof for LTL [3] and PLTL [5].

Theorem 2 (Termination). *Let φ be f-ALC(D)-LTL formula. When started on input φ, the tableau algorithm terminates.*

Proof. Let $l = |\text{sub}(\varphi)|$ be the length (number) of *f-ALC*(D)-LTL formula φ.

(1) In tableau construction phase. Each node $n_i \in N$ labeled by $L(n_i)$ contains two types of formulas: closure $cl(\varphi)$ and $\circ$-$cl(\varphi)$. By Definition 8, the closure

---

Algorithm 2. *Eliminate*(*T*)

Input: a tableau graph *T*

Output: Boolean value

1: **while** ($\exists n \in N$ and *n* is *unsatisfiable*) **do**
2:    eliminate node *n* of *T*, and marked ($\times$)
3: **end while**
4: **while** ($\exists n \in N$ and all children of *n* are eliminated) **do**
5:    eliminate node *n* of *T*, and marked ($\times$);
6: **end while**
7: **while** ($\exists n \in N$ is *pre-state* and the eventualities
8: formulas in $L(n_i)$ is *unsatisfiable*) **do**
9:    eliminate node *n* of *T*, and marked ($\times$);
10: **end while**
11: **if** (root node $n_0 \in N$ is not marked ($\times$)) **then**
12:    *T* is *"open"*, return true;
13: **else**
14:    *T* is *"closed"*, return false;
15: **end if**

$cl(\varphi)$ is composed of all subformulas of φ and their negation. So, the number of $cl(\varphi)$ is equal to $2l$. In a similar way, the number of $\circ\text{-}cl(\varphi)$ is equal to $2l$. The number of the two type of formulas in $L(n_i)$ is then at most equal to $4l$. It follows that the upper bound of the set of formulas consisting of these two types of formulas is 2^{4l}. Again since each node contains a Hintikka set for formula φ and each Hintikka set is a subset of $cl(\varphi)$, the upper bound of the nodes is also 2^{4l}.

(2) In tableau elimination phase. Let *T* be tableau graph constructed by the algorithm *ConTab*(φ). For the algorithm *Eliminate*(*T*), we consider two possible cases: Case 1: *all nodes in T are unsatisfiable*. Since the upper bound of the nodes in *T* is 2^{4l}, the number of iterations of the whole *Eliminate*(*T*) algorithm is bounded by 2^{4l}. Case 2: *there is only unsatisfiable eventualities in T*. Since the number of eventualities in each node is at most equal to *l*, the number of eventualities is bounded by 2^l. It follows that the number of eventualities in the whole tableau graph is bounded by $2^l \times 2^{4l}$. Thus, the number of iterations of the whole *Eliminate*(*T*) algorithm is bounded by $2^l \times 2^{4l}$. $\square$

Theorem 3 (Soundness). *Let φ be an f-$\mathcal{ALC}$(D)-LTL formula to the tableau algorithm. If φ is satisfiable, then the algorithm returns "open".*

Proof. To show soundness, we have to show that if the algorithm returns "closed", then φ is unsatisfiable. Let $T = (N, n_0, \rightarrow, L)$ be a tableau graph obtained by performing elimination algorithm. Since the algorithm return "closed", the root node in *T* is not eliminated. Again since the root node is labeled by $L(n_0) = \varphi$, we have to show that if root node n_0 is eliminated, then $L(n_0) = \varphi$ is unsatisfiable. We prove by induction that if a node $n \in N$ labeled by

$L(n) = \{\varphi_1, \varphi_2, \ldots, \varphi_m\}$ is eliminated, then $L(n) = \{\varphi_1, \varphi_2, \ldots, \varphi_m\}$ is unsatisfiable. In the elimination algorithm, the elimination of the node is consequences of the properties of elimination rules. Combined with the *Eliminate* (*T*), we analyse different types of nodes eliminated.

Case 1: *an unsatisfiable node n is eliminated.* By Definition 15, the unsatisfiable node labeled by $L(n)$ must satisfy one of four conditions at least: $L(n)$ contains two fuzzy conjugated triples; $L(x)$ contains axiom $<C, <, 0>$; $L(x)$ contains axiom $<C, >, k>$ with $k > 1$; there is a formula ϕ such that $\circ\phi \in L(n)$ and $\neg\circ\neg\phi \notin L(n)$. Obviously, $L(n)$ is unsatisfiable.

Case 2: *a son of a state is eliminated.* According to algorithm *ConTab(φ)* (lines 11–15), a state n will produce a son n' labeled by $L(n') = \{\varphi_1, \varphi_2, \ldots, \varphi_m\}$ in each application of tableau rules, where φ_m denotes *f-ALC*(D) literals. Since a son n' of a state n is eliminated, $L(n') = \{\varphi_1, \varphi_2, \ldots, \varphi_m\}$ is unsatisfiable. It follows that $L(n) = \{\circ\varphi_1, \circ\varphi_2, \ldots, \circ\varphi_m\}$ is also unsatisfiable.

Case 3: *a son of a non-state is eliminated.* According to algorithm *ConTab(φ)* (lines 4–10), a non-state n may produce one or two sons in each application of tableau rules. That is, a non-state n has at least one son n' labeled by $L(n')$. Since one can obtain $L(n')$ is by decomposing $L(n)$, $L(n)$ is satisfiable iff there exists at least one son n' of n such that $L(n')$ is satisfiable. Again since all sons of non-state n are eliminated, $L(n')$ is unsatisfiable.

Case 4: *pre-state containing eventualities is eliminated.* Let n be a pre-state labeled by $L(n) = \{\varphi_1^1\varphi_2^1, \ldots, \varphi_1^m\varphi_2^m\}$ $(m \geq 1)$. By the definition of pre-state, there are some paths $n_i^1 \ldots n_j^1, \ldots, n_i^m \ldots n_j^m$ $(i \geq 1, j \geq i+1)$ such that $n_i^1 = n, \ldots, n_i^m = n, \ldots, \varphi_2^1 \in L(n_j^1), \ldots, \varphi_2^m \in L(n_j^m)$. Since node n is eliminated, there is at least one eventuality in $L(n)$ that is unsatisfiable.   □

Theorem 4 (Completeness). *Let φ be an f-ALC(D)-LTL formula to the tableau algorithm. If the tableau algorithm returns "open", then φ is satisfiable.*

Proof. Suppose that we have constructed a tableau $T = (N, n_0, \rightarrow, L)$ when the tableau algorithm returns "open". To show completeness, we first show that the constructed tableau graph has a Hintikka structure.

Construct a Hintikka structure $\mathcal{H} = (\Sigma, \mathcal{L})$ from T, where $\Sigma = \{\omega_0, \omega_1, \ldots, \omega_n\}$ is nonempty set of states, ω_0 denotes an initial state, $\mathcal{L}$ is a mapping function from Σ to L. The construction steps are as follows:

(1) *Find a path.* Find a path leading from root node n_0 to leaf node n_k that is not marked (×). It can be defined as: a path $P = n_0, n_1, \ldots, n_k$ $(k \geq 0)$, where n_0 is root node, $L(n_0) = \{\varphi\}$, n_k is a state node containing only *f-ALC*(D) literal ϕ_m, $L(n_k) = \{\phi_1, \ldots, \phi_m\}$ $(m \geq 1)$. Please note that there is a special case that if $n_0 = n_k$, then $n_1, \ldots, n_{k-1}$ are empty and $L(n_0) = L(n_k) = \{\varphi\}$.
(2) *Construct a Hintikka structure $\mathcal{H}$ from a path P.*
   Step 1: the root node n_0 in P can be defined as an initial state ω_0 of $\mathcal{H}$ such that $\mathcal{L}(\omega_0) = L(n_0) = \{\varphi\}$. Step 2: starting from root node n_0 in P to find out states n_i $(0 \leq i \leq k)$ in turn. If $L(n_i)$ contains only elementary formulas, i.e., $L(n_i) = \{\phi_1, \ldots, \phi_m, \circ\varphi_1, \ldots, \circ\varphi_m\}$, then $\mathcal{L}(\omega_j) = L(n_j) =$

$\{\phi_1, \ldots, \phi_m\}$. If $n_i = n_k$, i.e., $L(n_i) = \{\phi_1, \ldots, \phi_m\}$, then $\mathcal{L}(\omega_j) = L(n_k) = \{\phi_1, \ldots, \phi_m\}$. Finally, until all the state nodes are found.

By the definition of Hintikka structure, the constructed structure $\mathcal{H} = (\Sigma, \mathcal{L})$ is a Hintikka structure for φ. According to the Theorem 1, if there is a Hintikka structure for φ, then φ is satisfiable. Thus, it follows that φ is satisfiable.   □

6   Conclusion and Future Work

In this work, we have presented a fuzzy spatio-temporal description logic f-$\mathcal{ALC}(D)$-LTL, which is a temporal extension of the f-$\mathcal{ALC}(D)$. Our description logic enables us to reason fuzzy RCC topological relationships that change over time. We first define concept syntax, formulas syntax, and semantic interpretation. Then, we present a tableau-based algorithm for f-$\mathcal{ALC}(D)$-LTL to decide satisfiability problem of f-$\mathcal{ALC}(D)$-LTL formula. Also, we show the termination, soundness, and completeness of the tableau algorithm using Hintikka structure. As a result, the satisfiability problem of f-$\mathcal{ALC}(D)$-LTL formula φ is decidable.

Acknowledgements. The work is supported by the National Natural Science Foundation of China (61370075, 61572118, 61672139).

References

1. Allen, J.F.: Maintaining knowledge about temporal intervals. Commun. ACM **26**(11), 832–843 (1983)
2. Artale, A., Franconi, E.: A temporal description logic for reasoning about actions and plans. J. Artif. Intell. Res. **9**(1), 463–506 (1998)
3. BenAri, M.: Mathematical Logic for Computer Science. Springer, London (2012). doi:10.1007/978-1-4471-4129-7
4. Cristani, M., Gabrielli, N.: Practical issues of description logics for spatial reasoning. In: Proceedings of the 2009 AAAI Spring Symposium: Benchmarking of Qualitative Spatial and Temporal Reasoning Systems, pp. 5–10 (2009)
5. Gaintzarain, J., Hermo, M., Lucio, P., Navarro, M.: Systematic semantic tableaux for PLTL. Electron. Notes Theor. Comput. Sci. **206**, 59–73 (2008)
6. Galton, A.: Spatial and temporal knowledge representation. Earth Sci. Inf. **2**(3), 169–187 (2009)
7. Goranko, V., Shkatov, D.: Tableau-based decision procedures for logics of strategic ability in multiagent systems. ACM Trans. Comput. Logic **11**(1), 1–51 (2009)
8. Haarslev, V., Lutz, C., Moller, R.: A description logic with concrete domains and a role-forming predicate operator. J. Logic Comput. **9**(3), 351–384 (1999)
9. Hudelot, C., Atif, J., Bloch, I.: $\mathcal{ALC}(\mathbf{F})$: a new description logic for spatial reasoning in images. In: Agapito, L., Bronstein, M.M., Rother, C. (eds.) ECCV 2014. LNCS, vol. 8926, pp. 370–384. Springer, Cham (2015). doi:10.1007/978-3-319-16181-5_26
10. Kaplunova, A., Haarslev, V., Moller, R.: Adding ternary complex roles to ALCRP(D). In: DL Workshop, pp. 112–126 (2002)
11. Klamma, R., Cao, Y., Spaniol, M., Leng, Y.: Spatiotemporal knowledge visualization and discovery in dynamic social networks. In: CIKM, pp. 384–391 (2007)

12. Lutz, C., Milicic, M.: A tableau algorithm for description logics with concrete domains and general tboxes. J. Autom. Reason. **38**(1), 227–259 (2007)
13. Lutz, C., Wolter, F., Zakharyaschev, M.: Temporal description logics: a survey. In: TIME, pp. 3–14 (2008)
14. Manna, Z., Pnueli, A.: The Temporal Logic of Reactive and Concurrent. Springer, New York systems (1992). doi:10.1007/978-1-4612-0931-7
15. Randell, D.A., Cui, Z., Cohn, A.G.: A spatial logic based on regions and connection. In: KR, pp. 165–176 (1992)
16. Ribaric, S., Hrkac, T.: A model of fuzzy spatio-temporal knowledge representation and reasoning based on high-level petri nets. Inf. Syst. **37**(3), 238–256 (2012)
17. Schockaert, S., Cock, M.D., Cornelis, C., Kerre, E.E.: Fuzzy region connection calculus: an interpretation based on closeness. Int. J. Approx. Reason. **48**(1), 332–347 (2008)
18. Schockaert, S., De Cock, M., Kerre, E.E.: Spatial reasoning in a fuzzy region connection calculus. Artif. Intell. **173**(2), 258–298 (2009)
19. Seylan, I., Jamroga, W.: Coalition description logic with individuals. Electron. Notes Theor. Comput. Sci. **262**, 231–248 (2010)
20. Sozer, A., Yazici, A., Oguztuzun, H.: Modeling and querying fuzzy spatiotemporal databases. Inf. Sci. **178**(19), 3665–3682 (2008)
21. Straccia, U.: Towards spatial reasoning in fuzzy description logics. In: FUZZ-IEEE, pp. 512–517 (2009)
22. Straccia, U.: All about fuzzy description logics and applications. Reason. Web **2015**, 1–31 (2015)
23. Sturm, H., Wolter, F.: A tableau calculus for temporal description logic: the expanding domain case. J. Logic Comput. **12**(5), 809–838 (2002)
24. Wang, S., Liu, D.: Spatial description logic and its application in geospatial semantic web. In: Proceedings of the 2008 International Multi-symposiums on Computer and Computational Sciences, pp. 214–221 (2008)
25. Wolper, P.: The tableau method for temporal logic: an overview. Log. Anal. **110**, 119–136 (1985)

R-Calculus for the Primitive Statements in Description Logic $\mathcal{ALC}$

Yuhui Wang[1,2]($\boxtimes$), Cungen Cao[1], and Yuefei Sui[1,2]

[1] Key Laboratory of Intelligent Information Processing,
Institute of Computing Technology, Chinese Academy of Sciences, Beijing, China
yhwang_ict@qq.com
[2] School of Computer and Control Engineering,
University of Chinese Academy of Sciences, Beijing, China

Abstract. The AGM postulates [1] are for the belief revision (revision by a single belief), and the DP postulates [14] are for the iterated revision (revision by a finite sequence of beliefs). Li [4] gave an **R**-calculus for **R**-configurations $\Delta|\Gamma$, where Δ is a set of atomic formulas or negations of atomic formulas, and Γ is a finite set of formulas. With an idea to delete the requirement that Δ is a set of atoms, we will give an **R**-calculus $\mathbf{S}^{\mathrm{DL}}$ (a set of deduction rules) with respect to $\subseteq$-minimal change such that for any finite consistent sets Γ, Δ of statements in the description logic $\mathcal{ALC}$, there is a consistent subset $\Theta \subseteq \Gamma$ of statements such that $\Delta|\Gamma \Rightarrow \Delta, \Theta$ is provable; and prove that $\mathbf{S}^{\mathrm{DL}}$ is sound and complete with the $\subseteq$-minimal change.

Keywords: Description logics · Belief revision · **R**-calculus · Set-inclusion minimal change · Soundness and completeness

1 Introduction

The AGM postulates [1–3], which are a set of requirements a revision operator should satisfy, are set for using a formula A to revise a theory K, so that if $K \circ A \Rightarrow K'$ then K' is a maximal consistent subset of $K \cup \{A\}$.

While the **R**-calculus [6–11], which is a revision operator which satisfies the AGM postulates, is a Gentzen-type deduction system to deduce a consistent one $\Gamma' \cup \Delta$ from an inconsistent theory $\Gamma \cup \Delta$ of the first-order logic. $\Gamma' \cup \Delta$ should be a maximal consistent subtheory of $\Gamma \cup \Delta$ that includes Δ as a subset, where Γ is a consistent set of formulas, and Δ is a consistent set of atomic formulas or negations of atomic formulas. Moreover, there is a concept $\Delta|\Gamma$ defined in **R**-calculus, called **R**-configuration. It is to use a theory Δ to revise another theory Γ by eliminating the formulas in Γ(or in the theory $Th(\Gamma)$ of Γ) which negations are deducible by Δ, to make the remain of Γ consistent with Δ. It was proved that if $\Delta|\Gamma \Rightarrow \Delta|\Gamma'$ is deducible and $\Delta|\Gamma'$ is an **R**-termination, i.e., there is no **R**-rule to reduce $\Delta|\Gamma'$ to another **R**-configuration $\Delta|\Gamma''$, then $\Delta \cup \Gamma'$ is a contraction of Γ by Δ.

© Springer International Publishing AG 2017
G. Li et al. (Eds.): KSEM 2017, LNAI 10412, pp. 106–116, 2017.
DOI: 10.1007/978-3-319-63558-3_9

The $\subseteq$-minimal change [5,12,13] is with respect to the set-inclusion, that is, let $\Delta|\Gamma \Rightarrow \Delta, \Theta$ be provable in an **R**-calculus. Then, Θ is a minimal change of Γ by Δ, if for any $A \in \Gamma, \Theta \cup \Delta \nvdash A$ implies $\Theta \cup \Delta \vdash \neg A$. Therefore, we also call Θ as a $\subseteq$-maximal consistent set of Γ by Δ.

Ontologies play a crucial role for the success of the Semantic Web [15]. One of the challenging problems for the development of ontology is ontology evolution, which is defined as the timely adaptation of an ontology to the arisen changes and the consistent management of these changes [16,17]. One of the center problem during ontology evolution is inconsistency handling. There are various forms of inconsistencies, such as structural inconsistency, logical inconsistency and user-defined inconsistency. Among them, logical inconsistency has received lots of attention, where ontologies are represented by logical theories, such as description logics (DLs) [16,18,19].

Description logics are different from traditional logics in that description logics are a set of logics, of which each is a fragment of the first-order logic, and the logical symbols are decomposed into two layers: one for concepts and another for statements. For example, in $\mathcal{ALC}$, the logical symbols are the concept constructors $\neg, \sqcap, \sqcup, \forall, \exists$; and the subsumption relation $\sqsubseteq, \not\sqsubseteq$. The revision for the subsumption statements can be transformed into that for the concept statements and it is also studied in the semantic networks so that we will consider only the statements $C(a)$ and $R(a, b)$ in the following.

R-calculus for description logics is between the one for propositional logic and the one for first-order logic, because the complexity of the deduction relation of description logics is between the ones of the deduction relation of the former and of the latter. Even though the quantifier $\forall$ occurs in the concept constructor $\forall R$, its equivalent form in the first-order logic is a guarded first-order formula, and the deduction relation in the guarded first-order logic is decidable. Therefore, for the **R**-calculi of description logics, we can decompose a statement $C(a)$ into atomic statements such that we will delete the requirement of the **R**-calculus [4] that Δ is a set of atoms and give two sets of deduction rules: one for $C(a)$ consistent with Δ, and another for $C(a)$ inconsistent with Δ.

Given two theories Δ and Γ, assume that Γ be a finite consistent set of statements such that $\Gamma = \{C_1(a), \ldots, C_n(a)\}$. The following procedure produces a minimal change of Γ by Δ. Let $\leq$ be an ordering on Γ such that $C_1(a) \leq C_2(a) \leq \cdots \leq C_n(a)$. For each $i \leq n$, define

$$\Theta_0 = \Delta;$$
$$\Theta_i = \begin{cases} \Theta_{i-1} \cup \{C_i(a)\} & \text{if } \Theta_{i-1} \cup \{C_i(a)\} \text{ is consistent} \\ \Theta_{i-1} & \text{otherwise.} \end{cases}$$

Then $\Gamma' = \Theta_n - \Delta$ is a maximal consistent subtheory of Γ such that $\Gamma' \cup \Delta$ is consistent.

In this paper we firstly consider a simple case of revision: $\Delta|C(a)$, where Δ is a set of statements to revise, and $C(a)$ is a statement to be revised; and then consider a general case of revision: $\Delta|\Gamma$, which is reduced to successive revisions $(\cdots ((\Delta|C_1(a))|C_2(a)) \cdots)|C_n(a)$, where $\Gamma = \{C_1(a), \ldots, C_n(a)\}$ and there is an ordering $\leq$ on Γ, so that we can make the revision $\Delta|\Gamma$ turned into an iterated revision of form $\Delta|C_i(a)$.

We will give an **R**-calculus $\mathbf{S}^{\mathrm{DL}}$ for the $\subseteq$-minimal change, which is a set of deduction rules for $\Delta|C(a)$, such that $\mathbf{S}^{\mathrm{DL}}$ is sound and complete, that is, for any theory Δ, and a statement $C(a)$, if $\Delta|C(a) \Rightarrow \Delta, C^i(a)$ is provable in $\mathbf{S}^{\mathrm{DL}}$ then $\Delta \cup \{C^i(a)\}$ is consistent, and if $\Delta \cup \{C(a)\}$ is consistent then $C^i(a) = C(a)$; and if $\Delta \cup \{C(a)\}$ is inconsistent then $C^i(a) = \lambda$; and conversely, if $\Delta \cup \{C(a)\}$ is consistent then $\Delta|C(a) \Rightarrow \Delta \cup \{C(a)\}$ is provable in $\mathbf{S}^{\mathrm{DL}}$; and if $\Delta \cup \{C(a)\}$ is inconsistent then $\Delta|C(a) \Rightarrow \Delta$ is provable in $\mathbf{S}^{\mathrm{DL}}$.

Moreover, $\mathbf{S}^{\mathrm{DL}}$ is sound and complete with respect to the $\subseteq$-minimal change for $\Delta|\Gamma$ too. That is, for any finite consistent sets Δ, Γ of statements in the description logic $\mathcal{ALC}$, there is a consistent subset $\Theta \subseteq \Gamma$ of statements such that $\Delta|\Gamma \Rightarrow \Delta, \Theta$ is *provable*, denoted by

$$\vdash_{\mathbf{S}\mathrm{DL}} \Delta|\Gamma \Rightarrow \Delta, \Theta,$$

and Θ is a minimal change of Γ by Δ. Therefore, $\mathbf{S}^{\mathrm{DL}}$ is sound with respect to the $\subseteq$-minimal change if for any consistent sets Γ, Δ, Θ of statements, if $\Delta|\Gamma \Rightarrow \Delta, \Theta$ is provable in $\mathbf{S}^{\mathrm{DL}}$ then Θ is a maximal consistent subset of Γ by Δ, i.e., $\vdash_{\mathbf{S}\mathrm{DL}}$ $\Delta|\Gamma \Rightarrow \Delta, \Theta$ implies $\models_{\mathbf{S}\mathrm{DL}} \Delta|\Gamma \Rightarrow \Delta, \Theta$; and $\mathbf{S}^{\mathrm{DL}}$ is complete with respect to the $\subseteq$-minimal change if for any maximal consistent subset Θ of Γ by $\Delta, \Delta|\Gamma \Rightarrow \Delta, \Theta$ is provable in $\mathbf{S}^{\mathrm{DL}}$, i.e., $\models_{\mathbf{S}\mathrm{DL}} \Delta|\Gamma \Rightarrow \Delta, \Theta$ implies $\vdash_{\mathbf{S}\mathrm{DL}} \Delta|\Gamma \Rightarrow \Delta, \Theta$. Here, Θ is a maximal consistent subset of Γ by Δ if (i) $\Theta \subseteq \Gamma$, (ii) $\Delta \cup \Theta$ is consistent, and (iii) for any Θ' with $\Theta \subset \Theta' \subseteq \Gamma, \Delta \cup \Theta'$ is inconsistent.

The paper is organized as follows: the next section gives the basic definitions of description logic $\mathcal{ALC}$, and defines the logical language and the semantics of the simplified $\mathcal{ALC}$; the third section gives a set $\mathbf{S}^{\mathrm{DL}}$ of deduction rules which is proved to be sound and complete with the $\subseteq$-minimal change, and the last section concludes the whole paper and discusses further works.

2   Description Logic $\mathcal{ALC}$

Let L_1 be the logical language for the description logic $\mathcal{ALC}$, which contains the following symbols:

- constant symbols: $c_0, c_1, \ldots$;
- atomic concepts: $A_0, A_1, \ldots$;
- roles: $R_0, R_1, \ldots$;
- concept constructors: $\neg, \sqcap, \sqcup, \forall, \exists$,

Concepts:
$$C := A|\neg A|C_1 \sqcap C_2|C_1 \sqcup C_2|\forall R.C|\exists R.C.$$

Primitive statements:
$$\varphi := C(c)|R(c, d)$$

A model $\mathbf{M}$ is a pair (Δ, I), where Δ is a non-empty set, and I is an interpretation, such that

o for any constant $c, I(c) \in \Delta$;

○ for any atomic concept $A, I(A) \subseteq \Delta$;
○ for any role $R, I(R) \subseteq \Delta^2$.
The interpretation C^I of C:

$$
C^I = \begin{cases}
I(A) & \text{if } C = A \\
\Delta - A & \text{if } C = \neg A \\
C_1^I \cap C_2^I & \text{if } C = C_1 \sqcap C_2 \\
C_1^I \cup C_2^I & \text{if } C = C_1 \sqcup C_2 \\
I(a) \in \Delta : \mathbf{A}b((I(a), I(b)) \in I(R) \Rightarrow I(b) \in C^I) & \text{if } C = \forall R.C \\
I(a) \in \Delta : \mathbf{E}b((I(a), I(b)) \in I(R) \ \& \ I(b) \in C^I) & \text{if } C = \exists R.C
\end{cases}
$$

where in syntax, we use $\neg, \wedge, \rightarrow, \forall, \exists$ to denote the logical connectives and quantifiers; and in semantics we use $\sim, \&, \Rightarrow, \mathbf{A}, \mathbf{E}$ to denote the corresponding connectives and quantifiers.

By the definition of $(\forall R.C)(a)$ and $(\exists R.C)(a)$, we have

$$\neg(\forall R.C)(a) \Leftrightarrow (\exists R.(\neg C))(a) \Leftrightarrow (R(a,d) \sqcap (\neg C)(d))$$
$$\neg(\exists R.C)(a) \Leftrightarrow (\forall R.(\neg C))(a) \Leftrightarrow (R(a,c) \rightarrow (\neg C)(c)) \Leftrightarrow ((\neg R)(a,c) \sqcup (\neg C)(c))$$

where d is an old constant and c is a new one.

The satisfaction $\mathbf{M} \models \varphi$ of statement φ:

$$
\mathbf{M} \models \varphi \text{ iff } \begin{cases}
I(c) \in C^I & \text{if } \varphi = C(c) \\
(I(c), I(d)) \in I(R) & \text{if } \varphi = R(c,d)
\end{cases}
$$

A sequent δ is of form $\Gamma \Rightarrow \Delta$, where Γ, Δ are sets of primitive statements. Given an interpretation I, we say that I satisfies δ, denoted by $I \models \delta$, if $I \vdash \Gamma$ implies $I \models \Delta$, where $I \models \Gamma$ if for each primitive statement $\varphi \in \Gamma, I \models \varphi$; and $I \models \Delta$ if for some primitive statement $\psi \in \Delta, I \models \psi$.

A sequent δ is valid, denoted by $\models \Gamma \Rightarrow \Delta$, if for any interpretation $I, I \models \Gamma \Rightarrow \Delta$.

3   R-Calculus for Subset-Minimal Change

Definition 3.1. *Given any consistent theories Γ and Δ, a theory Θ is a subset-minimal($\subseteq$-minimal) change of Γ by Δ, denoted by $\models_\mathbf{S} \Delta | \Gamma \Rightarrow \Delta, \Theta$, if (i) Θ is consistent, (ii) $\Theta \subseteq \Gamma$, and (iii) for any Θ' with $\Theta \subset \Theta' \subseteq \Gamma, \Theta' \cup \Delta$ is inconsistent.*

Here, if $C(a)$ is consistent with Δ we denote by $\models_\mathbf{SDL} \Delta | C(a) \Rightarrow \Delta, C(a)$; and if $C(a)$ is inconsistent with Δ we denote by $\models_\mathbf{SDL} \Delta | C(a) \Rightarrow \Delta$. For a set Γ of statements, if a theory Θ is a $\subseteq$-minimal change of Γ by Δ then we write

$$\models_\mathbf{SDL} \Delta | \Gamma \Rightarrow \Delta, \Theta.$$

3.1 $\mathbf{S}^{DL}$: R-Calculus for a Statement

The deduction rules in R-calculus $\mathbf{S}^{DL}$ are the composing rules, which compose substatements (e.g., $C_1(a), C_2(a)$) in the precondition of a rule into a complex statement (e.g., $(C_1 \sqcap C_2)(a)$) in the postcondition of the rule.

R-calculus $\mathbf{S}^{DL}$ for a statement $C(a)$:

Axioms:

$$(S^{\mathbf{A}}) \ \frac{\Delta \not\vdash \neg A(c)}{\Delta|A(c) \Rightarrow \Delta, A(c)} \qquad (S_{\mathbf{A}}) \ \frac{\Delta \vdash \neg A(c)}{\Delta|A(a) \Rightarrow \Delta}$$

$$(S^{\neg}) \ \frac{\Delta \not\vdash A(c)}{\Delta|\neg A(c) \Rightarrow \Delta, \neg A(c)} \qquad (S_{\neg}) \ \frac{\Delta \vdash A(c)}{\Delta|\neg A(a) \Rightarrow \Delta}$$

Deduction rules:

$$(S^{\sqcap}) \ \frac{\Delta|C_1(a) \Rightarrow \Delta, C(a)}{\Delta, C_1(a)|C_2(a) \Rightarrow \Delta, C_1(a), C_2(a)}{\Delta|(C_1 \sqcap C_2)(a) \Rightarrow \Delta, (C_1 \sqcap C_2)(a)} \qquad (S^1_{\sqcap}) \ \frac{\Delta|C_1(a) \Rightarrow \Delta}{\Delta|(C_1 \sqcap C_2)(a) \Rightarrow \Delta}$$

$$(S^2_{\sqcap}) \ \frac{\Delta, C_1(a)|C_2(a) \Rightarrow \Delta, C_1(a)}{\Delta|(C_1 \sqcap C_2)(a) \Rightarrow \Delta}$$

$$(S^{\sqcup}_1) \ \frac{\Delta|C_1(a) \Rightarrow \Delta, C_1(a)}{\Delta|(C_1 \sqcup C_2)(a) \Rightarrow \Delta, (C_1 \sqcup C_2)(a)} \qquad (S_{\sqcup}) \ \frac{\Delta|C_1(a) \Rightarrow \Delta}{\Delta|C_2(a) \Rightarrow \Delta}{\Delta|(C_1 \sqcup C_2)(a) \Rightarrow \Delta}$$

$$(S^{\sqcup}_2) \ \frac{\Delta|C_2(a) \Rightarrow \Delta, C_2(a)}{\Delta|(C_1 \sqcup C_2)(a) \Rightarrow \Delta, (C_1 \sqcup C_2)(a)}$$

$$(S^{\forall}) \ \frac{\Delta|C(d) \Rightarrow \Delta, C(d)}{\Delta|(\forall R.C)(a) \Rightarrow \Delta, (\forall R.C)(a)} \qquad (S_{\forall}) \ \frac{\Delta|C(c) \Rightarrow \Delta}{\Delta|(\forall R.C)(a) \Rightarrow \Delta}$$

$$(S^{\exists}) \ \frac{\Delta|C(c) \Rightarrow \Delta, C(c)}{\Delta|(\exists R.C)(a) \Rightarrow \Delta, (\exists R.C)(a)} \qquad (S_{\exists}) \ \frac{\Delta|C(d) \Rightarrow \Delta}{\Delta|(\exists R.C)(a) \Rightarrow \Delta}$$

where d is a constant and c is a new one, that is, c does not occur in Δ.

Remark. Because Δ can not contradict with $R(a,b)$, whether Δ contradicts with $(\forall R.C)(a)$ depends on both whether Δ has $R(a,d)$ and whether Δ contradicts with $C(d)$; and whether Δ contradicts with $(\exists R.C)(a)$ only depends on whether Δ contradicts with $C(c)$. □

Definition 3.2. $\Delta|C(a) \Rightarrow \Delta, C^i(a)$ *is provable in* $\mathbf{S}^{DL}$, *denoted by* $\vdash_{\mathbf{S}^{DL}}$ $\Delta|C(a) \Rightarrow \Delta, C^i(a)$, *if there is a sequence* $\theta_1, \dots, \theta_m$ *of statements such that*

$$\theta_1 = \Delta|C(a) \Rightarrow \Delta|C_1(a),$$
$$\dots$$
$$\theta_m = \Delta|C_{m-1}(a) \Rightarrow \Delta, C^i(a);$$

and for each $j < m, \Delta|C_j(a) \Rightarrow \Delta|C_{j+1}(a)$ *is an axiom or is deduced from the previous statements by a deduction rule, where* $i \in \{0,1\}, C^1(a) = C(a)$ *and* $C^0(a) = \lambda$, *the empty string.*

Intuitively, we first decompose $C(a)$ into literals according to the structure of $C(a)$, and delete/add literals by rule $(S^{\mathbf{A}}), (S_{\mathbf{A}})/(S^{\neg}), (S_{\neg})$.

Theorem 3.3 *(Soundness theorem). For any consistent theory Δ and statement $C(a)$, if $\vdash_{\mathbf{SDL}} \Delta|C(a) \Rightarrow \Delta, C^i(a)$ then if $i = 0$ then $\Delta \cup \{C(a)\}$ is inconsistent, i.e.,*

$$\vdash_{\mathbf{SDL}} \Delta|C(a) \Rightarrow \Delta \text{ implies } \models_{\mathbf{SDL}} \Delta|C(a) \Rightarrow \Delta;$$

otherwise, $\Delta \cup \{C(a)\}$ is consistent, i.e.,

$$\vdash_{\mathbf{SDL}} \Delta|C(a) \Rightarrow \Delta, C(a) \text{ implies } \models_{\mathbf{SDL}} \Delta|C(a) \Rightarrow \Delta, C(a).$$

Proof. Assume that $\Delta|C(a) \Rightarrow \Delta, C^i(a)$ is provable. We prove the theorem by induction on the structure of C.

Case $C(a) = B(a)$, where $B ::= A|\neg A$. Then $\vdash_{\mathbf{SDL}} \Delta|B(a) \Rightarrow \Delta, B^i(a)$ only if

$$\begin{cases} \Delta \vdash \neg B(a) \text{ if } i = 0 \\ \Delta \nvdash \neg B(a) \text{ if } i = 1; \end{cases}$$

that is,

$$\begin{cases} i = 0 \Rightarrow \text{incon}(\Delta, B(a)), \\ i = 1 \Rightarrow \text{con}(\Delta, B(a)). \end{cases}$$

where $\text{incon}(\Delta, B(a))$ denotes $\Delta \cup \{B(a)\}$ is inconsistent and $\text{con}(\Delta, B(a))$ denotes $\Delta \cup \{B(a)\}$ is consistent.

Case $C(a) = (C_1 \sqcap C_2)(a)$. If $\vdash_{\mathbf{SDL}} \Delta|(C_1 \sqcap C_2)(a) \Rightarrow \Delta, (C_1 \sqcap C_2)(a)$ then

$$\vdash_{\mathbf{SDL}} \Delta|C_1(a) \Rightarrow \Delta, C_1(a);$$
$$\vdash_{\mathbf{SDL}} \Delta, C_1(a)|C_2(a) \Rightarrow \Delta, C_1(a), C_2(a).$$

By the induction assumption, $\Delta \cup \{C_1(a)\}$ and $\Delta \cup \{C_1(a), C_2(a)\}$ are consistent, and so is $\Delta \cup \{(C_1 \sqcap C_2)(a)\}$;

If $\vdash_{\mathbf{SDL}} \Delta|(C_1 \sqcap C_2)(a) \Rightarrow \Delta$ then either

$$\vdash_{\mathbf{SDL}} \Delta|C_1(a) \Rightarrow \Delta$$

or

$$\vdash_{\mathbf{SDL}} \Delta, C_1(a)|C_2(a) \Rightarrow \Delta, C_1(a).$$

By the induction assumption, either $\Delta \cup \{C_1(a)\}$ is inconsistent or $\Delta \cup \{C_1(a), C_2(a)\}$ is inconsistent, which implies $\Delta \cup \{(C_1 \sqcap C_2)(a)\}$ is inconsistent.

Case $C(a) = (C_1 \sqcup C_2)(a)$. If $\vdash_{\mathbf{SDL}} \Delta|(C_1 \sqcup C_2)(a) \Rightarrow \Delta, (C_1 \sqcup C_2)(a)$ then either $\vdash_{\mathbf{SDL}} \Delta|C_1(a) \Rightarrow \Delta, C_1(a)$ or $\vdash_{\mathbf{SDL}} \Delta|C_2(a) \Rightarrow \Delta, C_2(a)$. By the induction assumption, either $\Delta \cup \{C_1(a)\}$ is consistent or $\Delta \cup \{C_2(a)\}$ is consistent, either of which implies $\Delta \cup \{(C_1 \sqcup C_2)(a)\}$ is consistent;

If $\vdash_{\mathbf{SDL}} \Delta|(C_1 \sqcup C_2)(a) \Rightarrow \Delta$ then $\vdash_{\mathbf{SDL}} \Delta|C_1(a) \Rightarrow \Delta$ and $\vdash_{\mathbf{SDL}} \Delta|C_2(a) \Rightarrow \Delta$. By the induction assumption, $\Delta \cup \{C_1(a)\}$ is inconsistent and $\Delta \cup \{C_2(a)\}$ is inconsistent, which imply $\Delta \cup \{(C_1 \sqcup C_2)(a)\}$ is inconsistent.

Case $C(a) = (\forall R.C_1)(a)$. If $\vdash_{\mathbf{SDL}} \Delta|(\forall R.C_1)(a) \Rightarrow \Delta, (\forall R.C_1)(a)$ then $\vdash_{\mathbf{SDL}}$ $\Delta|C_1(c) \Rightarrow \Delta, C_1(c)$, where c is a new constant not occurring in Δ and C_1. By the induction assumption, $\Delta \cup \{C_1(c)\}$ is consistent, and so is $\Delta \cup \{(\forall R.C_1)(a)\}$;

If $\vdash_{\mathbf{SDL}} \Delta|(\forall R.C_1)(a) \Rightarrow \Delta$ then $\vdash_{\mathbf{SDL}} \Delta|C_1(d) \Rightarrow \Delta$. By the induction assumption, $\Delta \cup \{C_1(d)\}$ is inconsistent, and so is $\Delta \cup \{(\forall R.C_1)(a)\}$.

Case $C(a) = (\exists R.C_1)(a)$. If $\vdash_{\mathbf{SDL}} \Delta|(\exists R.C_1)(a) \Rightarrow \Delta, (\exists R.C_1)(a)$ then $\vdash_{\mathbf{SDL}}$ $\Delta|C_1(d) \Rightarrow \Delta, C_1(d)$, where d is a constant. By the induction assumption, $\Delta \cup \{C_1(d)\}$ is consistent, and so $\Delta \cup \{(\exists R.C_1)(a)\}$;

If $\vdash_{\mathbf{SDL}} \Delta|(\exists R.C_1)(a) \Rightarrow \Delta$ then $\vdash_{\mathbf{SDL}} \Delta|C_1(c) \Rightarrow \Delta$, where c is a new constant not occurring in Δ and C_1. By the induction assumption, $\Delta \cup \{C_1(c)\}$ is inconsistent, and so is $\Delta \cup \{(\forall R.C_1)(a)\}$. □

The last theorem is soundness theorem for R-calculus $\mathbf{S}^{\mathrm{DL}}$, and completeness theorem is the following

Theorem 3.4 *(Completeness theorem). For any consistent theory Δ and statement $C(a)$, if $\Delta \cup \{C(a)\}$ is consistent then $\Delta|C(a) \Rightarrow \Delta, C(a)$ is provable in $\mathbf{S}^{\mathrm{DL}}$, i.e.,*

$$\models_{\mathbf{SDL}} \Delta|C(a) \Rightarrow \Delta, C(a) \text{ implies } \vdash_{\mathbf{SDL}} \Delta|C(a) \Rightarrow \Delta, C(a);$$

and if $\Delta \cup \{C(a)\}$ is inconsistent then $\Delta|C(a) \Rightarrow \Delta$ is provable in $\mathbf{S}^{\mathrm{DL}}$, i.e.,

$$\models_{\mathbf{SDL}} \Delta|C(a) \Rightarrow \Delta \text{ implies } \vdash_{\mathbf{SDL}} \Delta|C(a) \Rightarrow \Delta.$$

Proof. We prove the theorem by induction on the structure of C.

Case $C(a) = B(a)$, where $B ::= A|\neg A$. If $\Delta \cup \{B(a)\}$ is consistent then $\Delta \not\vdash \neg B(a)$, by $(S^{\mathbf{A}})$ and $(S^{\neg})$,

$$\vdash_{\mathbf{SDL}} \Delta|B(a) \Rightarrow \Delta, B(a);$$

and if $\Delta \cup \{B(a)\}$ is inconsistent then $\Delta \vdash \neg B(a)$, by $(S_{\mathbf{A}})$ and $(S_{\neg})$,

$$\vdash_{\mathbf{SDL}} \Delta|B(a) \Rightarrow \Delta.$$

Case $C(a) = (C_1 \sqcap C_2)(a)$. If $\Delta \cup \{(C_1 \sqcap C_2)(a)\}$ is consistent then $\Delta \cup \{C_1(a)\}$ and $\Delta \cup \{C_1(a), C_2(a)\}$ are consistent, and by induction assumption,

$$\vdash_{\mathbf{SDL}} \Delta|C_1(a) \Rightarrow \Delta, C_1(a)$$
$$\vdash_{\mathbf{SDL}} \Delta, C_1(a)|C_2(a) \Rightarrow \Delta, C_1(a), C_2(a).$$

By $(S^{\sqcap})$, we have that $\vdash_{\mathbf{SDL}} \Delta|(C_1 \sqcap C_2)(a) \Rightarrow \Delta, (C_1 \sqcap C_2)(a)$;

If $\Delta \cup \{(C_1 \sqcap C_2)(a)\}$ is inconsistent then either $\Delta \cup \{C_1(a)\}$ is inconsistent or $\Delta \cup \{C_1(a)\} \cup \{C_2(a)\}$ is inconsistent. By the induction assumption, either $\vdash_{\mathbf{SDL}} \Delta|C_1(a) \Rightarrow \Delta$, or $\vdash_{\mathbf{SDL}} \Delta, C_1(a)|C_2(a) \Rightarrow \Delta, C_1(a)$. By $(S_{\sqcap})$, we have that $\vdash_{\mathbf{SDL}} \Delta|(C_1 \sqcap C_2)(a) \Rightarrow \Delta$.

Case $C(a) = (C_1 \sqcup C_2)(a)$. If $\Delta \cup \{(C_1 \sqcup C_2)(a)\}$ is consistent then either $\Delta \cup \{C_1(a)\}$ or $\Delta \cup \{C_2(a)\}$ are consistent, and by induction assumption, either

$$\vdash_{\mathbf{SDL}} \Delta|C_1(a) \Rightarrow \Delta, C_1(a),$$

or

$$\vdash_{\textbf{SDL}} \Delta|C_2(a) \Rightarrow \Delta, C_2(a).$$

By $(S_1^{\sqcap})$ and $(S_2^{\sqcap})$, we have that $\vdash_{\textbf{SDL}} \Delta|(C_1 \sqcup C_2)(a) \Rightarrow \Delta, (C_1 \sqcup C_2)(a)$;

If $\Delta \cup \{(C_1 \sqcup C_2)(a)\}$ is inconsistent then $\Delta \cup \{C_1(a)\}$ and $\Delta \cup \{C_2(a)\}$ are inconsistent. By the induction assumption,

$$\vdash_{\textbf{SDL}} \Delta|C_1(a) \Rightarrow \Delta,$$
$$\vdash_{\textbf{SDL}} \Delta|C_2(a) \Rightarrow \Delta.$$

By $(S_{\sqcup})$, we have that $\vdash_{\textbf{SDL}} \Delta|(C_1 \sqcup C_2)(a) \Rightarrow \Delta$.

Case $C(a) = (\forall R.C_1)(a)$. If $\Delta \cup \{(\forall R.C_1)(a)\}$ is consistent then for any c, $\Delta \cup \{C_1(c)\}$ is consistent, and by induction assumption,

$$\vdash_{\textbf{SDL}} \Delta|C_1(c) \Rightarrow \Delta, C_1(c).$$

By $(S^{\forall})$, we have that $\vdash_{\textbf{SDL}} \Delta|(\forall R.C_1)(a) \Rightarrow \Delta, (\forall R.C_1)(a)$;

If $\Delta \cup \{(\forall R.C_1)(a)\}$ is inconsistent then there is a constant d such that $\Delta \cup \{C_1(d)\}$ is inconsistent. By the induction assumption, $\vdash_{\textbf{SDL}} \Delta|C_1(d) \Rightarrow \Delta$. By $(S_{\forall})$, we have that $\vdash_{\textbf{SDL}} \Delta|(\forall R.C_1)(a) \Rightarrow \Delta$.

Case $C(a) = (\exists R.C_1)(a)$. If $\Delta \cup \{(\exists R.C_1)(a)\}$ is consistent then there is a constant d such that $\Delta \cup \{C_1(d)\}$ is consistent, and by induction assumption,

$$\vdash_{\textbf{SDL}} \Delta|C_1(d) \Rightarrow \Delta, C_1(d).$$

By $(S^{\exists})$, we have that $\vdash_{\textbf{SDL}} \Delta|(\exists R.C_1)(a) \Rightarrow \Delta, (\exists R.C_1)(a)$;

If $\Delta \cup \{(\exists R.C_1)(a)\}$ is inconsistent then for any constant $c, \Delta \cup \{C_1(c)\}$ is inconsistent. By the induction assumption, $\vdash_{\textbf{SDL}} \Delta|C_1(c) \Rightarrow \Delta$. By $(S_{\exists})$, we have that $\vdash_{\textbf{SDL}} \Delta|(\exists R.C_1)(c) \Rightarrow \Delta$.   □

3.2 $\textbf{S}^{\text{DL}}$: R-Calculus for a Set of Statements

Let Γ be a finite consistent set of statements such that $\Gamma = \{C_1(a), \ldots, C_n(a)\}$. Define

$$\Delta|\Gamma = (\cdots ((\Delta|C_1(a))|C_2(a))|\cdots)|C_n(a).$$

Correspondingly, we have the following

Theorem 3.5. *For any consistent theories* Δ, Γ *and* Θ, *if* $\Delta|\Gamma \Rightarrow \Delta, \Theta$ *is provable in* $\textbf{S}^{\text{DL}}$ *then* Θ *is a* $\subseteq$-*minimal change of* Γ *by* Δ. *That is,*

$$\vdash_{\textbf{SDL}} \Delta|\Gamma \Rightarrow \Delta, \Theta \text{ implies } \models_{\textbf{SDL}} \Delta|\Gamma \Rightarrow \Delta, \Theta.$$

Proof. We prove the theorem by induction on n.

Assume that $\Delta|\Gamma \Rightarrow \Delta, \Theta$ is provable in $\textbf{S}^{\text{DL}}$.

Let $n = 1$. Then, either $\Theta = C_1(a_1)$ or $\Theta = \lambda$. If $\Theta = C_1(a_1)$ then $\Delta \cup \{C_1(a_1)\}$ is consistent, and Θ is a $\subseteq$-minimal change of $C_1(a_1)$ by Δ; otherwise, $\Delta \cup \{C_1(a_1)\}$ is inconsistent, and $\Theta = \lambda$ is a $\subseteq$-minimal change of $C_1(a_1)$ by Δ.

Assume that the theorem holds for n, that is, if $\Delta|\Gamma \Rightarrow \Delta, \Theta$ then Θ is a $\subseteq$-minimal change of Γ by Δ, where $\Gamma = (C_1(a_1), \ldots, C_n(a_n))$.

Let $\Gamma' = (\Gamma, C_{n+1}(a_{n+1})) = (C_1(a_1), \ldots, C_{n+1}(a_{n+1}))$. Then, if $\Delta|\Gamma' \Rightarrow \Delta, \Theta'$ is provable then $\Delta|\Gamma \Rightarrow \Delta, \Theta$ and $\Delta, \Theta|C_{n+1}(a_{n+1}) \Rightarrow \Delta, \Theta'$ are provable. By the case $n = 1$ and the induction assumption, Θ' is a $\subseteq$-minimal change of $C_{n+1}(a_{n+1})$ by $\Delta \cup \Theta$, and Θ is a $\subseteq$-minimal change of Γ by Δ, therefore, Θ' is a $\subseteq$-minimal change of Γ' by Δ.

$\square$

Theorem 3.6. *For any consistent theories Δ and Γ and any $\subseteq$-minimal change Θ of Γ by Δ, $\Delta|\Gamma \Rightarrow \Delta, \Theta$ is provable in $\mathbf{S}^{\mathrm{DL}}$. That is,*

$$\models_{\mathbf{S}_{\mathrm{DL}}} \Delta|\Gamma \Rightarrow \Delta, \Theta \text{ implies } \vdash_{\mathbf{S}_{\mathrm{DL}}} \Delta|\Gamma \Rightarrow \Delta, \Theta.$$

Proof. Assume that Θ is a $\subseteq$-minimal change of Γ by Δ. Then, there is an ordering $<$ of Γ such that $\Gamma = (C_1(a_1), C_2(a_2), \ldots, C_n(a_n))$, where $C_1(a_1) < C_2(a_2) < \cdots < C_n(a_n)$, and Θ is a maximal subset of Γ such that $\Delta \cup \Theta$ is consistent.

We prove the theorem by induction on n.

Let $n = 1$. By the last theorem, if $\Theta = \{C_1(a_1)\}$ then $\Delta|C_1(a_1) \Rightarrow \Delta, C_1(a_1)$ is provable; and if $\Theta = \emptyset$ then $\Delta|C_1(a_1) \Rightarrow \Delta$ is provable.

Assume that the theorem holds for n, that is, if Θ is a $\subseteq$-minimal change of Γ by Δ then $\Delta|\Gamma \Rightarrow \Delta, \Theta$ is provable.

Let $\Gamma' = (\Gamma, C_{n+1}(a_{n+1})) = (C_1(a_1), \ldots, C_{n+1}(a_{n+1}))$ and Θ' is a $\subseteq$-minimal change of Γ' by Δ. Then, Θ' is a $\subseteq$-minimal change of $C_{n+1}(a_{n+1})$ by $\Delta \cup \Theta$, and $\Delta, \Theta|C_{n+1}(a_{n+1}) \Rightarrow \Delta, \Theta'$ is provable. By the induction assumption, $\Delta|\Gamma \Rightarrow \Delta, \Theta$ is provable and so is $\Delta|\Gamma' \Rightarrow \Delta, \Theta|C_{n+1}(a_{n+1})$, and hence, $\Delta|\Gamma' \Rightarrow \Delta, \Theta'$ is provable in $\mathbf{S}^{\mathrm{DL}}$.

$\square$

4   Conclusions and Further Works

In this paper we gave an **R**-calculus $\mathbf{S}^{\mathrm{DL}}$ that is sound and complete with respect to the $\subseteq$-minimal change. However, for set-inclusion minimal change, if $A_1(a)$ is inconsistent with $\Delta \cup \Gamma$ or $A_2(a)$ is inconsistent with $\Delta \cup \{A_1(a)\} \cup \Gamma$ then $(A_1 \sqcap A_2)(a)$ is eliminated from $\{(A_1 \sqcap A_2)(a)\} \cup \Gamma$ revised by Δ, even though it may be the case that $\Delta \cup \{A_1(a)\} \cup \Gamma$ is consistent.

Specifically, the rule $(S_\sqcap)$ in $\mathbf{S}^{\mathrm{DL}}$ may eliminate too much information in Γ. For example,

$$\vdash_{\mathbf{S}_{\mathrm{DL}}} \neg havingarms(a)|(havingarms \sqcap havinglegs)(a) \Rightarrow \neg havingarms(a).$$

Intuitively, we should have

$$\neg havingarms(a)|(havingarms \sqcap havinglegs)(a) \Rightarrow \neg havingarms(a), havinglegs(a).$$

Therefore, a further work is to give a new **R**-calculus to preserve as much as possible information of statements to be revised such that we will have the above deduction in the new **R**-calculus $\mathbf{R}_*^{\mathrm{DL}}$. Formally, we have

$$\vdash_{\mathbf{S}^{\mathrm{DL}}} \neg C(a)|(C \sqcap D)(a) \Rightarrow \neg C(a);$$
$$\vdash_{\mathbf{R}_*^{\mathrm{DL}}} \neg C(a)|(C \sqcap D)(a) \Rightarrow \neg C(a), D(a).$$

Acknowledgments. This work was supported by the Open Fund of the State Key Laboratory of Software Development Environment under Grant No. SKLSDE-2010KF-06, Beijing University of Aeronautics and Astronautics, by the National Natural Science Foundation of China (Grant No. 91224006, 61173063, 61035004, 61203284), by the National Basic Research Program of China (973 Program) under Grant No. 2005CB321901, and by the CNSSF 10AYY003.

References

1. Alchourrón, C.E., Gärdenfors, P., Makinson, D.: On the logic of theory change: partial meet contraction and revision functions. J. Symb. Logic **50**, 510–530 (1985)
2. Hansson, S.O.: A Textbook of Belief Dynamics, Theory Change and Database Updating. Kluwer, Dordrecht (1999)
3. Hansson, S.O.: Ten philosophical problems in belief revision. J. Logic Comput. **13**, 37–49 (2003)
4. Li, W.: R-calculus: an inference system for belief revision. Comput. J. **50**, 378–390 (2007)
5. Katsuno, H., Mendelzon, A.O.: Propositional knowledge base revision and minimal change. Artif. Intell. **52**, 263–294 (1991)
6. Li, W., Sui, Y.: The R-calculus based-on addition instead of cancelation. In: Frontier in Computer Science (to appear)
7. Li, W., Sui, Y.: The sound and complete R-calculi with respect to pseudo-revision and pre-revision. I. J. Intell. Sci. **3**, 110–117 (2013)
8. Li, W., Sui, Y.: The set-based and inference-based R-calculus
9. Li, W., Sui, Y.: The sound and complete R-calculi with pseudo-subtheory minimal change property. Int. J. Softw. Inform. (to appear)
10. Li, W., Sui, Y., Sun, M.: The sound and complete R-calculus for revising propositional theories. Sci. China: Inf. Sci. **58**, 092101:1–092101:12 (2015)
11. Li, W., Sui, Y.: The set-based and inference-based r-calculus the sound and complete r-calculi with respect to set-inclusion and pseudo-subformula minimal change
12. Rott, H., Williams, M.-A. (eds.): Frontiers of Belief Revision. Kluwer, Dordrecht (2001)
13. Satoh, K.: Nonmonotonic reasoning by minimal belief revision. In: Proceedings of the International Conference on Fifth Generation Computer Systems, Tokyo, pp. 455–462 (1988)
14. Darwiche, A., Pearl, J.: On the logic of iterated belief revision. Artif. Intell. **89**, 1–29 (1997)
15. Berners-Lee, T., Hendler, J., Lassila, O.: The semantic web. Sci. Am. **284**(5), 3443 (2001)
16. Haase, P., Stojanovic, L.: Consistent evolution of OWL ontologies. In: Gómez-Pérez, A., Euzenat, J. (eds.) ESWC 2005. LNCS, vol. 3532, pp. 182–197. Springer, Heidelberg (2005). doi:10.1007/11431053_13

17. Stojanovic, L.: Methods and tools for ontology evolution. Ph.D. thesis, University of Karlsruhe (2004)
18. Flouris, G., Plexousakis, D., Antoniou, G.: On applying the AGM theory to DLs and OWL. In: Gil, Y., Motta, E., Benjamins, V.R., Musen, M.A. (eds.) ISWC 2005. LNCS, vol. 3729, pp. 216–231. Springer, Heidelberg (2005). doi:10.1007/11574620_18
19. Haase, P., van Harmelen, F., Huang, Z., Stuckenschmidt, H., Sure, Y.: A framework for handling inconsistency in changing ontologies. In: Gil, Y., Motta, E., Benjamins, V.R., Musen, M.A. (eds.) ISWC 2005. LNCS, vol. 3729, pp. 353–367. Springer, Heidelberg (2005). doi:10.1007/11574620_27

A Multi-objective Attribute Reduction Method in Decision-Theoretic Rough Set Model

Lu Wang[1], Weiwei Li[2], Xiuyi Jia[1,3]([✉]), and Bing Zhou[4]

[1] School of Computer Science and Engineering,
Nanjing University of Science and Technology, Nanjing 210094, China
njwxl28@hotmail.com
[2] College of Astronautics,
Nanjing University of Aeronautics and Astronautics, Nanjing 210016, China
liweiwei@nuaa.edu.cn
[3] Key Laboratory of Oceanographic Big Data Mining and Application
of Zhejiang Province, Zhoushan 316022, China
jiaxy@njust.edu.cn
[4] Department of Computer Science, Sam Houston State University,
Huntsville 77341, USA
zhou@shsu.edu

Abstract. Many attribute reduction methods have been proposed for decision-theoretic rough set model based on different definitions of attribute reduct, while an attribute reduct can be seen as an attribute subset that satisfies specific criteria. Most reducts are defined on the basis of a single criterion, which may result in the difficulty for users to choose appropriate reduct to design related reduction algorithm. To address this problem, we propose a multi-objective attribute reduction method based on NSGA-II for decision-theoretic rough set model. Three different definitions of attribute reduct based on positive region, decision cost and mutual information are considered and transferred to a multi-objective optimization problem. Experimental results show that the multi-objective reduction method can obtain a robust and better classification performance.

1 Introduction

Rough set theory is an important mathematical tool to handle imprecision, vagueness and uncertainty in data analysis [1]. The classical Pawlak rough set model is not suitable for noisy data as the set inclusion in the model is required to be fully correct or certain [2]. To handle this problem, Yao [3] proposed a decision-theoretic rough set model (DTRS) with a tolerance of errors by introducing Bayesian decision principle into rough set theory. As one of the most important concepts in DTRS or other rough set models, attribute reduction [3,4] plays a key role in many areas including machine learning and data mining [5,6]. By using attribute reduction, we can delete redundant attributes and induce a more simplified knowledge representation result.

© Springer International Publishing AG 2017
G. Li et al. (Eds.): KSEM 2017, LNAI 10412, pp. 117–128, 2017.
DOI: 10.1007/978-3-319-63558-3_10

Generally speaking, attribute reduction can be seen as a process of finding a minimum attribute subset that satisfies some specific criteria [7]. The minimum set of attributes is usually called an attribute reduct. In recent years, many different definitions of attribute reduct in DTRS have been proposed based on different criteria, such as region based attribute reducts [8–10], decision cost based attribute reducts [11–13] and entropy based attribute reducts [14–16]. Based on these different kinds of attribute reducts, the indiscernibility matrix algorithms [17], heuristic algorithms [16] and genetic algorithms [18] can be applied to design related attribute reduction methods. However, most attribute reduction methods in DTRS are single objective reduction, that is to say, only one kind of attribute reduct definition based on one specific criterion is considered in the process of attribute reduction. In this case, the obtained reduct may perform perfectly on the specific criterion but poorly on other criteria. For example, a positive region based attribute reduct could keep the same positive region, but it may generate a large decision cost. In the real word, it is not easy to clearly figure out the characteristics of the data sets, therefore, it will make a difficulty for users choosing the most appropriate attribute reduct to design attribute reduction method. The single objective attribute reduction methods prefer a certain criterion and may obtain a biased result.

To address the above problems, we propose a multi-objective attribute reduction method based on NSGA-II (Non-Dominated Sorting in Genetic Algorithm) [19] in DTRS. Firstly, we represent the attribute reduction as a multi-objective optimization problem by considering several different criteria, including positive region, decision cost, and the mutual information. The three kinds of criteria characterize the utility of an attribute reduct from different views. As rough set theory usually measures the classification ability by computing the size of positive region, the positive region criterion ensures classification ability unchanged based the obtained reduct, while the decision cost reflects the additional constraints on the problem, and the mutual information shows the characteristic of data itself (here is the uncertainty). Secondly, we apply NSGA-II on the multi-objective problem to obtain a set of candidate solutions, and adopt a wrapper method to output a final result as the attribute reduct. By incorporating the crowding distance and the elitist mechanism of NSGA-II, we can find a set of solutions with optimal performance on above three criteria. After obtaining a set of solutions, we use several classifiers to evaluate corresponding accuracy based on each candidate solution and output the best one as the final attribute reduct.

The rest of this paper is organized as follows. Section 2 summarizes some concepts about our work. Section 3 introduces the attribute reduct based on a multi-objective optimization problem, and gives the corresponding attribute reduction approach. Section 4 shows the experimental results. Section 5 concludes the paper.

2   Preliminaries

In this section, we will summarize some basic concepts about decision-theoretic rough set model.

2.1   Decision-Theoretic Rough Set Model

In DTRS, given a decision system $DS = (U, At = C \cup D, \{V_a | a \in At\}, \{I_a | a \in At\})$, where U is a finite nonempty set of objects, At is a finite nonempty set of attributes, C is a set of condition attributes describing the objects, and D is a set of decision attributes that indicates the classes of objects. V_a is a nonempty set of values of $a \in At$, and $I_a : U \to V_a$ is an information function that maps an object in U to exactly one value in V_a. In a decision table, an object x is described by its equivalence class under a set of attributes $A \subseteq At$: $[x]_A = \{y \in U | \forall a \in A(I_a(x) = I_a(y))\}$.

The set of states is denoted by $\Omega = \{X, X^C\}$, which represent an object is in a state X or not in X. The probabilities for these two complement states can be denoted as $p(X|[x]) = \frac{|X \cap [x]|}{|[x]|}$ and $p(X^C|[x]) = 1 - p(X|[x])$. The set of actions is denoted by $A = \{a_P, a_B, a_N\}$, where a_P, a_B and a_N represent the three actions in classifying an object x into the sets the positive region $POS(X)$, the boundary region $BND(X)$ and the negative region $NEG(X)$, respectively. When an object actually belongs to X, let $\lambda_{PP}, \lambda_{BP}, \lambda_{NP}$ denote the costs of taking actions of a_P, a_B, a_N, respectively. Similarly, let $\lambda_{PN}, \lambda_{BN}, \lambda_{NN}$ denote the costs of taking the same three actions, respectively, when an object actually belongs to X^C. $R(a_P|[x]), R(a_B|[x]), R(a_N|[x])$ represent the expected cost associated with each individual action:

$$R_P = R(a_P|[x]) = \lambda_{PP} \cdot p(X|[x]) + \lambda_{PN} \cdot p(X^C|[x]),$$
$$R_B = R(a_B|[x]) = \lambda_{BP} \cdot p(X|[x]) + \lambda_{BN} \cdot p(X^C|[x]), \tag{1}$$
$$R_N = R(a_N|[x]) = \lambda_{NP} \cdot p(X|[x]) + \lambda_{NN} \cdot p(X^C|[x]).$$

According to the Bayesian decision procedure, the minimum cost decision rules are suggested as follows:

(P) If $R_P \leq R_B$ and $R_P \leq R_N$, decide $x \in POS(X)$,

(B) If $R_B \leq R_P$ and $R_B \leq R_N$, decide $x \in BND(X)$,

(N) If $R_N \leq R_P$ and $R_N \leq R_B$, decide $x \in NEG(X)$.

Consider the reasonable loss condition $\lambda_{PP} \leq \lambda_{BP} < \lambda_{NP}, \lambda_{NN} \leq \lambda_{BN} < \lambda_{PN}$, that is, the cost of classifying an object x belonging to X into the positive region $POS(X)$ is less than or equal to the cost of classifying x into the boundary region $BND(X)$, and the both of these costs are strictly less than the cost of classifying x into the negative region $NEG(X)$. The reverse order of cost is used for classifying an object not in X. Then, we derive the following condition on cost functions [20]:

$$\frac{\lambda_{NP} - \lambda_{BP}}{\lambda_{BN} - \lambda_{NN}} > \frac{\lambda_{BP} - \lambda_{PP}}{\lambda_{PN} - \lambda_{BN}}. \tag{2}$$

Finally, the final minimum-risk decision rules (P)-(B) can be written as

(P1) If $p(X|[x]]) \geq \alpha$, decide $x \in POS(X)$,

(B1) If $\beta < p(X|[x]]) < \alpha$, decide $x \in BND(X)$,

(N1) If $p(X|[x]]) \leq \beta$, decide $x \in NEG(X)$.

Where the parameters α, β are defined as:

$$\alpha = \frac{\lambda_{PN} - \lambda_{BN}}{(\lambda_{PN} - \lambda_{BN}) + (\lambda_{BP} - \lambda_{PP})},$$

$$\beta = \frac{\lambda_{BN} - \lambda_{NN}}{(\lambda_{BN} - \lambda_{NN}) + (\lambda_{NP} - \lambda_{BP})}. \tag{3}$$

2.2 Three Kinds of Criteria in Decision-Theoretic Rough Set Model

In this paper, we consider three kinds of criteria characterizing the utility of an attribute reduct from different views, including positive region, decision cost and mutual information.

Positive Region. In rough set theory, a core concept is the positive region, which contains the objects that can be *certainly* classified under the current feature space. In general, the positive region can be applied to measure the classification ability, therefore, the positive region is the most commonly used criterion for defining an attribute reduct.

In DTRS, let $\pi_D = \{D_1, D_2, \ldots, D_m\}$ denote the partition of the universe U induced by D, and π_A denote the partition induced by the set of attributes $A \subseteq At$. Based on the threshold (α, β), one can divide the universe U into three regions of the decision partition π_D:

$$POS_{(\alpha,\beta)}(\pi_D|\pi_A) = \{x \in U | p(D_{max}([x]_A)|[x]_A) \geq \alpha\}$$

$$BND_{(\alpha,\beta)}(\pi_D|\pi_A) = \{x \in U | \beta < p(D_{max}([x]_A)|[x]_A) < \alpha\} \tag{4}$$

$$NEG_{(\alpha,\beta)}(\pi_D|\pi_A) = \{x \in U | p(D_{max}([x]_A)|[x]_A) \leq \beta\}$$

where $D_{max}([x]_A) = \arg\max_{D_i \in \pi_D}\{|[x]_A \cap D_i|/|[x]_A|\}$. $POS_{(\alpha,\beta)}(\pi_D|\pi_A)$ is the positive region by considering attribute set A. The greater the discriminative ability of attribute set A is, the larger the size of the corresponding positive region is. By using the positive region criterion, we can define an attribute reduct which keeps the positive region unchanged or expanded.

Decision Cost. DTRS is based on Bayesian decision procedure and the principle of making decisions is minimizing the decision cost. In DTRS, decision cost is a very important notion and it can be intuitively considered as the criterion for defining an attribute reduct.

For a given decision system $DS = (U, At = C \cup D, \{V_a | a \in At\}, \{I_a | a \in At\})$. The decision cost can be described as [10]:

$$dc = \sum_{p_i \geq \alpha} (p_i \cdot \lambda_{PP} + (1 - p_i) \cdot \lambda_{PN}) + \sum_{\beta < p_j < \alpha} (p_j \cdot \lambda_{BN} + (1 - p_j) \cdot \lambda_{BN})$$

$$+ \sum_{p_k \leq \beta} (p_k \cdot \lambda_{NP} + (1 - p_k \cdot \lambda_{NN})), \tag{5}$$

where $p_i = p(D_{max}([x_i]_A)|[x_i]_A)$ defines the maximum probability of an object belonging to the different decision class based on a set of attributes $A \subseteq At$.

Consider the real world where we assume zero cost of a correct classification, namely $\lambda_{PP} = \lambda_{NN} = 0$, then the decision cost can be rewritten as:

$$dc = \sum_{p_i \geq \alpha} ((1 - p_i) \cdot \lambda_{PN}) + \sum_{\beta < p_j < \alpha} (p_j \cdot \lambda_{BN} + (1 - p_j) \cdot \lambda_{BN})$$
$$+ \sum_{p_k \leq \beta} (p_k \cdot \lambda_{NP})). \tag{6}$$

Based on the decision cost criterion, a minimum cost attribute reduct in DTRS has been defined [11]. This kind of reduct definition ensures that its induced decision cost is minimum.

Mutual Information. Since mutual information in Shannon's information theory can be used to evaluate the relevance between attributes and class labels, this criterion is usually applied in feature selection or attribute reduction. Several definitions about mutual information present as follows [21]:

Definition 1. *Given a decision system $DS = (U, At, \{V_a | a \in At\}, \{I_a | a \in At\})$, if $A \subseteq At$, the entropy of A is*

$$H(A) = -\sum_{i=1}^{n} p(A_i) \cdot \log(p(A_i)), \tag{7}$$

where $p(A_i) = \frac{|A_i|}{|U|}, i = 1, 2, \cdots, n$, $U/IND(A) = \{A_1, A_2, \cdots, A_n\}$, and $IND(A)$ is the equivalence class under a set of attribute A.

Definition 2. *Given a decision system $DS = (U, At, \{V_a | a \in At\}, \{I_a | a \in At\})$, if $A \subseteq At$, $B \subseteq At$, $U/IND(A) = \{A_1, A_2, \cdots, A_m\}$ and $U/IND(B) = \{B_1, B_2, \cdots, B_n\}$, the conditional entropy of A with reference to B is*

$$H(B|A) = -\sum_{A_i \in U/IND(A)} p(A_i) \sum_{B_j \in U/IND(B)} p(B_j | Ai) \cdot \log(p(B_j | A_i)), \tag{8}$$

where $p(B_j | A_i) = \frac{|B_j \cap A_i|}{|A_i|}, i = 1, 2, \cdots, m, j = 1, 2, \cdots, n$.

Definition 3. *Given a decision system $DS = (U, At = C \cup D, \{V_a | a \in At\}, \{I_a | a \in At\})$, $R \subseteq C$, the mutual information between conditional attribute set R and the decision class D can be defined as*

$$I(R; D) = H(D) - H(D|R). \tag{9}$$

Usually, the mutual information between C and D has the maximum value, then an attribute reduct can be defined to find a minimal attribute set with the maximum mutual information value.

3   Multi-objective Attribute Reduction in Decision-Theoretic Rough Set Model

In this section, we will define a multi-objective attribute reduct in DTRS and propose a reduction method based on NSGA-II.

3.1   Multi-objective Attribute Reduct

Since most existing attribute reducts are defined based on single objective, it brings the difficulty of choosing which kind of attribute reduct for different applications. Therefore, to address this problem, we propose a multi-objective attribute reduct in this paper.

A classical multi-objective optimization problem consists of a set of objective functions with some related constraints, and it can be described as follows [22]:

$$\max_{Y \in \Omega} F(Y) = (f_1(Y), f_2(Y), \cdots, f_m(Y)),$$
$$s.t. \quad g_i(Y) \le 0, (i = 1, 2, \cdots, p),$$
$$h_j(Y) = 0, (j = 1, 2, \cdots, q),$$

$$(10)$$

where Ω is the decision (variable) space, $Y = (y_1, y_2, \cdots, y_n)$ is a decision vector and $F(Y)$ is the set of multiple objective functions. $g_i(Y) \le 0$ define inequality constraints and $h_j(Y) = 0$ define equality constraints.

In this paper, we consider three different sub objective functions in our definition of multi-objective attribute reduct, including maximizing the positive region size, minimizing decision cost and maximizing mutual information, which characterize the utility of an attribute reduct from different views. The definition of multi-objective attribute reduct is presented as follows.

Definition 4. *In a decision system, $DS = (U, At = C \cup D, \{V_a | a \in At\}, \{I_a | a \in At\})$; $R \subseteq C$ is a multi-objective attribute reduct if only if*

(1) $R = \arg\max_{R \subseteq C} F(R)$

(2) $\forall R' \subset R, \forall f_i(R') \le f_i(R)$, and $\exists f_i(R') < f_i(R), f_i(R) \in F(R)$

$F(R) = (f_1(R), f_2(R), f_3(R))$, is a multi-objective optimization function with three sub objective functions. $f_1(R)$, $f_2(R)$ and $f_3(R)$ are defined as follows.

$$f_1(R) = \frac{card(POS_R(U))}{card(POS_C(U))},$$
$$f_2(R) = -dc(U, R),$$
$$f_3(R) = H(D) - H(D|R),$$

$f_1(R)$, $f_2(R)$ and $f_3(R)$ expect that the attribute reduct can get a large positive region, a small decision cost and a large mutual information, respectively.

Generally speaking, sub objective functions in a multi-objective optimization problem may be conflicting with each other. That is to say, it is impossible to achieve the optimal performance on all sub objective functions at the same time. Instead, many multi-objective optimization problems output a set of Pareto optimal solutions for users. Therefore, we usually have more than one attribute reduct satisfying the definition of multi-objective attribute reduct.

Algorithm 1. Multi-objective attribute reduction algorithm

Input: A decision system DS with corresponding loss functions; The generations G and the population size N.

Output: The reduct R;

1: $R \leftarrow \emptyset$
2: Generate P_0 randomly;   //the initial parent population
3: $F = $ Non-dominated-sort(P_0) ;   // $F = (F_1, F_2, \ldots)$, all nondominated fronts of P_0
4: **for** $t = 0$ to G **do**
5:    $S_t = \emptyset, i = 1$;
6:    $Q_t = $ Genetic-operators(P_t);
7:    $R_t \leftarrow P_t \cup Q_t$;
8:    $F = $ Non-dominated-sort(R_t);   //$F = (F_1, F_2, \ldots)$, all nondominated fronts of R_t
9:    **repeat**
10:      $S_t = S_t \cup F_i$ and $i = i + 1$;
11:    **until** $|S_t| \geq N$
12:    Last front to be included: $F_l = F_i$;
13:    **if** $|S_t| = N$ **then**
14:      $P_{t+1} = S_t$;
15:    **else**
16:      $P_{t+1} = \bigcup_{j=1}^{l-1} F_i$;
17:      $K = N - |P_{t+1}|$;
18:      Crowding-distance-assignment(F_i);
19:      Sort(F_i);   // sort solutions in F_i with crowding distance in descending order
20:      SortPart(F_i);   //sort solutions with the length of reduct when crowing distance of solutions are same
21:      $P_{t+1} = P_{t+1} \cup F_i[1 : K]$;   //choose the first K elements of F_i
22:    **end if**
23: **end for**
24: Compute the accuracy of solutions in P_t based on several classifiers;
25: Select the chromosome with best performance from P_t to be R;
26: **return** R;

3.2   Multi-objective Attribute Reduction Algorithm

From the definition of multi-objective attribute reduct, we can find that the attribute reduction can be described as a process of solving the corresponding multi-objective optimization problem. In this regard, we introduce a multi-objective attribute reduction method based on NSGA-II algorithm, which is a kind of efficient genetic algorithm for multi-objective optimization problem.

Since NSGA-II usually gives a set of Pareto optimal solutions, in contrast, we design a wrapper method to output a solution with the highest classification accuracy as the final attribute reduct, while the accuracy is evaluated by several different classifiers ensemble.

In this algorithm, the basic framework of the proposed method is similar to the classical NSGA-II algorithm. Differing from the general genetic algorithm, NSGA-II proposed the crowding distance and the elitist mechanism. Let us consider tth generation of NSGA-II algorithm. Suppose the parent population at this generation is P_t and its size is N, while the offspring population created from P_t is Q_t having N members. After obtaining the offspring population, to preserve elite members of the combined parent and offspring population $R_t = P_t \cup Q_t$ (of size $2N$), R_t will be sorted according to different nondomination levels (F_1, F_2 and so on), and the solutions with less levels will be chosen. If the solutions have same levels, the solutions with less crowding distance will be chosen.

It is worth noting that, while generating the offspring population, differing from the classical NSGA-II, we check the offspring individuals whether repeated, and if they are repeated, then generate a new one. Meanwhile, to make the method more suitable for obtaining an attribute reduct, when the crowing distance of solutions are also same, we compare the length of attribute reduct and choose the solution with the shortest length. Finally, after obtaining a set of candidate solutions, we apply several classifiers to evaluate corresponding accuracy based on each candidate solution and output the best one as the final attribute reduct. More details can be found in Algorithm 1.

4   Experiments

4.1   Dataset

There are 8 UCI datasets [23] applied in our experiments. The basic information of all datasets are listed in Table 1, where $|C|$ is the number of conditional attributes, $|U|$ is the number of objects, and $|D|$ is the number of classes. We also replace the missing values and discretize all continuous attributes by using WEKA filters [24].

4.2   Experimental Setting

In our experiments, for each dataset, 10 times 10-fold cross-validation is applied and the average results are recorded. The population size N is set to 20, and the probabilities of crossing and mutation are 0.9 and 0.1, respectively. Four classifiers in WEKA [24] are considered to evaluate the classification accuracy, including NaiveBayes, J48(C4.5), RandomForest and SMO(SVM). For each data set, different loss functions are generated randomly in the interval of $(0,1)$, which satisfies the following constraint conditions: $\lambda_{BP} < \lambda_{NP}$, $\lambda_{BN} < \lambda_{PN}$, $\lambda_{PP} = \lambda_{NN} = 0$ and $(\lambda_{PN} - \lambda_{BN})(\lambda_{NP} - \lambda_{BP}) > (\lambda_{BP} - \lambda_{PP})(\lambda_{PN} - \lambda_{NN})$.

Table 1. Brief description of UCI data sets.

ID	Data sets	$|U|$	$|C|$	$|D|$
1	Spect-train	80	22	2
2	Hepatitis	155	21	2
3	Heart-statlog	270	13	2
4	Vote	435	16	2
5	Monks-1	124	6	2
6	Monks-2	169	6	2
7	Monks-3	122	6	2
8	Colic	368	22	2

4.3 Experimental Results

The experimental results are shown in Tables 2, 3, 4, 5 and 6. In these tables, PRER, MCR, MIR represent the positive region expanding reduct [8], the minimum decision cost reduct [11], the maximum mutual information reduct [21], respectively. MOAR is our multi-objective attribute reduct by considering positive region, decision cost and mutual information simultaneously. The best performance of each row in the tables is boldfaced, in addition, the average ranks of these four methods on all datasets are also recorded.

Table 2. Comparison of the average length of the reduct for different methods.

ID	PRER	MCR	MIR	MOAR
1	12.99 ± 0.12	18.00 ± 0.56	15.97 ± 0.46	$\mathbf{10.84 \pm 0.27}$
2	6.48 ± 0.07	11.00 ± 0.00	6.63 ± 0.10	$\mathbf{5.98 \pm 0.20}$
3	6.94 ± 0.25	8.00 ± 0.00	6.99 ± 0.10	$\mathbf{6.89 \pm 0.13}$
4	$\mathbf{1.00 \pm 0.00}$	12.34 ± 0.19	16.00 ± 0.00	12.38 ± 0.61
5	4.00 ± 0.00	$\mathbf{3.00 \pm 0.00}$	6.00 ± 0.00	4.20 ± 0.21
6	6.00 ± 0.00	6.00 ± 0.00	6.00 ± 0.00	$\mathbf{4.70 \pm 0.15}$
7	$\mathbf{2.23 \pm 0.17}$	4.00 ± 0.00	6.00 ± 0.00	4.73 ± 0.10
8	12.01 ± 0.27	11.33 ± 0.26	10.61 ± 0.25	$\mathbf{8.16 \pm 0.25}$
Rank	2	2.75	3.125	1.75

From Table 2, we can find that MOAR obtains the shortest reduct on most datasets, because we consider the length of attribute reduct as a measure to choose solutions, when their crowding distances are same.

Tables 3 and 4 show the classification accuracy of four classifiers based on different reducts. For NaiveBayes and SMO, MOAR can produce the best accuracy on all datasets. For J48 and RandomForest, MOAR could also get the best result on most datasets.

Table 3. Comparison of accuracy for different methods based on NaiveBayes and J48.

ID	NaiveBayes				J48			
	PRER	MCR	MIR	MOAR	PRER	MCR	MIR	MOAR
1	0.77 ± 0.02	0.75 ± 0.01	0.77 ± 0.01	**0.79 ± 0.02**	0.72 ± 0.03	0.73 ± 0.03	0.73 ± 0.04	**0.74 ± 0.03**
2	0.81 ± 0.02	0.80 ± 0.01	0.82 ± 0.01	**0.85 ± 0.02**	0.79 ± 0.01	**0.81 ± 0.01**	0.78 ± 0.01	**0.81 ± 0.01**
3	0.77 ± 0.02	0.76 ± 0.01	0.78 ± 0.01	**0.84 ± 0.01**	0.73 ± 0.03	0.74 ± 0.00	0.75 ± 0.02	**0.82 ± 0.01**
4	0.87 ± 0.00	**0.91 ± 0.01**	0.90 ± 0.00	**0.91 ± 0.01**	0.87 ± 0.00	**0.96 ± 0.00**	**0.96 ± 0.00**	**0.96 ± 0.00**
5	0.75 ± 0.02	0.75 ± 0.01	0.74 ± 0.02	**0.77 ± 0.02**	0.86 ± 0.03	**0.93 ± 0.03**	0.79 ± 0.02	0.88 ± 0.03
6	0.57 ± 0.02	0.57 ± 0.02	0.57 ± 0.02	**0.63 ± 0.01**	0.61 ± 0.02	0.61 ± 0.02	0.61 ± 0.02	**0.66 ± 0.01**
7	**0.93 ± 0.00**	**0.93 ± 0.00**	**0.93 ± 0.00**	**0.93 ± 0.00**	0.93 ± 0.01	0.93 ± 0.01	0.93 ± 0.01	0.93 ± 0.01
8	0.79 ± 0.01	0.80 ± 0.00	0.80 ± 0.00	**0.83 ± 0.01**	0.85 ± 0.01	0.85 ± 0.01	0.85 ± 0.00	**0.86 ± 0.00**
Rank	2.375	2.375	2	1	2.375	1.625	2.125	1.125

Table 4. Comparison of accuracy for different methods based on RandomForest and SMO.

ID	RandomForest				SMO			
	PRER	MCR	MIR	MOAR	PRER	MCR	MIR	MOAR
1	0.72 ± 0.03	0.71 ± 0.03	0.70 ± 0.03	**0.73 ± 0.03**	0.75 ± 0.02	0.72 ± 0.04	0.74 ± 0.02	**0.76 ± 0.01**
2	0.79 ± 0.02	0.76 ± 0.02	0.82 ± 0.02	**0.83 ± 0.03**	0.83 ± 0.02	0.80 ± 0.02	0.82 ± 0.02	**0.85 ± 0.01**
3	0.71 ± 0.02	0.69 ± 0.02	0.71 ± 0.02	**0.78 ± 0.02**	0.77 ± 0.02	0.78 ± 0.01	0.77 ± 0.02	**0.84 ± 0.01**
4	0.87 ± 0.00	0.95 ± 0.01	**0.96 ± 0.00**	**0.96 ± 0.00**	0.87 ± 0.00	**0.96 ± 0.00**	**0.96 ± 0.00**	**0.96 ± 0.00**
5	0.89 ± 0.02	**0.96 ± 0.01**	0.85 ± 0.04	0.93 ± 0.01	0.79 ± 0.03	0.79 ± 0.03	0.79 ± 0.03	**0.80 ± 0.03**
6	0.58 ± 0.03	0.58 ± 0.03	0.58 ± 0.03	**0.64 ± 0.03**	0.59 ± 0.01	0.59 ± 0.01	0.59 ± 0.01	**0.61 ± 0.01**
7	0.92 ± 0.01	**0.93 ± 0.01**	0.92 ± 0.02	0.92 ± 0.02	**0.93 ± 0.00**	**0.93 ± 0.00**	**0.93 ± 0.00**	**0.93 ± 0.00**
8	0.82 ± 0.01	0.83 ± 0.02	0.83 ± 0.02	**0.85 ± 0.01**	0.82 ± 0.01	0.82 ± 0.01	0.83 ± 0.01	**0.84 ± 0.01**
Rank	2.5	2.25	2.375	1.25	2.125	2.375	2.125	1

Table 5. Comparison of misclassification cost for different methods based on Naive-Bayes and J48.

ID	NaiveBayes				J48			
	PRER	MCR	MIR	MOAR	PRER	MCR	MIR	MOAR
1	0.74 ± 0.46	0.81 ± 0.47	0.74 ± 0.46	**0.69 ± 0.47**	0.86 ± 0.49	0.85 ± 0.44	0.84 ± 0.50	**0.82 ± 0.44**
2	1.48 ± 0.78	1.64 ± 0.80	1.40 ± 0.77	**1.18 ± 0.62**	1.45 ± 0.43	1.38 ± 0.52	1.56 ± 0.52	**1.34 ± 0.52**
3	3.81 ± 1.38	3.93 ± 1.29	3.63 ± 1.44	**2.57 ± 1.06**	4.37 ± 1.36	4.29 ± 1.20	4.09 ± 1.26	**3.04 ± 0.99**
4	1.12 ± 0.41	**0.77 ± 0.38**	0.89 ± 0.38	0.79 ± 0.36	1.12 ± 0.41	0.35 ± 0.26	0.33 ± 0.25	**0.32 ± 0.25**
5	2.26 ± 1.01	2.25 ± 1.05	2.35 ± 0.97	**2.11 ± 0.97**	1.26 ± 0.90	**0.60 ± 0.80**	1.94 ± 0.94	1.04 ± 0.86
6	4.64 ± 0.77	4.64 ± 0.77	4.64 ± 0.77	**4.36 ± 0.79**	4.08 ± 1.17	4.08 ± 1.17	4.08 ± 1.17	**3.79 ± 1.14**
7	0.52 ± 0.64	**0.50 ± 0.55**	**0.50 ± 0.55**	**0.50 ± 0.55**	0.57 ± 0.64	0.54 ± 0.62	0.58 ± 0.66	0.54 ± 0.62
8	1.81 ± 0.68	1.61 ± 0.75	1.71 ± 0.72	**1.59 ± 0.75**	1.80 ± 0.73	1.81 ± 0.81	1.77 ± 0.68	**1.74 ± 0.68**
Rank	2.875	2.375	2.375	1.125	3.125	2.375	2.625	1.125

From Tables 5 and 6, we can find that MOAR can also obtain the best average rank on the misclassification cost criterion. An interesting result is that MOAR is better than MCR, while MCR is a single-objective attribute reduct by minimizing decision cost only. The reason is that the Bayesian decision cost minimized in MCR is a kind of expected cost, while the misclassification cost is a kind of actual cost. Although MOAR considers the expected decision cost as well, it also considers other two criteria as the optimization objectives, the result shows the robustness of MOAR.

Table 6. Comparison of missclassification cost for different methods based on RandomForest and SMO.

ID	RandomForest				SMO			
	PRER	MCR	MIR	MOAR	PRER	MCR	MIR	MOAR
1	0.89 ± 0.89	0.92 ± 0.50	0.96 ± 0.46	**0.86 ± 0.48**	0.78 ± 0.46	0.87 ± 0.44	0.81 ± 0.46	**0.75 ± 0.46**
2	1.60 ± 0.69	1.89 ± 0.70	1.37 ± 0.70	**1.32 ± 0.72**	1.27 ± 0.72	1.54 ± 0.65	1.40 ± 0.64	**1.08 ± 0.57**
3	4.76 ± 1.41	5.11 ± 1.42	4.71 ± 1.48	**3.65 ± 1.27**	3.86 ± 1.42	3.65 ± 1.20	3.71 ± 1.21	**2.68 ± 0.97**
4	1.12 ± 0.41	0.47 ± 0.25	**0.38 ± 0.29**	0.39 ± 0.24	1.12 ± 0.41	0.40 ± 0.26	**0.36 ± 0.28**	0.38 ± 0.28
5	0.98 ± 0.91	**0.39 ± 0.62**	1.32 ± 0.96	0.62 ± 0.75	1.91 ± 1.14	1.86 ± 1.14	1.91 ± 1.13	**1.80 ± 1.11**
6	4.05 ± 1.01	4.05 ± 1.01	4.05 ± 1.01	**3.25 ± 0.95**	4.96 ± 0.61	4.96 ± 0.61	4.96 ± 0.61	5.03 ± 0.53
7	**0.61 ± 0.66**	0.63 ± 0.69	0.71 ± 0.68	0.69 ± 0.71	0.52 ± 0.58	**0.50 ± 0.55**	**0.50 ± 0.55**	**0.50 ± 0.55**
8	1.94 ± 0.66	1.86 ± 0.79	1.83 ± 0.66	**1.54 ± 0.70**	1.76 ± 0.69	1.80 ± 0.73	1.65 ± 0.66	**1.58 ± 0.68**
Rank	2.75	2.75	2.625	1.5	2.625	2.625	2.125	1.25

From the experimental results, we can conclude that the multi-objective attribute reduction is a better choice for decision-theoretic rough set model, as it can achieve a better accuracy with a lower misclassification cost.

5 Conclusion

In this paper, a multi-objective attribute reduct definition is proposed based on DTRS. This definition can consider positive region, decision cost and mutual information at the same time. Meanwhile, it solves the problem that the attribute reduct based on one criterion perform perfectly on the specific criterion but poorly on other criterion. In addition, a multi-objective attribute reduction method corresponding the definition is proposed, and we can find an optimal attribute reduct with optimal classification accuracy from a set of Pareto solutions. Experiments results show the effectiveness and robustness of our proposed method.

Acknowledgment. This paper is supported by the National Natural Science Foundations of China (Grant Nos. 61403200, 71671086), the Natural Science Foundation of Jiangsu Province (Grant No. BK20140800), and Key Laboratory of Oceanographic Big Data Mining & Application of Zhejiang Province (Grant No. OBDMA201602).

References

1. Pawlak, Z.: Rough sets. Int. J. Comput. Inf. Sci. **11**(5), 341–356 (1982)
2. Li, W., Huang, Z., Jia, X., Cai, X.: Neighborhood based decision-theoretic rough set models. Int. J. Approx. Reason. **69**(C), 1–17 (2016)
3. Yao, Y.: Decision-theoretic rough set models. In: Yao, J.T., Lingras, P., Wu, W.-Z., Szczuka, M., Cercone, N.J., Ślęzak, D. (eds.) RSKT 2007. LNCS, vol. 4481, pp. 1–12. Springer, Heidelberg (2007). doi:10.1007/978-3-540-72458-2_1
4. Yao, Y., Zhao, Y.: Attribute reduction in decision-theoretic rough set models. Inf. Sci. **178**(17), 3356–3373 (2008)
5. Krawczak, M., Szkatuła, G.: An approach to dimensionality reduction in time series. Inf. Sci. **260**, 15–36 (2014)

6. Mac Parthaláin, N., Jensen, R.: Unsupervised fuzzy-rough set-based dimensionality reduction. Inf. Sci. **229**, 106–121 (2013)
7. Jia, X., Shang, L., Zhou, B., Yao, Y.: Generalized attribute reduct in rough set theory. Knowl.-Based Syst. **91**, 204–218 (2016)
8. Li, H., Zhou, X., Zhao, J., Liu, D.: Non-monotonic attribute reduction in decision-theoretic rough sets. Fundamenta Informaticae **126**(4), 415–432 (2013)
9. Ma, X., Wang, G., Hong, Y., Li, T.: Decision region distribution preservation reduction in decision-theoretic rough set model. Inf. Sci. **278**, 614–640 (2014)
10. Zhang, X., Miao, D.: Region-based quantitative and hierarchical attribute reduction in the two-category decision theoretic rough set model. Knowl.-Based Syst. **71**, 146–161 (2014)
11. Jia, X., Liao, W., Tang, Z., Shang, L.: Minimum cost attribute reduction in decision-theoretic rough set models. Inf. Sci. **219**, 151–167 (2013)
12. Min, F., Qinghua, H., Zhu, W.: Feature selection with test cost constraint. Int. J. Approx. Reason. **55**(1), 167–179 (2012)
13. Liao, S., Zhu, Q., Min, F.: Cost-sensitive attribute reduction in decision-theoretic rough set models. Math. Probl. Eng. **2014**(2), 1–9 (2014)
14. Qian, Y., Liang, J., Pedrycz, W., Dang, C.: Positive approximation: an accelerator for attribute reduction in rough set theory. Artif. Intell. **174**(9–10), 597–618 (2010)
15. Feifei, X., Bi, Z., Lei, J.: Cost minimization attribute reduction based on mutual information. In: International Conference on Fuzzy Systems and Knowledge Discovery, vol. 2015, pp. 215–219 (2015)
16. Wang, B., Li, X., Zhang, S.: An improved heuristic minimal attribute reduction algorithm based on condition information entropy. In: International Conference on Machinery, Materials and Information Technology Applications, vol. 2015, pp. 538–543 (2015)
17. Yang, M.: A novel algorithm for attribute reduction based on consistency criterion. Chin. J. Comput. **33**(2), 231–239 (2010)
18. Fang, Y., Liu, Z.-H., Min, F.: A PSO algorithm for multi-objective cost-sensitive attribute reduction on numeric data with error ranges. Soft Comput. (2016). doi:10.1007/s00500-016-2260-5
19. Deb, K., Agrawal, S., Pratap, A., Meyarivan, T.: A fast and elitist multiobjective genetic algorithm: NSGA-II. IEEE Trans. Evol. Comput. **6**(2), 182–197 (2002)
20. Yao, Y.: Three-way decisions with probabilitic rough sets. Inf. Sci. **18**, 341–353 (2010)
21. Miao, D., Hu, G.: A heuristic algorithm for reduction of knowledge. J. Comput. Res. Dev. **36**(6), 681–684 (1999)
22. Deb, K., Kalyanmoy, D.: Multi-objective optimization using evolutionary algorithms, vol. 2. Wiley, Hoboken (2001)
23. Blake, C.L., Merz, C.J.: UCI repository of machine learning databases. http://www.ics.uci.edu/mlearn/MLRepository.html
24. Hall, M., Frank, E., Holmes, G., Pfahringer, B., Reutemann, P., Witten, I.H.: The WEKA data mining software: an update. ACM SIGKDD Explor. Newsl. **11**(1), 10–18 (2009)

A Behavior-Based Method for Distinction of Flooding DDoS and Flash Crowds

Degang Sun, Kun Yang, Zhixin Shi[✉], and Bin Lv

Institute of Information Engineering, Chinese Academy of Sciences,
Beijing, China
shizhixin@iie.ac.cn

Abstract. DDoS and Flash Crowds are always difficult to distinguish. In order to solve this issue, this paper concluded a new feature set to profile the behaviors of legitimate users and Bots, and proposed an idea employed Random Forest to distinguish DDoS and FC on two widely-used datasets. The results show that the proposed idea can achieve distinguishing accuracy more than 95%. With comparison with traditional methods-Entropy, it still has a high accuracy.

Keywords: Flooding DDoS · Flash crowds · Random Forest · User behavior analysis · Entropy

1 Introduction

Flooding Distributed Denial of Service (DDoS) attacks have been wreaking havoc on the Internet and no signs show the trend of fading [1]. What is worse, DDoS start to simulate the behaviors of normal users and hidden in the traffic of Flash Crowds (FC) [2] in order to avoid the detecting by defending systems.

Due to normal and abnormal users have different characteristics of behaviors, therefore, it can be used to detect anomalies by modeling legitimate users behaviors. Xie and Yu [3] proposed a novel method to detect anomaly events based on the hidden Markov model. This approach used the entropy of document popularity as the input feature to establish model. But the input parameters are difficult to achieve through training. Oikonomou and Mirkovic [4] tried to discriminate mimicking attacks from real flash crowds by modeling human behavior. However, it is difficult to be employed for large-scale dynamic web pages. Thapngam et al. [5] proposed a discriminating method based on the packet arrival patterns. Pearson's correlation coefficient was used to measure the packet patterns. Patterns are defined by using the repeated properties observed from the traffic flow and also calculated the packet delay. However, defining the packet patterns are difficult. Yu et al. [6,7] employed flow similarities to discriminate DDoS and FC and achieved a better effect. The author mainly used fixed thresholds needed craft design and professional field knowledge, so this would be little deficiencies for deploying it in practice. Saravanan et al. [8] combined multi-parameters with weights to discriminate DDoS and FC, and achieved better results than the single

© Springer International Publishing AG 2017
G. Li et al. (Eds.): KSEM 2017, LNAI 10412, pp. 129–136, 2017.
DOI: 10.1007/978-3-319-63558-3_11

parameter. Those weights were fixed and could not be updated automatically. What's more, Turing test also are good methods to distinguish normal users and abnormal ones before years. However, the development of Reverse Turing Test makes the hacker could identify the pictures more easily without the participation in humans [9], which means Turing Test could not defend DDoS and distinguish DDoS and FC in a good way any more. Although, various methods have been proposed to detect and defend DDoS, however, there are still few researches on distinguishing Flooding DDoS and FC and there are also some deficiencies of the existing methods. In response to the above case, we propose a method of behavior analysis to solve this issue.

The paper is structured as follows: Sect. 2 shows analyzes and summarizes the traffic differences and the proposed idea. Section 3 furthers evaluation of the proposed method and comparison with traditional methods-Entropy. Section 4 is the summary of this paper.

2   Proposed Method

In order to solve the problem of discrimination of DDoS and FC, a few assumptions should be considered. In this paper, our assumption is that: the number of Bots simultaneously launched attacks (active Bots) have a limitation and usually do no more than the number of legitimate clients. Otherwise, it is impossible to distinguish Bots and legitimate users by statistical approaches, according to the authors' researches [6]. DDoS are usually launched by Botnets while FC derive from legitimate clients, there will be some traffic difference among network layer between DDoS and FC. In order to profile those differences of each Bot or client, we proposed a concept called Behavior Vector.

Definition (Behavior Vector). For a given time interval-Δt, aggregate packets which have the same source and destination IP as one flow, calculate the related statistical features, such as packets' size, packet arrival time difference, and label the behavior class: normal or abnormal, according to the priori knowledge. All of those could be used to make up of Behavior Vector:

$$(TS, srcIP, dstIP, nPkts, uPktsSize, stdPktsSize, uArrivalTime,$$
$$stdArrivalTime, behaviorClass)$$

(1) Timestamp (TS): To locate the accurate attack time of each Bot.
(2) Source IP (srcIP): To trace the Bot took part in the attack.
(3) Destination IP (dstIP): To confirm which target is the victim.
   Based on the above-mentioned assumption, the number of active Bots is limit, so Bots have to spend packets as many as they can in order to attain the expected attack effect, and DDoS mainly derived from Botnets, FC derived from legitimate users, each user will have a different behavior. Therefore, the following features are different: the number of packets sent by each Bot will larger than that of each legitimate user (nPkts), the average packet size of each Bot and that of legitimate user (uPktsSize) will be different, the standard deviation of the packet size between each Bots will be much smaller than that of legitimate user (stdPktsSize).

(4) The Number of Packets Sent by Each IP or Client In Each Interval (nPkts).
(5) The Average Packet Size Sent By Each Source IP in Each Interval (uPkts-Size).
(6) The Standard Deviation of Packet Size Sent by Each Source IP in Each Interval (stdPktsSize).
(7) The Average Packet Arrival Time Difference of Packets Sent by Each Source IP In Each Interval (uArrivalTime): the reason like 'uPktsSize'.
(8) The Standard Deviation of Packet Arrival Time Difference of Packets Sent By Each Source IP In Each Interval (stdArrivalTime): the reason like 'stdPktsSize'.
(9) The classification of Behavior Vector (Behavior Class): label the Behavior Vector into one class: *DDoS_Flow* or *FC_Flow*, which could be achieved according to the priori knowledge.

In this paper, we will evaluate our method on two public datasets. However, different datasets are usually derived from different situations, such as different topology structure, different network bandwidth. In order to conquer those dilemmas and obtain the feature set, we do some pre-processings to eliminate the existing difficulties:

(1) Different Topology: It means different IP masks and different IP addresses in each dataset. In order to solve this issue, we apply relevant statistical features instead of using IPs directly.
(2) Network Bandwidth: In order to cope with this issue, we scale the interval to ensure that different datasets have the same network bandwidth in each interval on the whole. For example, assuming that the bandwidth of first dataset is 10 m/s, while the second dataset is 100 m/s. In order to achieve the same traffic volume in each interval, we enlarge the first dataset to be 100 m/s, that means to increase 100/10(ten) times of the first interval. We believe that the impact of the network bandwidth is minimized through the scaling.

Through these pre-processings, we could make different datasets mixed to extract behavior features to evaluate our method further. Based on Behavior Vector, we could map the traffic behavior of each Bot and legitimate user into points in Euclid space and lots of methods have been proposed to measure points, so we could translate the issue of distinguishing DDoS and FC into how to identify abnormal points-outliers in Euclid space. What's more, we propose an idea employed Random Forest [10], which is an excellent and efficient Data Mining method, to classify abnormal points and normal points, that is to distinguish DDoS and FC. The whole procedure can be shown in Algorithm 1 directly.

3 Experiments

In this section, we do some experiments to estimate Behavior Vector and the idea proposed on two public real-world datasets: CAIDA "DDoS Attack 2007"

Algorithm 1. Proposed Idea.

Input:
   DDoS and FC datasets
Output:
   The discriminating results: $DDoS\_Flow$ or FC_Flow
1: Do pre-processings, such as scaling the interval of each dataset in order to eliminate the effect of the difference between datasets.
2: Cacluate features in every interval, label the class: $DDoS\_Flow$ or FC_Flow, according to the priori knowledge. Finally, achieve Behavior Vector of each dataset.
3: Mix Behavior Vector came from each dataset and normalize the mixed results.
4: Take the normalized data to train the model with Random Forest.
5: Once achieved the model fitted, then use it to test on other sets.
6: Return the final distinguished results.

Dataset (CAIDA 2007 Dataset) [11] used as the DDoS data and World Cup 1998 Dataset (WC 1998 Dataset) [12] used as the FC data. In our experiments, we select 1 min DDoS data-2007-08-05 05:30:00 to 2007-08-05 05:31:00- from CAIDA 2007 Dataset, And select 1 h FC data-1998-06-10 16:00:00 to 1998-06-10 17:00:00-from WC 1998 Dataset.

It could be found that the bandwidth of DDoS is much larger than that of FC, so we should enlarge the FC interval to eliminate the effect of bandwidth. In this paper, we employ the minimal interval Δt is 1 s and after analyzed the selected data, we can find when the FC interval scale is 100 times of the DDoS interval, they will have the nearly same traffic in each interval, so we choose the scale 100. After pre-processings, we calculate on each scaled dataset to achieve the features required and label the behavior Class-$DDoS\_Flow$ or FC_Flow in each interval to achieve Behavior Vector, more details can be seen in Sect. 2. Finally, we could achieve input data for the latter process.

Once achieved the input data, we should firstly make selection of features to eliminate the redundancy because we could not assure that all the latter five features in Behavior Vector are necessary and there are not any redundancy between features. To solve this problem, we adopt Pearson's Correlation Method (CorrelationAttributeEval), Gain Ratio Method (GainRatioAttributeEval) and Symmetrical Uncertainty Attribute Method (SymmetricalUncertAttributeEval) to make evaluation of features. As a result, we could conclude that the order of importance of whole features is as follows:

$$uPktsSize > stdPktsSize > uArrivalTime > stdArrivalTime > nPkts$$

According to the order of features importance, we integrate the different features with the sequence of features' importance to find which combination is the most optimal one, and all of the combinations are as follows:

Then we choose Random Forest to evaluate the different combination. The results can be seen in Table 1, which shows that the more features are combined, the more accuracy can be achieved. However, as the increasing of the number of the features, the more complexity of calculation it will take and will not achieve obvious performance improvement. Finally, we found that the combination 2 is the optimal one with high accuracy and less complexity of calculation, therefore, we use it as the input data for the evaluation of method in the future.

Combination 1 : $(uPktsSize)$

Combination 2 : $(uPktsSize, stdPktsSize)$

Combination 3 : $(uPktsSize, stdPktsSize, uArrivalTime)$

Combination 4 : $(uPktsSize, stdPktsSize, uArrivalTime, stdArrivalTime)$

Combination 5 : $(uPktsSize, stdPktsSize, uArrivalTime, stdArrivalTime, nPkts)$

Table 1. The distinguished result of different features combination between features.

Combination ways	Random forest					
	Confusion matrix			Accuary	FPR	FNR
		FC_Flow	DDoS_Flow			
Combination 1	FC_Flow	291777	53	99.918%	0.138%	0.018%
	DDoS_Flow	461	332900			
Combination 2	FC_Flow	291783	47	99.954%	0.071%	0.016%
	DDoS_Flow	238	333123			
Combination 3	FC_Flow	291803	27	99.951%	0.084%	0.009%
	DDoS_Flow	280	333081			
Combination 4	FC_Flow	291805	25	99.953%	0.080%	0.009%
	DDoS_Flow	268	333093			
Combination 5	FC_Flow	291812	18	99.955%	0.078%	0.006%
	DDoS_Flow	261	333100			

In order to fully evaluate our method, we divide the 1 min DDoS data into different scales by sampling technique. According to attack strength, it refers to the rate of packet sent by each Bot in each interval, while the 1 h FC data keep unchanged.

In each sampled data, the number of Bots is kept unchanged, the only change is the amounts of traffic produced by Bots, that is the attack strength, which have become bigger with different scales from 10% to 100% by sample. Finally, when the sampled scale is 100%, it means to use the whole 1 min DDoS data to analyze, The amount of traffic by Bots launched in DDoS is basically equal to the amount of traffic by legitimate users produced in FC.

In this paper, we integrate with 10 fold cross validation, Confusion Matrix, Accuracy, TPR and FPR all together as measure standards to evaluate the trained model and the estimated result can be seen in Table 2 which shows that the different sampled scales will affect the discriminated result, it means attack strength play a vital role to distinguish DDoS and FC. The more attack strength, the better discriminated accuracy, this is reasonable in the real environment. The reason is that more attack strength will lead to more obvious traffic abnormalities, which will enlarge the difference of different behavior between Bots and legitimate clients. However, there are a little bit difference of our results. The reason is that we sample on the DDoS data which has some noises [11] by different scales, the bigger sample rate is, the more noises will bring in. As a result, the discriminated result will have a slight change, but it will not cause a huge impact on our idea.

Table 2. The distinguished result with the optimal features by random forest.

Attack strength	Random forest					
	Confusion matrix			Accuary	FPR	FNR
		FC_Flow	DDoS_Flow			
10%	FC_Flow	291829	1	99.996%	0.008%	0.000%
	DDoS_Flow	21	266318			
20%	FC_Flow	291823	7	99.990%	0.018%	0.002%
	DDoS_Flow	52	294673			
30%	FC_Flow	291815	15	99.985%	0.025%	0.005%
	DDoS_Flow	75	305567			
50%	FC_Flow	291811	19	99.971%	0.049%	0.007%
	DDoS_Flow	159	327457			
80%	FC_Flow	291800	30	99.966%	0.056%	0.010%
	DDoS_Flow	184	327432			
100%	FC_Flow	291780	50	99.954%	0.072%	0.017%
	DDoS_Flow	239	333122			

In order to make further evaluation of our method, we make additional experiments by using another test set, 1 min DDoS test data-2007-08-05 05:34:00 to 2007-08-05 05:35:00-randomly selected from the CAIDA 2007 Dataset and 1 h FC test data-1998-07-03 16:00:00 To 1998-07-03 17:00:00-came from the Quarter of WC 1998 Dataset to test our trained model based on the previous train datasets. Similarly, we should some pre-processings to the new dataset in order to eliminate the impact of different dataset like the previous pre-processings-scaling the interval, the only difference is the scale which is 80:1, that is when the FC interval scale is 80 times of the DDoS interval, they will have almost the same traffic in each interval. After pre-processings, we could extract features and use the fitted model to evaluate our method further.

Ultimately, the test result is shown in Table 3 which shows that our idea can obtain more than 95.000% accuracy, further with less FPR around 0.005% and FNR around 5.000%. In addition, we use the whole features-the combination 5-to further evaluation, the result also shows the high accuracy, which means our method could also have a good effect on the new dataset.

Table 3. The distinguished results by random forest on the other test data

Attack strength	Combination ways	Random forest					
		Confusion matrix			Accuary	FPR	FNR
			FC_Flow	DDoS_Flow			
100%	Combination 2	FC_Flow	344890	14768	97.810%	0.002%	4.106%
		DDoS_Flow	5	314970			
100%	Combination 5	FC_Flow	354763	4895	99.273%	0.003%	1.361%
		DDoS_Flow	10	314965			

In order to compare with traditional methods further, we take another experiment on the test dataset with Shannon Entropy, which is a classic and common method. We calculate the value of Shannon Entropy of srcIPs in each interval on the test dataset. The result can be seen in Fig. 1, which indicates that it is not easy to distinguish DDoS and FC based on Shannon Entropy of srcIPs, because their entropy values are very close, so it is very difficult to select proper thresholds to distinguish each other.

Based on the above analysis and experiments, we could conclude that traditional methods are mainly based on thresholds which usually need to be selected elaborately but are considerably difficult to be obtained in reality. Our idea employed Random Forest which could avoid the selection of thresholds in a good way. It could adjust the value of threshold flexibly but basically need no human interventions and could achieve better accuracy for distinguishing DDoS and FC. Furthermore, the Behavior Vector we proposed could also locate precisely the Bot participating attacks and time of attack and the victims with the first features of Behavior Vector. In addition, our idea is deployed on end-victim, therefore, it is easy to deploy without modifying the existing network protocols, just to deploy at the front of end-victim.

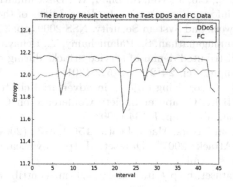

Fig. 1. The entropy result between the test DDoS and FC data

4   Conclusion

In this paper, an extensive analysis is made into two widely-used datasets and a new feature set is concluded to form Behavior Vector, which is used to profile the traffic behavior of each Bot and legitimate user. Based on it, experiments have been taken to evaluate our method of distinguishing DDoS and FC. The results show that our idea could achieve high accuracy, less FPR and FNR. In addition, compared with the Entropy-based method, the results also indicate that our method can achieve not only better accuracy, but also can locate the Bot, the time of attack and victims by the first features of Behavior Vector.

References

1. Mansfield-Devine, S.: The growth and evolution of DDoS. Netw. Secur. **2015**(10), 13–20 (2015)
2. Jung, J., Krishnamurthy, B., Rabinovich, M.: Flash crowds and denial of service attacks: characterization and implications for CDNs and web sites. In: Proceedings of the 11th International Conference on World Wide Web, pp. 293–304. ACM (2002)
3. Xie, Y., Yu, S.-Z.: A large-scale hidden semi-Markov model for anomaly detection on user browsing behaviors. IEEE/ACM Trans. Netw. (TON) **17**(1), 54–65 (2009)
4. Oikonomou, G., Mirkovic, J.: Modeling human behavior for defense against flash-crowd attacks. In: 2009 IEEE International Conference on Communications, pp. 1–6. IEEE (2009)
5. Thapngam, T., Yu, S., Zhou, W., Beliakov, G.: Discriminating DDoS attack traffic from flash crowd through packet arrival patterns. In: 2011 IEEE Conference on Computer Communications Workshops (INFOCOM WKSHPS), pp. 952–957. IEEE (2011)
6. Yu, S., Guo, S., Stojmenovic, I.: Fool me if you can: mimicking attacks and anti-attacks in cyberspace. IEEE Trans. Comput. **64**(1), 139–151 (2015)
7. Yu, S., Thapngam, T., Liu, J., Wei, S., Zhou, W.: Discriminating DDoS flows from flash crowds using information distance. In: Proceedings of the third International Conference on Network and System Security, NSS 2009, pp. 351–356. IEEE (2009)
8. Saravanan, R., Shanmuganathan, S., Palanichamy, Y.: Behavior-based detection of application layer distributed denial of service attacks during flash events. Turkish J. Electr. Eng. Comput. Sci. **24**(2), 510–523 (2016)
9. Mori, G., Malik, J.: Recognizing objects in adversarial clutter: breaking a visual captcha. In: 2003 IEEE Computer Society Conference on Computer Vision and Pattern Recognition, vol. 1, pp. I–134 (2003)
10. Breiman, L.: Random forests. Mach. Learn. **45**(1), 5–32 (2001)
11. CAIDA "DDoS Attack 2007" Dataset. http://www.caida.org/data/passive/ddos-20070804_dataset.xml
12. World Cup 1998 dataset. http://ita.ee.lbl.gov/html/contrib/WorldCup.html

Knowledge Management

Analyzing Customer's Product Preference Using Wireless Signals

Na Pang[1,2(✉)], Dali Zhu[1,2], Kaiwen Xue[3], Wenjing Rong[1,2],
Yinlong Liu[1,2], and Changhai Ou[1,2]

[1] Institute of Information Engineering, Chinese Academy of Sciences,
Beijing, China
{pangna,zhudali,rongwenjing,liuyinlong,ouchanghai}@iie.ac.cn
[2] School of Cyber Security, University of Chinese Academy of Sciences,
Beijing, China
[3] The Second High School Attached to Beijing Normal University,
Beijing, China
Kevin3358Kevin@163.com

Abstract. Customer's product preference provides how a customer collects products or prefers one collection over another. Understanding customer's product preference can provide retail store owner and librarian valuable insight to adjust products and service. Current solutions offer a certain convenience over common approaches such as questionnaire and interviews. However, they either require video surveillance or need wearable sensor which are usually invasive or limited to additional device. Recently, researchers have exploited physical layer information of wireless signals for robust device-free human detection, ever since *Channel State Information* (CSI) was reported on commodity WiFi devices. Despite of a significant amount of progress achieved, there are few works studying customer's product preference. In this paper, we propose a customer's product preference analysis system, *PreFi*, based on *Commercial Off-The-Shelf* (COTS) WiFi-enabled devices. The key insight of *PreFi* is to extract the variance features of the fine-grained time-series CSI, which is sensitively affected by customer activity, to recognize what is the customer doing. First, we conduct *Principal Component Analysis* (PCA) to smooth the preprocessed CSI values since general denoising method is insufficient in removing the bursty and impulse noises. Second, a sliding window-based feature extraction method and majority voting scheme are adopted to compare the distribution of activity profiles to identify different activities. We prototype our system on COTS WiFi-enabled devices and extensively evaluate it in typical indoor scenarios. The results indicate that *PreFi* can recognize a few representative customer activity with satisfied accuracy and robustness.

Keywords: Channel state information · Customer's product preference · COTS WiFi devices

© Springer International Publishing AG 2017
G. Li et al. (Eds.): KSEM 2017, LNAI 10412, pp. 139–148, 2017.
DOI: 10.1007/978-3-319-63558-3_12

1 Introduction

Analyzing customer's product preference using WiFi infrastructure should ideally meet the following four goals:

Privacy protection: The system should not result in privacy leakage for customer. Using of video surveillance should be avoided since it reveals the iden-
tity of customer.

Device-free: It should not require the customer to carry on-body sensors or
smartphones since these devices decreases the overall usability, expecially for
the customer who does not carry the device as expected.

High accuracy: It should not only do coarse grained customer activity recognition but also do fine-grained recognition with high accuracy even in complex
indoor environment.

Deployment: It should be easily deployed on *Commercial Off-The-Shelf*
(COTS) WiFi-enabled devices without changing any firmware at the *access
point* (AP).

If the above four requirements are satisfied, accurate analysis on customer's
product preference is available to provide valuable insights to retail store owner
or librarian in terms of arrangement of products and efficiency of services. It can
be used to recommend customer with some of their potentially interesting items.

However, no previous studies simultaneously satisfy the capability of privacy protection, device-free, high accuracy, and deployment. Computer vision
based systems such as *Microsoft Kinect* [18], *Leap* [1], and *Pharos* [10], require
the *line-of-sight* (LOS) with adequate lighting, and bring the new concern in
privacy disclosure. Wearable sensor based systems cause inconvenience in occa-
sions such as *bathing* and swimming, and it leads to extra cost [21]. Recently,
researchers exploit the characteristics of wireless signals to "hear" talking [22],
track position [13], detect moving human through-the-wall [29], detect abnormal
activity [28], and "see" keystrokes [5]. *WiTrack* [3] designs *Frequency Modulated
Carrier Wave* (FMCW) signals to achieve 3D motion tracking. *Wi-Vi* [4] leverages an *inverse synthetic aperture radar* (ISAR) to track moving user through
walls. *WiHear* [22] adopts specialized directional antennas and stepper motors to
obtain variations of CSI for hearing talking. However, those specialized devices
used in these systems imply high start-up costs.

In this paper, we focus on micro-movement activity sensing to analyze customer's product preference. We extract the fine-grained *Channel State Information* (CSI) to analyze the multipath channel features of OFDM subcarriers.
The time series of CSI values can reflect the unique characteristics of different
activities. To translate these ideas into a practical system involves the following
challenges:

- How to derive the CSI affected by customer activity under the subtle signal
  changes?
- Which features should be extracted from the time series of CSI values?
- How to differentiate between different customer activities during the various
  statistical features from the CSI values?

We address the above challenges and propose the WiFi-based customer's product preference analysis system, called *PreFi* (Customer's product P̲reference using WiF̲i̲). After calculating various statistical features to create CSI profiles of the activity, K-nearest neighbour (kNN) classifier and majority voting scheme are adopted to analyze customer activity under both LOS and *non-line-of-sight* (NLOS) situations. It does not require any dedicated sensors or any firmware change at the access point. We implement *PreFi* on *Commercial Off-The-Shelf* (COTS) WiFi-enabled devices and evaluate its performance in retail store and laboratory. Experiment results indicate that *PreFi* can achieve satisfied performance. More specifically, the average accuracy of *PreFi* is observed to be as high as 88.4%, and the average false positive rate is as low as 3.2%. The main contributions are as follows:

- By exploiting *Channel State Information* (CSI), we validate the feasibility of non-invasive customer's product preference analysis using WiFi signals.
- We extract CSI features to characterize the difference between multiple customers.
- We implement *PreFi* on COTS WiFi devices and evaluate it in different scenarios. The experiments show that *PreFi* can analyze customer's product preference in real time, and it is robust to environmental change.

This paper is organized as follows. In Sect. 2 we state the background of *Channel State Information*. Section 3 presents the details of analyzing customer's product preference using CSI. Experiment and evaluation are presented in Sect. 4. Section 5 discusses the related work. Finally, in Sect. 6 we conclude our paper.

2   Channel State Information

COTS WiFi device supports IEEE 802.11n/ac wireless standards which employ *Multiple-Input Multiple-Output* (MIMO). MIMO adopts multiple antennas for transmitting and receiving in physical layer to increase data transmission performance and bandwidth efficiency via spatial multiplexing and improve the reliability using spatial diversity. Specifically, *Orthogonal Frequency Division Multiplexing* (OFDM), which is a high data rate wireless transmission multicarrier modulation scheme, is used in IEEE 802.11n/ac standards. In a wireless network, *channel state information* (CSI) reveals channel properties depicting the phase and amplitude of each OFDM subcarrier, and it represents the wireless phenomenon of fading, scattering, and power decay.

We consider a flat-fading communications system with N_c and N_r antennas of the transmitter and the receiver, respectively. The received signals $\mathbf{x}(t)$ at symbol time t can be expressed in matrix form as

$$\mathbf{x}(t) = \mathbf{H}(t)\mathbf{s}(t) + \mathbf{w}(t),$$

where $\mathbf{s}(t)$ is N_c vector of transmit signals at time t, and $\mathbf{w}(t)$ is N_r vector of noise. The estimated $\mathbf{H}$ for N sub-carriers can be expressed as

$$\mathbf{H} = [\mathbf{H}_1, \mathbf{H}_2, \ldots, \mathbf{H}_i, \ldots, \mathbf{H}_N]^T, \quad \mathbf{H}_i = \mid \mathbf{H}_i \mid e^{j\sin(\angle \mathbf{H}_i)},$$

where $\angle \mathbf{H}_i$ is the phase information and $\mid \mathbf{H}_i \mid$ is the amplitude information, respectively. Specifically, we collect the CSI values of 30 *Orthogonal Frequency Division Multiplexing* (OFDM) subcarriers for every antenna from IEEE 802.11 data frames using a modified driver as described in [9].

3   Analyzing Customer's Product Preference Using CSI

3.1   CSI Preprocessing

Due to the fact that complex indoor environment affects the stability of commercial WiFi devices, the extracted CSI values are inherently very noisy from electromagnetic interference, or even the environment change such as air pressure. *PreFi* firstly removes the high-frequency noise through a low-pass filter. Then *Principal Component Analysis* (PCA) denoising scheme is applied to smooth out noisy signals. It's observed that the sensitivity of the human activity for different subcarriers is different. We perform PCA to correlate the CSI fluctuations of different filtered subcarriers for unveiling the most common variations.

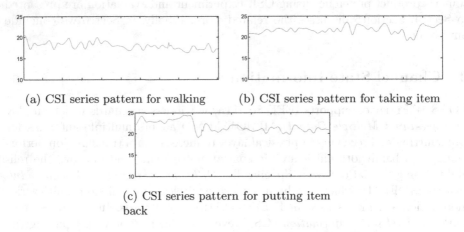

(a) CSI series pattern for walking   (b) CSI series pattern for taking item

(c) CSI series pattern for putting item back

Fig. 1. CSI profiles when a customer performs different activities.

To observe that filtered CSI values distinguish between different activity, a customer is asked to walk, take item, and put item back while the CSI value is being captured. Figure 1 shows the filtered CSI values of walking, taking item, and putting item back. We experimentally observed that different customer activity results in different CSI variance for each transmit-receive antenna pair.

3.2   Feature Extraction

Considering that the length of normal customer activity is no longer than 2 s. We choose a sliding window of 4 s in length, moving with a step of 0.2 s. There would be 21 subsegments in a 8 s CSI variance, which will be further represented as a set of features as shown in Fig. 2. A number of notable features are extracted to characterize the activity as shown in Table 1. The instances could be represented as $S_i = \{F_{i,1}, F_{i,2}, F_{i,3}, \ldots, F_{i,l}\}(l \leq 21)$, in which each instance $F_{i,j}$ represents a 15-D feature vector as defined above.

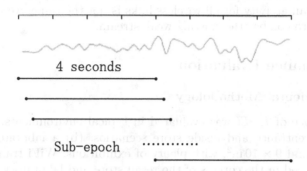

Fig. 2. Bag decomposition of an CSI segment.

Table 1. Extracted features from CSI waveforms.

Feature	Value	Description
f_1	x_{max}	The max amplitude
f_2	$t_{x_{max}}$	The time when the amplitude is max
f_3	x_{min}	The min amplitude
f_4	$t_{x_{min}}$	The time when the amplitude is min
f_5	$|(f_1 - f_3)/(f_2 - f_4)|$	The sharpness of the CSI from max point to min point
f_6	Mean	The mean CSI
f_7	Median	The median CSI
f_8	$|t_{start} - t_{middle}|$	The duration of the positive curve
f_9	$|t_{end} - t_{middle}|$	The duration of the negative curve
f_{10}	STD	The normalized standard deviation
f_{11}	MAD	The median absolute deviation
f_{12}	STD	The normalized standard deviation
f_{13}	NE	The normalized entropy
f_{14}	PDR	The power decline ratio
f_{15}	IR	Interquartile range

3.3 Classification

PreFi uses the obtained notable features to characterize customer activity. *Dynamic time warping* (DTW) method is adopted to calculate the distance between features of different activities. DTW handles waveforms with different lengths and provides a non-linear mapping by minimizing the distance between different waveforms. Compared with Eu-clidean distance, DTW determines the minimum distance even though they are shifted or distorted versions of each other [17]. *PreFi* trains an ensemble of k-nearest neighbour (kNN) classifiers using those features from all transmitter receiver antenna pairs. Since the sensitivity to human activity for all wireless links is not the same, we use majority voting scheme to combat the existing weak stream.

4 Performance Evaluation

4.1 Experimental Methodology

The performance of *PreFi* was evaluated in typical environments: (a) a retail store with an entrance and inside store scenarios. (b) a laboratory covering an area of around $9 \times 10\,m^2$, with plenty of exhibitions. WiFi transmitter and receiver are placed in the corners of the retail store and laboratory as shown in Fig. 3. The blue rectangle represents the receiver, and the green circular represents the transmitter, while the red pentagram represents the location in which the user performs activity. A total of 7 typical daily customer activities in retail store scenario and 5 walking activities in laboratory scenario are performed by 3 users. These activities are listed in Table 2.

We did experiments on *Think-pad X200* laptop with Intel 5300 wireless NIC installed, and the router is set as an AP which runs on 5 GHz with 20 MHz bandwidth. The receiver has 2 antennas while the AP has 3. We collect the CSI of 30 *Orthogonal Frequency Division Multiplexing* (OFDM) subcarriers from IEEE 802.11 data frames using tools developed by Halperin *et al.* [9]. The pair of WiFi devices communicate with 100 *pkts/s*. MATLAB is used to process the CSI value.

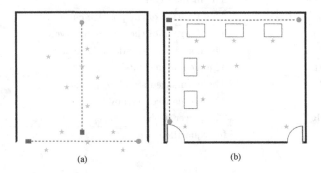

Fig. 3. Experiment settings: retail store (a), laboratory (b). (Color figure online)

Table 2. Codes for different activities in two test scenarios.

Code	Retail store activities	Code	Laboratory activities
A	Walk outside	a	Walk in
B	Walk inside	b	Taking item to observe
C	Walk away	c	Put item back
D	No person	d	Walk away
E	Take item	e	No person
F	Put the item back		
G	Put the item in basket		

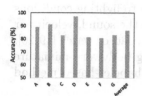 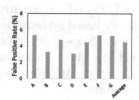 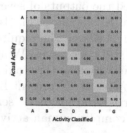

(a) Accuracy of customer activity classification

(b) False positive rate of classification

(c) Confusion matrix of classification

Fig. 4. Performance around the retail store.

4.2 Feasibility of Customer's Product Preference Analysis

The accuracy of *PreFi*'s customer's product preference analysis is evaluated by asking the customer to vary location and activity as shown in Fig. 3. The results in Fig. 4 show that *PreFi* can get average accuracy of 86.5% around the retail store, with an average false positive rate of 4.5%. The results in Fig. 5 indicate that *PreFi* has an average accuracy of 88.4% in laboratory, with an average

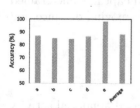 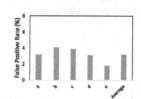 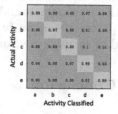

(a) Accuracy of customer activity classification

(b) False positive rate of classification

(c) Confusion matrix of classification

Fig. 5. Performance in the laboratory.

false positive rate of 3.2%. Figures 4c and 5c plot the confusion matrix in two scenarios. The results indicate that *PreFi* can distinguish a set of retail store and laboratory activities with satisfied accuracy. The laboratory environment is in higher accuracy than the retail store environment.

5   Related Work

5.1   Device-Based Activity Sensing

Popular vision-based sensing systems such as *Xbox Kinect* [18] and *Leap* [1] use depth cameras to achieve human computer interactions. Ijsselmuiden et al. [11] fused the outputs of several perceptual components for human activity recognition. Kushwaha et al. [14] present an object classification and recognition system. However, the requirement of LOS and the sensitivity to lighting conditions are still open issues. Wearable sensor based systems sense the sound, velocity or acceleration on three axises by embedding sensors in belt, coat, or watch [19,20]. Zappi et al. [6] present human activity classification using bodyworn miniature inertial and magnetic sensors. Liu et al. [16] present a shapelet-based method for complex human activity recognition.

5.2   Device-Free Activity Sensing Using WiFi

Doppler radar can be used to measure the speeds of human body. *WiTrack* [3] designs *Frequency Modulated Carrier Wave* (FMCW) signals to achieve 3D motion tracking. *Wi-Vi* [4] leverages an *inverse synthetic aperture radar* (ISAR) to track moving user through walls. However, those specialized devices imply high start-up costs. *Received Signal Strength* (RSS) fingerprints can be utilized for activity sensing. *WiGest* [2] leverages RSS values to sense in-air human hand motions. However, they can only provide coarse-grained recognition due to multi-path fading. Recently, *Channel State Information* (CSI) are increasingly adopted to fine-grained activity sensing since it discriminates multi-path characteristics [15,23,26]. *WiseFi* [27] combines CSI with Angle-of-arrival (AOA) to recognize and localize human activity. *WiFigure* [7,8,15] have demonstrated the viability of micro-movement sensing with high accuracy. *WiWho* [25] exploits CSI to identify walking gait and steps of human body. *WifiU* [24] captures fine-grained gait patterns to identify humans. *C2IL* [12] estimates the moving distance and speed in complex environment.

6   Conclusion

In this paper, we presented *PreFi*, a non-invasive WiFi-based abnormal activity sensing system with only two *commodity off-the-shelf* (COTS) WiFi devices. We first employ *Principal Component Analysis* (PCA) to smooth the preprocessed CSI values. Then, a sliding window-based feature extraction method and majority voting scheme are adopted to compare the distribution of activity profiles to identify different activities. *PreFi* does not require dedicated wearable sensors, additional hardware, and works under both LOS and NLOS conditions. In the future, we will pay more attention to the device-free static human activity detection.

Acknowledgement. This work was supported by Research of life cycle management and control system for equipment household registration, No. J770011104. We also thank the anonymous reviewers and shepherd for their valuable feedback.

References

1. Leap motion. https://www.leapmotion.com
2. Abdelnasser, H., Youssef, M., Harras, K.A.: Wigest: a ubiquitous wifi-based gesture recognition system. In: 2015 IEEE Conference on Computer Communications (INFOCOM), pp. 1472–1480. IEEE (2015)
3. Adib, F., Kabelac, Z., Katabi, D., Miller, R.C.: 3D tracking via body radio reflections. In: 11th USENIX Symposium on Networked Systems Design and Implementation (NSDI 14), pp. 317–329 (2014)
4. Adib, F., Katabi, D.: See through walls with wifi!, vol. 43. ACM (2013)
5. Ali, K., Liu, A.X., Wang, W., Shahzad, M.: Keystroke recognition using wifi signals. In: Proceedings of the 21st Annual International Conference on Mobile Computing and Networking, pp. 90–102. ACM (2015)
6. Altun, K., Barshan, B.: Human activity recognition using inertial/magnetic sensor units. In: Salah, A.A., Gevers, T., Sebe, N., Vinciarelli, A. (eds.) HBU 2010. LNCS, vol. 6219, pp. 38–51. Springer, Heidelberg (2010). doi:10.1007/978-3-642-14715-9_5
7. Bagci, I.E., Roedig, U., Martinovic, I., Schulz, M., Hollick, M.: Using channel state information for tamper detection in the internet of things. In: ACSAC 2015 - The Computer Security Applications Conference, pp. 131–140 (2015)
8. Chang, J.Y., Lee, K.Y., Wei, Y.L., Lin, C.J., Hsu, W.: We can "see" you via wi-fi - an overview and beyond (2016)
9. Halperin, D., Hu, W., Sheth, A., Wetherall, D.: Tool release: gathering 802.11n traces with channel state information. ACM SIGCOMM Comput. Commun. Rev. **41**(1), 53 (2011)
10. Hu, P., Li, L., Peng, C., Shen, G., Zhao, F.: Pharos: enable physical analytics through visible light based indoor localization. In: Twelfth ACM Workshop on Hot Topics in Networks, p. 5 (2013)
11. Ijsselmuiden, J., Stiefelhagen, R.: Towards high-level human activity recognition through computer vision and temporal logic. In: Proceedings of KI 2010: Advances in Artificial Intelligence, German Conference on AI, Karlsruhe, Germany, 21–24 September 2010, pp. 426–435 (2010)
12. Jiang, Z.P., Xi, W., Li, X., Tang, S., Zhao, J.Z., Han, J.S., Zhao, K., Wang, Z., Xiao, B.: Communicating is crowdsourcing: wi-fi indoor localization with CSI-based speed estimation. J. Comput. Sci. Technol. **29**(4), 589–604 (2014)
13. Kotaru, M., Katti, S.: Position tracking for virtual reality using commodity wifi (2017)
14. Kushwaha, A.K.S., Kolekar, M., Khare, A.: Vision based method for object classification and multiple human activity recognition in video survelliance system. In: Cube International Information Technology Conference, pp. 47–52 (2012)

15. Li, H., Yang, W., Wang, J., Xu, Y., Huang, L.: Wifinger: talk to your smart devices with finger-grained gesture. In: Proceedings of the 2016 ACM International Joint Conference on Pervasive and Ubiquitous Computing, pp. 250–261. ACM (2016)

16. Liu, L., Peng, Y., Liu, M., Huang, Z.: Sensor-based human activity recognition system with a multilayered model using time series shapelets. Knowl.-Based Syst. **90**(C), 138–152 (2015)

17. Long, X., Fonseca, P., Foussier, J., Haakma, R., Aarts, R.M.: Sleep and wake classification with actigraphy and respiratory effort using dynamic warping. IEEE J. Biomed. Health Inform. **18**(4), 1272–1284 (2014)

18. Microsoft. X-box kinect. http://www.xbox.com

19. Muhammad, S., Stephan, B., Durmaz, I.O., Hans, S., Havinga, P.J.M.: Complex human activity recognition using smartphone and wrist-worn motion sensors. Sensors **16**(4), 426 (2016)

20. Radhakrishnan, M., Eswaran, S., Misra, A., Chander, D., Dasgupta, K.: IRIS: tapping wearable sensing to capture in-store retail insights on shoppers. In: IEEE International Conference on Pervasive Computing and Communications, pp. 1–8 (2016)

21. Rallapalli, S., Ganesan, A., Chintalapudi, K., Padmanabhan, V.N., Qiu, L.: Enabling physical analytics in retail stores using smart glasses. In: International Conference on Mobile Computing and Networking, pp. 115–126 (2014)

22. Wang, G., Zou, Y., Zhou, Z., Wu, K., Ni, L.M.: We can hear you with wi-fi!. In: ACM International Conference on Mobile Computing and Networking, pp. 593–604 (2014)

23. Wang, H., Zhang, D., Wang, Y., et al.: RT-Fall: a real-time and contactless fall detection system with commodity WiFi devices. IEEE Trans. Mob. Comput. **16**(2), 1 (2017)

24. Wang, W., Liu, A.X., Shahzad, M.: Gait recognition using wifi signals. In: Proceedings of the 2016 ACM International Joint Conference on Pervasive and Ubiquitous Computing, pp. 363–373. ACM (2016)

25. Zeng, Y., Pathak, P.H., Mohapatra, P.: WiWho: wifi-based person identification in smart spaces. In: Proceedings of the 15th International Conference on Information Processing in Sensor Networks, p. 4. IEEE Press (2016)

26. Zheng, X., Wang, J., Shangguan, L., Zhou, Z., Liu, Y.: Smokey: ubiquitous smoking detection with commercial wifi infrastructures. In: IEEE INFOCOM 2016-The 35th Annual IEEE International Conference on Computer Communications, pp. 1–9. IEEE (2016)

27. Zhu, D., Pang, N., Li, G., Liu, S.: WiseFi: activity localization and recognition on commodity off-the-shelf wifi devices. In: IEEE International Conference on High Performance Computing and Communications; IEEE International Conference on Smart City; IEEE International Conference on Data Science and Systems (2017)

28. Zhu, D., Pang, N., Li, G., Rong, W., Fan, Z.: Win: non-invasive abnormal activity detection leveraging fine-grained wifi signals. In: Trustcom/BigDataSE/ISPA (2017)

29. Zhu, H., Xiao, F., Sun, L., et al.: R-TTWD: robust device-free through-the-wall detection of moving human with WiFi. IEEE J. Sel. Areas Commun. **35**(5), 1090–1103 (2017)

Improved Knowledge Base Completion
by the Path-Augmented TransR Model

Wenhao Huang[1,2(✉)], Ge Li[1,2], and Zhi Jin[1,2]

[1] Key Laboratory of High Confidence Software Technologies,
Ministry of Education, Peking University, Beijing, China
wenhao.huang@pku.edu.cn, {lige,zhijin}@sei.pku.edu.cn
[2] Software Institute, Peking University, Beijing, China

Abstract. Knowledge base completion aims to infer new relations from existing information. In this paper, we propose path-augmented TransR (PTransR) model to improve the accuracy of link prediction. In our approach, we build PTransR based on TransR, which is the best one-hop model at present. Then we regularize TransR with information of relation paths. In our experiment, we evaluate PTransR on the task of entity prediction. Experimental results show that PTransR outperforms previous models.

Keywords: Knowledge base completion · Relation path · Link prediction

1 Introduction

Large scale knowledge bases such as WordNet [1] and FreeBase [2] are important resources for natural language processing (NLP) applications like web searching [3], automatic question answering systems [4], and even medical informatics [5]. Formally, a *knowledge base* is a dataset containing triples of two entities and their relation. A triplet (h, r, t), for example, indicates that the *head entity* h and the *tail entity* t have a relation r. Despite massive triplets a knowledge base contains, evidence in the literature suggests that existing knowledge bases are far from complete [6,7].

In the past decades, researchers have proposed various methods to automatically construct or populate knowledge bases from plain texts [6,8], semi-structured data on the Web [9,10], etc. Recently, studies have shown that embedding the entities and relations of a knowledge base into a continuous vector space is an effective way to integrate the global information in the existing knowledge base and to predict missing triplets without using external resources (i.e., additional text or tables) [7,11–14].

Bordes et al. [11] propose the TransE approach, which translates entities' embeddings by that of a relation, to model knowledge bases. That is to say, the relation between two entities can be represented as a vector offset, similar to

G. Li et al. (Eds.): KSEM 2017, LNAI 10412, pp. 149–159, 2017.
DOI: 10.1007/978-3-319-63558-3_13

word analogy tasks for word embeddings [15] and sentence relation classification by sentence embeddings [16]. For one-to-many, many-to-one, and many-to-many relations, however, such straightforward vector offset does not make much sense. Considering a head entity `China` and the `country-city` relation, we can think of multiple plausible tail entities like `Beijing`, `Tianjin`, and `Shanghai`. These entities cannot be captured at the same time by translating the head entity and relation embeddings. Therefore, researchers propose to map entities to a new space where embedding translation is computed, resulting in TransH [7], TransR [12], and other variants. Among the above approaches, TransR achieves the highest performance on established benchmarks.

One shortcoming of the above methods is that only the direct relation (i.e., one-hop relation) between two entities is considered. In a knowledge base, some entities and relations only appear a few times; they suffer from the problem of data sparsity during training. Fortunately, the problem can be alleviated by using multi-hop information in a knowledge base. Guu et al. [13] present a random walk approach to sample entities with composited relations. Likewise, Lin et al. [14] propose a path-augmenting approach that uses multi-hop relations between two entities to regularize the direct relation between the same entity pair. Their experiments show the path-augmented TransE model (denoted as PTransE) outperforms the one-hop TransE model.

In this paper, we are curious whether we can combine the worlds, i.e., whether the path-augmenting technique is also useful for a better one-hop "base" model. Therefore, we propose to leverage TransR [12] as our cornerstone, but enhance it with path information as in [14], resulting a new variant, PTransR. We evaluate our model on the FreeBase dataset. Experimental results show that modeling relation paths is beneficial to the base model TransR, and that PTransR also outperforms PTransE in entity prediction. In this way, we achieve the state-of-the-art link prediction performance in the category that uses only the knowledge base itself (i.e., without additional textual information).

The rest of this paper is organized as follows. In Sect. 2, we describe the base model TransR and then discuss the path-augmented variant PTransR. In Sect. 3, we compare our PTransR model with other baselines in an entity prediction experiment; we also have in-depth analysis regarding different groups of relations, namely 1-to-1, 1-to-n, n-to-1, and n-to-n relations. In Sect. 4, we briefly review previous work in information extraction. Finally, we conclude our paper in Sect. 5.

2   Our Approach

In this section, we present our PTransR model in detail. In Subsect. 2.1, we introduce the TransE model and explain how TransR overcomes the weakness of TransE. Then, we augment TransR model with path information in Subsect. 2.2.

2.1 Base Model: TransR

As said in Sect. 1, embedding entities and their relation into vector spaces can effectively exploit internal structures that a knowledge base contains, and thus is helpful in predicting missing triplets without using additional texts.

The first model in such research direction is TransE [11]. It embeds entities and their relation in a same low-dimensional vector space; the two entities' embeddings are translated by a relation embedding, which can be viewed as an offset vector. In other words, for a triplet (h, r, t), we would like $\mathbf{h} + \mathbf{r} \approx \mathbf{t}$. (Here, bold letters refer to the embeddings of head/tail entities and the relation.) The plausibility of a triplet (h, r, t) is then evaluated by a scoring function

$$f_r(\mathbf{h}, \mathbf{t}) \triangleq \|\mathbf{h} + \mathbf{r} - \mathbf{t}\|, \qquad (1)$$

where $\| \cdot \|$ denotes either ℓ_1-norm or ℓ_2-norm. $f_r(\mathbf{h}, \mathbf{t})$ is expected to be small if (h, r, t) is a positive triplet.

To further analyze the performance of TransE, Bordes et al. [11] divide relations into four groups, namely 1-to-1, 1-to-n, n-to-1, and n-to-n, according to the mapping properties of a relation. For example, country-city is a 1-to-n relation, because a country may have multiple cities, but a city belongs to only one country. The weakness of TransE is that entity embeddings on the many side tend to be close to each other, which is the result of expecting $f_r(\mathbf{h}, \mathbf{t})$ to be small for all positive triplets. Therefore, it is hard for TransE to distinguish among the entities which are on the many side.

To solve the above problem, TransR [12] embeds entities and relations into two separate spaces: the *entity space* and the *relation space*. It uses relation-specific matrices $\mathbf{M}_r$ to map an entity from its own space to the relation space, given by $\mathbf{h}_r = \mathbf{M}_r \mathbf{h}$ and $\mathbf{t}_r = \mathbf{M}_r \mathbf{t}$, so that translation can be accomplished by regarding relation embedding as an offset vector, i.e., $\mathbf{h}_r + \mathbf{r} \approx \mathbf{t}_r$. To achieve this goal, TransR defines the scoring function as

$$f_r(\mathbf{h}, \mathbf{t}) \triangleq \|\mathbf{M}_r \mathbf{h} + \mathbf{r} - \mathbf{M}_r \mathbf{t}\|_2^2. \qquad (2)$$

To train the model, we shall generate negative samples and use the hinge loss. The overall cost function of the TransR model is

$$\mathcal{L}_{\text{TransR}} = \sum_{(h,r,t) \in S} \sum_{(h',r,t') \in S'} \max\left\{0, \gamma + f_r(\mathbf{h}, \mathbf{t}) - f_r(\mathbf{h}', \mathbf{t}')\right\}, \qquad (3)$$

where negative samples are constructed as

$$S' = \{(h', r, t) | (h', r, t) \notin S\} \bigcup \{(h, r, t') | (h, r, t') \notin S\}.$$

Results in link prediction show that TransR outperforms other models on established benchmarks, indicating TransR is the best one-hop model at present. However, TransR fails to utilize the rich path information, which will be dealt with in the following subsection.

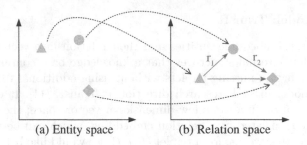

(a) Entity space   (b) Relation space

Fig. 1. An illustration of the PTransR model.

2.2   Path-Augmented TransR: PTransR

Using path information to regularize one-hop models can be beneficial [13,14]. Here, we adopt the path modeling method in PTransE, and extend the TransR model to path-augmented TransR (denoted as PTransR) (Fig. 1).

A relation path is a set of relations that connect head entity and tail entity in succession. An n-hop relation path from h to t is defined as $p = \{r_1, r_2, \cdots, r_n\}$, satisfying $h \xrightarrow{r_1} e_1 \xrightarrow{r_2} \cdots \xrightarrow{r_n} t$. If $n = 1$, then $p = r_1$ is a direct (1-hop) relation. To enhance TransR model with multi-hop information, we follow the treatment in PTransE [14] and represent a relation path as an embedding vector by additive compositional methods. Then such multi-hop information is used to regularize one-hop direct relation between the same entity pair. A reliability score is computed to address the strength of regularization by a particular path. The details are described as follows.

To compute the representation of a relation path p that composites primitive relations $r_1, r_2, \cdots, r_n$, i.e., $p = r_1 \circ r_2 \circ \cdots \circ r_n$ (where $\circ$ denotes the composition operation), we add the embeddings of these primitive relations, given by

$$\mathbf{p} = \mathbf{r}_1 + \mathbf{r}_2 + \cdots + \mathbf{r}_n, \tag{4}$$

where bold letters denote the vector of a relation or a path.

The choice of addition as the composition operation is reasonable, because the vector representation of path p should be close to that of direct relation r if it is likely to infer r from p. For example, the representation of path $\xrightarrow{father} \xrightarrow{mother}$ is expected to be close to that of direct relation $\xrightarrow{grandmother}$.

Although a knowledge base may contain a variety of relation paths between two entities, not every path is equally useful for inferring direct relations. For example, the relation path $John \xrightarrow{friend} Tim \xrightarrow{gender} male$ gives little contribution to inferring the gender of $John$.

To evaluate the reliability of a path, PTransE uses a path-constraint resource algorithm (PCRA) [14], which is also applied in our approach. This algorithm first assigns a certain amount of resource (i.e., a value of 1) to the head entity h; then each node distributes resource evenly to its direct child nodes (Fig. 2). The value of p along an entity pair h, t is denoted as $v(p|h, t)$.

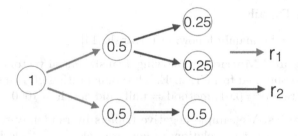

Fig. 2. An illustration of the path-constraint resource algorithm (PCRA).

However, $v(p|h,t)$ alone does not embody the relatedness between a relation path p and a direct relation r. To address this problem, a relatedness measure is defined as $P_r(r|p) = P_r(r,p)/P_r(p)$, where $P_r(p)$ is the sum of $v(p|h,t)$ for every training triplet (h,r,t) with p as a relation path from h to t. $P_r(r,p)$ is the sum of $v(p|h,t)$ for every training triplet (h,r,t) with r being a direct relation and p as a relation path. The overall reliability of a path p on a triplet (h,r,t) is given by

$$R(p|h,r,t) = P_r(r|p) \cdot v(p|h,t). \qquad (5)$$

PTransR's scoring function $f_{\text{PTransR}}(h,r,t)$ is composed of two scores:

$$f_{\text{PTransR}}(h,r,t) = E(h,r,t) + E(\mathbf{P}|h,r,t). \qquad (6)$$

$E(h,r,t)$ is the same as the scoring function of TransR (Eq. 2) which evaluates the plausibility of (h,r,t) without considering relation paths from h to t. Following PTransE, $E(\mathbf{P}|h,r,t)$ is defined as

$$E(\mathbf{P}|h,r,t) = \frac{1}{Z}\sum_{p\in\mathbf{P}} E(p|h,r,t), \qquad (7)$$

$$E(p|h,r,t) = R(p|h,r,t)\|\mathbf{p}-\mathbf{r}\|_2^2 = P_r(r|p)v(p|h,t)\|\mathbf{p}-\mathbf{r}\|_2^2, \qquad (8)$$

where Z is a normalizing factor for $R(p|h,t)$ and $\mathbf{P}$ is the set of all paths from h to t. $E(\mathbf{P}|h,r,t)$ evaluates the plausibility of (h,r,t) with the consideration of relation paths from h to t. The overall loss function of PTransR is

$$\mathcal{L}_{\text{PTransR}} = \sum_{(h,r,t)\in S} \left[L(h,r,t) + \frac{1}{Z}\sum_{p\in\mathbf{P}} L(p|h,r,t) \right], \qquad (9)$$

$$L(h,r,t) = \sum_{(h',r,t')\notin S} \max\left\{0, \gamma_1 + E(h,r,t) - E(h',r,t')\right\}, \qquad (10)$$

$$L(p|h,r,t) = \sum_{(h,r',t)\notin S} \max\left\{0, \gamma_2 + E(p|h,r,t) - E(p|h,r',t)\right\}, \qquad (11)$$

PTransR learns entity and relation embeddings by minimizing $\mathcal{L}_{\text{PTransR}}$.

2.3   Training Details

We train PTransR by mainly following PTransE [14].

Initial Vectors and Matrices. Following TransR, initial vectors and matrices for PTransR are obtained from TransE. The configuration of TransE is: margin $\gamma = 1$, learning rate $\alpha = 0.01$, method = unif, and epoch = 1000.

Negative Samples. We sample negative triplets by randomly replacing head entity h or tail entity t or relation r. For example, (h, r, t)-derived negative triplets are (h', r, t), (h, r, t'), and (h, r', t), where $(h, r, t) \in S$ and (h', r, t), (h, r', t), $(h, r, t') \notin S$.

Vector Representation Constraints. Following TransR, to regularize the representations, we impose the following constraints on the entity and relation embeddings.

$$\| \mathbf{h} \| = \| \mathbf{t} \| = \| \mathbf{r} \| = 1, \| \mathbf{M}_r \mathbf{h} \| \leq 1, \| \mathbf{M}_r \mathbf{t} \| \leq 1. \tag{12}$$

Path Selection. PTranE restricts the length of path to less than 3. Its results show that 3-hop paths do not make significant improvement, compared to 2-hop paths. For efficiency, we only consider 2-hop relation paths.

Inverse Relation. As inverse relations sometimes contain useful information, for each training triplet (h, r, t), (t, r^{-1}, h) is added to the training set.

3   Evaluation

In this section, we present results of our experiment. We first briefly introduce the dataset and the task of entity prediction. Then we show the experimental results and analyze the performance.

3.1   Dataset

FB15k is a commonly used dataset in knowledge base completion. Table 1 shows statistics of FB15k. FB15k dataset contains factual information in our world, e.g., location/country/language_spoken. As FB15k has various kinds of relations, it is suitable for the evaluation of PTransR. Therefore, we choose FB15k as our experimental dataset.

Table 1. Statistics of the FB15k dataset.

Dataset	Relation	Entity	Train	Valid	Test
FB15k	1,345	14,951	483,142	50,000	59,071

3.2   Experimental Settings

We evaluate PtransR on the task of entity prediction. Entity prediction aims at predicting the missing entity in an incomplete triplet, i.e., predicting h given r and t, or predicting t given h and r. Following the settings in TransE, for a triplet (h, r, t), we replace the head entity h with every entity e and compute the score of (e, r, t). Entity candidates are ranked according to their scores. We repeat the same process to predict the tail entity t. Then we use the two metrics in TransE to evaluate the performance: MeanRank (average rank of the expected entity) and Hits@10 (proportion of triplets whose head/tail entity is among top-10 in the ranking). However, there could be several entities that are plausible for the same incomplete triplet. The plausible entities which are ranked before h or t may cause underestimation of performance. One solution is to remove other plausible entities in the ranking, which is referred to as a *filter*. In comparison, the results without removing other plausible entities are referred to as *raw*. A good model should achieve low MeanRank and high Hits@10.

To utilize the inverse relation, instead of only using the score $f_{\text{PTransR}}(h, r, t)$, we use the sum of $f_{\text{PTransR}}(h, r, t)$ and $f_{\text{PTransR}}(t, r^{-1}, h)$ to rank the candidates, i.e.,

$$score(h, r, t) = f_{\text{PTransR}}(h, r, t) + f_{\text{PTransR}}(t, r^{-1}, h). \tag{13}$$

To accelerate the testing process, we use the reranking method in PTransE. We first rank all candidates according to their scores which are computed by the scoring function of TransR, which means that path information is not considered in the first ranking. Then we rerank the top-500 candidates according to the scores computed by the scoring function mentioned above, namely $score(h, r, t)$.

The configurations for experiments are given as follows: learning rate α for SGD among $\{0.01, 0.001, 0.0001\}$, dimension of entity space $\mathbb{R}^k$ and relation space $\mathbb{R}^d$ between $\{20, 50\}$, γ_1 and γ_2 among $\{1, 2, 4\}$, batch size B among $\{480, 960, 4800\}$. The optimal configuration on valid set is $\alpha = 0.001$, $k = d = 50$, $\gamma_1 = \gamma_2 = 1$, and $B = 4800$. The training process is limited to less than 500 epochs.

3.3   Overall Performance

Table 2 shows the experimental results. By comparing the results of PTransR with the results of previous models, we have the following main observations: (1) PTransR outperforms TransR on every metric to a large margin, which shows that path-augmented model can achieve better results than one-hop base model. (2) PTransR outperforms PTransE in MeanRank and is comparable to PTransE in Hits@10, which shows that path-augmented model with a better one-hop base model can achieve better performance.

Table 3 presents the performance on the four relation categories 1-to-1, 1-to-n, n-to-1, and n-to-n, with Hits@10(*filter*) as the metric. From Table 3, we find that, compared to TransR, PTransR shows consistent improvement on all four relation categories. Also, compared to PTransE, PTransR performs better on 1-to-1, 1-to-n and n-to-1 relations, especially on 1-to-n and n-to-1.

Table 2. Evaluation results of entity prediction on FB15k.

Metric	Mean rank		Hits@10(%)	
	Raw	Filter	Raw	Filter
Unstructured (Bordes et al. 2012)	1,074	979	4.5	6.3
RESCAL (Nickel et al. 2011)	828	683	28.4	44.1
SE (Bordes et al. 2011)	273	162	28.8	39.8
SME (linear) (Bordes et al. 2012)	274	154	30.7	40.8
SME (bilinear) (Bordes et al. 2012)	284	158	31.3	41.3
LFM (Jenatton et al. 2012)	283	164	26.0	33.1
TransE (Bordes et al. 2013)	243	125	34.9	47.1
TransH (unif) (Wang et al. 2014)	211	84	42.5	58.5
TransH (bern) (Wang et al. 2014)	212	87	45.7	64.4
TransR (unif) (Lin et al. 2015)	226	78	43.8	65.5
TransR (bern) (Lin et al. 2015)	198	77	48.2	68.7
CTransR (unif) (Lin et al. 2015)	233	82	44.0	66.3
CTransR (bern) (Lin et al. 2015)	199	75	48.4	70.2
PTransE (2-hop) (Lin et al. 2015)	200	54	51.8	83.4
PTransE (3-hop) (Lin et al. 2015)	207	58	51.4	**84.6**
PTransR (2-hop)	**171**	**47**	**53.0**	84.3

Table 3. Evaluation results of different relation catogories.

Tasks	Predicting head (Hits@10)				Predicting tail (Hits@10)			
Relation category	1-to-1	1-to-N	N-to-1	N-to-N	1-to-1	1-to-N	N-to-1	N-to-N
Unstructured	34.5	2.5	6.1	6.6	34.3	4.2	1.9	6.6
SE	35.6	62.6	17.2	37.5	34.9	14.6	68.3	41.3
SME (linear)	35.1	53.7	19.0	40.3	32.7	14.9	61.6	43.3
SME (bilinear)	30.9	69.6	19.9	38.6	28.2	13.1	76.0	41.8
TransE	43.7	65.7	18.2	47.2	43.7	19.7	66.7	50.0
TransH (unif)	66.7	81.7	30.2	57.4	63.7	30.1	83.2	60.8
TransH (bern)	66.8	87.6	28.7	64.5	65.5	39.8	83.3	67.2
TransR (unif)	76.9	77.9	38.1	66.9	76.2	38.4	76.2	69.1
TransR (bern)	78.8	89.2	34.1	69.2	79.2	37.4	90.4	72.1
CTransR (unif)	78.6	77.8	36.4	68.0	77.4	37.8	78.0	70.3
CTransR (bern)	81.5	89.0	34.7	71.2	80.8	38.6	90.1	73.8
PTransE (2-hop)	91.0	92.8	60.9	83.8	**91.2**	74.0	88.9	86.4
PTransE (3-hop)	90.1	92.0	58.7	**86.1**	90.7	70.7	87.5	**88.7**
PTransR (2-hop)	**91.4**	**93.4**	**65.5**	84.2	**91.2**	**74.5**	**91.8**	86.8

3.4 In-Depth Analysis and Discussion

As pointed out in Sect. 1, despite the massive train set of FB15k, some relations cannot be properly captured due to the problem of data sparsity. We separate relations into five groups according to their frequency in the train set, as shown in Table 4. MeanRank(raw) of TransR and PTransR is compared in Table 4 and the improvement from TransR to PTransR is presented. First of all, we see PTransR outperforms TransR in all five groups of relations. Second, as relation frequency decreases, the improvement goes up, which means that path information is useful for dealing with the problem of data sparsity.

Table 4. Evaluation results concerning relations of different frequency in train set.

Relation frequency in train set	1–3	4–15	16–50	51–300	>300
Relation number	291	305	243	271	235
MeanRank of TransR	159	98	54	81	202
MeanRank of PTransR	85	63	41	63	182
Improvement (%)	46.5	35.7	24.1	22.2	9.9

4 Related Work

Relation extraction is an important research topic in NLP. It can be roughly divided into two categories based on the source of information.

Text-based approaches extraction entities and/or relations from plain text. For example, Hearst [17] uses "is a|an" pattern to extract hyponymy relations. Banko et al. [6] proposes to extract open-domain relations from the Web. Fully supervised relation extraction, which classify two marked entities into several predefined relations, has become a hot research arena in the past several years [8,18,19].

Knowledge base completion/population, on the other hand, does not use additional text. Socher et al. [20] propose a tensor model to predict missing relations in an existing knowledge base, showing neural networks' ability of entity-relation inference. Then, translating embeddings approaches are proposed for knowledge base completion [7,11,12,14]. Recently, Wang et al. [21] use additional information to improve knowledge base completion by using textual context.

In this paper, we focus on pure knowledge base completion, i.e., we do not use additional resources. We combine the state-of-the-art one-hop TransR model [12] and path augmentation method [14], resulting in the new PTransR variant.

5 Conclusion

In this paper, we augment one-hop TransR model with path modeling method, resulting in PTransR model. We evaluate PTransR on the task of entity prediction and compare the performance of PTransR with that of previous models.

Experimental results show that path information is useful in solving the problem of data sparsity, and that PTransR outperforms previous models, which makes PTransR the state-of-the-art model in the field that populates knowledge base without using additional text.

Acknowledgments. This research is supported by the National Basic Research Program of China (the 973 Program) under Grant No. 2015CB352201 and the National Natural Science Foundation of China under Grant Nos. 61421091, 61232015 and 61502014.

References

1. Miller, G.A.: WordNet: a lexical database for English. Commun. ACM **38**(11), 39–41 (1995)
2. Bollacker, K., Evans, C., Paritosh, P., Sturge, T., Taylor, J.: Freebase: a collaboratively created graph database for structuring human knowledge. In: Proceedings of the 2008 ACM SIGMOD International Conference on Management of Data, pp. 1247–1250 (2008)
3. Jiang, X., Tan, A.H.: Learning and inferencing in user ontology for personalized semantic web search. Inf. Sci. **179**(16), 2794–2808 (2009)
4. Yao, X., Van Durme, B.: Information extraction over structured data: question answering with freebase. In: Proceedings of the 52nd Annual Meeting of the Association for Computational Linguistics, pp. 956–966 (2014)
5. Ashburner, M., Ball, C.A., Blake, J.A., Botstein, D., Butler, H., Cherry, J.M., Davis, A.P., Dolinski, K., Dwight, S.S., Eppig, J.T., et al.: Gene ontology: tool for the unification of biology. Nat. Genet. **25**(1), 25–29 (2000)
6. Banko, M., Cafarella, M.J., Soderland, S., Broadhead, M., Etzioni, O.: Open information extraction from the web. In: Proceedings of the 20th International Joint Conference on Artificial Intelligence, pp. 2670–2676 (2007)
7. Wang, Z., Zhang, J., Feng, J., Chen, Z.: Knowledge graph embedding by translating on hyperplanes. In: Proceedings of the 28th AAAI Conference on Artificial Intelligence, pp. 1112–1119 (2014)
8. Xu, Y., Mou, L., Li, G., Chen, Y., Peng, H., Jin, Z.: Classifying relations via long short term memory networks along shortest dependency paths. In: Proceedings of the Conference on Empirical Methods in Natural Language Processing, pp. 1785–1794 (2015)
9. Chang, C.H., Kayed, M., Girgis, M.R., Shaalan, K.F.: A survey of web information extraction systems. IEEE Trans. Knowl. Data Eng. **18**(10), 1411–1428 (2006)
10. Zesch, T., Müller, C., Gurevych, I.: Extracting lexical semantic knowledge from Wikipedia and Wiktionary. In: Proceedings of the 6th International Conference on Language Resources and Evaluation, pp. 1646–1652 (2008)
11. Bordes, A., Usunier, N., Garcia-Duran, A., Weston, J., Yakhnenko, O.: Translating embeddings for modeling multi-relational data. In: Advances in Neural Information Processing Systems, pp. 2787–2795 (2013)
12. Lin, Y., Liu, Z., Sun, M., Liu, Y., Zhu, X.: Learning entity and relation embeddings for knowledge graph completion. In: Proceedings of the 29th AAAI Conference on Artificial Intelligence, pp. 2181–2187 (2015)

13. Guu, K., Miller, J., Liang, P.: Traversing knowledge graphs in vector space. In: Proceedings of the 2015 Conference on Empirical Methods in Natural Language Processing, Lisbon, Portugal, pp. 318–327. Association for Computational Linguistics, September 2015

14. Lin, Y., Liu, Z., Luan, H., Sun, M., Rao, S., Liu, S.: Modeling relation paths for representation learning of knowledge bases. In: Proceedings of the Conference on Empirical Methods in Natural Language Processing, pp. 705–714 (2015)

15. Mikolov, T., Yih, W.T., Zweig, G.: Linguistic regularities in continuous space word representations. In: Proceedings of the 2013 Conference of the North American Chapter of the Association for Computational Linguistics: Human Language Technologies, pp. 746–751. Association for Computational Linguistics (2013)

16. Mou, L., Men, R., Li, G., Xu, Y., Zhang, L., Yan, R., Jin, Z.: Natural language inference by tree-based convolution and heuristic matching. In: The 54th Annual Meeting of the Association for Computational Linguistics, p. 130 (2016)

17. Hearst, M.A.: Automatic acquisition of hyponyms from large text corpora. In: Proceedings of the 14th Conference on Computational Linguistics-Volume 2, pp. 539–545. Association for Computational Linguistics (1992)

18. dos Santos, C.N., Xiang, B., Zhou, B.: Classifying relations by ranking with convolutional neural networks. arXiv preprint arXiv:1504.06580 (2015)

19. Xu, Y., Jia, R., Mou, L., Li, G., Chen, Y., Lu, Y., Jin, Z.: Improved relation classification by deep recurrent neural networks with data augmentation. In: COLING (2016)

20. Socher, R., Chen, D., Manning, C.D., Ng, A.: Reasoning with neural tensor networks for knowledge base completion. In: Advances in Neural Information Processing Systems, pp. 926–934 (2013)

21. Wang, Z., Li, J.: Text-enhanced representation learning for knowledge graph. In: International Joint Conference on Artificial Intelligence, pp. 1293–1299 (2016)

Balancing Between Cognitive and Semantic Acceptability of Arguments

Hiroyuki Kido[1(✉)] and Keishi Okamoto[2]

[1] Institute of Logic and Cognition, Sun Yat-sen University, Guangzhou, China
kido@mail.sysu.edu.cn
[2] Department of Information Systems, National Institute of Technology,
Sendai College, Sendai, Japan
okamoto@sendai-nct.ac.jp

Abstract. This paper addresses the problem concerning approximating human cognitions and semantic extensions regarding acceptability status of arguments. We introduce three types of logical equilibriums in terms of satisfiability, entailment and semantic equivalence in order to analyse balance of human cognitions and semantic extensions. The generality of our proposal is shown by the existence conditions of equilibrium solutions. The applicability of our proposal is demonstrated by the fact that it detects a flaw of argumentation actually taking place in an online forum and suggests its possible resolution.

1 Introduction

Argumentation is a verbal, social and rational activity [17]. It is also a daily activity in the sense that thinking hard has almost the same meaning as arguing oneself carefully. Meanwhile, it is unclear how one should engage in and facilitate rational argumentation. This is because there seems to be no universal evaluation standard telling what rational argumentation is. However, in this paper, we argue that one aspect of such standards is given by acceptability semantics, e.g., Dung's acceptability semantics [13], as well as more advanced ones, e.g., stage semantics [18], semi-stable semantics [10], ideal semantics [14], CF2 semantics [5] and prudent semantics [12].

These and most acceptability semantics have the language independence principle [4] meaning that they intrinsically refer not to contents, e.g., sentences and words, existing *in* arguments but to relations, e.g., attack and support relations, existing *between* arguments when they define acceptability status of arguments. We thus call them *(formal) semantic acceptability* in the sense that such contents do not directly affect evaluation.[1] By contrast, we call human judgment based on contents of arguments *cognitive acceptability*. Both acceptability might be the same in some situations, but different and moreover incompatible

---

[1] We recognise that sentences existing in arguments indirectly affect semantic acceptability in the sense that they define relations among arguments.

© Springer International Publishing AG 2017
G. Li et al. (Eds.): KSEM 2017, LNAI 10412, pp. 160–173, 2017.
DOI: 10.1007/978-3-319-63558-3_14

in other situations. As an example, let us consider the following two conflicting arguments A and B attached with their votes.

Argument A (having 1001 positive votes). *Veterinarians should have a right to apply animal euthanasia to pets because they must respect pet owners' will.*

Argument B (having 1002 positive votes). *It is not acceptable to allow veterinarians to kill innocent pets because they cannot confirm pets' own will.*

In terms of semantic acceptability, e.g., Dung's semantics, A and B cannot be acceptable at the same time because they are in conflict, i.e., attack each other. However, in terms of cognitive acceptability, A and B should be acceptable at the same time because they have a lot of positive votes. Now, does this situation show that agents who vote for both A and B are irrational or the truth is that there is no attack relation between these arguments? We think neither is true. The conflict is a result of rational judgment caused by focusing on different sources of information: on the one hand, a relation between arguments, and on the other hand, contents of arguments. Then, would it be reasonable to leave the conflict as it is, as the existing contradiction? We think that the acceptability should be at least compatible each other. Therefore, in this paper we investigate what kind of difference exists between cognitive and semantic acceptability, and how we should approximate and resolve them.

A difficulty associated with these questions is how to make a detailed analysis of the difference in a unified way. Our novel approach is to introduce equilibriums to analyse balances between cognitive and semantic acceptability using an entailment relation of propositional logic. In this paper, we divide the difference into three types: satisfiability, entailment and semantic equivalence. We formalise a detection of unsatisfiability, and resolutions of satisfiability, entailment and semantic equivalence. We show that there always exist contraction-based satisfiability, entailment and semantic-equivalence resolutions, and there always exist their expansion-based resolutions under the condition that agent's perceptions do not violate conflict-freeness of acceptability semantics.

The contributions of this paper are as follows. Firstly, to the best of our knowledge, this is the first paper arguing for the need for balancing between agent's perceptions and semantic extensions. Although some recent work covers a change of agent's perceptions and semantic extension, it handles a change of either of them and their balanced change is outside its scope (See Sect. 5). By benefitting from expression and analytical power of propositional logic, this paper makes it possible to handle sensitively balanced contractions and expansions of them (See Sect. 3). Secondly, this paper relates to both theory and practice of computational argumentation. In fact, we demonstrate how our theoretical concepts, i.e., equilibrium notions, highlight flaws of argument actually taking place in an online forum, and suggest their possible resolutions (See Sect. 4).

2   Preliminaries

Dung's acceptability semantics [13] is a general and abstract theory of formal argumentation. It is general in the sense that the semantics reinterprets consequence relations of various approaches for nonmonotonic reasoning. It is abstract in the sense that the semantics is defined on a directed graph, called an abstract argumentation framework, denoted by AF. AF is defined as a pair $\langle Args, Atts \rangle$ where $Args$ is a set of arguments and $Atts \subseteq Args \times Args$ is a binary relation on $Args$ ((A, B) $\in Atts$ means "A attacks B"). Suppose $A \in Args$ and $S \subseteq Args$. S attacks A iff some member of S attacks A. S is conflict-free iff S attacks none of its members. S defends A iff S is conflict-free and S attacks all arguments attacking A. Given AF, Dung's acceptability semantics defines four kinds of sets, called extensions, of acceptable arguments. S is a complete extension iff S is a fixed point of the function $F : Pow(Args) \to Pow(Args)$ where $F(S) = \{a | S$ defends $a\}$ and $Pow(Args)$ is the power set of $Args$. S is a grounded (resp. preferred) extension iff it is the minimum (resp. a maximal) complete extension with respect to set inclusion. S is a stable extension iff it is a complete extension attacking all members in $Args \setminus S$. We assume functions $arg(AF)$ and $att(AF)$ to refer to the set of arguments $Args$ and attack relations $Atts$ of AF, respectively.

Example 1. Let AF be the argumentation framework where $arg(AF) = \{A, B, C, D\}$, and $att(AF) = \{(B, C), (C, B), (C, D), (D, D)\}$. The graph representation of AF and four types of extensions are described in the left and right below, respectively.

AF: $A \quad B \leftrightarrow C \to D$

- Preferred extensions: $\{A, B\}, \{A, C\}$
- Stable extension: $\{A, C\}$
- Grounded extension: $\{A\}$
- Complete extensions: $\{A\}, \{A, B\}, \{A, C\}$

3   Equilibrium-Based Resolutions

3.1   Semantic and Cognitive Acceptabilities

We introduce a propositional language associated with an abstract argumentation framework. It is used to describe acceptability status of arguments in the framework.

Definition 1 (Language). *Let AF be an abstract argumentation framework. A propositional language L_{AF} associated with AF is defined as follows. For all arguments $X \in arg(AF)$, x is an atomic formula of L_{AF}. When x and y are formulas of L_{AF}, $(x \wedge y)$, $(x \vee y)$, $(x \to y)$ and $\neg x$ are formulas of L_{AF}.*

For all atomic formulas $x \in L_{AF}$, "x is true" intuitively means "argument X is acceptable." Given an extension, arguments are acceptable if and only if they are members of the extension. Different two extensions define alternative memberships. Therefore, a logical expression of extensions is defined as follows.

Definition 2 (Extensions). *Let AF be an abstract argumentation framework and $\mathcal{E}$ be a set of extensions of AF. A logical expression of $\mathcal{E}$ is given as follows.*[2]

$$\{\bigvee_{E \in \mathcal{E}} (\bigwedge_{X \in E} x \wedge \bigwedge_{X \in Args \setminus E} \neg x)\}$$

Example 1 (continued). Logical expressions of all types of extensions of AF are given as follows.

- Preferred extensions: $\{(a \wedge b \wedge \neg c \wedge \neg d) \vee (a \wedge \neg b \wedge c \wedge \neg d)\}$
- Stable extension: $\{a \wedge \neg b \wedge c \wedge \neg d\}$
- Grounded extension: $\{a \wedge \neg b \wedge \neg c \wedge \neg d\}$
- Complete extensions: $\{(a \wedge \neg b \wedge \neg c \wedge \neg d) \vee (a \wedge b \wedge \neg c \wedge \neg d) \vee (a \wedge \neg b \wedge c \wedge \neg d)\}$

In what follows, we assume a fixed and arbitrary abstract argumentation framework AF and acceptability semantics ε. We use function $\varepsilon(AF)$ to refer to the logical expression of the set of extensions of AF with respect to ε. We assume a fixed and arbitrary consistent set $\Sigma \subseteq L_{AF}$ of formulas. On the one hand, $\varepsilon(AF)$ is used to represent semantic extensions, and on the other hand, Σ is used to represent cognitive extensions.

3.2 Satisfiability Resolution

This subsection discusses how to detect and resolve incompatibility between cognitive acceptability Σ and semantic acceptability $\varepsilon(AF)$. A minimally unsatisfiable set is a minimal subset of cognitive acceptability that causes incompatibility with semantic acceptability.

Definition 3 (Minimally unsatisfiable set). *$\Sigma_\downarrow \subseteq L_{AF}$ is a minimally unsatisfiable set with respect to AF iff $\Sigma_\downarrow$ is a minimal subset of Σ with respect to $\supseteq$ such that $\varepsilon(AF) \cup \Sigma_\downarrow$ is unsatisfiable.*

An order relation between abstract argumentation frameworks is introduced to refer to maximality and minimality of frameworks.

Definition 4 (Subframework/superframework). *Let AF_i and AF_j be abstract argumentation frameworks. AF_i is a subframework of AF_j (or AF_j is a superframework of AF_i), denoted by $AF_i \sqsubseteq AF_j$, iff $arg(AF_i) \subseteq arg(AF_j)$ and $att(AF_i) \subseteq att(AF_j)$.*

An unsatisfiable core is intuitively a minimal subframework preserving incompatibility existing between AF and Σ.

Definition 5 (Unsatisfiable core). *An abstract argumentation framework $AF_\downarrow$ is an unsatisfiable core of AF with respect to Σ iff there is a minimally unsatisfiable set $\Sigma_\downarrow$ with respect to AF, and $AF_\downarrow$ is a minimal subframework of AF such that $\Sigma_\downarrow$ is a minimally unsatisfiable set with respect to $AF_\downarrow$.*

---

[2] The exclusive OR is strictly appropriate. However, we use OR because of their equivalence.

Example 2. Let us consider the following argumentation framework AF and the satisfiable set $\Sigma = \{a, a \rightarrow b, \neg c\} \subseteq L_{AF}$.

$$AF:\ A \leftrightarrow B \leftrightarrow C$$

Complete extensions of AF is $\{(a \wedge \neg b \wedge c) \vee (\neg a \wedge b \wedge \neg c) \vee (\neg a \wedge \neg b \wedge \neg c)\}$. In this situation, $\Sigma_1 = \{a, a \rightarrow b\}$ and $\Sigma_2 = \{a, \neg c\}$ are minimally unsatisfiable sets with respect to AF. The following AF_1 and AF_2 are the unsatisfiable cores of AF with respect to Σ.

$$AF_1:\ A \leftrightarrow B \qquad\qquad AF_2:\ A \leftrightarrow B \leftrightarrow C$$

A question here is how to resolve incompatibility between cognitive and semantic acceptability. When we consider the situation where semantic acceptability imposes cognitive acceptability to change, a possible resolution is to change the cognitive acceptability. A maximally satisfiable set is a maximal subset of the cognitive acceptability that causes no incompatibility with the semantic acceptability.

Definition 6 (Maximally satisfiable set). *$\Sigma_\downarrow \subseteq \Sigma$ is a maximally satisfiable set with respect to AF iff $\Sigma_\downarrow$ is a maximal subset of Σ with respect to $\supseteq$ such that $\varepsilon(AF) \cup \Sigma_\downarrow$ is satisfiable.*

Meanwhile, when we consider the situation where cognitive acceptability imposes on semantic acceptability to change, a possible resolution is to contract or expand AFs. Each maximally and minimally satisfiable framework represents a contraction and expansion of AFs, respectively.

Definition 7 (Maximally/minimally satisfiable framework). *Let $AF_{\updownarrow}$ be an abstract argumentation framework. $AF_{\updownarrow}$ is a maximally (resp. minimally) satisfiable framework with respect to Σ iff $AF_{\updownarrow}$ is a maximal subframework (resp. minimal superframework) of AF with respect to $\supseteq$ such that $\varepsilon(AF_{\updownarrow}) \cup \Sigma$ is satisfiable.*

A contraction is appropriate when an AF is uncertain in the sense that existence of arguments and attacks is unclear. An expansion is appropriate when an AF is incomplete in the sense that there can exist additional arguments or attacks. A satisfiability resolution is defined as an equilibrium point between cognitive and semantic acceptability.

Definition 8 (Satisfiability resolutions). *Let $AF_{\updownarrow}$ be an abstract argumentation framework and $\Sigma_\downarrow \subseteq \Sigma$ be a set. The pair $(AF_{\updownarrow}, \Sigma_\downarrow)$ is a satisfiability resolution of (AF, Σ) iff $\Sigma_\downarrow$ is a maximally satisfiable set with respect to $AF_{\updownarrow}$ and $AF_{\updownarrow}$ is a maximally or minimally satisfiable framework with respect to $\Sigma_\downarrow$.*

In particular, we call satisfiability resolutions contraction-based when they are based on maximally satisfiable frameworks, and expansions-based when they are based on minimally satisfiable frameworks.

Example 2 (Continued). All possible contraction-based and expansion-based satisfiability resolutions of (AF, Σ) are described in the left and right below, respectively.

$$(A \leftrightarrow B \leftrightarrow C \; , \{a\} \qquad)$$
$$(A \leftrightarrow B \leftrightarrow C \; , \{a \to b, \neg c\} \quad)$$
$$(A \qquad B \leftrightarrow C \; , \{a, a \to b, \neg c\} \;)$$

$$(A \leftrightarrow B \leftrightarrow C \qquad , \{a\} \qquad)$$
$$(A \leftrightarrow B \leftrightarrow C \qquad , \{a \to b, \neg c\} \;)$$
$$(A \overset{\frown}{\leftrightarrow} B \leftrightarrow C \qquad , \{a, \neg c\} \qquad)$$
$$(A \leftrightarrow B \leftrightarrow C \leftarrow D \; , \{a, \neg c\} \qquad)$$

Here, the domain of discourse for possible expansions of AF is assumed to be given by the complete graph AF_D as follows.

$$AF_D: \; A \overset{\longleftrightarrow}{} B \leftrightarrow C \overset{\longleftrightarrow}{} D$$

3.3 Entailment Resolution

This subsection deals with the issue that semantic acceptability cannot entail cognitive acceptability. The first approach for solving this problem is to change cognitive acceptability. We define a maximally entailed set as follows.

Definition 9 (Maximally entailed set). $\Sigma_\downarrow \subseteq \Sigma$ *is a maximally entailed set with respect to AF iff $\Sigma_\downarrow$ is a maximal subset of Σ with respect to $\supseteq$ such that $\varepsilon(AF) \models \Sigma_\downarrow$ holds.*[3]

The second approach is to contract or expand argumentation frameworks defining semantic acceptability. Both maximally and minimally entailing frameworks are minimal changes of AFs whose semantic acceptability entails cognitive acceptability.

Definition 10 (Maximally/minimally entailing framework). *Let $AF_\uparrow$ be an abstract argumentation framework. $AF_\uparrow$ is a maximally (resp. minimally) entailing framework with respect to Σ iff $AF_\uparrow$ is a maximal subframework (resp. minimal superframework) of AF with respect to $\supseteq$ such that $\varepsilon(AF_\uparrow) \models \Sigma$ holds.*

An entailment resolution is defined as an equilibrium point between the acceptabilities.

Definition 11 (Entailment resolutions). *Let $AF_\uparrow$ be an abstract argumentation framework and $\Sigma_\downarrow \subseteq \Sigma$ be a set. The pair $(AF_\uparrow, \Sigma_\downarrow)$ is an entailment resolution of (AF, Σ) iff $\Sigma_\downarrow$ is a maximally entailed set with respect to $AF_\uparrow$ and $AF_\uparrow$ is a maximally or minimally entailing framework with respect to $\Sigma_\downarrow$.*

We call entailment resolutions contraction-based (resp. expansion-based) when they are based on maximally (resp. minimally) entailing frameworks.

Example 2 (Continued). All possible contraction-based and expansion-based entailment resolutions of (AF, Σ) are described in the left and right below, respectively.

---
[3] $\varepsilon(AF) \models \Sigma_\downarrow$ denotes that, for all $x \in L_{AF}$, if $x \in \Sigma_\downarrow$ then $\varepsilon(AF) \models x$.

$$(A \leftrightarrow B \leftrightarrow C \quad , \emptyset \quad)$$
$$(A \leftrightarrow B \leftrightarrow C \quad , \emptyset \quad) \qquad (A \leftrightarrow B \leftrightarrow C \leftarrow D \quad , \{\neg c\} \quad)$$
$$(A \rightarrow B \leftrightarrow C \quad , \{a\} \quad) \qquad (A \leftrightarrow B \leftrightarrow C \quad D \quad , \{a \rightarrow b\} \quad)$$
$$(A \leftrightarrow B \leftarrow C \quad , \{a\} \quad) \qquad (A \leftrightarrow B \leftrightarrow C \leftarrow D \quad , \{a \rightarrow b, \neg c\} \quad)$$
$$(A \leftarrow B \rightarrow C \quad , \{a \rightarrow b, \neg c\} \quad) \qquad (A \leftrightarrow B \leftrightarrow C \quad D \quad , \{a\} \quad)$$
$$(A \quad B \rightarrow C \quad , \{a, a \rightarrow b, \neg c\} \quad) \qquad (A \leftrightarrow B \leftrightarrow C \leftarrow D \quad , \{a, \neg c\} \quad)$$

3.4   Semantic Equivalence Resolution

This subsection asks how to make cognitive and semantic acceptability coincide with each other. The first approach is to change cognitive acceptability. A maximally (resp. minimally) equivalent set is a maximal subset (resp. minimal superset) of the cognitive acceptability that is equivalent to the semantic acceptability.

Definition 12 (Maximally/minimally equivalent set). $\Sigma_\downarrow \subseteq L_{AF}$ *is a maximally (resp. minimally) equivalent set with respect to AF iff $\Sigma_\downarrow$ is a maximal subset (resp. minimal superset) of Σ with respect to $\supseteq$ such that $\Sigma_\downarrow \Leftrightarrow \varepsilon(AF)$ holds.*[4]

The second approach is to change AFs defining semantic acceptability. Both maximally and minimally equivalent frameworks are minimal changes of *AF* whose semantic acceptability is equivalent to the cognitive acceptability.

Definition 13 (Maximally/minimally equivalent framework). *Let $AF_\uparrow$ be an abstract argumentation framework. $AF_\uparrow$ is a maximally (resp. minimally) equivalent framework with respect to Σ iff $AF_\uparrow$ is a maximal subframework (resp. minimal superframework) of AF with respect to $\sqsupseteq$ such that $\Sigma \Leftrightarrow \varepsilon(AF_\uparrow)$ holds.*

A semantic equivalence resolution is formally defined as an equilibrium point between cognitive and semantic acceptability.

Definition 14 (Semantic equivalence resolutions). *Let $AF_\uparrow$ be an abstract argumentation framework and $\Sigma_\downarrow \subseteq L_{AF}$ be a set. The pair $(AF_\downarrow, \Sigma_\downarrow)$ is a semantic equivalence resolution of (AF, Σ) iff $\Sigma_\downarrow$ is a maximally or minimally equivalent set with respect to $AF_\uparrow$ and $AF_\uparrow$ is a maximally or minimally equivalent framework with respect to $\Sigma_\downarrow$.*

Example 2 (Continued). The followings are all contraction-based semantic equivalence resolutions of (AF, Σ).

$$(A \quad B \rightarrow C \quad , \{a, a \rightarrow b, \neg c\} \quad) \qquad (A \qquad , \{a\} \qquad)$$
$$(A \quad B \quad , \{a, a \rightarrow b\} \quad) \qquad (\qquad , \emptyset \qquad)$$

---

[4] $\Sigma_\downarrow \Leftrightarrow \varepsilon(AF)$ denotes that $\Sigma_\downarrow \models \varepsilon(AF)$ and $\varepsilon(AF) \models \Sigma_\downarrow$ hold.

Note that, for example, $(\langle \{A, B, C\}, \{(B, C)\}\rangle, \{a, a \rightarrow b, \neg c, b\})$ is not a contraction-based semantic equivalence resolution. Meanwhile, there is no expansion-based semantic equivalence resolution of (AF, Σ). So, we here consider another example with $\Sigma^* = \{a \lor b, c\}$. We assume a set $\Sigma_D \subseteq L_{AF_D}$ of a domain discourse for possible expansions of Σ^* where AF_D is the complete graph as shown in the previous example. There is the following expansion-based semantic equivalence resolution of (AF, Σ^*).

$$(A \leftrightarrow B \overleftrightarrow{} C \quad D \qquad \{a \lor b, c, \neg b \land d\})$$

4 Generality and Applicability

4.1 Characterising Existence of Resolutions

In this subsection, we show the necessary and sufficient conditions of the existence of our equilibrium-based resolutions. We use the model checking techniques [8] to analyse argumentation theoretic properties of equilibriums. According to the literature, a set $S \subseteq arg(AF)$ of arguments is conflict-free iff the set $\{x \in L_{AF}|X \in S\}$ is a model of the formula Φ_{AF} defined as follows.

$$\Phi_{AF} = \bigwedge_{A \in arg(AF)} (a \rightarrow \bigwedge_{B:(B,A)\in att(AF)} \neg b)$$

We divide expansion-based and contraction-based resolutions into three types: Σ fixation, Σ expansion and Σ contraction. The first theorem supposes the situation where Σ is fixed in expansion-based resolutions.

Theorem 1. *For any AF and consistent set $\Sigma \subseteq L_{AF}$, there exist abstract argumentation frameworks $AF_\uparrow$ such that $(AF_\uparrow, \Sigma)$ are expansion-based satisfiability (entailment, respectively) resolutions of (AF, Σ) iff Σ do not violate the conflict-free property, i.e., $\Sigma \not\models \neg\Phi_{AF}$.*

There does not always exist an expansion-based semantic equivalence resolution $(AF_\uparrow, \Sigma)$ even when Σ do not violate the conflict-free property. An example case is when $AF = \langle\{A\}, \emptyset\rangle$ and $\Sigma = \emptyset$. The next two theorems suppose the situations where Σ can be expanded and contracted in expansion-based resolutions, respectively.

Theorem 2. *For any AF and consistent set $\Sigma \subseteq L_{AF}$, there exist abstract argumentation frameworks $AF_\uparrow$ and supersets $\Sigma_\uparrow$ of Σ such that $(AF_\uparrow, \Sigma_\uparrow)$ are expansion-based satisfiability (entailment and semantic equivalence, respectively) resolutions of (AF, Σ) iff Σ do not violate the conflict-free property, i.e., $\Sigma \not\models \neg\Phi_{AF}$.*

Theorem 3. *For any AF and consistent set $\Sigma \subseteq L_{AF}$, there exist abstract argumentation frameworks $AF_\uparrow$ and subsets $\Sigma_\downarrow$ of Σ such that $(AF_\uparrow, \Sigma_\downarrow)$ are expansion-based satisfiability (entailment, respectively) resolutions of (AF, Σ).*

Note that Theorems 2 and 3 include Theorem 1, i.e., the case of $\Sigma_\downarrow = \Sigma$ and $\Sigma_\uparrow = \Sigma$, as special cases, respectively. The next theorem regarding contraction-based resolutions supposes the situation where Σ is fixed.

Theorem 4. *For any AF and consistent set $\Sigma \subseteq L_{AF}$, there exists an abstract argumentation framework $AF_\downarrow$ such that $(AF_\downarrow, \Sigma)$ is a contraction-based satisfiability resolution of (AF, Σ).*

There does not always exist an entailment and semantic equivalence resolution. For example, when $AF = \langle \{A\}, \emptyset \rangle$ and $\Sigma = \{\neg a\}$, there is neither entailment nor semantic equivalence resolution. The next two theorems suppose the situations where Σ can be expanded and contracted in contraction-based resolutions, respectively.

Theorem 5. *For any AF and consistent set $\Sigma \subseteq L_{AF}$, there exist an abstract argumentation framework $AF_\downarrow$ and a superset $\Sigma_\uparrow$ of Σ such that $(AF_\downarrow, \Sigma_\uparrow)$ is a contraction-based satisfiability resolution of (AF, Σ).*

Theorem 6. *For any AF and consistent set $\Sigma \subseteq L_{AF}$, there exist abstract argumentation frameworks $AF_\downarrow$ and subsets $\Sigma_\downarrow$ of Σ such that $(AF_\downarrow, \Sigma_\downarrow)$ are contraction-based satisfiability (entailment and semantic equivalence, respectively) resolutions of (AF, Σ).*

Table 1 shows the summary of the above mentioned theorems in terms of change of Σ. Symbol "✓" means that there exists a resolution. Meanwhile, "(cond.)" means that there exists a resolution if and only if Σ does not violate the conflict-free property. These facts show generality of equilibriums-based approaches in the sense that they have the ability to give resolutions for all types of equilibriums.

Table 1. Summary of the existence of resolutions.

	Expansion-based resolutions			Contraction-based resolutions		
	Σ fixation (Theorem 1)	Σ expansion (Theorem 2)	Σ contraction (Theorem 3)	Σ fixation (Theorem 4)	Σ expansion (Theorem 5)	Σ contraction (Theorem 6)
Satisfiability	✓ (cond.)	✓ (cond.)	✓	✓	✓	✓
Entailment	✓ (cond.)	✓ (cond.)	✓			✓
Equivalence		✓ (cond.)				✓

4.2 Application Illustration in Online Forum

Application potential is one of important criteria for evaluating accuracy of research proposals. This subsection illustrates applicability of our equilibrium based resolutions to argumentation analysis in online debate forums.

CreateDebate [1] is an online forum preserving a number of written arguments, casted votes, and relationships, e.g., dispute, support and clarification relations, among arguments. On the left in Fig. 1, we show a whole dispute

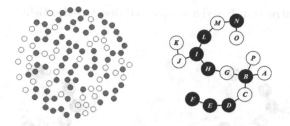

Fig. 1. The left graph shows a whole dispute relations and a summary of positive and negative votes. The right graph shows an argumentation framework constructed from the left graph.

structure and a summary of users' votes that are both extracted from argument on abortion actually taking place in CreateDebate. The edges represent the whole dispute relations and the white (resp. black) nodes represent that they get user's positive votes above (resp. below) average. On the right in Fig. 1, we show a partial graph of the left that especially focuses on discussing a specific subject on abortion. We represent it as an abstract argumentation framework, denoted by AF, with a symmetric attack relation, i.e., if $(X, Y) \in att(AF)$ then $(Y, X) \in att(AF)$, for all $X, Y \in arg(AF)$. We assume that $\varepsilon(AF)$ represents the logical expression of the set of preferred extensions. We define agents' cognitions Σ from the votes to individual arguments in such a way that $x \in \Sigma$ holds when node X is white and $\neg x \in \Sigma$ holds when node X is black.

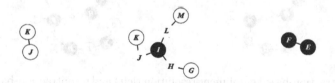

Fig. 2. The inconsistent cores of the argumentation framework shown in the right in Fig. 1.

Figure 2 shows all unsatisfiable cores of AF with respect to Σ.[5] It is observed that $\{k, j\}$, $\{g, \neg i, k, m\}$ and $\{\neg e, \neg f\} \subseteq \Sigma$ are minimally unsatisfiable subsets of Σ with respect to AF. The inconsistent cores verify consistency between acceptance of arguments coming from the argumentation structure and user's votes. They are useful in the sense that they tell where people should pay attention to resolve incompatibility between agent cognitions and semantic extensions.

A consistent core is obtained by eliminating attacks (K, J), (J, K), (E, F), $(F, E) \in att(AF)$ from AF. Given the AF, the left and right graphs in

---

[5] The rightmost graph is not an unsatisfiable core when complete extensions are assumed.

Fig. 3 show a contraction-based and an expansion-based satisfiability resolutions, respectively.

Fig. 3. A contraction-based (left) and an expansion-based (right) satisfiability resolutions.

Note that Σ has no change in the contraction-based resolution, although it changes to $\Sigma \setminus \{k, \neg e\}$ in the expansion-based resolution. The satisfiability resolutions facilitate rational argumentation in the sense that they suggest simultaneous updates on both argumentation frameworks and users' cognitions.

Given the expansion-based resolution, Fig. 4 shows a contraction-based entailment resolution and a contraction-based semantic equivalence resolution.

Fig. 4. A contraction-based entailment resolution (left) and a contraction-based semantic equivalence resolution (right).

Both resolutions change their attack relations by changing some mutual attacks to one-directional. Note that the entailment resolution has the contraction $\Sigma \setminus \{\neg f\}$, but the semantic equivalence resolution has the expansion $\Sigma \cup \{e, \neg k, q\}$. Note that a semantic equivalence resolution is stricter than an entailment resolution, and an entailment resolution is stricter than a satisfiability resolution. Each type of resolution corresponds to agent's attitude of how much closeness between cognitive and semantic acceptability they require.

5   Conclusions and Discussion

Our motivation of this work is that, as with us human beings, agent's perceptive development can be achieved by resolving the gaps between cognitive and semantic acceptability. In this paper, we introduced equilibrium-based satisfiability resolutions, entailment resolutions and semantic equivalence resolutions. We showed generality of our proposal by proving existence conditions of their resolutions. We demonstrated its applicability to online forums. The limitations of this paper is that our equilibrium-based resolutions mention nothing about sentences additional arguments should have. Our expansion-based resolutions only mention about an attack relation that additional arguments should have with other arguments.

This paper relates to research studies of belief revisions on abstract argumentation frameworks (AFs). [11] proposes a typology of revisions of AFs in terms of changes on extensions. [6] addresses a problem of how to modify AFs so that a desired set of arguments becomes an extension of revised AFs. [7] handles changes of AFs by introducing several revision operators that satisfy AGM (Alchourrón, Gärdenfors, and Makinson) postulates [2]. These research studies commonly focus on changes of AFs. In contrast, [16] focuses on belief revisions on agents' perceptions about acceptability status of arguments, and provide web-platform *ArgTech* that alerts users to irrational agents' perceptions through incompatibility checking with the complete labelling [3]. [9,15] focus on changes of both AFs and agents' perceptions, and deal with resolutions of incoherence of agent's beliefs by expanding AF and constraints on its outcomes. In contrast, we primarily focus on balancing between perceptions (or constraints) and outcomes of argumentation. We thus introduce equilibrium mechanisms to handle simultaneous changes of them although the related work handles, at most, each of them. Notice that the resolutions analysed in Sect. 4.2 can only be achieved as an equilibrium, and a change of either AF or agents' perceptions is a special case of our equilibrium-based resolutions. Moreover, although the related work focuses on expansions and incoherence ("satisfiability" in this paper), this paper further deals with contraction and introduces entailment and semantic equivalence resolutions.

Acknowledgments. The authors thank Dr. Martin Caminada for his valuable discussion. This work was supported by JSPS KAKENHI Grant Number 15KT0041.

A   Proofs

Proof. (Theorem 1) ($\Rightarrow$) Assume $\Sigma \models \neg\Phi_{AF}$. $\Phi_{AF_\uparrow} \models \Phi_{AF}$ holds because $att(AF_\uparrow) \supseteq att(AF)$ holds, for all $AF_\uparrow \sqsupseteq AF$. Thus, $\Sigma \models \neg\Phi_{AF_\uparrow}$ holds. Since $\varepsilon(AF_\uparrow) \models \Phi_{AF_\uparrow}$ holds, $\{\Phi_{AF_\uparrow}\} \cup \Sigma$ is unsatisfiable and $\varepsilon(AF_\uparrow) \not\models \Sigma$ holds. Thus, there is neither minimally satisfiable framework nor minimally entailing framework with respect to Σ. ($\Leftarrow$) It is sufficient to show entailment resolutions. Assuming that $\Sigma \not\models \neg\Phi_{AF}$ holds, there is a model M of Σ and Φ_{AF}, i.e., $M \models \Sigma \cup \{\Phi_{AF}\}$. We want to show that there is $AF_\uparrow \sqsupseteq AF$ such that, for all $x \in M$ (resp. $x \notin M$), $\varepsilon(AF_\uparrow) \models x$ (resp. $\varepsilon(AF_\uparrow) \models \neg x$) hold. Consider $AF_\Uparrow$ adding a new argument Y attacking only all arguments X where $x \notin M$ holds. Obviously, $\varepsilon(AF_\Uparrow) \models \neg x$, for all $x \notin M$. Consider $AF_\uparrow$ adding another new argument Z attacking only all arguments attacking X where $x \in M$ holds. Here, since $M \models \Phi_{AF}$ holds, for all $(X, Y) \in att(AF)$, if $y \in M$ then $x \notin M$, and vice versa. Thus, $\varepsilon(AF_\uparrow) \models x$, for all $x \in M$. □

Proof. (Theorems 2, 3, 4, 5 and 6 (Sketch)) Similar to Theorem 1 or obvious from the case $AF = \langle \emptyset, \emptyset \rangle$ or $\Sigma = \emptyset$. □

References

1. CreateDebate: Debate forum, online debate community (2006). www.createdebate. com/
2. Alchourrón, C.E., Gärdenfors, P., Makinson, D.C.: On the logic of theory change: partial meet contraction and revision functions. J. Symbolic Logic **50**(2), 510–530 (1985)
3. Baroni, P., Caminada, M., Giacomin, M.: An introduction to argumentation semantics. Knowl. Eng. Rev. **26**, 365–410 (2011)
4. Baroni, P., Giacomin, M.: On principle-based evaluation of extension-based argumentation semantics. Artif. Intell. **171**(10–15), 675–700 (2007)
5. Baroni, P., Giacomin, M., Guida, G.: Scc-recursiveness: a general schema for argumentation semantics. Artif. Intell. **168**(1–2), 162–210 (2005)
6. Baumann, R., Brewka, G.: Expanding argumentation frameworks: enforcing and monotonicity results. In: Proceedings of the 3rd International Conference on Computational Models of Argument, pp. 75–86 (2010)
7. Baumann, R., Brewka, G.: AGM meets abstract argumentation: expansion and revision for dung frameworks. In: Proceedings of the 24th International Conference on Artificial Intelligence, pp. 2734–2740 (2015)
8. Besnard, P., Doutre, S.: Checking the acceptability of a set of arguments. In: Proceedings of the 10th International Workshop on Non-Monotonic Reasoning, pp. 59–64 (2004)
9. Booth, R., Kaci, S., Rienstra, T., van der Torre, L.: A logical theory about dynamics in abstract argumentation. In: Liu, W., Subrahmanian, V.S., Wijsen, J. (eds.) SUM 2013. LNCS (LNAI), vol. 8078, pp. 148–161. Springer, Heidelberg (2013). doi:10.1007/978-3-642-40381-1_12
10. Caminada, M.: Semi-stable semantics. In: Proceedings of the 1st International Conference on Computational Models of Argument, pp. 121–130 (2006)
11. Cayrol, C., de Saint-Cyr, F.D., Lagasquie-Schiex, M.C.: Change in abstract argumentation frameworks: adding an argument. J. Artif. Intell. Res. **38**, 49–84 (2010)
12. Coste-Marquis, S., Devred, C., Marquis, P.: Prudent semantics for argumentation frameworks. In: Proceedings of the 17th International Conference on Tools with Artificial Intelligence, pp. 568–572 (2005)
13. Dung, P.M.: On the acceptability of arguments and its fundamental role in non-monotonic reasoning, logic programming, and n-person games. Artif. Intell. **77**, 321–357 (1995)
14. Dung, P.M., Mancarella, P., Toni, F.: A dialectic procedure for sceptical, assumption-based argumentation. In: Proceedings of the 1st International Conference on Computational Models of Argument, pp. 145–156 (2006)
15. Rienstra, T.: Argumentation in flux: modelling change in the theory of argumentation. Dissertations and theses: Doctoral thesis. University of Luxembourg, Luxembourg, Luxembourg (2014)
16. Schulz, C., Dumitrache, D.: The ArgTeach web-platform. In: Proceedings of the 6th International Conference on Computational Models of Argument, pp. 475–476 (2016)

17. van Eemeren, F.H., Grootendorst, R., Henkemans, F.S.: Fundamentals of Argumentation Theory: A Handbook of Historical Backgrounds and Contemporary Developments. Routledge, Abingdon (1996)
18. Verheij, B.: Two approaches to dialectical argumentation: admissible sets and argumentation stages. In: Proceedings of the 8th Dutch Conference on Artificial Intelligence, pp. 357–368 (1996)

Discovery of Jump Breaks in Joint Volatility for Volume and Price of High-Frequency Trading Data in China

Xiao-Wei Ai[1(✉)], Tianming Hu[2], Gong-Ping Bi[1], Cheng-Feng Lei[1], and Hui Xiong[3]

[1] Nanchang Hangkong University, Nanchang, China
axw010306@126.com
[2] Dongguan University of Technology, Dongguan, China
[3] Rutgers University, New Brunswick, USA

Abstract. Recent years have witnessed more and more frequent abnormal fluctuations in stock markets and thus it is important to real-time monitor dynamically such fluctuations. To that end, this paper first proposes a realized trading volatility (RTV) model and analyzes its properties. Next, based on the RTV model, it develops a critical jump point test for the joint volatility of volume and price using matrix singular values. Finally, the proposed models are evaluated on the minute transaction data of China's Shanghai and Shenzhen A-share stock markets over 2009.01.05–2009.03.31. With the PV, VV and RTV sequence values extracted from the transaction data, case studies are performed on certain stocks and empirical suggestions are offered for the maintenance of the stability of the market index.

Keywords: Realized trading volatility · Matrix singular values · Matrix norm · Multivariate statistical

1 Introduction

Volatility in stock markets refers to the variance range of stock trading. It is characteristic of stock market. Volatility within certain range keeps stock market running properly. However, if share price changes sharply, from a market macro-structure point of view, the risk of the whole financial system will be increased. From a micro-structure point of view, the investors will face a larger risk and lose confidence in the market in the long run. Therefore, many countries emphasize on the monitoring of stock market.

The volatility behavior of the stock return rates is characterized by its continuity and jumps. Continuity means the stock price and return change stably most of time. However, at certain times they may abruptly change radically upward or downward, which is called jumps in financial econometrics. Although jumps happen rarely, when they do happen, they make huge impact on the stock market. Hence, the detection and regulation of jumps is an important problem for both theory and practice.

Recent studies on jumps of high frequency financial data can be divided into two classes. One views financial sequences as random processes and separates the continuous diffusion from the pure jumps. For instance, the RRV model was proposed in [1, 2], the

© Springer International Publishing AG 2017
G. Li et al. (Eds.): KSEM 2017, LNAI 10412, pp. 174–182, 2017.
DOI: 10.1007/978-3-319-63558-3_15

RBV model was proposed in [3] and improved in [4]. [5] offered an empirical analysis for jumping volatility. [6] compared jumping volatility for a few models. The other class directly tests the jump point by improving the classic RV model. [7] proposed the HAR-RV-J, which was later improved in [8] to detect change points with penalty functions. [9, 10] developed the point-by-point methods for jump test.

Although these models are able to offer interpretation for volatility jumps of stock return, all of them used stock index. In contrast, this paper focuses on specific stocks and analyzes their trading data.

With the advent of network, it has been easier for the institutions and large traders to cover their manipulation of stock trading and insider trading. However, as long as it is abnormal trading, there will be traces left in the transaction data. Take large traders for example. Their operation can be divided into four phases: accumulation, washing, elevating, and shipping. Usually the jump happens on the eve of washing and shipping.

The remainder of the paper is organized as follows. First we introduce a Realized Trading Volatility (RTV) model for price and volume of high frequency trading data. Based on the RTV model, we develop a point-by-point detection method for critical jump points of return volatility using matrix singular values. Finally, with the minute transaction data of the Shanghai and Shenzhen A-share stock market from Jan 5 2009 to Mar 31 2009, we perform empirical analysis of certain stocks with high volatility.

2 Bivariate Normal Distribution

Suppose bivariate normal random variable $X - \begin{pmatrix} X_1 \\ X_2 \end{pmatrix} \sim N_2(\mu, \Sigma)$, with density function:

$$f(x) = \frac{1}{2\pi} \cdot \frac{1}{\sqrt{|\Sigma|}} \cdot \exp\left\{ -\frac{1}{2}(x-\mu)' \cdot \sum^{-1} \cdot (x-\mu) \right\}$$

where $x = \begin{pmatrix} x_1 \\ x_2 \end{pmatrix}$, $\mu = \begin{pmatrix} \mu_1 \\ \mu_2 \end{pmatrix}$ covariance matrix $\sum = \begin{pmatrix} \sigma_{11} & \sigma_{12} \\ \sigma_{21} & \sigma_{22} \end{pmatrix}$, expectation $\mu_i = E(X_i)$, covariance $\sigma_{ij} = \text{cov}(X_i, X_j)$, $i,j = 1,2$, correlation coefficient (between X_1 and X_2): $\rho = \frac{\sigma_{12}}{\sqrt{\sigma_{11}} \cdot \sqrt{\sigma_{22}}}$.

Lemma 1. For $x \sim N_2(\mu, \Sigma)$, hyper-elliptic

$$D = \{x : (x-\mu)' \cdot \Sigma^{-1} \cdot (x-\mu) \le \chi^2_{1-\alpha}(2)\}$$

we have

1. $(x-\mu)' \cdot \Sigma^{-1} \cdot (x-\mu) \sim \chi^2(2)$.
2. $P\{x \in D\} = 1 - \alpha$.
3. For any area D*, if $P\{x \in D^*\} = 1 - \alpha$, then hyper-elliptic D's volume $VD \le$ D*'s volume V*. Among all D for which $P\{x \in D\} = 1 - \alpha$ holds, the hyper-elliptic D's volume is the smallest.

4. Suppose V^* is the volume of the unit hypersphere, $D_* = \left\{ (x_1, x_2) : x_1^2 + x_2^2 \leq 1 \right\}$, we have $V_D = \int_D dx = V^* \cdot \chi_{1-\alpha}^2(2) \cdot \sqrt{|\Sigma|}$. It means a smaller $|\Sigma|$ leads to a smaller V_D and a denser distribution of X.

3 Realized Trading Volatility of Price and Volume

3.1 Price Volatility

Let P(t) denote the settlement price of some stock on t-th day, $p_j(t)$ the share price for the j-th transaction on t-th day, $r_j(t) = \log p_j(t) - \log P(t-1)$ the logarithmic yield of the j-th transaction on t-th day. Let T denote the number of transactions on the t-th day, then the average logarithmic yield on the t-th day is:

$$\overline{r(t)} = \frac{1}{T} \sum_{j=1}^{T} r_j(t)$$

Definition 1. The stock's Price Volatility (PV) on the t-th day is defined as the variance of its logarithmic yield, as in Eq. (1).

$$PV(t) = \frac{1}{T-1} \sum_{j=1}^{T} [r_j(t) - \overline{r(t)}]^2 \tag{1}$$

3.2 Volume Volatility

Let Q denote the number of circulating shares of some stocks, $q_j(t)$ the turnover of the j-th transaction, $h_j(t) = q_j(t)/Q$ the turnover rate, then the average turnover rate is:

$$\overline{h(t)} = \frac{1}{T} \sum_{j=1}^{T} h_j(t)$$

Definition 2. The stock's Volume Volatility (VV) is defined as the variance of the turnover rate, as in Eq. (2).

$$VV(t) = \frac{1}{T-1} \sum_{j=1}^{T} [h_j(t) - \overline{h(t)}]^2 \tag{2}$$

3.3 Volatility Rate of Price and Volume of Realized Trading

To reflect the joint volatility between trading price and volume, we propose a new model: Realized Trading Volatility.

Definition 3. The stock's realized trading volatility is defined as RTV = $c \cdot f(x, y)$,

$$c = 1/[2(1 - \rho2) \ln 20], f(x, y) = \frac{(x - \mu_1)^2}{\sigma_1^2} - \frac{2\rho(x - \mu_1)(y - \mu_2)}{\sigma_1\sigma_2} + \frac{(y - \mu_2)^2}{\sigma_2^2} \quad (3)$$

the random variable x, y denote the price volatility and the volume volatility respectively, $\mu_1 = E(x), \mu_2 = E(y), \sigma_1^2 = D(x), \sigma_2^2 = D(y), \rho$ is the correlation coefficient.

From Lemma 1, let RTV(t) = $c \cdot f[$ PV(t), VV(t)$]$, then $P\{(x, y) \in D\} = 95\%$ i.e. PV and VV fall in D 95% of the time. This means RTV is suitable to describe the trading volatility.

Below we seek extreme values of f(x, y).

$$\begin{cases} \frac{\partial f}{\partial x} = \frac{2(x - \mu_1)}{\sigma_1^2} - \frac{2\rho(y - \mu_2)}{\sigma_1\sigma_2} = 0 \\ \frac{\partial f}{\partial y} = \frac{2(y - \mu_2)}{\sigma_2^2} - \frac{2\rho(x - \mu_1)}{\sigma_1\sigma_2} = 0 \end{cases}$$

Solving the above equation, we have extreme values of f(x, y) at $\rho^2 = 1$. At this time, Y is linearly dependent on X. Let $Y = aX + b$, $\mu_1 = E(X), \mu_2 = E(Y) = a\mu_1 + b$, $\sigma_1^2 = D(X), \sigma_2^2 = D(Y) = a^2\sigma_1^2$. Substituting them into f(x, y) gives the following theorem:

Theorem 1. When $\rho = 1, f_{\min}(x, y) = 0$;
When When $\rho = -1, f_{\max}(x, y) = 4(x - \mu_1)^2/\sigma_1^2$

This shows that when $\rho = 1$ the price volatility is positively correlated with the volume volatility. At this time, there is no trading at daily limit. When $\rho = -1$ the price volatility is negatively correlated with the volume volatility. At this time, the price goes from limit up to limit down.

4 The Jump Point Model for High-Frequency Trading Volatility Break

Suppose m-dimensional random vector $X_1, X_2, \cdots, X_m$ has been observed n times and we record the results in matrix

$$A = (a_{ij})_{m \times n} = (\alpha_1, \cdots, \alpha_n) \in R^{m \times n}$$

We perform singular decomposition for A:

$$A = U \cdot \begin{pmatrix} \Sigma_r \\ & 0 \end{pmatrix} \cdot V^T$$

Then the singular values of A is, $\sigma_1 \geq \cdots \geq \sigma_r > 0$, r = Rank(A) ,F norm of A is:

$$\|A\|_F^2 = \|\alpha_1\|_2^2 + \cdots + \|\alpha_n\|_2^2 = \sigma_1^2 + \cdots + \sigma_r^2$$

We have

$$r \cdot \sigma_r^2 \leq \|\alpha_1\|_2^2 + \cdots + \|\alpha_n\|_2^2 \leq r \cdot \sigma_1^2$$

If $X_1, X_2, \cdots, X_m \sim N(0, 1)$, then $X_1^2 + \cdots + X_m^2 \sim \chi^2(m)$
For the given significance value $1 - \varepsilon$, we have

$$P\{\sum_{i=1}^{m} X_i^2 \leq \chi_{1-\varepsilon/2}^2 \text{ or } \sum_{i=1}^{m} X_i^2 \geq \chi_{\varepsilon/2}^2\} = \varepsilon$$

Let the realized trading volatility for a certain stock on day t be denoted by $X(t) = \begin{pmatrix} PV(t) \\ VV(t) \\ RTV(t) \end{pmatrix}$, Observing n times for X gives:

$$A(t) = (a_{ij})_{3 \times n} = \begin{pmatrix} PV(t-n+1) & \cdots & PV(t) \\ VV(t-n+1) & \cdots & VV(t) \\ RTV(t-n+1) & \cdots & RTV(t) \end{pmatrix}_{3 \times n}$$

where PV(t), VV(t), RTV(t) denote the price volatility, volume volatility, and the realized trading volatility, respectively. Since all of them follow the normal distribution, we make the following conclusion:

Theorem 2.

$$P\{\|X(t)\|_2^2 \leq \frac{r}{n} \cdot \sigma_r^2 \text{ or } \|X(t)\|_2^2 \geq \frac{r}{n} \cdot \sigma_1^2\} = \frac{r}{n}$$

From Theorem 2, we get two jump points for the volatility change of high-frequency trading stocks. When $\|X(t)\|_2^2 \geq r \cdot \sigma_1^2/n$, X(t) is called jump upward point, and it is on the eve of washing. When $\|X(t)\|_2^2 \leq r \cdot \sigma_r^2/n$, X(t) is called jump downward point, and it is on the eve of shipping.

5   The Algorithm of Point-by-Point Test for Jump Critical Points

It takes a long time, e.g., one to a few years, for the large traders to manipulate the stocks, from accumulation, washing, elevating to shipping. In such a period, there are only a few jump critical points, e.g., 5–10 points. To test the points accurately, we take 30 days as sample size (n = 30).

Algorithm 1: point-by-point test for jump critical points based on singular values

---

Input: From day t, take trading volatility values for the successive n days:

$$A(t) = (a_{ij})_{3 \times n}$$

The trading volatility values for day t+1 are:

$$X(t+1) = \left(PV(t+1), VV(t+1), RTV(t+1) \right)'$$

Output: iReturn   /* return value: 0 indicates not jump point for day t+1*/

/* 1 indicates jump upward critical point for day t+1*/

/* -1 indicates jump downward critical point for day t+1*/

Step:

1. Initialization: set_day = n   /* number of days for point-by-point test */

$\overline{X(t)}=0$ /* the average of the successive n days for day t */

iReturn = 0   /*return value: 0 indicates not jump point*/

2. For t =1 to set_day

Compute eigen-values for $A(t)' \cdot A(t)$: $\lambda_1 \geq \cdots \geq \lambda_r > 0, \lambda_{r+1} = \cdots = \lambda_n = 0$

Compute singular values for $A(t)$: $\sigma_1 \geq \sigma_2 \geq \cdots \geq \sigma_r > 0$, where $\sigma_i = \sqrt{\lambda_i}$

$$\overline{X(t)} = \sum_{i=1}^{3} \sum_{j=1}^{n} a_{ij} / n \quad \text{/* the average of the successive n days for day t*/}$$

$S(t+1) = X(t+1)' \cdot X(t+1)$ /*norm of trading volatility vector for day t+1*/

If $S(t+1) \geq 3\sigma_1^2 / n$ Then

iReturn = 1   /*jump upward critical point*/

Return iReturn   /*end of algorithm*/

End

If $S(t+1) \leq 3\sigma_r^2 / n$ Then

iReturn = -1   /*jump downward critical point*/

Return iReturn   /*end of algorithm*/

End

Next

---

6 The Empirical Analysis of Jump Critical Points

We collect the minute trading data of 888 stocks from the Shanghai and Shenzhen A-share stock markets in China from Jan 5 2009 to Mar 31 2009. Then we select a few stocks with high volatility and perform simulation analysis. In detail, from the

3-dimensional volatility data for 30 consecutive days, we compute the singular values, compare them with the values of the 31st day, and determine if the 31st day is the jump critical point. This operation is repeated point-by-point for every point in the trading sequence.

Below is the case for stock SZ000718. From Jan 5, 2009 to Mar 31, 2009, despite small volatility in 2 days, it is stable most of the time. On Mar 23, 2009, however, there was huge volatility, apparently indicating a jump critical point.

To find the reasons for the abnormal fluctuation of SZ000718 on Mar 23, 2009, we checked the details of its minute transaction data. Over the previous month, neither the price volatility nor the volume volatility of its trading is significant, both ranging from 0.01% to 0.1%. On Mar 23, 2009, however, there came a sudden increase in the volume volatility. At 10:38, a trading volume of 2347.1 K was observed with 21506.9 K in value. At 14:00, we saw another huge trading volume of 5768.6 K with 52897.6 K in value. As a consequence, the stock price was raised 9.95% that day, indicating an obvious price lift. This case shows that due to the daily price limit, for the manipulators to evade regulation, it is easier to manipulate trading volume that has no limit. However, as the trading volume goes high, even a small amount of price volatility will magnify the value of RTV, which is beyond the manipulators' control (Fig. 1 and Table 1).

Table 1. The realized trading volatility for stock SZ000718

Date	PV	VV	RTV
2009.01.05	0.00482	0.000081	0.10591
2009.01.06	0.00277	0.000001	0.00826
...	...	...	...
2009.03.20	0.02301	0.00024	0.02602
2009.03.23	0.13051	0.16905	8.70686
2009.03.24	0.01402	0.00217	0.42850
...	...	...	...
2009.03.31	0.03499	0.00016	0.19356
Avg	0.02151	0.00061	0.32795

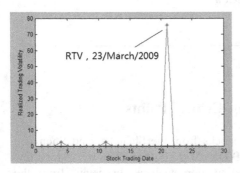

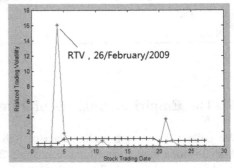

Fig. 1. The jump critical point for SZ000718   **Fig. 2.** The jump critical point for SH600537

Figure 2 illustrates the case for stock SH600537. Within the test period, its price and turnover rate varied in [4, 5] and [1%, 2.5%], respectively. The jump point is detected on Feb 26 2009. The stock price was elevated drastically half a year later to 60 and went down slowly afterward.

Our method works for stocks with varying marketable volume. Figure 3 shows the result for stock SH600146 whose marketable volume is 0.2 billion. Figure 4 shows the result for stock SH600028 whose marketable volume is 95.5 billion.

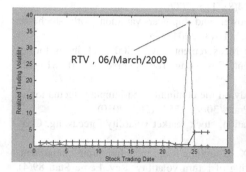
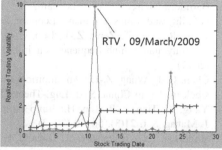

Fig. 3. The jump critical point for SH600146   **Fig. 4.** The jump critical point for SH600028

7 Conclusion

It is a challenge in both theory and practice to recognize and monitor dynamically abnormal volatility caused by anomalous trading. This paper first proposed the RTV model, and then developed a test for jump critical point of high-frequency trading volatility based on α percentiles of matrix singular values. They were evaluated over high-frequency trading data of Shanghai and Shenzhen A-share stock markets over half a year. The following empirical conclusions can be drawn from the case studies.

- It is suitable to use 30 consecutive trading days as the time window.
- In point-by-point test in a period (e.g., a month), a stock may exhibit more than one jump critical points, as shown in Fig. 4. Although some signals are of moderate volatility, those jump points with abrupt radical change must be of significant information that will be reflected in later trading.
- Our evaluation only identified critical points for jump upward and no jump downward. This may be dependent on the specific data we chose and remains a question for future study.

In summary, the case studies demonstrated that due to the power to capture the joint volatility of price and volume, the RTV model and the resultant test using matrix singular values are capable of capturing the jump critical points for both blue-chip and small-cap stocks. In the future, we plan to improve the model for clearer signals of jump points.

Acknowledgments. This work was supported by Department of Education of Jiangxi Province, PR China, through Grant No. GJJ14525.

References

1. Martens, M., Dijk, D.: Measuring volatility with the realized range. J. Econom. **138**(1), 181–207 (2007)
2. Tang, Y., Zhang, S.-Y.: Weighted realized range-based volatility based on high-frequency data and its empirical analysis. Syst. Eng. **24**(8), 1–5 (2006)
3. Barndorff-Nielsen, O.E., Shephard, N.: Power and bipower variation with stochastic volatility and jumps. J. Financ. Econom. **2**(1), 1–48 (2004)
4. Long, R., Xie, C., Zeng, Z.-J., Luo, C.-Q.: Measurement of CSI 300 stock index futures volatility under high-frequency environment. Syst. Eng.-Theory Pract. **31**(5), 813–822 (2011)
5. Chen, G.-J., Wang, Z.-H.: An empirical study on the continuity and jumping fluctuation of stock market in China. Syst. Eng.-Theory Pract. **30**(9), 1554–1562 (2010)
6. Yang, K., Tian, F.-P., Lin, H.: Jump estimation, stock market volatility forecasting. Chin. J. Manag. Sci. **21**(3), 50–60 (2013)
7. Andersen, T.G., Bollerslev, T., Diebold, F.X.: Roughing it up: including jump components in the measurement, modeling and forecasting of return volatility. Rev. Econ. Stat. **89**(4), 701–720 (2007)
8. Zhang, H., Li, W., Yu, T.-T.: Detection research on the structural breaks of the volatility of financial data high-frequency. J. Quant. Tech. Econ. **7**, 50–63 (2011)
9. Lee, S., Mykland, P.A.: Jumps in financial markets: a new nonparametric test and jump dynamics. Rev. Financ. Stud. **21**(6), 2535–2561 (2008)
10. Shen, G.-X.: Jump test on time points and jump dynamics empirical study on CSI 300. Chin. J. Manag. Sci. **20**(1), 43–50 (2012)

Device-Free Intruder Sensing Leveraging Fine-Grained Physical Layer Signatures

Dali Zhu[1,2], Na Pang[1,2(✉)], Weimiao Feng[1,2], Muhmmad Al-Khiza'ay[3], and Yuchen Ma[1,2]

[1] Institute of Information Engineering, Chinese Academy of Sciences,
Beijing, China
{zhudali,pangna,fengweimiao,mayuchen}@iie.ac.cn
[2] School of Cyber Security, University of Chinese Academy of Sciences,
Beijing, China
[3] School of Information Technology, Deakin University, Geelong, Australia
malkhiza@deakin.edu.au

Abstract. With the development of smart indoor spaces, intruder sensing has attracted great attention in the past decades. Realtime intruder sensing in intelligent video surveillance is challenging due to the various covariate factors such as walking surface, clothing, carrying condition. Gait recognition provides a feasible approach for human identification. Pioneer systems usually rely on computer vision or wearable sensors which pose unacceptable privacy risks or be limited to additional devices. In this paper, we present *CareFi*, a device-free intruder sensing system that can identify a stranger or a burglar based on *Commercial Off-The-Shelf* (COTS) WiFi-enabled devices. *CareFi* extracts the fine-grained physical layer *Channel State Information* (CSI) to analyze the distinguishing gait characteristics for intruder sensing. *CareFi* can identify the intruder under both *line-of-sight* (LOS) and *non-line-of-sight* (NLOS) situations. *CareFi* does not require any dedicated sensors or lighting and works in dark just as well as in light. We prototype *CareFi* using commercial off-the-shelf WiFi devices and experimental results in typical indoor scenarios show that it achieves more than 87.2% detection rate for intruder sensing.

Keywords: Intruder sensing · Channel State Information · Privacy · COTS WiFi devices

1 Introduction

Intruder sensing system, also called security monitoring system, has attracted much interest in both research and industry community in the past decades. If an intruder sensing system can satisfy the following requirements: device-free, non-intrusive, deployable, and low cost, we can imagine that this system can automatically give an alarm when a stranger enters the office and automatically

© Springer International Publishing AG 2017
G. Li et al. (Eds.): KSEM 2017, LNAI 10412, pp. 183–194, 2017.
DOI: 10.1007/978-3-319-63558-3_16

open the door when the staff arrives. Gait, a pattern of walking, reflect a unique biometric trait as a result of habitat differences [2,20]. It can be extracted for intruder sensing without subject cooperation.

Traditional methods use camera [4,6,28], wearable sensor [10,13,14,18], and radar [19,25]. Camera based method tracks human motion in video and then does the feature extraction phase and pattern classification phase. Wearable sensor based method leverages the signal characteristics, such as acceleration, temperature, and gyroscope, produced by walking for gait recognition. Radar based method measures the change of Doppler shift in wireless signals. However, they are faced with the fundamental limitations respectively.

- Camera based method needs enough lighting with *line-of-sight* (LOS). At the same time, privacy intrusion is one of the most crucial concerns.
- Wearable sensor based method is inconvenient since it requires that all sensors are worn and connected correctly, which brings additional burden especially for elderly.
- Radar based method is with high cost since it requires very specialized hardware.

In a wireless network, *Channel State Information* (CSI) is used to estimate the channel properties of wireless communication. It has been recently employed to recognize some simple activities. Zheng et al. [31] proposed *Smokey* that leverages the CSI variation to automatically detect smoking. Wu et al. [27] presented *WiDir* that estimates walking direction of human by analyzing the phase change dynamics of CSI. Qian et al. [16] proposed to detect and recognize human motions by correlating Doppler shifts with human activity directions. Kotaru and Katti [11] presented a virtual reality position tracking method. Zhu et al. [33] proposed an isolation-based abnormal activity detection on commercial WiFi devices. Wang et al. [26] proposed *WiFall* that employs the wireless signal properties CSI to achieve device-free fall detection. The goal of intruder sensing system is to automatically identify whether there is any stranger or burglar at home or not. This system should meet the following requirements: device-free, non-intrusive, deployable, and low cost.

In this paper, we propose *CareFi*, a device-free intruder sensing system leveraging fine-grained physical layer signatures. As shown in Fig. 1, *CareFi* consists of only two COTS WiFi devices. The sender (such as a router) continuously emits signals to the receiver (such as a laptop). The first challenge for *CareFi* is to extract the features of detailed *Channel State Information* (CSI) affected by different person. The most challenging task is to accurately characterize the walking pattern of human. The gait cycle time, walking speed, and other gait metrics should be precisely measured for intruder sensing. We show that the gait information is available to sense an intruder under both LOS and NLOS situations. It will save the human interventions needed.

Experiment results in different scenarios including *meeting room* and *apartment* demonstrate that *CareFi* can achieve great performance. More specifically, *CareFi* accurately senses intruder with an average accuracy of 87.2%. In summary, the main contributions are as follows:

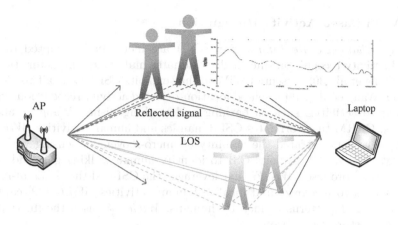

Fig. 1. *CareFi* system overview.

- By exploiting *Channel State Information* (CSI), we validate the feasibility of non-invasive WiFi-based intruder sensing.
- We propose a walking pattern estimation method based on sliding window of step analysis to accurately measure the gait characteristic such as the gait cycle time and walking speed.
- We implement *CareFi* on COTS WiFi devices, and extensive evaluations show that human gait information extracted by physical layer signatures is rich enough to identify an intruder in typical meeting room and apartment environments.

The rest of this paper is organized as follows. Section 2 motivates the problem of related work. Section 3 presents the design details of *CareFi*. Section 4 presents our experimental results. Finally, we conclude the whole paper in Sect. 5.

2   Related Work

We review the existing research work from two perspectives: related work on gait based human identification and related work on WiFi based activity recognition.

2.1   Gait Based Human Identification

Computer vision based method records human walking and extracts step information from depth cameras. It brings serious privacy disclosure and introduces higher cost by deploying depth camera. Yu et al. [29] evaluated the performance of different gait recognition methods. Hu et al. [9] presented to detect, track and recognize gait by employing *Local Binary Pattern* (LBP). Wearable sensor based systems sense the sound, velocity or acceleration on three axises by embedding sensors in belt, coat, or watch [17]. Pan et al. [15] utilize footstep induced structural vibration to identify different people. However, additional sensors need to be installed or worn, which is difficult for many people to comply with.

2.2 WiFi Based Activity Recognition

Recently, *Channel State Information* (CSI) are increasingly adopted to fine-grained activity recognition since it discriminates multi-path characteristics [12, 21,31]. Zeng et al. [30] presented WiWho that exploits CSI-based gait to identify a person. Wang et al. [24] presented location-oriented activity recognition system to recognize activities such as bathing and washing dishes. Wang et al. [23] presented CARM that correlates CSI dynamics and human activities. *WiFigure* [5,7,12] have demonstrated the viability of micro-movement sensing with high accuracy. *WiWho* [30] exploits CSI to identify human walking gait and steps. Zhu et al. [32] proposed *WiseFi* that leverages the CSI and the *Angle-of-arrival* (AOA) values to recognize and localize human activities. *WifiU* [22] captures fine-grained gait patterns to identify humans. *WiGer* [3] used the fluctuations in CSI for gesture recognition.

3 System Design

CareFi is a device-free intruder sensing system leveraging fine-grained physical layer signatures. Figure 2 shows the system architecture of *CareFi*. It consists of four main phases: CSI extraction, data preprocessing, step analysis, and intruder sensing phase.

– *Channel State Information* (CSI) Extraction
  The transmitter propagates wireless beacon signals continuously to the receiver. *CareFi* extracts CSI measurements at the receiver of a WiFi link.

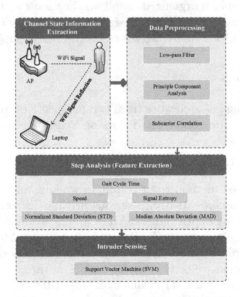

Fig. 2. System architecture of *CareFi*.

– **Data Preprocessing**
  The raw CSI values are too noisy to be directly used in intruder sensing. Low-pass filter and *Principal Component Analysis* (PCA) are adopted to smooth CSI.
– **Step Analysis**
  This is the most challenging phase in this work. Accurately characterize the walking pattern including the gait cycle time and walking speed is critical for intruder sensing.
– **Intruder Sensing**
  Based on the walking features extracted from physical layer signatures, machine learning methods are employed to build an intruder sensing system.

3.1 *Channel State Information* Extration

The channel can be represented as

$$Y = \mathbf{H}x + N,$$

in which x is transmitted signal vector, $\mathbf{H}$ is channel matrix, N represents channel noise vector, and y is received signal vector. The estimated $\mathbf{H}$ for N subcarriers can be expressed as

$$\mathbf{H}_i =\mid \mathbf{H}_i \mid e^{j\sin(\angle\mathbf{H}_i)},$$

where $\angle\mathbf{H}_i$ is the phase information and $\mid \mathbf{H}_i \mid$ is the amplitude information, respectively. We extract CSI channel matrix $\mathbf{H}_{i,j}$ from *Network Interface Card* (NIC) based on the developed CSI-tools [8] as following.

$$\mathbf{H}_{i,j} = \begin{bmatrix} h_{1,1} & h_{1,2} & h_{1,3} & \cdots & h_{1,30} \\ h_{2,1} & h_{2,2} & h_{2,3} & \cdots & h_{2,30} \\ \vdots & \vdots & \vdots & \ddots & \vdots \\ h_{6,1} & h_{6,2} & h_{6,3} & \cdots & h_{6,30} \end{bmatrix}, \tag{1}$$

where $h_{i,j}$ is the CSI value for the ith stream of the jth subcarrier. Each CSI stream is the CSI values captured from each transmitter-receiver pair. 30 represents the number of readable subcarriers.

3.2 Data Prepocessing

The CSI values extracted from commercial WiFi cards include too much noise from electromagnetic interference or environment change. Low-pass filter is not enough to remove most of the noise for intruder sensing. The filtered CSI-based gait across different subcarriers has the property that the impacts of human activity on CSI vary dynamically on a single subcarrier [31]. *Principal Component Analysis* (PCA) can automatically correlate CSI streams and recombine CSI streams to extract components that represent the variation caused by human activity. The first 15 PCA components from the 180 CSI streams are extracted and the rest, which are mostly noisy components, are discarded. We captured CSI values when three persons was asked to walk on the same path in one room. Figure 3 shows the initial CSI values and filtered CSI-based gait of the three persons. The results show that the step lengths and the other differences between the CSI-based gait shape are quite clear.

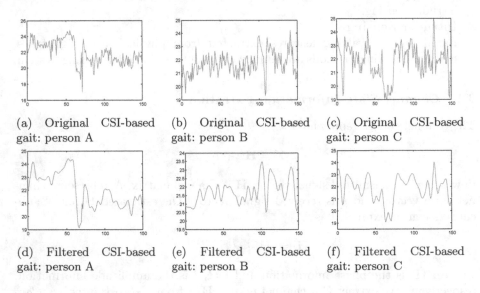

(a) Original CSI-based gait: person A

(b) Original CSI-based gait: person B

(c) Original CSI-based gait: person C

(d) Filtered CSI-based gait: person A

(e) Filtered CSI-based gait: person B

(f) Filtered CSI-based gait: person C

Fig. 3. CSI-based gait information of 3 persons.

3.3   Step Analysis

Step analysis is used to accurately characterize the walking pattern of human. The gait cycle time, walking speed, and other gait metrics should be precisely measured for intruder sensing. *CareFi* involves following steps as shown in Fig. 4:

- (1) A sliding window with a bandwidth w_i is created for CSI waveforms. We set w_1 to 0.90. $w_i < 1.5$.
- (2) The sum of CSI amplitude in the sliding window is calculated as $s_j (0 < j < l)$. We set l to 15 in this paper.
- (3) The sliding window moves k seconds. We set k to 0.1 in our experiment. The variance between s_j is calculated as q_i.
- (4) Starting from $w_i + 0.01$, step (2) and (3) are repeated until $q_i < r$.

The gait cycle time is w_i when the variance $q_i < r$. To verify that the gait cycle time extracted by *CareFi* is accurate, a person is asked to wear smartphone in pocket as the ground-truth, the average accuracy where the deviation is lower than 0.03 seconds is more than 96%.

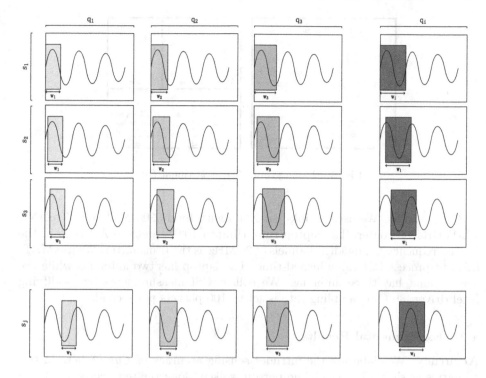

Fig. 4. Sliding window based step analysis.

3.4 Device-Free Intruder Sensing

To extract more obvious features affected by human walking, we then adopt the *Short-Time Fourier Transform* (STFT) to construct the spectrogram. In each walking instance, we extract features including gait cycle time, speed, the normalized standard deviation (STD), signal entropy, and the median absolute deviation (MAD). Support vector machine (SVM) classifier has been applied to distinguish different human or sense intruder. We took the data from testbed in four hours, which consists of 16 volunteers walking straight, and each sample lasts 6 s. The human subjects are 10 male and 6 female students, with similar ages in the range of 26–28 years. We collect 50 walking instances for each human.

4 Experimentation Evaluation

4.1 Equipment

We conducted *CareFi* in a typical meeting room and an apartment as shown in Fig. 5: (a) a meeting room covering an area of around $9 \times 9 \ m^2$ with one sofa and two tables; (b) an apartment covering about $8 \times 9 \ m^2$ area with two bedrooms. In addition to evaluating *CareFi* in different test environments, we extensively evaluate *CareFi*'s performance under two typical scenarios: LOS and

Fig. 5. Meeting room (a), Apartment (b).

NLOS scenarios. We used Think-pad X200 with Intel 5300 NIC and TP-LINK TL-WDR4300 router. The laptop runs Ubuntu 10.04, and the AP runs in the 5 GHz frequency bandwidth channels of 20 MHz as the transmitter. We use MAT-LAB to process CSI value in real-time. The laptop has two antennas while the access point has three antennas. We collect CSI measurements by modifying Intel driver [8]. Our sampling rate is set to 100 packets per second.

4.2   Experimental Results

Accuracy. We elaborate the intruder sensing accuracy of *CareFi* here in two scenarios as shown in Fig. 5. The person walked along the red five-pointed star area. The receiver in Fig. 5 is the blue rectangle, while the transmitter is the green circular. For each person over the randomly selected 5 persons, we calculate the mean accuracy of 50 instance. Figure 6 plots the average accuracy of *CareFi* in LOS and NLOS. The overall intruder sensing accuracy for *CareFi* is 87.2% in meeting room LOS, 83.8% in meeting room NLOS, 86.7% in apartment LOS, and 82.4% in apartment NLOS. For person 13, the result shows that the average accuracy is around 80%, which is obviously lower than others. The reason is that the walking posture is distinctly different from others.

Impact of Multiple APs. We change the number of APs up to 4 to evaluate the performance impact. The Fig. 7 indicates that *CareFi* can achieve average accuracy as high as 90.8% when there is four APs. The accuracy 87.2% in the above evaluation is satisfied in some scenes. The performance of *CareFi* is robust in different scenarios.

Impact of Different Group Size. We evaluate the performance of intruder sensing with different group size. Take the meeting room for an example, we consider there are totally 8 persons as normal. The Fig. 8 plots the performance of intruder sensing in meeting room where the persons are with different height,age, and weight. The average accuracy is as high as 93% when the group size is less than 4, which is appropriate for whole-home intruder sensing. The result also shows the accuracy decreases as the group size increases. It is because that more persons will have similar gait information.

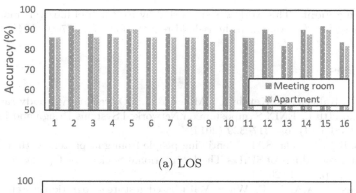

(a) LOS

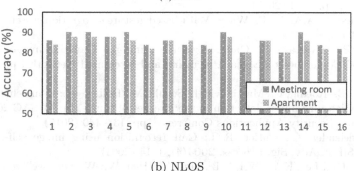

(b) NLOS

Fig. 6. Intruder sensing accuracy for each person in different scenarios.

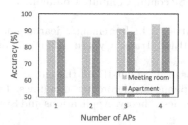

Fig. 7. Impact of multiple APs.

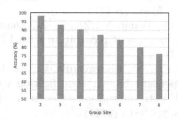

Fig. 8. The accuracy of different group size.

5   Conclusion

Intruder sensing has been an important component in various applications ranging from health care and building surveillance. In this paper, we presented *CareFi*, a non-invasive device-free intruder sensing system leveraging fine-grained physical layer signatures. *CareFi* can work under both LOS and NLOS conditions. This current implementation of *CareFi*, however, has the limitation. When there are multiple strangers moving at the same time, the CSI variance patterns captured by *CareFi* are complex mixtures of multiple steps. For future work, we will be observed in separating concurrent steps and then track them respectively [1].

Acknowledgement. This work was supported by Research of life cycle management and control system for equipment household registration, No. J770011104.

References

1. Adib, F., Kabelac, Z., Katabi, D., Miller, R.C.: 3D tracking via body radio reflections. In: 11th USENIX Symposium on Networked Systems Design and Implementation (NSDI 14), pp. 317–329 (2014)
2. Ailisto, H.J., Makela, S.M.: Identifying people from gait pattern with accelerometers. In: Proceedings of SPIE - The International Society for Optical Engineering, vol. 5779, pp. 7–14 (2005)
3. Al-Qaness, M.A.A., Li, F.: Wiger: WiFi-based gesture recognition system, vol. 5(6), p. 92 (2016)
4. Arora, P., Srivastava, S.: Gait recognition using gait Gaussian image. In: International Conference on Signal Processing and Integrated Networks, pp. 791–794 (2015)
5. Bagci, I.E., Roedig, U., Martinovic, I., Schulz, M., Hollick, M.: Using channel state information for tamper detection in the internet of things. In: ACSAC 2015 - The Computer Security Applications Conference, pp. 131–140 (2015)
6. Benabdelkader, C., Cutler, R.G.: Gait recognition using image self-similarity. EURASIP J. Adv. Sig. Process. **2004**(4), 1–14 (2004)
7. Chang, J.Y., Lee, K.Y., Wei, Y.L., Lin, C.J., Hsu, W.: We can "see" you via WiFi - an overview and beyond (2016)
8. Halperin, D., Wenjun, H., Sheth, A., Wetherall, D.: Tool release: gathering 802.11n traces with channel state information. ACM Sigcomm Comput. Commun. Rev. **41**(1), 53–53 (2011)
9. Hu, M., Wang, Y., Zhang, Z., Zhang, D., Little, J.J.: Incremental learning for video-based gait recognition with LBP flow. IEEE Trans. Cybern. **43**(1), 77–89 (2013)
10. Juefei-Xu, F., Bhagavatula, C., Jaech, A., Prasad, U.: Gait-ID on the move: pace independent human identification using cell phone accelerometer dynamics. In: IEEE Fifth International Conference on Biometrics: Theory, Applications and Systems, pp. 8–15 (2012)
11. Kotaru, M., Katti, S.: Position tracking for virtual reality using commodity WiFi (2017). arXiv preprint arXiv:1703.03468
12. Li, H., Yang, W., Wang, J., Xu, Y., Huang, L.: WiFinger: talk to your smart devices with finger-grained gesture. In: Proceedings of the 2016 ACM International Joint Conference on Pervasive and Ubiquitous Computing, pp. 250–261. ACM (2016)
13. Mantyjarvi, J., Lindholm, M., Vildjiounaite, E., Makela, S.M.: Identifying users of portable devices from gait pattern with accelerometers. In: Proceedings of IEEE International Conference on Acoustics, Speech, and Signal Processing, vol. 2, pp. ii/973–ii/976 (2005)
14. Ngo, T.T., Makihara, Y., Nagahara, H., Mukaigawa, Y., Yagi, Y.: The largest inertial sensor-based gait database and performance evaluation of gait-based personal authentication. Pattern Recogn. **47**(1), 228–237 (2014)
15. Pan, S., Wang, N., Qian, Y., Velibeyoglu, I., Noh, H.Y., Zhang, P.: Indoor person identification through footstep induced structural vibration. In: International Workshop on Mobile Computing Systems and Applications, pp. 81–86 (2015)

16. Qian, K., Wu, C., Zhou, Z., Zheng, Y., Yang, Z., Liu, Y.: Inferring motion direction using commodity WiFi for interactive exergames. In: Proceedings of the 2017 CHI Conference on Human Factors in Computing Systems, pp. 1961–1972. ACM (2017)
17. Radhakrishnan, M., Eswaran, S., Misra, A., Chander, D., Dasgupta, K.: Iris: Tapping wearable sensing to capture in-store retail insights on shoppers. In: IEEE International Conference on Pervasive Computing and Communications, pp. 1–8 (2016)
18. Schwesig, R., Leuchte, S., Fischer, D., Ullmann, R., Kluttig, A.: Inertial sensor based reference gait data for healthy subjects. Gait Posture **33**(4), 673–678 (2011)
19. Tahmoush, D., Silvious, J.: Radar micro-doppler for long range front-view gait recognition. In: IEEE International Conference on Biometrics: Theory, Applications, and Systems, pp. 1–6 (2009)
20. Wang, C., Zhang, J., Wang, L., Jian, P., Yuan, X.: Human identification using temporal information preserving gait template. IEEE Trans. Softw. Eng. **34**(11), 2164 (2011)
21. Wang, H., Zhang, D., Wang, Y., et al.: RT-Fall: a real-time and contactless fall detection system with commodity WiFi devices. IEEE Trans. Mob. Comput. **16**(2), 1 (2017)
22. Wang, W., Liu, A.X., Shahzad, M.: Gait recognition using WiFi signals. In: Proceedings of the 2016 ACM International Joint Conference on Pervasive and Ubiquitous Computing, pp. 363–373. ACM (2016)
23. Wang, W., Liu, A.X., Shahzad, M., Ling, K., Lu, S.: Understanding and modeling of WiFi signal based human activity recognition. In: Proceedings of the 21st Annual International Conference on Mobile Computing and Networking, pp. 65–76. ACM (2015)
24. Wang, Y., Liu, J., Chen, Y., Gruteser, M., Yang, J., Liu, H.: E-eyes: device-free location-oriented activity identification using fine-grained WiFi signatures. In: Proceedings of the 20th Annual International Conference on Mobile Computing and Networking, pp. 617–628. ACM (2014)
25. Wang, Y., Fathy, A.E.: Micro-doppler signatures for intelligent human gait recognition using a UWB impulse radar. In: IEEE International Symposium on Antennas and Propagation (2011)
26. Wang, Y., Wu, K., Ni, L.M.: WiFall: Device-free fall detection by wireless networks. IEEE Trans. Mobile Comput. **16**(2), 581–594 (2017)
27. Wu, D., Zhang, D., Xu, C., Wang, Y., Wang, H.: Widir: walking direction estimation using wireless signals. In: Proceedings of the 2016 ACM International Joint Conference on Pervasive and Ubiquitous Computing, pp. 351–362. ACM (2016)
28. Xu, D., Yan, S., Tao, D., Lin, S., Zhang, H.J.: Marginal fisher analysis and its variants for human gait recognition and content-based image retrieval. IEEE Trans. Image Process. **16**(11), 2811–2821 (2007)
29. Yu, S., Tan, D., Tan, T.: A framework for evaluating the effect of view angle, clothing and carrying condition on gait recognition. In:18th International Conference on Pattern Recognition, ICPR 2006, vol. 4, pp. 441–444. IEEE (2006)
30. Zeng, Y., Pathak, P.H., Mohapatra, P.: WiWho: WiFi-based person identification in smart spaces. In: Proceedings of the 15th International Conference on Information Processing in Sensor Networks, p. 4. IEEE Press (2016)
31. Zheng, X., Wang, J., Shangguan, L., Zhou, Z., Liu, Y.: Smokey: ubiquitous smoking detection with commercial WiFi infrastructures. In: IEEE INFOCOM 2016-The 35th Annual IEEE International Conference on Computer Communications, pp. 1–9. IEEE (2016)

32. Zhu, D., Pang, N., Li, G., Liu, S.: WiseFi: Activity localization and recognition on commodity off-the-shelf wi-Fi devices. In: 2016 IEEE 18th International Conference on High Performance Computing and Communications, IEEE 14th International Conference on Smart City, IEEE 2nd International Conference on Data Science and Systems (HPCC/SmartCity/DSS), pp. 562–569. IEEE (2016)
33. Zhu, D., Pang, N., Li, G., Rong, W., Fan, Z.: WiN: Non-invasive abnormal activity detection leveraging ne-grained wi signals. In: Trustcom/BigDataSE/I SPA, 2016 IEEE, pp. 744–751. IEEE (2016)

Understanding Knowledge Management in Agile Software Development Practice

Yanti Andriyani[1(⊠)], Rashina Hoda[1], and Robert Amor[2]

[1] Department of Electrical and Computer Engineering, SEPTA Research,
The University of Auckland, Building 903, 386 Khyber Pass, New Market,
Auckland 1023, New Zealand
yand610@aucklanduni.ac.nz, r.hoda@auckland.ac.nz
[2] Department of Computer Science, The University of Auckland,
Auckland, New Zealand
trebor@cs.auckland.ac.nz

Abstract. Knowledge management in agile software development has typically been treated as a broad topic resulting in major classifications of its schools and concepts. What inherent knowledge is involved in everyday agile practice and how agile teams manage it is not well understood. To address these questions, we performed a Systematic Literature Review of 48 relevant empirical studies selected from reputed databases. Using a thematic analysis approach to the synthesis, we discovered that (a) agile teams use three knowledge management strategies: discussions, artifacts and visualisations to manage knowledge (b) there are three types of software engineering knowledge: team progress as project knowledge; requirements as product knowledge; and coding techniques as process knowledge. (c) this knowledge is presented in several everyday agile practices. A theoretical model describing how knowledge management strategies and knowledge types are related to agile practices is also presented. These results will help agile practitioners become aware of the specific knowledge types and knowledge management strategies and enable them to better manage them in everyday agile practices. Researchers can further investigate and build upon these findings through empirical studies.

Keywords: Agile software development · Knowledge type · Knowledge management strategies

1 Introduction

Agile Software Development (ASD) methods such as Scrum and Extreme Programming (XP) ushered in an era of lightweight software development [1]. Unlike traditional software methods such as waterfall, which was driven by detailed specifications and design upfront and involved rigorous documentation [2] as a process of managing knowledge, agile methods emphasize social interactions and collaboration among team members in applying and sharing knowledge.

Prior reviews and investigations of knowledge management in ASD have typically treated it as a broad topic resulting in major classifications of its schools, concepts [3], strengths, weaknesses, opportunities and threats [4]. However, what inherent knowledge

© Springer International Publishing AG 2017
G. Li et al. (Eds.): KSEM 2017, LNAI 10412, pp. 195–207, 2017.
DOI: 10.1007/978-3-319-63558-3_17

is involved in everyday agile practice and how agile teams manage this knowledge is not well understood. To address this gap, we performed a Systematic Literature Review (SLR) involving 48 relevant empirical studies [5] filtered from an initial pool of 2317 articles selected from the reputed academic databases of Springer, Scopus, and IEEE Xplore.

This SLR aims to provide an overview of the studies on this particular topic by answering the following research questions (RQ):

RQ1: Which specific agile practices support knowledge management?
RQ2: What is the inherent knowledge involved in these agile practices and how do agile teams manage that knowledge?

The next section provides an overview of background and related work of this research. Section 3 gives the research method used and Sect. 4 discusses the results of this study. Section 5 provides the discussions of the findings and Sect. 6 provides the conclusion.

2   Background and Related Work

The relevant ASD and knowledge management concept definitions and summaries of prior reviews are presented in this section to aid in the understanding of the findings described later.

2.1   Knowledge Classifications

Knowledge is defined as: *"A fluid mix of framed experience, values, contextual information, and expert insight that provides a framework for evaluating and incorporating new experiences and information"* [6]. Knowledge is described in two basic forms: tacit knowledge or the knowledge that is implicit and not clarified in an accessible form; and explicit knowledge or the knowledge that is visualised or clarified in accessible forms such as writing on sticky notes, drawing pictures to describe the processes or feelings, drawing progress charts, etc. [7].

A classification of knowledge in software engineering describes three types of knowledge: project, product and process knowledge [8]. Project knowledge is defined as *"the knowledge about resources, functional and attributes requirements, work products, budget, timing, milestones, deliverables, increments, quality targets and performance parameters"* [8]. Documentation and resources include contracts and project plans based on requirement designs; and the parameters are analysed through comparisons between the planned and actual cost, effort and time [9].

Product knowledge is defined as *"the knowledge about [the] product features and how they relate to other products, standards, protocols and the like"* [8]. Specifically, product knowledge is related to the product features, its interface and its dependency on technology (e.g. specific programming language or platform), network configurations, standards (related to components of the product) and protocols.

Process knowledge is defined as *"the knowledge about business processes, work-flows, responsibilities, supporting technologies and interfaces between processes"* [8]. In other words, process knowledge comprises of work targets and task information retrieved from the business model; the workflow and responsibilities referring to how the various artifacts are produced and who is responsible for specific tasks and how they accomplish them based on the milestones [8].

2.2 Prior Reviews on Knowledge Management in ASD

Knowledge management in organizations is defined as: *"A method that simplifies the process of sharing, distributing, creating, capturing and understanding the company knowledge"* [6]. Knowledge management in traditional software development involved the use of various documents to capture and represent the knowledge related to the various stages of software development life cycles [3]. In contrast to traditional methods, agile methods emphasize tacit knowledge over explicit knowledge, relying on individual, team and customer communications and interactions [4].

Prior reviews and investigations on knowledge management in software engineering in general and ASD, in particular, have classified knowledge schools [10] and concepts [3]. Bjørnson and Dingsøyr [10] classified two types of knowledge management schools in software engineering. The first category is the *technocratic* school, which refers to knowledge management strategies that focus on explicating knowledge and its flows. The second category is the *behavioural* school, which focuses on collaboration and communication as knowledge management strategies.

Yanzer et al. [3] presented several concept maps about knowledge management in agile projects. The first map covers ways of communication that focus on techniques and tools to manage the conversation. The second is about human and social factors, which discusses knowledge management adoption in agile projects and knowledge artifact usage in the projects. Other concepts include tools for knowledge management and knowledge representation forms related to managing tacit knowledge, and emphasize the managing of tacit knowledge by using tools to clarify the knowledge.

Analysing the prior reviews on knowledge management in ASD, we could see that there was a need for a specific explanation of knowledge management in ASD from the viewpoint of daily agile practice.

3 Review Method

A Systematic Literature Review (SLR) aims to collect evidence from the prior literature based on research questions to provide guidelines for practitioners [11]. We followed the specific steps of performing a SLR as recommended by Kitchenham [11], such as planning, study selection, data extraction and synthesis, and reporting the review.

3.1 Planning the Review and Identifying Relevant Literature

We constructed a search strategy to identify relevant literature by deriving major terms from the research questions, listing the keywords and developing search strings from these major terms and their synonyms using AND/OR operators. The search string that we used was: *("knowledge manag*" OR "learning manag*") AND ("agile" OR "scrum" OR "XP" OR "Lean") AND ("software" AND "team").*

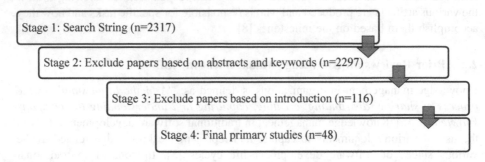

Stage 1: Search String (n=2317)

Stage 2: Exclude papers based on abstracts and keywords (n=2297)

Stage 3: Exclude papers based on introduction (n=116)

Stage 4: Final primary studies (n=48)

Fig. 1. SLR screening process

The search string was used to filter articles from three reputed academic databases:

- Springer, http://www.springer.com
- Scopus, www.scopus.com
- IEEE Xplore, http://www.ieeexplore.ieee.org

A total of 2317 papers were obtained initially, of which 195 papers were from Scopus based on title and abstracts while Springer Link resulted in 2097 papers and 25 results were from IEEE Xplore.

3.2 Publication Selection

Inclusion and exclusion criteria (described in [5]) were designed to help select from the initial pool of papers resulting from the searches. The results from the search string generated 2317 papers, which were screened in the next stage (see Fig. 1). The second stage generated 2297 papers, which screened the papers based on how the abstract and keywords related to the RQs and the inclusion and exclusion criteria [5]. Stage 3 involved reading the introduction section of the 2297 papers checked against the inclusion and exclusion criteria and resulted in 116 papers being filtered. All 116 papers were read in stage 4 resulting in a total of 48 papers selected as the primary studies based on the inclusion and exclusion criteria. Most papers were excluded because they only provided theoretical explanations and no empirical evidence.

3.3 Data Extraction and Synthesis

Data extraction involved extracting detailed information from the 48 primary studies, such as the paper citation details, answers to the review questions and the main stufy findings into an excel sheet. The emphasis was placed on extracting evidence to support the review questions. The first author was responsible for data extraction overseen by the co-authors who provided guidance throughout the process and helped reach consensus in certain cases.

Data was analysed by determining themes from the selected papers using thematic analysis [12]. Initial codes summarizing key ideas were selected after reading the primary studies thoroughly. For example, the specific artifacts used in ASD, such as product backlog, user stories etc. were collectively classified as *artifacts* since they contained useful knowledge about the software requirements. Similarly, UML modeling, burn down charts etc. were classified as *visualisations*.

Themes were derived as the next level of abstraction, gathering similar codes together [12]. For example, the codes artifacts, visualisation and discussion collectively formed the theme knowledge management strategies since each of them is a type of knowledge management strategy. Figures 2 and 3 depict the emergence of the themes

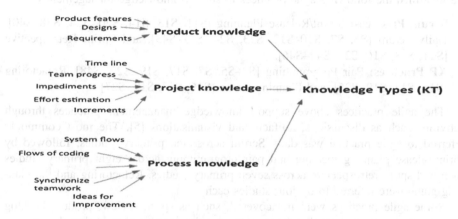

Fig. 2. Emergence of knowledge types (KT) theme categories

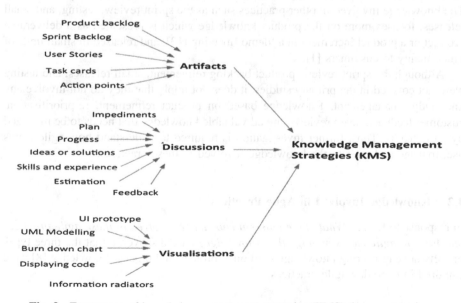

Fig. 3. Emergence of knowledge management strategies (KMS) theme categories

'*knowledge management strategies*' *(KMS)* and '*knowledge types*' *(KT)* from the underlying codes. Another theme that emerged naturally from the codes was '*agile practices*' *(AP)* which included the specific agile practices reported in the primary studies.

4 Results

4.1 Agile Practices Supporting Knowledge Management

In response to *RQ1 "Which specific agile practices support knowledge management?"* we identified the following agile practices to support knowledge management [5]:

- **Scrum Practices:** Sprint/Release Planning [S11, S13, S14, S20, S24, S38–S40], Daily Scrum [S6, S7, S10–S12, S23, S24, S41–S43], and Sprint Retrospective [S11, S13, S16, S24, S44–S46].
- **XP Practices:** Pair Programming [S3–S5, S7, S17, S19, S22, S48], Refactoring [S15, S17, S19, S25], and Planning Game [S16, S17, S19, S27].

The agile practices above support knowledge management practices through activities such as discussions, artifacts and visualisations [5]. The most commonly referred to agile practice was daily Scrum across ten primary studies, followed by sprint/release planning meeting and pair programming across eight primary studies each and sprint retrospective across seven primary studies. Refactoring and the planning game were referred to by four studies each.

Some agile practices were not covered, such as sprint review, product backlog refinement, testing and small releases. One possible explanation is that the product backlog refinement is a relatively new, optional and lesser known Scrum practice [13]. The knowledge involved in other practices such as the sprint review, testing, and small releases, focuses more on the product knowledge which is presented as a deliverable product or a product increment in a 'demo' meeting [13], and released in small units of functionality to customers [1].

Although the sprint review, product backlog refinement, small releases, and testing were not covered in the primary studies, it does not imply that they do not involve any knowledge management. Knowledge based on product refinement, re-prioritisation, customer feedback, team reviews, are all valuable knowledge that needs to be managed by agile teams. Thus, further investigation is required to understand how agile teams use, manage and refer to the knowledge involved in these practices.

4.2 Knowledge Involved in Agile Practices

In response to RQ2: *"What is the inherent knowledge involved in these agile practices and how do agile teams manage that knowledge?"* we discovered that the three types of software engineering knowledge – product, project, and process knowledge [8] were captured in everyday agile practices.

Table 1 presents a summarized view of the knowledge types involved in agile practices. These knowledge types are managed in six agile practices. For example, product knowledge found to be involved in the release and sprint planning meetings included domain context [S6] and product features (systems requirements) [S12], which facilitates agile teams to gain knowledge about the product to be developed. In daily stand-up, a wall or Scrum board [S12] is used to stick up story cards that contain several tasks, which are broken-down from the product backlog.

Table 1. Knowledge types (KT) in ASD (based on [8])

Knowledge types (KT) based on [8]	Description [8]	Examples in ASD practices
Product knowledge	*"The knowledge about [the] product features and how they relate to other products, standards, protocols and the like"*	Domain context [S6]; product features on user stories [S12]; coding [S12], testing [S11]
Project knowledge	*"The knowledge about resources, functional and attributes requirements, work products, budget, timing, milestones, deliverables, increments, quality targets and performance parameters"*	Project/daily goals [S6]; timeline [S10]; progress line; lack of time for testing and work targets [S10]
Process knowledge	*"The knowledge about business processes, workflows, responsibilities, supporting technologies and interfaces between processes"*	Systems flows [S11]; business process [S11]; other team member's role and their interdependencies [S6]; synchronizing teamwork; ideas of improvement [S24]; workflow of coding and working code [S19]

Project knowledge is managed in several agile practices, such as sprint/release planning, daily stand-up, sprint retrospective, and planning game. For example, in a daily stand-up meeting agile teams clarify their cumulative work done by the team in a burn-down chart [S10]. In a sprint retrospective project knowledge is shared when agile teams share issues about lack of time to accomplish some tasks and uncompleted tasks in the last sprint which can lead to some changes in the timeline [S6, S11].

Process knowledge is managed in several agile practices, such as sprint/release planning, daily stand-up, sprint retrospective, pair programming and refactoring [S6, S11, S19, S24]. Process knowledge in these practices includes the knowledge about the system flows that are visualised in UML modeling or other visualisations that represent coding flow, features and business processes [S11].

With regards to knowledge management strategies applied, agile teams were seen to use: discussions (e.g. sharing requirements), artifacts (e.g. user stories) and visualisations (e.g. burn-down charts), to manage the project, product and process knowledge [5]. See summarized description in Table 2.

Table 2. Knowledge management strategies (KMS) in ASD

Knowledge management strategies (KMS)	Description	Examples in ASD practices
Discussions	Verbal communication that involves interaction among agile team members which aims to share knowledge	Sharing requirements [S20]; progress; plan and impediments [S13]; feedback; ideas/solutions [S24]; system flow; coding; techniques; design problems [S11]; coding problems; techniques; analysing; estimating and negotiate to agree; communication over video conference (e.g. Skype) [S47]
Artifacts	Physical forms that contain specific product features and project information	Story cards from Product backlog and Sprint backlog [S20]; user stories; task card [S46]; the card of code; code repositories [S13], JIRA [S47], Wiki [S6]
Visualisations	Strategies that clarify the resources about product, process and project into a visualised form	Information radiators; UI prototyping [S13]; UML modelling [S11]; Burn down chart [S25]; story cards on the Scrum board; wall; action points [S46]; showing the code/working code [S5], JIRA, Wiki [S47], Microsoft Excel [S8]

Discussion was the most commonly used knowledge management strategy across all agile practices (e.g. release/sprint planning, daily stand-up and retrospective). This strategy facilitates agile teams to share knowledge, in particular process and project knowledge [S11, S13, S20, S24, S47]. Process knowledge was shared during discussions where agile teams share issues, ideas, solutions, new techniques in solving problems [S13] (e.g. coding, testing) and feedback, and negotiate with team members in making plans or decisions [S24]. Project knowledge was also included in the discussions where the content of the discussion related to the project, such as blockers in the last sprint which can affect the project timeline and other team members' progress [S11].

Artifacts in agile practices were commonly used to share product knowledge which included product requirements (e.g. in the form product backlog) and were further broken down into user stories. The product backlog also helped capture product knowledge through task cards for coding [S46], design and user interface development [S20].

Visualisation is the technique used to manage knowledge in a visible and accessible form. This strategy helped agile teams to support tacit knowledge sharing among team members. For example, the Scrum board was used to show progress based on the story cards; the whiteboard for information such as feedback, feelings and action points in the retrospective meetings [S46]; the burn-down chart for showing team progress, achievements and performance [S25] and software code on display screens for showing working code in pair programming [S48]. The most commonly used visualisation strategy in agile practices was the Scrum board, which contained all user stories and presented teamwork progress through the work status of team members [S11].

5 Discussion

In this section, we discuss a theoretical model of knowledge management in ASD which emerged from analyzing the knowledge types (KT), knowledge management strategies (KMS) and agile practices (AP) as described in the previous section.

Table 3 presents a summarized view of the agile practices, knowledge types, and knowledge management strategies identified in this review. The first column lists the agile practices (AP) identified in the literature (Sect. 4.1); the 'Knowledge Types (KT) Supported' column lists the knowledge types as related to agile practices (Sect. 4.2 and Table 1); 'Knowledge Management Strategies (KMS)' column lists the knowledge management strategies inherent in the agile practices (Sect. 4.2 and Table 2); and the primary studies, which can be seen in [5], supporting these findings. We found that all three types of knowledge were addressed in all the Scrum practices in varying degrees.

Table 3. Agile practices that support knowledge management

Agile practices (AP)		Knowledge types (KT) supported	Knowledge management strategies (KMS)	Supporting primary studies
Scrum practices	Sprint/release planning	Product knowledge; project knowledge; process knowledge	Discussions; artifacts; visualisations	S11, S13, S14, S20, S24, S38–S40
	Daily Scrum	Product knowledge; project knowledge; process knowledge	Discussion; artifacts; visualisations	S6, S7, S10–S12, S23, S24, S41–S43
	Sprint retrospective	Product knowledge; project knowledge; process knowledge	Discussion; artifacts; visualisations	S11, S13, S16, S24, S44–S46
XP practices	Pair programming	Product knowledge; process knowledge	Discussion; artifacts; visualisations	S3, S5, S7, S17, S19, S22, S48
	Refactoring	Product knowledge; process knowledge	Discussion; artifacts; visualisations	S15, S17, S19, S25
	Planning game	Product knowledge; project knowledge	Discussions; artifacts; visualisations	S16, S17, S19, S27

Furthermore, in synchronising teamwork through practices such as the daily stand-up and inspecting the process through retrospectives, agile teams combine product, project and process knowledge and build a framework to solve the problems or find ways to improve. Dingsøyr [14] explained that agile teams gather and link the knowledge through several processes referring to Nonaka and Takeuchi's knowledge creation theory. We expand on this work by highlighting the knowledge types and management strategies involved in these processes: Socialization, a process of sharing tacit knowledge (e.g. sharing mental models and technical skills) is achieved through discussion (identified as a KM strategy in our study). Externalization occurs when agile teams gain the shape of metaphors, concepts and models in written form, such as documentation, diagrams or artifacts (i.e. types of product and project knowledge). Agile teams gain and process the externalized knowledge to understand about "know-how" (i.e. a type of process knowledge) as part of the internalization process. The final process is a combination, where agile teams compile the knowledge from different sources (e.g. artifacts, meetings, board) in order to transform it into action.

Despite the inclusion of product, process and project knowledge in daily agile practices, some knowledge management related challenges were also identified in other reviews. Ringstad et.al. [15] and Stray et.al. [16] mention some challenges including: lack of focus on what was working well; no specific discussion about improving teamwork; and difficulty in transforming lessons learned into action [17].

The results of this review indicate that there is a discrepancy between Scrum theory and real practice, which could be attributed to the effectiveness of using knowledge in each practice. In Scrum practices (e.g. sprint/release planning meeting, daily stand-up and retrospective meeting) where product, process and project knowledge were involved, agile teams do not fully pay attention to the knowledge at the same degree, which means that they do not use the knowledge effectively. Another interesting finding from Scrum practice was that agile teams also tend to discuss product and project knowledge in the retrospective, which is theoretically meant to focus on the process alone.

In XP practices, product and process knowledge are discussed in pair programming and refactoring (see the description in Table 3). These findings are aligned with the theoretical aims of these practices. Pair programming aims to enhance agile team skills by working in pairs [1]. In practice, agile teams discuss coding issues, integrate with design and learn from other team member's skills. Similarly, refactoring focuses on maintaining coding and design, which involves product knowledge and process knowledge. In addition, Table 3 shows that the planning game in XP refers to product and project knowledge, being consistent with the theory [1] as this practice aims to capture and analyse overall product requirements and build plans to accomplish the tasks.

The list of agile practices in Table 3 shows that most primary studies pay more attention to practices such as pair programming, daily Scrum and release/sprint planning meetings. Because meetings embody interaction and communication, these practices emphasize tacit knowledge sharing rather than explicit knowledge; however, there is evidence about some ways to transform tacit knowledge into explicit knowledge, such as storing it on sticky notes, paper or online documentation. Thus, important information or knowledge-related artifacts could be managed and used as a reference for team members.

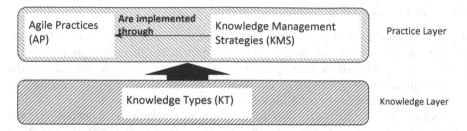

Fig. 4. A theoretical model of knowledge management in agile software development

Based on the findings of this SLR, a theoretical model of knowledge management in ASD was developed. The theory consists of Knowledge Types (KT), Knowledge Management Strategies (KMS) and Agile Practices (AP). Figure 4 depicts the theoretical model that illustrates two layers involved in knowledge management in ASD: a knowledge layer and a practice layer. The knowledge layer includes the three knowledge types and is a fundamental layer on which the practice layer functions. The practice layer includes agile practices and knowledge management strategies (or practices). The knowledge types from the knowledge layer are required and used in the agile practices and knowledge management strategies in the practice layer.

Our hypothesis is that the three knowledge types are managed by performing agile practices and knowledge management strategies, and the practices work using the knowledge types involved. For example, in daily stand-up (an agile practice), discussion, artifacts and visualisations (knowledge management strategies) are implemented to manage product, project and process knowledge (knowledge types). It can be seen that artifacts, visualisation and discussion in daily stand-up require particular knowledge to be managed. Without using artifacts on the Scrum board and discussing the stories, the knowledge in daily stand-up cannot be managed effectively.

5.1 Implications

For practitioners, this SLR shows that there are three specific types of knowledge involved in everyday agile practice, which suggests that knowledge is necessary to be embedded and referred to by agile teams. It also seems important for agile teams to identify the three types of knowledge included in each knowledge management strategy (e.g. discussions, artifacts and visualisations), which must be applied by agile teams in everyday practice. Thus, in order to gain benefits of knowledge management, we suggest that practitioners need to pay more attention to the types of knowledge described in this review, focus on how to manage that knowledge using the strategies discussed and implement the knowledge management concepts consciously in agile practices.

In terms of research, the SLR results suggest that instead of managing knowledge in ASD by classifying it as tacit and explicit knowledge, the specific explanation of knowledge management in ASD based on software engineering would be more relevant for agile teams, such as product, project and process knowledge involved in each agile practice. It also seems there is a need for further study into the knowledge types and management strategies involved in agile practices.

5.2   Limitations

We now discuss some of the limitations of this SLR. First is related to completeness. Despite our best efforts, and as common with most SLRs, there is a possibility that some articles published in some journals and conferences, which can address the research questions, were missed in this SLR. Thus, our results must be classified as applying to the papers selected from the three major digital libraries.

Furthermore, we are aware that some questions might arise about the selected articles. Thus, we defined inclusion and exclusion criteria (listed in [5] due to space constraints) as a protocol in selecting primary studies guided by the research questions. As well as some papers that described knowledge management theories in ASD, the implementation of knowledge management theories was reported but not discussed in detail in the primary studies, and are therefore not included in this SLR. We also assumed that selecting primary studies based on industrial empirical results would be more valuable for agile practitioners than summarizing findings from educational settings.

6   Conclusion

This systematic literature review set out to analyse and summarize the empirical research on knowledge management in agile software development (ASD). We analysed 48 primary studies filtered from an initial pool of 2317 papers using inclusion and exclusion criteria, presenting agile practices that support knowledge management in ASD.

The results of this SLR describe the knowledge types and knowledge management strategies in everyday agile practices. The most important contribution of this SLR is providing a new understanding of knowledge management in ASD that involves managing three different types of knowledge – process, project and product – by implementing three knowledge management strategies – discussions, artifacts and visualisations – during every day agile practices. Understanding these specific dimensions of knowledge and specific knowledge management strategies will help agile practitioners become aware of, and enable them to manage, the knowledge in everyday agile practices effectively.

Acknowledgement. This research is supported by the Indonesia Endowment Fund for Education (LPDP) S-669/LPDP/2013 as scholarship provider from the Ministry of Finance, Indonesia.

References

1. Beck, K.: Extreme Programming Explained: Embrace Change. Addison- Wesley Professional, Boston (1999)
2. Royce, W.: Managing the development of large software systems. In: IEEE WESCON, vol. 26, pp. 328–338. IEEE (1970)

3. Yanzer Cabral, A.R., Ribeiro, M.B., Noll, R.P.: Knowledge management in agile software projects: a systematic review. J. Inf. Knowl. Manag. **13**, 1450010 (2014)

4. Neves, F.T., Rosa, V.N., Correia, A.M.R., de Castro Neto, M.: Knowledge creation and sharing in software development teams using agile methodologies: key insights affecting their adoption. In: 6th Iberian Conference on Information Systems and Technologies (CISTI 2011), pp. 1–6 (2011)

5. Andriyani, Y., Hoda, R., Amor, R.: Research literature of knowledge management in agile software development (ASD). Technical report (2017)

6. Davenport, T.H., Prusak, L.: Working Knowledge-How Organizations Manage What They Know, vol. 5, pp. 193–211. Harvard Business School Press, Brighton (1998)

7. Ikujirō, N., Takeuchi, H.: The Knowledge-Creating Company: How Japanese Companies Create the Dynamics of Innovation. Oxford University Press, New York (1995)

8. Ebert, C.D.M.: J: Effectively utilizing project, product and process knowledge. Inf. Softw. Technol. **50**(6), 579–594 (2008)

9. Lindvall, M., Rus, I.: Knowledge management for software organizations. In: Aurum, A., Jeffery, R., Wohlin, C., Handzic, M. (eds.) Managing Software Engineering, pp. 73–94. Springer, Heidelberg (2003). doi:10.1007/978-3-662-05129-0_4

10. Bjørnson, F.O., Dingsøyr, T.: Knowledge management in software engineering: a systematic review of studied concepts, findings and research methods used. Inf. Softw. Technol. **50**, 1055–1068 (2008)

11. Kitchenham, B.A., Pfleeger, S.L., Pickard, L.M., Jones, P.W., Hoaglin, D.C., El Emam, K., Rosenberg, J.: Preliminary guidelines for empirical research in software engineering. IEEE Trans. Softw. Eng. **28**, 721–734 (2002)

12. Braun, V., Clarke, V.: Using thematic analysis in psychology. Qual. Res. Psychol. **3**, 77–101 (2006)

13. Deemer, P., Benefield, G., Larman, C., Vodde, B.: A lightweight guide to the theory and practice of Scrum version 2.0, vol. 2015 (2012)

14. Dingsøyr, T.: Value-based knowledge management: the contribution of group processes. In: Biffl, S., Aurum, A., Boehm, B., Erdogmus, H., Grünbacher, P. (eds.) Value-Based Software Engineering, pp. 309–325. Springer, Heidelberg (2006). doi:10.1007/3-540-29263-2_15

15. Ringstad, M.A., Dingsøyr, T., Brede Moe, N.: Agile process improvement: diagnosis and planning to improve teamwork. In: O'Connor, Rory V., Pries-Heje, J., Messnarz, R. (eds.) EuroSPI 2011. CCIS, vol. 172, pp. 167–178. Springer, Heidelberg (2011). doi:10.1007/978-3-642-22206-1_15

16. Gulliksen Stray, V., Moe, N.B., Dingsøyr, T.: Challenges to teamwork: a multiple case study of two agile teams. In: Sillitti, A., Hazzan, O., Bache, E., Albaladejo, X. (eds.) XP 2011. LNBIP, vol. 77, pp. 146–161. Springer, Heidelberg (2011). doi:10.1007/978-3-642-20677-1_11

17. Andriyani, Y., Hoda, R., Amor, R.: Reflection in agile retrospectives. In: Baumeister, H., Lichter, H., Riebisch, M. (eds.) XP 2017. LNBIP, vol. 283, pp. 3–19. Springer, Cham (2017). doi:10.1007/978-3-319-57633-6_1

Knowledge Integration

Multi-view Unit Intact Space Learning

Kun-Yu Lin, Chang-Dong Wang$^{(\boxtimes)}$, Yu-Qin Meng, and Zhi-Lin Zhao

School of Data and Computer Science, Sun Yat-sen University, Guangzhou, China
kunyulin14@outlook.com, changdongwang@hotmail.com, yuqinmeng@outlook.com,
zhaozhl7@mail2.sysu.edu.cn

Abstract. Multi-view learning is a hot research topic in different research fields. Recently, a model termed multi-view intact space learning has been proposed and drawn a large amount of attention. The model aims to find the latent intact representation of data by integrating information from different views. However, the model has two obvious shortcomings. One is that the model needs to tune two regularization parameters. The other is that the optimization algorithm is too time-consuming. Based on the unit intact space assumption, we propose an improved model, termed multi-view unit intact space learning, without introducing any prior parameters. Besides, an efficient algorithm based on proximal gradient scheme is designed to solve the model. Extensive experiments have been conducted on four real-world datasets to show the effectiveness of our method.

Keywords: Multi-view learning · Unit intact space · Clustering

1 Introduction

Due to the rapid development of information technology, an increasing amount of multi-view data has been generated from diverse domains [21]. Different domains can be taken as different views. For example, in web page classification task, the description of a web page can be partitioned into two views, namely the words occurring on that page and the words occurring in hyperlinks pointing to that page [3]. Although different views exhibit heterogeneous properties of data, they are coupled by some underlying relations and exhibit some common structures. Properly combining the information from different views by exploring the relations will improve the learning performance. This leads to the emergence of multi-view learning, which is a challenging task in the field of machine learning.

Many multi-view learning approaches have been proposed from different perspectives. In [9,10], two simple schemes are introduced to combine multiple views. One is feature concatenation, which concatenates variables in all views into one view and applies the single-view learning method on the combined view. The other is to combine the single-view learning objective functions of all views into a multi-view learning model. However, both two schemes do not work quite well in improving the learning performance, especially when the multiple views

© Springer International Publishing AG 2017
G. Li et al. (Eds.): KSEM 2017, LNAI 10412, pp. 211–223, 2017.
DOI: 10.1007/978-3-319-63558-3_18

are from different representation spaces [17]. An alternating improved variant is the weighted combination scheme. Instead of naive convex combination of objective functions of different views, a hyperparameter is used to control the distribution of weight parameters [5,18,19]. Furthermore, an auto-weighted multi-view learning framework is proposed to learn an optimal weight without introducing any additional parameter [11]. The weighted scheme is effective in some relatively simple applications but may easily degenerate in some complex tasks.

Apart from the weighted combination based approaches, the co-training based approach has also been utilized [3]. Three main assumptions are made, namely sufficiency, compatibility and conditional independence. Based on these assumptions, the co-training approaches can produce relatively good performance by iteratively maximizing the mutual agreement between two distinct views. The scheme is attractive due to its effectiveness and some variants have been developed such as co-EM [4,12] and co-regularization [10,15]. However, the co-training approaches may produce poor results in some cases where the three rigorous assumptions cannot be satisfied.

Subspace learning approach has also been used for multi-view learning, which is based on the idea that all the views share a latent subspace. One representative technique is canonical correlation analysis (CCA), which models the relationships between several sets of variables [6]. Multi-view Fisher discriminant analysis is another subspace learning approach based on Fisher discriminant analysis [23]. Additionally, some efforts have been made to find low-dimensional embedding of data to fulfill the multi-view learning tasks, such as spectral analysis [19] and stochastic neighbor embedding [20].

Recently, an interesting model has been proposed termed *multi-view intact space learning*, which has drawn a large amount of attention [22]. The model is based on the view insufficiency assumption, i.e. each individual view only captures partial information. It aims to integrate the information from different views and find a latent intact space of data which contains all information. It is demonstrated to be effective on several real-world applications. Despite the success, this method has two shortcomings. First, the model adopts two regularization terms to control the scale of the latent feature vectors. Due to this reason, two regularization parameters are needed for trade-off between the reconstruction error and regularization terms. Determining these parameters by cross validation costs extra computation resources and does not work well in unsupervised learning. The second drawback is that the optimization process for solving the problem is too time-consuming because it involves matrix inversion.

To address the above two shortcomings, we propose a variant model, which is based on an assumption that the latent intact space of data is actually the surface of a unit hypersphere, namely *unit intact space assumption*. Based on the assumption, the latent intact space representations of objects are points on the surface of the unit hypersphere. Therefore, there is no need of regularization terms for controlling the scale of the latent feature vectors. An efficient optimization algorithm is designed to solve the model based on the proximal gradient method [13]. Extensive experiments have been conducted on four real-world datasets to demonstrate the effectiveness of our model.

2   The Proposed Model

2.1   Background

Recently, an interesting multi-view learning model termed *multi-view intact space learning* was proposed in [22] and has drawn a large amount of attention. Based on the view insufficiency assumption, the model aims to find a latent intact space which integrates information of data from all views. Given a dataset containing N objects collected from M views, the data representation in the v-th view can be represented using a data matrix $\boldsymbol{X}^v = [\boldsymbol{x}_1^v, \ldots, \boldsymbol{x}_N^v] \in \mathbb{R}^{D^v \times N}$ where D^v is the dimensionality of the v-th view. In this work, the view generation functions are assumed to be linear which extracts the feature from the intact space $\mathcal{Z}$ to a D^v-dimensional feature space $\mathcal{X}^v$. Therefore, a view generation function from the intact space to the v-th view can be expressed as $f^v(\boldsymbol{z}_i) = \boldsymbol{W}^v \boldsymbol{z}_i$ where $\boldsymbol{W}^v \in \mathbb{R}^{D^v \times d}, \forall v = 1, \ldots, M$ is the view generation matrix and $\boldsymbol{z}_i$ is the feature representation of an object in the d-dimensional intact space. The feature representation of the latent intact space for the whole dataset can be expressed in matrix form as $\boldsymbol{Z} = [\boldsymbol{z}_1, \ldots, \boldsymbol{z}_N] \in \mathbb{R}^{d \times N}$. Adopting the Cauchy loss to measure the reconstruction error, the model is formulated as

$$\min_{\boldsymbol{Z}, \boldsymbol{W}^v} \frac{1}{MN} \sum_{v=1}^{M} \sum_{i=1}^{N} \log\left(1 + \frac{\|\boldsymbol{x}_i^v - \boldsymbol{W}^v \boldsymbol{z}_i\|_2^2}{c^2}\right) + C_1 \sum_{v=1}^{M} \|\boldsymbol{W}^v\|_F^2 + C_2 \sum_{i=1}^{N} \|\boldsymbol{z}_i\|_2^2$$

(1)

where $\|\cdot\|_F$ is the Frobenius norm and $\|\cdot\|_2$ is the L_2 norm. In the model, two regularization terms are introduced to control the scale of the latent feature vectors and two regularization parameters are needed for trade-off between reconstruction error and regularization, which can be determined by cross validation. An iteratively reweighted residuals (IRR) algorithm is designed to solve the model.

2.2   Multi-view Unit Intact Space Learning

Despite the success, the aforementioned method has two shortcomings. First, the model adopts two regularization terms to control the scale of the latent feature vectors. Due to this reason, two regularization parameters are needed for trade-off between the reconstruction error and regularization terms. Determining these parameters by cross validation costs extra computation resources and does not work well in unsupervised learning. The second drawback is that the optimization process for solving the problem is too time-consuming because it involves matrix inversion. Therefore, we propose an improved model to solve the two shortcomings.

Our model is based on an assumption that the latent intact space of data is actually the surface of a unit hypersphere, namely *unit intact space assumption*. Based on the assumption, the latent intact representations of objects are points on the surface of the unit hypersphere so that all the feature representation vectors in the latent intact space have the same length, i.e. all the vectors are unit vectors. The assumption is reasonable for data clustering or classification problem when measuring the similarity among data points by the angles between two feature vectors [7]. Given two objects with unit feature vectors x and y, the Euclidean distance between them is $\|x - y\|_2 = \sqrt{2 - 2x^T y}$ and the cosine distance is $1 - x^T y$. It is easy to find that the two distance measures are positively correlated, i.e. the Euclidean distance is equivalent to the angles between the two unit vectors. Therefore, after obtaining the unit intact space representation of data, the Euclidean distance can be used to measure the similarity between data objects. For illustration purpose, Fig. 1 shows the main idea of the model.

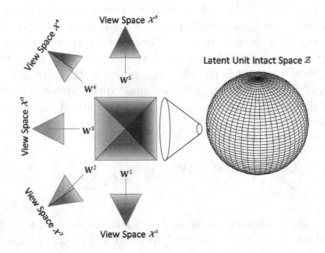

Fig. 1. Illustration of *unit intact space assumption*: the red point on the hypersphere stands for an object in the unit intact space $\mathcal{Z}$. Applying the view generation matrix W^v, we get the view representations of all the objects in feature space $\mathcal{X}^v$. (Color figure online)

According to the assumption, instead of introducing two regularization terms, the scale of latent intact feature vectors is controlled by directly adding a unit-length constraint for all the objects. The model is called *Multi-View Unit Intact space Learning* (MVUIL). The formulation of the model is

$$\min_{Z,W^v} \sum_{v=1}^{M} \|X^v - W^v Z\|_F^2 \qquad (2)$$
$$\text{s.t. } \|z_i\|_2 = 1, \forall i = 1, \dots, N.$$

Compared with the multi-view intact space learning, our model is much simpler where no prior parameters are needed.

3   Optimization

By using alternating optimization scheme, the optimization problem in Eq. (2) can be decomposed into two subproblems over the view generation matrices

$\{\boldsymbol{W}^v\}$ and the latent intact feature vectors $\{\boldsymbol{z}_i\}$ respectively. In particular, an iterative optimization method based on proximal gradient scheme [13] is proposed to solve the problem.

3.1   Update Latent Feature Vectors in Unit Intact Space

In this subsection, by fixing the view generation matrices $\{\boldsymbol{W}^v\}$, the latent intact feature vectors $\{\boldsymbol{z}_i\}$ are updated. By introducing a penalty function

$$g(\boldsymbol{z}_i) = \begin{cases} 0, & \|\boldsymbol{z}_i\|_2 = 1, \\ \infty, & \text{otherwise}, \end{cases} \tag{3}$$

the subproblem with respect to $\boldsymbol{z}_i$ is equivalent to minimizing the following

$$\mathcal{L}_z(\boldsymbol{z}_i) = f_z(\boldsymbol{z}_i) + g(\boldsymbol{z}_i) \tag{4}$$

where $f_z(\boldsymbol{z}_i) = \sum_{v=1}^M \|\boldsymbol{x}_i^v - \boldsymbol{W}^v \boldsymbol{z}_i\|_2^2$. Denoted by $\boldsymbol{z}_i^{(k)}$ the solution during the k-th iteration, we consider a quadratic model to approximate $f_z(\boldsymbol{z}_i)$ in the form

$$q_{t_k}\left(\boldsymbol{z}_i, \boldsymbol{z}_i^{(k)}\right) = f_z\left(\boldsymbol{z}_i^{(k)}\right) + \left\langle \boldsymbol{z}_i - \boldsymbol{z}_i^{(k)}, \nabla f_z\left(\boldsymbol{z}_i^{(k)}\right) \right\rangle + \frac{1}{2t_k}\left\|\boldsymbol{z}_i - \boldsymbol{z}_i^{(k)}\right\|_2^2 \tag{5}$$

where $\langle \cdot, \cdot \rangle$ is the inner product of vectors, t_k is the step size in the k-th iteration, and $\nabla f_z\left(\boldsymbol{z}_i^{(k)}\right) = 2\sum_{v=1}^M \boldsymbol{W}^{vT}(\boldsymbol{W}^v \boldsymbol{z}_i^{(k)} - \boldsymbol{x}_i^v)$ is the partial gradient of f_z with respect to $\boldsymbol{z}_i$ at $\boldsymbol{z}_i^{(k)}$. In Eq. (5), the first two terms stand for the linearized part of f_z at $\boldsymbol{z}_i^{(k)}$ and the last term measures the local error. Then an approximation model of $\mathcal{L}_z$ is given by

$$Q_{t_k}\left(\boldsymbol{z}_i, \boldsymbol{z}_i^{(k)}\right) = q_{t_k}(\boldsymbol{z}_i, \boldsymbol{z}_i^{(k)}) + g(\boldsymbol{z}_i). \tag{6}$$

We can get the $(k+1)$-th solution in following closed form

$$\begin{aligned}
\boldsymbol{z}_i^{(k+1)} &= \arg\min_{\boldsymbol{z}_i} Q_{t_k}\left(\boldsymbol{z}_i, \boldsymbol{z}_i^{(k)}\right) \\
&= \arg\min_{\boldsymbol{z}_i}\left\{g(\boldsymbol{z}_i) + \frac{1}{2t_k}\left\|\boldsymbol{z}_i - \left(\boldsymbol{z}_i^{(k)} - t_k\nabla f_z\left(\boldsymbol{z}_i^{(k)}\right)\right)\right\|_2^2\right\} \\
&= \mathbf{prox}_{t_k g}\left(\boldsymbol{z}_i^{(k)} - t_k\nabla f_z\left(\boldsymbol{z}_i^{(k)}\right)\right)
\end{aligned} \tag{7}$$

where $\mathbf{prox}_{t_k g}(\cdot)$ is the proximal operator of g with parameter t_k. Accordingly, the updating formula of $\boldsymbol{z}_i$ is given by

$$\boldsymbol{z}_i^{(k+1)} = \begin{cases} \boldsymbol{b}_i^{(k)}, & \left\|\boldsymbol{b}_i^{(k)}\right\|_2 = 1, \\ \boldsymbol{b}_i^{(k)} / \left\|\boldsymbol{b}_i^{(k)}\right\|_2, & \text{otherwise}, \end{cases} \tag{8}$$

where $\boldsymbol{b}_i^{(k)} = \boldsymbol{z}_i^{(k)} - t_k\nabla f_z\left(\boldsymbol{z}_i^{(k)}\right)$.

Algorithm 1. Multi-view unit intact space learning

Input: Data of M views $\{X^1, \ldots, X^M\}$, dimensionality of latent intact space d
1: Randomly initialize view generation matrices $W^{v(0)}$
2: **repeat**
3:    For each object i, update z_i by Eq. (8)
4:    For each view v, update W^v by Eq. (10)
5: **until** Convergence or reaching the maximum number of iterations
Output: Latent intact feature vectors $\{z_i\}$ for all objects

---

3.2   Update View Generation Matrices

In this subsection, by fixing the latent intact feature vectors $\{z_i\}$, the view generation matrices $\{W^v\}$ are updated. The subproblem with respect to W^v becomes minimizing the following

$$\mathcal{L}_w(W^v) = \|X^v - W^v Z\|_F^2 \tag{9}$$

which is an unconstrained minimization problem. We can solve the problem by the following updating formula

$$W^{v(k+1)} = W^{v(k)} - \eta_k \nabla \mathcal{L}_w\left(W^{v(k)}\right) \tag{10}$$

where $W^{v(k)}$ is the solution during the k-th iteration, η_k is the step size in the k-th iteration, and $\nabla \mathcal{L}_w\left(W^{v(k)}\right) = 2\left(W^{v(k)} Z - X^v\right) Z^T$ is the partial gradient of $\mathcal{L}_w$ with respect to W^v at $W^{v(k)}$.

Using the above alternating update scheme, we can get the latent unit intact feature representations of data from their multi-view feature representations. The pseudo code of the optimization algorithm is given in Algorithm 1.

3.3   Convergence Analysis

Since we utilize the proximal gradient method to solve the optimization problem, the convergence analysis about $\{z_i\}$ can be obtained as follows. Assume that the gradient ∇f_z satisfies the Lipschitz condition and $L(f_z)$ is the Lipschitz constant. We can get the convergence condition of z_i according to the theorems in [2]. Let $\{z_i^{(k)}\}$ be the sequence generated by the proximal gradient method with either a constant or a backtracking step-size rule, then by applying the theorems in [2] for every $n > 1$ we have

$$\min_{2 \leq k \leq n} \left\| z_i^{(k-1)} - z_i^{(k)} \right\|_2 \leq \frac{1}{\sqrt{n}} \left(\frac{2\mathcal{L}_z(z_i^{(1)}) - \mathcal{L}_z^*}{\beta L(f_z)} \right)^{1/2} \tag{11}$$

where $\mathcal{L}_z^*$ is the optimal value of $\mathcal{L}_z$, $\beta = 1$ for the constant step-size setting and $\beta = L_0/L(f_z)$ for the backtracking case[1]. Moreover, $\left\|z_i^{(k-1)} - z_i^{(k)}\right\|_2 \to 0$ as $k \to \infty$. Therefore, the value of $\{z_i\}$ will gradually converge in the iteration.

3.4   Complexity Analysis

In order to demonstrate the efficiency of our algorithm, we give computational complexity analysis and make comparison with the multi-view intact space learning algorithm. We first analyze the time complexity of our method. Since the optimization procedure is iterative, we begin with analyzing the complexity in each iteration and then obtain the total complexity of the whole algorithm. For simplicity, we denote $D_1 = \sum_{v=1}^{M} D^v$. In each iteration, the computation time of calculating the gradient during updating unit intact feature vectors $\{z_i\}$ is $O(D_1 dN)$. The computation time of calculating the gradient during updating view generation matrices $\{W^v\}$ is $O(D_1 dN)$. The updating time for $\{z_i\}$ and $\{W^v\}$ are respectively $O(dN)$ and $O(D_1 d)$. Therefore, the total time complexity of our algorithm is $O(T_1 D_1 dN)$ where T_1 is the number of iterations.

Similarly, the computational complexity of multi-view intact space learning in [22] can be analyzed as follows. Like the analysis before, we first consider the complexity in each iteration. According to the description in [22], the computational complexity for calculating the weight function is $O(D_1 dN)$. The time complexity for evaluating the latent intact feature for one object is $O(D_1 d^2 + d^3)$ since the complexity for calculating the matrix inversion is $O(d^3)$. For N objects in the whole dataset, the complexity is $O(D_1 d^2 N + d^3 N)$. For the v-th view, the complexity of evaluating the view generation function is $O(D^v d^2 + D^v dN + d^2 N + d^3)$. For all views, the computation complexity is $O(D_1 d^2 + D_1 dN + d^2 N + d^3)$ in one iteration. Therefore, assuming that the number of iterations is T_2, the total complexity of the algorithm is $O(T_2 D_1 d^2 N + T_2 d^3 N)$, which is several orders of magnitude larger than our method.

4   Experiments

In this section, extensive experiments are conducted to demonstrate the effectiveness of our method by regarding it as the preprocessing step of the multi-view clustering task. That is, k-means is applied on the unit intact space representations of the multi-view data, the performance of which is compared with several start-of-the-art clustering algorithms.

4.1   Datasets and Evaluation Measures

Handwritten Numeral Dataset: Multiple Features (Mfeat) dataset is a handwritten numeral image dataset from UCI machine learning repository [1]. The

---

[1] L_0 is the initial Lipschitz constant in backtracking step-size rule. More detailed description about the theorem for proof and step-size setting can be found in [2].

dataset contains 2000 images from 10 classes. In our experiment, three kinds of features are used to represent each image. The fac feature stands for 216 profile correlations, the fou feature stands for 76 Fourier coefficients and the kar feature stands for 64 Karhunen-Love coefficients.

Multi-source News Dataset: 3Sources dataset[2] is a multi-source news dataset whose objects are all news stories collected from three news sources, namely BBC, The Guardian and Reuters. The original dataset contains 984 new articles covering 416 distinct news stories. However, not all news are reported by all three medias. In our experiment, only 169 stories reported by all three sources are used so that each object has three views. All the stories from three sources use TF-IDF representaion.

Multi-view Text Datasets: BBC and BBCSport[3] are two multi-view text datasets constructed from the single-view BBC and BBCSport corpora by splitting news articles into related segments of text. The construction is done by first separating the raw documents into segments and then assigning segments randomly to views. Both datasets are split into four views and every view uses TF-IDF features to represent the data. For the BBC dataset 685 news objects are collected and for the BBCSport dataset 116 news objects are collected.

In our experiment, three measures are used for evaluating clustering performance namely normalized mutual information (NMI), clustering accuracy (ACC) and purity (PUR), which are all widely used. We can see [16] for detailed information.

4.2 Parameter Analysis

In this subsection, we analyze the effect of the dimensionality d of the latent unit intact space on the learning performance. By using different dimensionality d, we analyze the clustering performance on three evaluation measures. The results are shown in Fig. 2. From the figure, our algorithm performs well when the dimensionality d lies in some relatively wide range. For example, it produces good results with $d \in [100, 160]$ on Mfeat and with $d \in [3500, 5000]$ on 3Sources. The empirical results show that a value closed to the average dimensionality of all views is a suitable choice for generating good results.

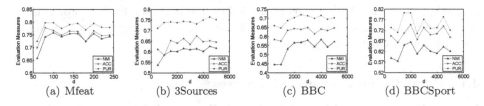

| (a) Mfeat | (b) 3Sources | (c) BBC | (d) BBCSport |

Fig. 2. Parameter analysis on four real-world datasets.

---

[2] http://mlg.ucd.ie/datasets/3sources.html.
[3] http://mlg.ucd.ie/datasets/segment.html.

4.3   Comparison Results

In this subsection, comparison experiments are conducted on the clustering performance between the proposed *Multi-View Unit Intact space Learning* (MVUIL) for clustering and several state-of-the-art algorithms, including clustering algorithms and the original multi-view intact space learning for clustering. We will first briefly introduce the compared algorithms and then analyze the comparison results on the four real-world datasets.

Two types of clustering algorithms are used to conduct our comparison experiments. The first type is the traditional single-view clustering algorithm, including k-means (KM) [24], affinity propagation (AP) [8] and normalized cut (NC) [14]. The other type is the state-of-the-art multi-view clustering algorithm, including multi-view k-means (MVKM) [5], co-training spectral clustering (CTMS) [9], co-regularized spectral clustering (CRMS) [10] and multi-view affinity propagation (MVAP) [17]. Besides, we also compare the performance of our method with multi-view intact space learning (MVIL) [24]. After obtaining the intact representation of data, we utilize k-means to get clustering results.

In the experiment, for the methods requiring pre-specifying the number of clusters, the ground-truth cluster number is used. For the single-view methods, different features from several views are concatenated to generate a combined view for each dataset, on which the single-view methods are applied. For single-view and multi-view affinity propagation, the similarity graphs are set as recommended in the original papers. For single-view and multi-view spectral clustering methods, we use Gaussian kernel to measure the similarity between two objects x and y in the dataset, i.e. $\mathrm{Sim}(x, y) = \exp{(-\|x - y\|_2^2/(2\sigma^2))}$ where the standard deviation σ is set as the median of Euclidean distances between all pairs of objects [16]. For both MVIL and the proposed MVUIL, the same dimensionality of intact space is used for each dataset. And for MVIL, the two regularization parameters are tuned to generate the best clustering performance. The results on three evaluation measures are given in Table 1, Tables 2 and 3. In what follows, we analysis the comparison results.

Mfeat: From the tables, we can see that the AP and MVAP algorithms achieve the highest purity. However, the two algorithms produce the worst performance in terms of ACC. It is because that the true number of clusters are not given for the two algorithms so that the two algorithms tend to separate some ground-truth classes into several small clusters. Considering all the three measures, we find that CRMS produces the best results. This is mainly due to the fact that spectral methods outperform other clustering methods on image datasets. Nevertheless, our method still achieves relatively good performance on this dataset.

3Sources: Unlike the Mfeat dataset, it is more difficult for single-view clustering algorithms to generate good clustering results. Besides, the multi-view learning approach based on weighted combination (i.e. MVKM) is also not applicable for this dataset. When comparing KM and MVKM, we find MVKM produces worse results than KM in terms of all the three measures. The phenomenon implies that the weighted combination approach is not suitable for improving the

Table 1. Comparison on NMI over 100 runs. For each dataset, the first line represents mean and the second line represents standard deviation. The highest mean NMIs are highlighted in bold.

Datasets	KM	NC	AP	MVKM	CRMS	CTMS	MVAP	MVIL	MVUIL
Mfeat	0.616	0.596	0.628	0.719	**0.769**	0.755	0.643	0.720	0.758
	0.024	0.013	0.000	0.036	0.035	0.023	0.000	0.020	0.034
3Sources	0.390	0.385	0.274	0.321	0.513	0.565	0.437	0.475	**0.626**
	0.065	0.051	0.000	0.073	0.019	0.028	0.000	0.054	0.055
BBC	0.305	0.032	0.271	0.233	0.229	0.296	0.287	0.459	**0.583**
	0.149	0.063	0.000	0.058	0.031	0.008	0.000	0.060	0.068
BBCSport	0.317	0.158	0.214	0.235	0.273	0.407	0.389	0.566	**0.665**
	0.110	0.065	0.000	0.033	0.025	0.044	0.000	0.089	0.105

Table 2. Comparison on ACC over 100 runs. For each dataset, the first line represents mean and the second line represents standard deviation. The highest mean ACCs are highlighted in bold.

Datasets	KM	NC	AP	MVKM	CRMS	CTMS	MVAP	MVIL	MVUIL
Mfeat	0.613	0.549	0.161	0.717	**0.803**	0.778	0.294	0.761	0.767
	0.054	0.036	0.000	0.072	0.072	0.042	0.000	0.046	0.067
3Sources	0.498	0.470	0.515	0.495	0.562	0.626	0.254	0.535	**0.652**
	0.066	0.046	0.000	0.069	0.029	0.065	0.000	0.054	0.082
BBC	0.472	0.330	0.310	0.417	0.457	0.454	0.168	0.571	**0.641**
	0.080	0.087	0.000	0.038	0.015	0.020	0.000	0.087	0.083
BBCSport	0.453	0.360	0.310	0.406	0.461	0.591	0.224	0.603	**0.746**
	0.075	0.043	0.000	0.029	0.033	0.057	0.000	0.103	0.110

Table 3. Comparison on PUR over 100 runs. For each dataset, the first line represents mean and the second line represents standard deviation. The highest mean PURs are highlighted in bold.

Datasets	KM	NC	AP	MVKM	CRMS	CTMS	MVAP	MVIL	MVUIL
Mfeat	0.663	0.605	0.906	0.751	0.817	0.794	**0.909**	0.771	0.798
	0.038	0.025	0.000	0.055	0.061	0.034	0.000	0.033	0.051
3Sources	0.574	0.563	0.527	0.535	0.691	0.732	0.757	0.655	**0.764**
	0.048	0.045	0.000	0.068	0.015	0.019	0.000	0.041	0.049
BBC	0.479	0.333	0.577	0.434	0.483	0.543	0.639	0.642	**0.716**
	0.081	0.077	0.000	0.040s	0.028	0.007	0.000	0.074	0.058
BBCSport	0.458	0.371	0.405	0.409	0.532	0.636	0.716	0.740	**0.802**
	0.076	0.049	0.000	0.027	0.029	0.052	0.000	0.087	0.095

clustering performance. We also find that CTMS and CRMS are both effective approaches to combine information from different views. MVAP has relatively better performance that it improves the values on NMI and purity. It is shown that MVIL produces worse result than CTMS but our method produces better result, implying that MVIL fails to find the intact space but our method succeeds. Overall, our method achieves the best clustering result compared with other methods in terms of all three measures.

BBC: Similar to the 3Sources dataset, the multi-view learning approach based on weighted combination (i.e. MVKM) is not applicable, i.e. it produces worse results than single-view KM. Besides, we find MVAP produces lower ACC and higher PUR than AP, implying that it tends to separate the true clusters into small clusters. CTMS also performs relatively well on this dataset. Compared with these methods, both our method and MVIL have made significant improvement on clustering performance. Our method performs better and it nearly doubles the NMI produced by CTMS. All the other algorithms, including both single-view and multi-view algorithms, produce poor results whose NMI values are all below 0.31. The main reason may be that, on this dataset, each original news article is split into four segments, taken as four different but complementary views. All the compared clustering methods fail to recover a complete feature space, leading to poor clustering performance. However, our method finds a unit intact space for data, which provides a relatively complete feature representation for clustering.

BBCSport: Since the dataset is obtained by using the same data extraction method as BBC, the dataset has the similar property. Therefore, the multi-view approaches mentioned above have similar performance to those on the BBC dataset. Our method has also made significant improvement on clustering performance. The underlying reason is the same as the previous BBC dataset.

5 Conclusion

In this paper, we propose a multi-view learning method termed *multi-view unit intact space learning*. It is an improved model of *multi-view intact space learning* which aims to learn a latent intact space integrating the information from different views. Although *multi-view intact space learning* succeeds in finding intact space, the model has two obvious shortcomings, namely the adoption of two regularization parameters and the high time complexity of optimization. Based on the *unit intact space assumption*, our model can find the intact feature representation without introducing any prior parameters. An efficient algorithm is designed to solve the model based on proximal gradient method. Experiments on real-world datasets demonstrate the effectiveness of our method.

Acknowledgments. This work was supported by Key Research and Development Program of Guangdong (2015B010108001), NSFC (61502543), Guangdong Natural Science Funds for Distinguished Young Scholar (2016A030306014) and Tip-top Scientific and Technical Innovative Youth Talents of Guangdong special support program (No. 2016TQ03X542).

References

1. Bache, K., Lichman, M.: UCI machine learning repository (2013)
2. Beck, A., Teboulle, M.: Gradient-based algorithms with applications to signal recovery. Convex Optim. Signal Process. Commun. 42–88 (2009)
3. Blum, A., Mitchell, T.: Combining labeled and unlabeled data with co-training. In: Proceedings of The Eleventh Annual Conference on Computational Learning Theory, pp. 92–100. ACM (1998)
4. Brefeld, U., Scheffer, T.: Co-EM support vector learning. In: Proceedings of The Twenty-first International Conference on Machine Learning, p. 16. ACM (2004)
5. Cai, X., Nie, F., Huang, H.: Multi-view k-means clustering on big data. In: IJCAI, pp. 2598–2604. Citeseer (2013)
6. Chaudhuri, K., Kakade, S.M., Livescu, K., Sridharan, K.: Multi-view clustering via canonical correlation analysis. In: Proceedings of the 26th Annual International Conference on Machine Learning, pp. 129–136. ACM (2009)
7. Dhillon, I.S., Modha, D.S.: Concept decompositions for large sparse text data using clustering. Mach. Learn. 42(1–2), 143–175 (2001)
8. Frey, B.J., Dueck, D.: Clustering by passing messages between data points. Science 315(5814), 972–976 (2007)
9. Kumar, A., Daumé, H.: A co-training approach for multi-view spectral clustering. In: Proceedings of the 28th International Conference on Machine Learning (ICML 2011), pp. 393–400 (2011)
10. Kumar, A., Rai, P., Daume, H.: Co-regularized multi-view spectral clustering. In: Advances in Neural Information Processing Systems, pp. 1413–1421 (2011)
11. Nie, F., Li, J., Li, X., et al.: Parameter-free auto-weighted multiple graph learning: a framework for multiview clustering and semi-supervised classification. In: International Joint Conferences on Artificial Intelligence (2016)
12. Nigam, K., Ghani, R.: Analyzing the effectiveness and applicability of co-training. In: Proceedings of the Ninth International Conference on Information and Knowledge Management, pp. 86–93. ACM (2000)
13. Parikh, N., Boyd, S., et al.: Proximal algorithms. Found. Trends Optim. 1(3), 127–239 (2014)
14. Shi, J., Malik, J.: Normalized cuts and image segmentation. IEEE Trans. Pattern Anal. Mach. Intell. 22(8), 888–905 (2000)
15. Sindhwani, V., Niyogi, P., Belkin, M.: A co-regularization approach to semi-supervised learning with multiple views. In: Proceedings of ICML Workshop on Learning with Multiple Views, pp. 74–79 (2005)
16. Von Luxburg, U.: A tutorial on spectral clustering. Stat. Comput. 17(4), 395–416 (2007)
17. Wang, C.D., Lai, J.H., Philip, S.Y.: Multi-view clustering based on belief propagation. IEEE Trans. Knowl. Data Eng. 28(4), 1007–1021 (2016)
18. Wang, M., Hua, X.S., Yuan, X., Song, Y., Dai, L.R.: Optimizing multi-graph learning: towards a unified video annotation scheme. In: Proceedings of the 15th ACM International Conference on Multimedia, pp. 862–871. ACM (2007)
19. Xia, T., Tao, D., Mei, T., Zhang, Y.: Multiview spectral embedding. IEEE Trans. Syst. Man Cybern. Part B (Cybern.) 40(6), 1438–1446 (2010)
20. Xie, B., Mu, Y., Tao, D., Huang, K.: m-SNE: Multiview stochastic neighbor embedding. IEEE Trans. Syst. Man Cybern. Part B (Cybern.) 41(4), 1088–1096 (2011)
21. Xu, C., Tao, D., Xu, C.: A survey on multi-view learning (2013). arXiv preprint arXiv:1304.5634

22. Xu, C., Tao, D., Xu, C.: Multi-view intact space learning. IEEE Trans. Pattern Anal. Mach. Intell. **37**(12), 2531–2544 (2015)
23. Xu, J., Han, J., Nie, F.: Discriminatively embedded k-means for multi-view clustering. In: Proceedings of the IEEE Conference on Computer Vision and Pattern Recognition, pp. 5356–5364 (2016)
24. Xu, R., Wunsch, D.: Survey of clustering algorithms. IEEE Trans. Neural Netw. **16**(3), 645–678 (2005)

A Novel Blemish Detection Algorithm
for Camera Quality Testing

Kun Wang, Kwok-Wai Hung$^{(\boxtimes)}$, and Jianmin Jiang

College of Computer Science and Software Engineering,
Shenzhen University, Shenzhen, Guangdong Province, China
2161230417@email.szu.edu.cn,
{kwhung,jianmin.jiang}@szu.edu.cn

Abstract. In the camera manufacturing, there exist dusts, fingerprints, and water spots on the image sensor and lens. Hence, the resultant effects of darker region is called blemish, which causes a significant reduction in camera quality. The shapes of blemishes are diverse and irregular. Traditional method detects blemishes using image median filtering in a single direction which leads to false alarm and mis-detection for images with high level of noises. Thus, we present a novel filtering method for blemish detection, which utilizes four directional filters in the 0, 45, 90, 135 degrees directions. Compared to the conventional single direction filter, the multidirectional filters take into account more spatial information to more accurately detect blemishes for both weak and strong blemishes. Moreover, the proposed method uses a new adaptive threshold to better accommodate different image noise levels automatically. Experimental results on two batches of production samples (600 images) show the effectiveness of the proposed method over the conventional method.

Keywords: Blemish detection · Camera manufacturing · Multidirectional filter

1 Introduction

Image capture devices, including cell phone cameras, handheld digital cameras and video cameras, are prone to lens smudges and blemishes due to the industrial design and the production environment of the devices [1]. It is an important task to control the quality in the camera manufacturing and one of the pivotal issues is blemish detection. Inspection by human operator is depended on their physical and psychological state [2] and thus is inconsistent. Human visual inspection is time-consuming, tedious and highly dependent on inspector experience, conditions or mood [3]. Compared with human inspection, the automated blemish detection system is more accurate and cuts down the costs. Currently, the use of various filtering techniques is common in automated visual inspection tasks, for example, by using various texture filters [4]. One growing area of the blemish detection is the manufacturing process of flat liquid-crystal device displays [5] and light-emitting device chips [6].

The appearance of blemish will affect the quality of the camera in camera manufacturing. Blemish is a darker area than the surrounding pixels. Most blemishes are caused by dust on sensor, blemishes can be caused also by water spot or fingerprint on

© Springer International Publishing AG 2017
G. Li et al. (Eds.): KSEM 2017, LNAI 10412, pp. 224–236, 2017.
DOI: 10.1007/978-3-319-63558-3_19

lens as shown in Fig. 1. The majority of blemishes are not only very small but also they are extremely diverse and can assume various forms [7]. Blemishes appear as low contrast, non-uniform brightness regions, typically larger than single pixels [8]. A blemish defect in a camera may be manifested by one or more blurred spots as seen when a resulting digital image produced by the camera is displayed.

Blemishes may be contrasted with other defects such as vignettes. A photograph or drawing whose edges gradually fade into the surrounding paper is called a vignette. The vignetting means a phenomenon of radial falloff of the image intensity from the image center [9]. Vignettes are usually minor defects that are present in almost every manufactured specimen, and may be alleviated by post-capture image processing that can correct for vignetting and less shading defects. But the blemishes are difficult to detect because their edges are smooth and slow changes in the image signal. The number of pixels in high resolution complementary metal oxide semiconductor camera sensors has increased rapidly in recent years. Hence, a physically small dust can cause a large amount of blemish pixels. Blemishes may be severe enough so as to result in a particular specimen being flagged as a failed unit.

In this paper, we propose an efficient filtering method for blemish detection. The method is based on image scaling, filtering, difference calculation and thresholding. The novel method performs well on testing 600 production samples. The rest of organization of this paper is as follows. Section 2 describes the related works, and Sect. 3 gives the details of the proposed method. Section 4 shows the experimental results and Sect. 5 conclude the paper.

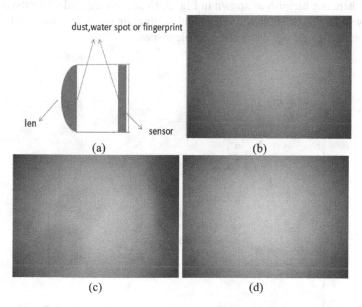

Fig. 1. (a) Blemish caused by dust on sensor, or by water spot or fingerprint on lens (b) blemishes caused by dust on sensor (c) blemishes caused by water spot on lens (d) blemishes caused by fingerprint on lens.

2 Related Works

2.1 Traditional Methods Based on Image Filtering

There are several methods to detect blemishes [10, 11] by using various filtering techniques to detect blemish. These traditional methods are based on an image taken on a bright flat surface by the camera being tested. The raw image may contain certain artifacts such as noise and vignetting. Specifically, the traditional method [11] uses the single direction median filter to detect blemishes. The method has four major steps: (*i*) image size reduction by scaling, (*ii*) blemish detection by filtering, (*iii*) image subtraction, and (*iv*) thresholding. Let us describe the details of the traditional method as follows.

2.2 Image Size Reduction by Scaling

First, the raw image is scaled down to a smaller size scaled image. Scaling the raw image speeds up the processing and reduces the noise. Image size reduction can speed up the processing and reduce the influence by noise to some extent.

2.3 Median Filtering

Next, a median filtering operation is performed on the scaled image. Since blemish is a darker area than the surrounding pixels, the intensity profile of the scaled image will drop when there is a blemish as shown in Fig. 2. Hence, we can calculate the difference image between the median filtered image and original image, where the positive values of the difference image are considered as blemishes.

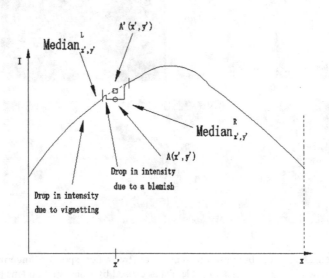

Fig. 2. Intensify profile for line x

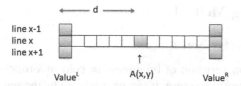

Fig. 3. Single directional median filter

After scaling down the raw image, a traditional approach for this would be to use a filter, presented in Fig. 3, to reduce the effect of blemish and then calculate a difference image between the filtered image and original image. The filter is applied on the rows x of the scaled image, where the width d is experimentally tuned, as follows,

$$Value^L = median[A(x-1,y-d), A(x,y-d), A(x+1,y-d)]$$
$$Value^R = median[A(x-1,y+d), A(x,y+d), A(x+1,y+d)] \qquad (1)$$
$$A'(x,y) = (Value^L + Value^R)/2$$

However, shape of blemishes is so diversified and irregular that it is difficult to detect all kinds of blemishes in only one direction.

2.4 Image Subtraction

Then we calculate the difference between the scaled image and the filtered image,

$$A_{difference} = A'(x,y) - A(x,y) \qquad (2)$$

Blemish is a region darker than the surrounding pixels. So, the positive values in the difference image between the filtered image and the scaled image will be considered as blemishes.

2.5 Thresholding

Once the difference is calculated, a thresholding operation will be applied to the difference image to retain the blemish pixels from background noises. Pixels which are greater than the threshold will be considered as blemishes. The traditional threshold value can be written as

$$T = Ms \qquad (3)$$

where the parameter M can be founded according to experiments, and the variable s stands for standard deviation of the difference image. Traditional threshold formula (3) can hardly deal with both low and high noisy images with one parameter M.

Due to the single direction filter and the traditional threshold formula with one parameter, the detection result is not promising by using traditional method which cannot alleviate the interference by noise but considers noise as blemishes. In order to overcome this difficulty, a new type of filter which can detect blemishes accurately in both high and low noisy images with a new threshold formula is needed.

3   Novel Filtering Method

3.1   Influences of Image Noises

Image noise is random variation of brightness or color information in images, and is usually an aspect of electronic noise. It can be produced by the sensor and circuitry of a scanner or digital camera. Image noise can also originate in the unavoidable shot noise of an ideal photon detector. Image noise is an undesirable by-product of image capture that adds spurious and extraneous information [12]. Intensity profile will drop when image is noisy as shown in Fig. 4. Blemish is a region darker than the surrounding pixels. Traditional approach aims at detecting the drops of intensity profile and

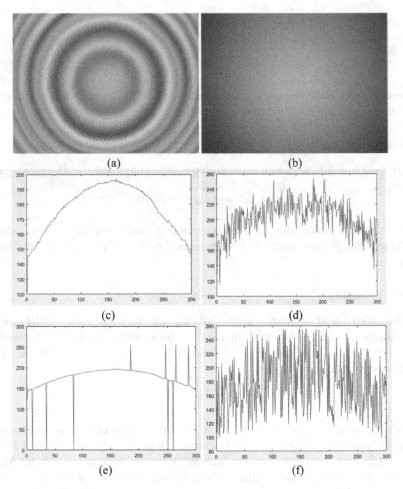

Fig. 4. A low noise image for test with size 300 × 400. (a) Raw image (b) RGB image (c) intensity profile for row 140 (d)–(f) intensity profile for row 140 after adding (d) Gaussian noise (e) salt and pepper noise (f) multiplicative noise.

considers any drops as blemishes. This is the major cause of the unsatisfactory detection results by using traditional method. In order to better separate noises from blemish, the filter needs to take into account more spatial information.

Drops caused by image noise are random and it is uncertain that drops appear in which directions. Detection results in different directions can be seen in Fig. 5. In the test image, pixels located within the red rectangle are in a noise region, pixels located within red circle are in a blemish region. It is obvious that the drops caused by noise region can only be detected in vertical direction and is almost non-existent in horizontal direction, but the drops caused by blemish can be detected in both horizontal and vertical directions. This implies detecting blemishes in multiple directions and

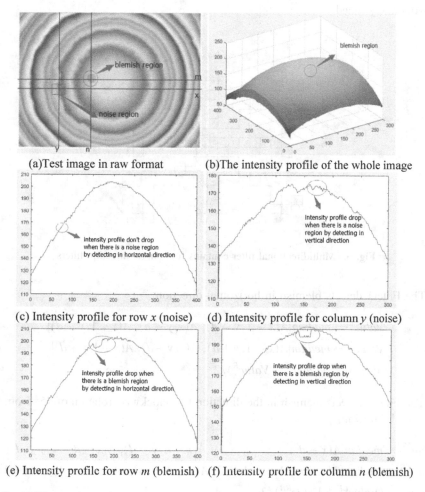

(a)Test image in raw format   (b)The intensity profile of the whole image

(c) Intensity profile for row x (noise)   (d) Intensity profile for column y (noise)

(e) Intensity profile for row m (blemish)   (f) Intensity profile for column n (blemish)

Fig. 5. (a) The test image in raw format with size 300×400, line x is 170th row, line y is 80th column, line m is 150th row, line n is 160th column (b) intensity profile of the whole image (c) intensity profile for line x (d) intensity profile for line y (e) intensity profile for line m (f) intensity profile for line n (Color figure online)

averaging all filtering results in different directions can alleviate the influence caused by noise. Besides, the shapes of blemishes are irregular, the proposed multidirectional filter can detect blemishes with various shapes more accurately.

3.2 Proposed Multi-directional Median Filter

In order to improve the detection results of the blemishes in noisy images, we proposed a novel filtering method. The filter we used contains four filters in different directions, respectively, in the horizontal direction, clockwise rotation of 45° from the horizontal direction, clockwise rotation of 90° from the horizontal direction, clockwise rotation of 135° from the horizontal direction as shown in Fig. 6. The parameters d and f are experimentally tuned.

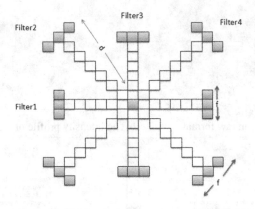

Fig. 6. Multidirectional filter contains four single direction filters

The Filter1 detects blemish in horizontal direction,

$$Value^L = median[A(x-1,y-d), A(x,y-d), A(x+1,y-d)]$$
$$Value^R = median[A(x-1,y+d), A(x,y+d), A(x+1,y+d)] \qquad (4)$$
$$A(x,y)_1 = (Value^L + Value^R)/2$$

The Filter2 detects blemish in the direction that clockwise rotation of 45° from the horizontal direction,

$$Value^{LT} = median[A(x-d+1,y-d-1), A(x-d,y-d), A(x-d-1,y-d+1)]$$
$$Value^{RB} = median[A(x+d+1,y+d-1), A(x+d,y+d), A(x+d-1,y+d+1)]$$
$$A(x,y)_2 = (Value^{LT} + Value^{RB})/2$$

$$(5)$$

The Filter3 detects blemish in the direction that clockwise rotation of 90° from the horizontal direction,

$$Value^B = median[A(x+d,y-1), A(x+d,y), A(x+d,y+1)]$$
$$Value^T = median[A(x-d,y-1), A(x-d,y), A(x-d,y+1)] \tag{6}$$
$$A(x,y)_3 = (Value^B + Value^T)/2$$

The Filter4 detects blemish in the direction that clockwise rotation of 135° from the horizontal direction,

$$Value^{LB} = median[A(x+d-1,y-d-1), A(x+d,y-d), A(x+d+1,y-d+1)]$$
$$Value^{RT} = median[A(x-d-1,y+d-1), A(x-d,y+d), A(x-d+1,y+d+1)]$$
$$A(x,y)_4 = (Value^{LB} + Value^{RT})/2$$

$$\tag{7}$$

Finally, the value of $A'(x, y)$ is the average of the results of the four filters in different directions,

$$A'(x,y) = (A(x,y)_1 + A(x,y)_2 + A(x,y)_3 + A(x,y)_4)/4 \tag{8}$$

Then we calculate the difference between the scaled image and the filtered image,

$$A_{difference} = A'(x,y) - A(x,y) \tag{9}$$

3.3 Adaptive Threshold with Bias

After that, a thresholding operation will be applied to the difference image. Compared to the traditional thresholding method, we proposed a novel thresholding method, and the thresholding value can be written as

$$T = Ms + N \tag{10}$$

where the parameters M and N can be found empirically, and the variable s stands for standard deviation of the difference image. Because of blemish effects, the standard deviation is not sufficient to describe the image noise in all scenarios, the bias N is added to make the threshold formula more robust to image noise and more accurate than traditional threshold formula. The adaptive threshold formula can get more appropriate threshold according to the noise level of difference image.

Compared with the traditional method, the multidirectional filter with new threshold can reduce the interference caused by noise effectively as shown in Fig. 7.

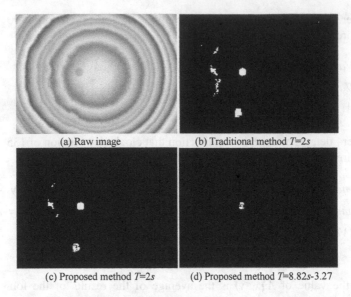

(a) Raw image   (b) Traditional method T=2s

(c) Proposed method T=2s (d) Proposed method T=8.82s-3.27

Fig. 7. (a) A noisy test image in raw format (b) the single direction filter detection result by using the thresholding value $T = 2s$ (c) the proposed multidirectional filter detection result by using the thresholding value $T = 2s$ (d) the proposed multidirectional filter detection result by using the new thresholding value $T = 8.82s - 3.27$

4 Results and Discussion

Experiments were conducted on images taken by two batches of camera samples at different production periods to evaluate the performance of proposed approach and traditional approach. Due to different production environments, 300 low noise raw images and 300 noisy raw images were captured for experiment. We have evaluated the detection results on low noise images and noisy images by using traditional method [11] and the method proposed in this paper. For traditional method, we adopted different thresholds ($T = 2s$, $T = 3.4s$) for low noise and noisy images for better performance. For the proposed method, the same threshold ($T = 8.82s - 3.27$) was adopted throughout the experiments. The experimental results are as follows.

4.1 Low Noise Samples

The filter we proposed takes into more spatial information and can detect the blemish with various shapes. By using the novel method, blemishes in low noise image can be detected accurately. The low noise image detection result can be seen in Figs. 8 and 9.

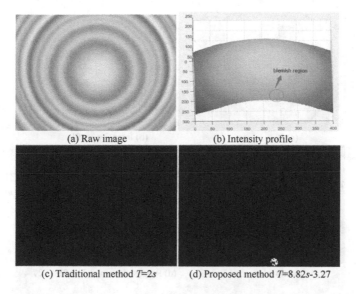

(a) Raw image   (b) Intensity profile

(c) Traditional method $T=2s$ (d) Proposed method $T=8.82s$-3.27

Fig. 8. (a) A low noise test image in raw format (b) the intensity profile of the whole image (c) the single direction filter detection result by using the thresholding value $T = 2s$ (d) the proposed multidirectional filter detection result by using the thresholding value $T = 8.82s - 3.27$.

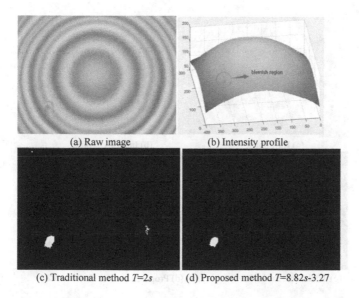

(a) Raw image   (b) Intensity profile

(c) Traditional method $T=2s$ (d) Proposed method $T=8.82s$-3.27

Fig. 9. (a) A low noise test image in raw format (b) the intensity profile of the whole image (c) the single direction filter detection result by using the thresholding value $T = 2s$ (d) the proposed multidirectional filter detection result by using the thresholding value $T = 8.82s - 3.27$.

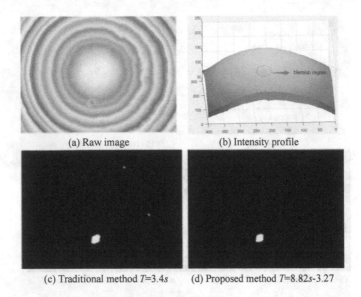

(a) Raw image   (b) Intensity profile

(c) Traditional method T=3.4s   (d) Proposed method T=8.82s-3.27

Fig. 10. (a) A noisy test image in raw format (b) the intensity profile of the whole image (c) the single direction filter detection result by using the thresholding value $T = 3.4s$ (d) the proposed multidirectional filter detection result by using the thresholding value $T = 8.82s - 3.27$.

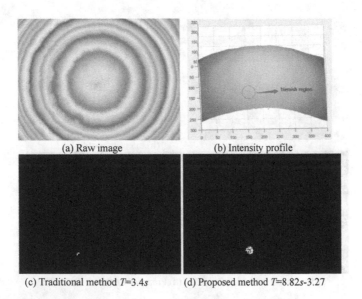

(a) Raw image   (b) Intensity profile

(c) Traditional method T=3.4s   (d) Proposed method T=8.82s-3.27

Fig. 11. (a) A noisy test image in raw format (b) the intensity profile of the whole image (c) the single direction filter detection result by using the thresholding value $T = 3.4s$ (d) the proposed multidirectional filter detection result by using the thresholding value $T = 8.82s - 3.27$.

4.2   High Noise Samples

In addition, by using the blemish detection method proposed in this paper, the blemishes in noisy images can be detected accurately and alleviate the interference caused by noise. The detection results in noisy images can be seen in Figs. 10 and 11.

Experimental results show that the proposed novel method can detect blemishes in low noise images more accurately than traditional method. Besides, the novel method also can detect blemishes in noisy images and alleviate the influence caused by noise.

5   Conclusion

In this paper, we proposed a novel method for blemish detection. The method goes through image scaling, filtering, difference calculation and thresholding to detect blemish. Compared to the traditional filter presented in Fig. 3, the proposed multidirectional filter considers more spatial information to detect the blemish in four directions. Besides, we proposed a novel threshold formula which makes the thresholding value to be more flexible and accurate to accommodate various noise levels. The novel detection method proposed in this paper can detect blemishes with various shapes and reduce the influence caused by noise, as verified by extensive experiments.

Acknowledgments. This work was supported in part by the Shenzhen Emerging Industries of the Strategic Basic Research Project (No. JCYJ20160226191842793), and the National Natural Science Foundation of China (No. 61602312, 61602314, 61620106008).

References

1. Cazier, R., Voss, S.D., Yost, J.: Blemish detection and notification in an image capture device. United States Patent No. 8982262 B2 (2015)
2. Kim, S., Kang, T., Jeong, D.: Region mura detection using efficient high pass filtering based on fast average operation. In: Proceedings of the 17th World Congress, International Federation of Automatic Control, vol. 17 (2008)
3. Lin, H.D., Chiu, Y.S.P.: Computer-aided quality system for visual blemish inspection of epoxy packages. Sci. Iranica **18**, 1591–1599 (2011)
4. Pietikainen, M., Ojala, T., Silven, O.: Approaches to texture-based classification, segmentation and surface inspection. In: Handbook of Pattern Recognition and Computer Vision, pp. 711–736 (1999)
5. Tsai, D.M., Chuang, S.T.: ID-based defect detection in patterned TFT-LCD panels using characteristic fractal dimension and correlations. Mach. Vis. Appl. **20**, 423–434 (2009)
6. Lin, H.-D., Chung, C.-Y., Chiu, S.W.: Computer-aided vision system for surface blemish detection of LED chips. In: Beliczynski, B., Dzielinski, A., Iwanowski, M., Ribeiro, B. (eds.) ICANNGA 2007. LNCS, vol. 4432, pp. 525–533. Springer, Heidelberg (2007). doi:10.1007/978-3-540-71629-7_59
7. Chiu, Y.S., Lin, H.D.: An innovative blemish detection system for curved LED lenses. Expert Syst. Appl. **40**, 471–479 (2013)

8. Pratt, W.K., Sawkar, S.S., O'Reilly, K.: Automatic blemish detection in liquid crystal flat panel displays. In: Proceedings of SPIE-The International Society (1998)
9. Reza, M.T.: Vignetting artifact reduction in an efficient way for digital camera image. Tampere University of Technology (2008)
10. Zhang, H., Shen, S., McAllister, I.A.: Camera blemish defects detection. United States Patent No. 8797429 B2 (2013)
11. Lepisto, L., Nikkanen, J., Suksi, M.: Blemish detection in camera production testing using fast difference filtering. J. Electron. Imaging **18**, 020501 (2009)
12. Boncelet, C., Bovik, A.C.: Handbook of Image and Video Processing. Academic Press, Cambridge (2005)

Learning to Infer API Mappings
from API Documents

Yangyang Lu[1,2], Ge Li[1,2(✉)], Zelong Zhao[1,2], Linfeng Wen[1,2],
and Zhi Jin[1,2(✉)]

[1] Key Lab of High-Confidence Software Technology, Ministry of Education,
Peking University, Beijing 100871, China
{luyy,lige,zhaozl,wenlf,zhijin}@pku.edu.cn
[2] School of Electronics Engineering and Computer Science, Peking University,
Beijing 100871, China

Abstract. To satisfy business requirements of various platforms and
devices, developers often need to migrate software code from one platform to another. During this process, a key task is to figure out API
mappings between API libraries of the source and target platforms. Since
doing it manually is time-consuming and error-prone, several code-based
approaches have been proposed. However, they often have the issues of
availability on parallel code bases and time expense caused by static or
dynamic code analysis.

In this paper, we present a document-based approach to infer API
mappings. We first learn to understand the semantics of API names and
descriptions in API documents by a word embedding model. Then we
combine the word embeddings with a text similarity algorithm to compute semantic similarities between APIs of the source and target API
libraries. Finally, we infer API mappings from the ranking results of API
similarities. Our approach is evaluated on API documents of *JavaSE*
and *.NET*. The results outperform the baseline model at precision@k
by 41.51% averagely. Compared with code-based work, our approach
avoids their issues and leverages easily acquired API documents to infer
API mappings effectively.

Keywords: API mappings · API similarity · API documents

1 Introduction

To support business requirements of different platforms or devices, software
developers often need to release several corresponding versions of their software projects or products. Open source projects, like Lucene and JUnit, often
support different versions for Linux, Windows and Mac OS X. Then with the
development of mobile devices, application products also release versions corresponding to iOS, Android and Windows Phone. Usually, developers often write
code under one platform first, then migrate them from the current platform

G. Li et al. (Eds.): KSEM 2017, LNAI 10412, pp. 237–248, 2017.
DOI: 10.1007/978-3-319-63558-3_20

(annotated as "the source platform") to another (annotated as "the target plat-form"). Compared with developing different versions independently, it is usually more economical to make the migration shown in Fig. 1. Since the program-ming languages used in the source and target platforms may be different, this migration process is often called language migration or code migration.

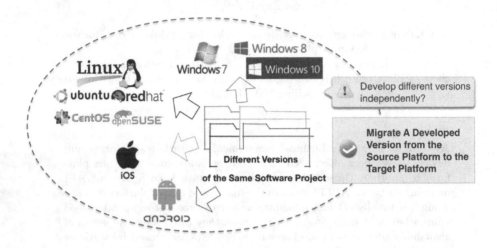

Fig. 1. Migrating software projects from one platform to another platform

During the process of code migration, a key task is to figure out API map-pings between API libraries of the source and target platforms [1,4,10]. Here API mappings are defined as the mappings of APIs that belong to different libraries but implement the same or similar functionality. Since modern software develop-ment heavily relies on API libraries [9], code migration usually need to replace APIs according to the knowledge of API mappings. For example, *JavaSE* is the basic library for Java projects and *.NET* for C# projects. Then API map-pings between the *JavaSE* library and the *.NET* library are important for the migration process of software code from platforms supporting Java to platforms supporting C#.

Since it is time-consuming and error-prone to discover API mappings manu-ally [10], researchers proposed several code-based approaches to mine API map-pings automatically. These approaches mostly built parallel code bases from software projects' different versions and took API calling information parsed by static [4,7,8,10] or dynamic [1] code analysis to figure out API mappings. How-ever, they have the issues that there may be no available corresponding versions of software projects and it is usually slow to parse API calling information with the static and dynamic code analysis.

In this paper, we propose a document-based approach to infer API mappings. We find that API documents provide functional information of APIs in API names and descriptions, which can be leveraged to infer APIs of the same or similar func-tionality. Our assumption is that two APIs have the mapping relation if their names or descriptions express similar semantics in functionality. So our approach tries to understand the semantics of API names and descriptions first by learning embeddings of words in them. On this basis, we use a text similarity algorithm to compute the semantic similarity of APIs and infer API mappings. Our experiments on *JavaSE* and *.NET* libraries outperform the baseline model at precision@k $(1, 5, 10, 20)$ by 41.5% averagely. Compared with current code-based work, it is easy to access API documents. Then the process of word embedding learning and similarity computation is fast and effective to infer API mappings.

The rest of this paper is organized as follows: Sect. 2 introduces related work; Sect. 3 illustrates our approach to infer API mappings from API documents; Sect. 4 presents details of the dataset and the experiment results; Sect. 5 gives the conclusion and discusses future work.

2   Related Work

The state-of-art work of discovering API mappings mostly takes code-based approaches.

Zhong et al. [10] proposed the MAM model to mine API mappings from different versions of software client code. They first constructed API transformation graphs (ATGs) based on aligned client code snippets, then leveraged heuristic rules to infer API mappings between *JavaSE* and *.NET* libraries.

Gokhale et al. [1] developed a prototype tool named Rosetta to infer API mappings between *JavaME* and *Android* libraries. They crawled mobile graphic applications using the above two libraries and recorded the information parsed by dynamic code analysis to mine API mappings.

Nguyen et al. [4,5] proposed the StaMiner model which takes the task of discovering API mappings as a machine translation task. They also built parallel code bases from different version of client code. Then they extracted API usage sequences from source code with static analysis and applied the IBM translation model to align API usages and infer API mappings.

After the work of StaMiner, Nguyen et al. [6] applied word2vec model on API sequences extracted from Java and C# source code and proposed the API2VEC model to learn APIs' vector representations. Then based on API2VEC's vectors and a large amount of known API mappings between *JavaSE* and *.NET*, they trained a transformation classifier (implemented as a multilayer perceptron) to find API mappings.

In the above code-based approaches, MAM, Rosetta and StaMiner need the alignment relations between method-level code snippets. These relations acquire parallel code bases of different client versions which may be unavailable for parts of projects. Then API2VEC needs a large amount of known API mappings. These approaches also have the issue of high time expense caused by static or dynamic code analysis.

Our work presents a document-based approach to infer API mappings on the basis of understanding words in API documents. API documents are available from official websites of API libraries. Compared with code analysis, it is fast to processing documents, learning word embeddings and ranking potential API mappings via similarity in our approach. Also, our approach is unsupervised. It can avoid issues of the above code-based approaches and be effective for any two API libraries without available code bases.

3 Approach

3.1 Overview

In API documents, API names and their descriptions provide useful semantic information of APIs' functionality. Because API documents are originally published to help developers understand what functions APIs have implemented. If we could understand the semantics of API names and descriptions, it seems to be feasible to infer API mappings based on their similarity.

Table 1 gives examples of API mappings between *JavaSE* and *.NET* from the literature [4]. We can find that mapped APIs share words in their names and descriptions, such as "io", "exception", "length" and "xml". Besides the same words in API names and descriptions, there are also words of relevant semantics in them, such as "length" and "number", "sequence" and "string". So we proposed the following approach to capture semantic relevance of API names and descriptions based on word embeddings and evaluate API similarity to infer API mappings.

Table 1. Examples of API mappings (left: *JavaSE*, right: *.NET*)

java.io.IOException	System.IO.IOException
Exception	Class
Signals that an I/O exception of some sort has occurred	The exception that is thrown when an I/O error occurs
java.lang.CharSequence.length	System.String.Length
Methods	Properties
Returns the length of this character sequence	Gets the number of characters in the current String object
org.w3c.dom	System.Xml
Package	Namespace
Provides the interfaces for the Document Object Model (DOM) which is a component API of the Java API for XML processing	The System.Xml namespaces contain types for processing XML. Child namespaces support serialization of XML documents or streams, XSD schemas, XQuery 1.0 and XPath 2.0, and LINQ to XML, which is an in-memory XML programming interface that enables easy modification of XML documents

Figure 2 shows the overview of our approach. We crawl raw documents of the source and target libraries from their official website and use HTML parser to extract names, descriptions and types of APIs as the preliminary corpus. Then our approach uses the following four steps to infer API mappings:

- We clean the preliminary corpus with text processing operations, which mainly contain removing analphabetic characters, removing stopwords, splitting words and lowercasing. Specifically, we add a decomposition operation of CamelCase words and package prefixes before lowercasing. This is to dense the semantic space and capture semantic relevance implicated in shared or similar words of API names and descriptions.
- Then we learn word embeddings so that we can capture semantic relevance between words. Here we train a CBOW model [3] on the processed corpus of the source and target libraries. We concatenate the name and description as training sentences, then merge vocabularies of the source and target libraries as one vocabulary.
- Before matching APIs based on the names and descriptions, we need to group APIs based on their types. Because the API in the source library should find candidates with the same type from the target library to avoid wrong mappings and reduce time expense of computation. However, API types in the source library are usually different from the ones in the target library (e.g. "Package" and "Namespace", "Methods" and "Properties" in Table 1). So in this step, we transfer their API types into a unified list with a predefined transfer map. Details are introduced in Sect. 4.1.
- Finally, for a given API in the source library, we take APIs of the same (unified) type as its candidates. Then we compute and rank APIs' semantic similarity based on word embeddings and a text similarity algorithm. Based on the ranking results, the top ones are inferred as potential API mappings.

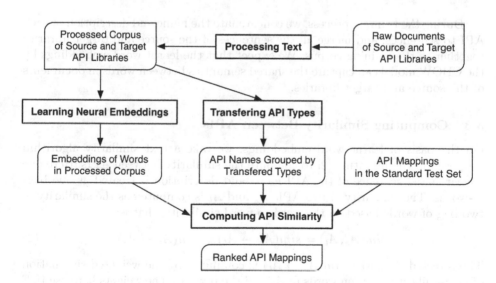

Fig. 2. Overview of our approach

3.2 Understanding API Documents

In this paper, we leverage the CBOW model to learn and represent the semantics of words in API documents. The CBOW model is proposed by Mikolov [3] to learn the language model from the natural language corpus. As shown in Fig. 3, its basic idea is predicting the intermedia word w_t through the previous k words $\{w_{t-k}, \ldots, w_{t-1}\}$ and posterior k words $\{w_{t+1}, \ldots, w_{t+k}\}$ to capture semantic relevance implicated in co-occurrence of words. Its training objective is to maximize the following log-likelihood function:

$$\frac{1}{T} \sum_{t=k}^{T-k} \log p(w_t | w_{t-k}, \ldots, w_{t-1}, w_{t+1}, w_{t+k}) \tag{1}$$

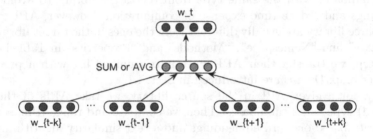

Fig. 3. CBOW model of learning word embeddings

During the training process, we concatenate the name and description of each API to sentences, and merge all the sentences of the source and target libraries together as the training corpus. We expect that the learnt word embeddings by the CBOW model can capture the shared semantics between words in documents of the source and target libraries.

3.3 Computing Similarity Between APIs

On the basis of learnt word embeddings, we take a text similarity algorithm proposed in the literature [2] to compute the similarity between APIs. Here one concatenated sentence of the API name and description is treated as one bag-of-words. The similarity of two APIs A_s and A_t is computed as the similarity of two bag-of-words based on the word-to-word similarity, that is:

$$sim(A_s, A_t) = sim(A_s \rightarrow A_t) + sim(A_t \rightarrow A_s) \tag{2}$$

The directed similarity $sim(A_s \rightarrow A_t)$ is computed by the weighted summation of the similarity between words in A_s and the text A_t. The weights here use IDF (Inverse Document Frequency,) values.

$$sim(A_s \rightarrow A_t) = \frac{\sum_{w_s \in A_s} sim(w_s, A_t) * IDF(w_s)}{\sum_{w_s \in A_s} IDF(w_s)} \tag{3}$$

As for the similarity $sim(w_s, A_t)$ between A_s's word w_s and the text A_t, it is defined as the maximal similarity between word w_s and all the words in A_t.

$$sim(w_s, A_t) = \max_{w_t \in A_t} sim(w_s, w_t) \qquad (4)$$

Then the key point of this algorithm is a reasonable measure of the similarity between words. Here we use the cosine similarity between learnt embeddings of words from Sect. 3.2, which is used widely in natural language processing tasks.

4 Evaluation

4.1 Dataset

We use the documents of *JavaSE* and *.NET* libraries to evaluate our approach. The text processing operations have been introduced in Sect. 3.1. Here we present details of transferring API types. Ideally, given an API A_s in the source library, its matching candidate A_t in the target library should have the same type so that mappings with mis-matched types will not be inferred. For example, APIs of the "Package" type should not be matched with APIs of the "Class" type. However, API types of different libraries are usually different, especially when they support different programming languages. Thus we do statistics on API types of *JavaSE* and *.NET* (the "Original" column in Table 3) and predefine a transfer map of API types based on their definitions (Table 2). Finally we get 8 unified API types for these two libraries (the "Transfered" column in Table 3).

4.2 Experimental Settings

We use the Word2Vec module in gensim 0.13.3[1] to implement the CBOW model. Its basic parameters are set as: embedding dimension = 128, the context window = 10 words, training epoch = 200. The training process of CBOW is fast

Table 2. The transfer map of API types

JavaSE		*.NET*	
Original types	Transfered types	Original types	Transfered types
Annotation type	Class	Attached properties	Method
Error	Class	Enumeration	Enum
Exception	Class	Events	Other
Nested classes	Class	Namespace	Package
Enum constants	Other	Properties	Method
Optional elements	Other	Attached events	Other
Required elements	Other	Delegate	Other
		Operators	Other
		Structure	Other

---

[1] http://radimrehurek.com/gensim.

Table 3. Dataset information of *JavaSE* and *.NET* API documents

	# Sample	Original		Filtered	
		49,150		44,762	
JavaSE	Grouped by API type	Original		Transfered	
		Annotation type	78	Class	3,370
		Class	2,261	Constructor	4,088
		Constructors	4,088	Enum	71
		Enum	71	Field	5,096
		Enum constants	430	Interface	1,026
		Error	33	Method	30,350
		Exception	506	Package	197
		Fields	5,096	Other	564
		Interface	1,026		
		Methods	30,350		
		Nested classes	492		
		Optional elements	113		
		Package	197		
		Required elements	21		
	# Sample	Original		Filtered	
		366,772		366,609	
.NET	Grouped by API type	Original		Transfered	
		Attached events	21	Class	9,005
		Attached properties	80	Constructor	12,464
		Class	9,005	Enum	1,645
		Constructors	12,464	Field	5,332
		Delegate	592	Interface	925
		Enumeration	1,645	Method	306,887
		Events	28,220	Package	95
		Fields	5,332	Other	30,256
		Interface	925		
		Methods	206,815		
		Namespace	95		
		Operators	1,127		
		Properties	99,992		
		Structure	296		

which takes less than 30 min on the dataset of Table 3 in the Core i7 CPU. Then the process of computing API similarity is also quick which outputs results averagely for 5–10 given APIs per minute of the source libraries. The speed is related to the number of candidate APIs with the same type in the target library.

We use the one-hot model as the baseline. Each API is represented as a vector, for which the dimension is the same as the vocabulary size. If the name or description of the API contains the i_{th} word of the vocabulary, then the value at the i_{th} position of the one-hot vector is 1, otherwise 0. API similarity in the baseline is computed by the cosine similarity of one-hot vectors.

The standard test set comes from the literature [4]. After filtering meaningless mappings like "@1", we finally get 145 standard mappings from *JavaSE* to *.NET*. The inferred API mappings is evaluated by precision@k. Given an API A_s from the source library, if the standard mapped API $A_{s \to t}$ of the target library is at the top-k results of inferred matched A_t, then we record it as one hit at top-k. The final precision@k is the ratio of the hitting count at top-k to the total number of test mappings.

4.3 Results

Table 4 shows precision@k ($k = 1, 5, 10, 20$) results of the baseline and our work (embedding dimension $= 128$, the context window $= 10$ words, training epoch $= 200$). It shows that our work outperforms the baseline at all the k-values, which improves precision@k ($1, 5, 10, 20$) by 61.95%, 47.60%, 34.63%, 21.86% respectively. The average promotion is 41.51%.

Furthermore, we make groups of experiments to analysis the effect brought by training parameters of word embeddings. First, we compare results with different embedding dimensions (Table 5). We find that precision@1 and precision@5 decrease distinctly with the growth of dimension from 128 to 256 then to 512. It shows that 128 dimension is enough for the current corpus to avoid over-fitting. Then we analyze results with different training epochs (Table 6).

Table 4. precision@k(%) of API migration

Model	One-hot	Our work
precision@1	14.48	23.45
precision@5	28.97	42.76
precision@10	35.86	48.28
precision@20	44.14	53.79

Table 5. precision@k(%) with different embedding dimensions

Dimension	128	256	512
precision@1	23.45	23.45	21.38
precision@5	42.76	40.00	40.00
precision@10	48.28	50.34	48.28
precision@20	53.79	53.79	53.79

Table 6. precision@k(%) with different training epochs

Train iteration	200	300	400	500
precision@1	23.45	24.83	25.51	24.83
precision@5	42.76	43.45	42.76	41.38
precision@10	48.28	48.28	48.90	50.34
precision@20	53.79	52.41	53.10	53.79

These experiments use 128 dimension of embeddings. We find that the growth of training epochs after 200 has little effect on the precision@k results.

Finally, we want to figure out the contribution of API names and descriptions in the process of inferring API mappings. We concatenated the name and description of an API as a sample in the above experiments. Here we try to divide these two parts from the corpus. Table 7 shows the statistical information

Table 7. Corpus information of different types

API library	*JavaSE*			*.NET*		
	NAME	DESC	FULL	NAME	DESC	FULL
MinLength	1	1	4	1	1	5
MaxLength	21	84	88	36	80	88
AvgLength	7.14	7.19	14.32	8.82	10.89	19.71
VocabSize	4,250	6,489	7,372	4,960	7,457	8,232
MergedVocabSize	NAME: 6,882;		DESC: 9,588;		FULL: 11,019	

Table 8. precision@k(%) with different kinds of corpus

Corpus	Iteration	precision@1	precision@5	precision@10	precision@20
NAME	200	20.69	32.41	38.62	44.83
	300	20.00	35.17	40.00	45.52
	400	20.69	31.72	41.37	45.52
	500	21.38	34.48	39.21	45.52
DESC	200	7.59	16.55	20.69	28.97
	300	7.59	14.49	19.31	28.28
	400	7.59	15.86	20.69	26.80
	500	7.59	15.17	17.24	28.97
FULL	200	23.45	42.76	48.28	53.79
	300	24.83	43.45	48.28	52.41
	400	25.51	42.76	48.90	53.10
	500	24.83	41.38	40.34	53.79

about different kinds of training corpus. "NAME" and "DESC" means use only API names and only API descriptions respectively. "FULL" means the concatenated corpus of names and descriptions. Table 8 presents precision@k results with different kinds of corpus. We can find that the results only using names is better then the results only using descriptions, but both of them can not achieve the results of concatenated corpus. It shows that the name part contributes most semantics in the process of inferring API mappings and the description part provides supplementary information for this process.

5   Conclusion

Software developers often need to release different versions of projects or products to support different platforms or devices. They usually migrate developed code from one platform to another. API mappings provide the key knowledge acquired in the process.

In this paper, we propose a document-based approach of understanding words in API documents and computing API similarity to infer API mappings. Our approach achieves averagely 41.51% improvement of precision@k ($k = 1, 5, 10, 20$) to the baseline model. While existed code-based approaches may come across the unavailability of parallel code and high time expense of code analysis, API documents is easy to access and our approach can infer API mappings more quickly by learning word embedding and computing API similarity. Then our approach only leverages weak-supervised knowledge of API types rather than strong supervision of method-level alignment on code or many known API mappings.

As for future work, we hope to measure similarity better with other algorithms or other networks, such as recurrent neural networks or convolutional neural networks. Then the available number of known API mappings for us is too small to provide supervision knowledge. We expect to collect more available API mappings and extend our approach by training supervised modules in the future.

Acknowledgement. This research is supported by the National Basic Research Program of China (the 973 Program) under Grant No. 2015CB352201 and the National Natural Science Foundation of China under Grant Nos. 61421091, 61232015 and 61502014.

References

1. Gokhale, A., Ganapathy, V., Padmanaban, Y.: Inferring likely mappings between APIs. In: 2013 35th International Conference on Software Engineering (ICSE), pp. 82–91. IEEE (2013)
2. Mihalcea, R., Corley, C., Strapparava, C., et al.: Corpus-based and knowledge-based measures of text semantic similarity. In: AAAI, vol. 6, pp. 775–780 (2006)
3. Mikolov, T., Sutskever, I., Chen, K., Corrado, G.S., Dean, J.: Distributed representations of words and phrases and their compositionality. In: Advances in Neural Information Processing Systems, pp. 3111–3119 (2013)

4. Nguyen, A.T., Nguyen, H.A., Nguyen, T.T., Nguyen, T.N.: Statistical learning approach for mining API usage mappings for code migration. In: Proceedings of the 29th ACM/IEEE International Conference on Automated Software Engineering, pp. 457–468. ACM (2014)
5. Nguyen, A.T., Nguyen, H.A., Nguyen, T.T., Nguyen, T.N.: Statistical learning of API mappings for language migration. In: Companion Proceedings of the 36th International Conference on Software Engineering, pp. 618–619. ACM (2014)
6. Nguyen, A.T., Nguyen, T.T., Nguyen, T.N.: Migrating code with statistical machine translation. In: Companion Proceedings of the 36th International Conference on Software Engineering, pp. 544–547. ACM (2014)
7. Nguyen, T.D., Nguyen, A.T., Nguyen, T.N.: Mapping API elements for code migration with vector representations. In: Proceedings of the 38th International Conference on Software Engineering Companion, pp. 756–758. ACM (2016)
8. Nguyen, T.D., Nguyen, A.T., Phan, H.D., Nguyen, T.N.: Exploring API embedding for API usages and applications. In: Proceedings of the 39th International Conference on Software Engineering. ACM (2017)
9. Thung, F., David, L., Lawall, J.: Automated library recommendation. In: Proceedings of the 2013 20th Working Conference on Reverse Engineering (WCRE 2013), Koblenz, Germany, 14–17 October 2013, pp. 182–191 (2013)
10. Zhong, H., Thummalapenta, S., Xie, T., Zhang, L., Wang, Q.: Mining API mapping for language migration. In: Proceedings of the 32nd ACM/IEEE International Conference on Software Engineering, vol. 1, pp. 195–204. ACM (2010)

Super-Resolution for Images with Barrel Lens Distortions

Mei Su, Kwok-Wai Hung[✉], and Jianmin Jiang

College of Computer Science and Software Engineering, Shenzhen University,
Shenzhen, Guangdong Province, China
kwhung@szu.edu.cn

Abstract. Camera lens distortions are widely observed in different applications for achieving specific optical effects, such as wide angle captures. Moreover, the image with lens distortion is often limited in resolution due to the cost of camera, limited bandwidth, etc. In this paper, we present a learning-based image super-resolution method for improving the resolution of images captured by cameras with barrel lens distortions. The key to the significant improvement of the resolution loss due to lens distortions is to learn a sparse dictionary with a post-processing step. During the training stage, the training images are used to learn the sparse dictionary and projection matrixes. During the testing stage, the observed low-resolution image uses the projection matrixes for two step super-resolution reconstructions of the final high-resolution image. Experimental results show that the proposed method outperforms the conventional learning-based super-resolution methods in terms of PSNR and SSIM values using the same set of training images for algorithm trainings.

Keywords: Super-resolution · Learning-based · Barrel distortion

1 Introduction

Recently, due to the rapid development of wide-angle lens for cameras, the research community is paying more attention to the camera nonlinear lens distortions such as radial barrel distortions [1, 2]. The lens distortion is that when the picture passes through the optical lens, it results in the loss of perspective, and forms the distortion effects. Barrel distortion is a distortion phenomenon that the image appears to be a barrel shape and it is caused by the physical properties and structure of the lens. Barrel distortion is the most common and the most widely used one. Hence, in our paper, we discuss the image enhancement algorithm for images with barrel distortions.

In general, cameras with fisheye lens produce images with barrel distortions. The image formation procedure of a typical camera with fisheye lens includes two major processes: barrel distortion and un-distortion (distortion correction) processes. However, the current research directions of enhancing the image quality captured from the wide-angle lens are limited to improving the accuracy of the un-distortion process [1, 2]. There are very limited researches on enhancing the resolutions of the images with nonlinear lens distortions. In [1], the authors proposed a reconstruction-based super- resolution algorithm for improving the resolutions of fisheye cameras; however,

G. Li et al. (Eds.): KSEM 2017, LNAI 10412, pp. 249–257, 2017.
DOI: 10.1007/978-3-319-63558-3_21

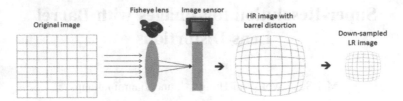

Fig. 1. Image formation process.

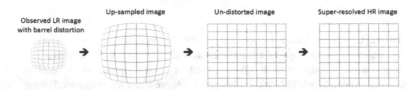

Fig. 2. Image restoration process.

this method requires multi-frame captures and the quality of reconstruction-based methods is generally worse than learning-based methods in various applications [3–6].

In this paper, a new learning-based image super-resolution method is proposed to address the resolution enhancement of the images captured from the barrel lens. Our method includes two processes: image formation process (Fig. 1) and image restoration process (Fig. 2). During image formation process, the original high-resolution (HR) image is initially distorted through fisheye lens and image sensor. Then, we get a low-resolution (LR) distorted image due to down-sampling (such as limited resolution of cameras, limited network bandwidth, etc.). During image restoration process, the observed LR image with barrel distortion is initially interpolated using bicubic method. In the next step, we restore the rectilinear image through the un-distortion process and then carry out super-resolution process to improve image quality.

The major novelty of this paper is to make use of the state-of-the-art learning-based method [3] to train a sparse dictionary for the images affected by the nonlinear barrel distortion and add a spatially varying post-processing refinement, in order to restore the image details of the distorted images. Due to the sophisticated performance of the super-resolution algorithms [3, 4] through adapting the training images to learn the sparse dictionary [4, 7] for training the ridge regressors, the resolution of the testing images is significantly improved. Meanwhile, the post-processing process is used to classify the image patches to compute coefficient matrixes, where these coefficients are then used to refine the HR image for producing the final resulting image.

Experimental results show that the proposed learning-based super-resolution method can improve the objective and subjective quality of the distorted images in PSNR and SSIM values compared with the bicubic interpolation and various state-of-the-art super-resolution algorithms [3–5] using the same training and testing images.

The rest of this paper is organized as follows. In Sect. 2, we describe the camera distortion and un-distortion processes using classical barrel lens models for simulating

$a_1 = 1, a_2 = 0.22$ $a_1 = 1, a_2 = 0.3$ $a_1 = 1, a_2 = 0.4$

Fig. 3. Physical effects of barrel distortion using different parameters a_1 and a_2 in Eq. (2).

the image formation and image restoration processes. Section 3 presents the proposed method in details. We conduct experiments and demonstrate the performance of our method in Sect. 4. Finally, in Sect. 5, we conclude this paper.

2 Related Work

For an image with an arbitrarily shaped lens radial distortion, the classical distortion method can be described as polar angle transformation [8–10]. Given an image with Cartesian coordinates (x, y), the polar coordinates are transformed as

$$r, \theta = T(x, y) \tag{1}$$

where the radius r and angle θ are the polar coordinates of (x, y) in the polar domain.

One of the commonly used nonlinear distortion model for wide-angle cameras is to transform the radius of the polar coordinates in a nonlinear manner from radius r to the new radius r_{new}, through low-order polynomial functions, as follows,

$$r_{new} = \sum_{i=1}^{n} a_i r^i \tag{2}$$

where the radius r is modified to r_{new} but the angle θ is kept constant, and the parameter a_i represents the distortion coefficients. Figure 3 shows the physical effects of this distortion model for various values of a_i, after inverse transformation to the Cartesian coordinates. To restore the radial distortion, the new radius r_{new} is restored back to the original radius r through forward and inverse polar transformations.

3 Proposed Method

In this paper, we propose a super resolution method used for images with barrel lens distortions. Our method first utilizes sparse dictionary to reconstruct the high-resolution image, and then use a post-processing procedure to reconstruct the image again. We will introduce the detail process of our method in this section.

3.1 Pretreatment Process

For the given high-resolution training images, the distortion process in Eqs. (1) and (2) is used to obtain the high-resolution barrel distortion images. Then, we utilize the given up-scaling factor u to obtain the corresponding low-resolution distortion images from the HR distortion images by down sampling. The LR distortion images are then interpolated to middle images by bicubic method with the same factor. Note that the high frequency information of these middle images is missing. After that, the middle images are un-distorted to recover the rectilinear shapes using polar transformations.

We follow the same feature extraction process as A+ [3] and TLA [11]. In order to extract local features and obtain high frequency contents, we use R high pass filters to convolute all the un-distorted middle images (for simplicity of notation, let us call the middle images as LR images in the following contexts). Thus, each LR image forms R filtered images. Gridding method is then used to extract patches from these images, and the patches from the same position are aggregated as a feature. High-resolution features from the original HR image are extracted by the same gridding method.

3.2 Training Stage

Dictionary and Projection Matrix
Since we have computed the LR and HR features, at this stage, we will compute low-resolution dictionary and projection matrixes. We use the same computational process as A+ [3]. The LR dictionary $\mathbf{D}_L = \left\{ \mathbf{d}_{L,k} \right\}_{k=1}^{K}$, (where $\mathbf{d}_{L,k}$ denotes the dictionary atom and K is the dictionary size), is computed as follows,

$$\min_{\delta} \|\mathbf{D}_L \delta - \mathbf{x}\|_2^2 + \lambda \|\delta\|_1 \tag{3}$$

where δ is the sparse representation, $\mathbf{x}$ is low resolution features and λ is a weighting factor. The general super-resolution methods use training samples in training pool to generate neighbors for regression. The LR and HR features extracted from the original high-resolution images and the un-distorted low-resolution images form the training pool. Hence, the optimization problem can be transformed into

$$\min_{\gamma} \|\mathbf{x}_i - \mathbf{N}_{L,k}\gamma\|_2^2 + \lambda \|\gamma\|_2 \tag{4}$$

where γ is a coefficient vector, $\mathbf{N}_{L,k}$ includes m neighbor samples from training pool. These m training samples lie closest to the dictionary atom to which the input feature x_i is matched. The closed-form solution is given as follows

$$\gamma_i = \left(\mathbf{N}_{L,k}^{\mathrm{T}} \mathbf{N}_{L,k} + \lambda \mathbf{I} \right)^{-1} \mathbf{N}_{L,k}^{\mathrm{T}} \mathbf{x}_i \tag{5}$$

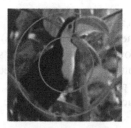

Fig. 4. An image divides into three and four classes.

The corresponding high-resolution neighbors are defined as $N_{H,k}$. Hence, for each input LR feature x_i, we compute the nearest atom in dictionary, and the HR patch $y_{H,i}$ can be reconstructed by the following equation,

$$y_{H,i} = N_{H,k}\left(N_{L,k}^T N_{L,k} + \lambda I\right)^{-1} N_{L,k}^T x_i, k = 1, 2, \ldots, K \tag{6}$$

Therefore, we define the projection matrix P_k in Eq. (7) since it can be computed offline. Each dictionary atom $d_{L,k}$ has a corresponding projection matrix P_k.

$$Pk = N_{H,k}\left(N_{L,k}^T N_{L,k} + \lambda I\right)^{-1} N_{L,k}^T \tag{7}$$

Post-processing Refinement

In the above process, the high-resolution patch has been computed, but the spatially varying characteristics of barrel distortion image are ignored in the former procedure. The information loss in the edge region and the center region of the image is different. Hence, we add a spatially varying post-processing procedure to refine the HR image.

First, for the original HR features y and the reconstructed HR features y_H from the reconstructed HR image, we divided them into n_{group} classes according to the geometric distance to image center. We denote the class of y and y_H as y_c and $y_{H,c}$ respectively (where $c = 1, \ldots, n_{group}$). Figure 4 shows an image divides into three and four classes.

We use H_c represents the coefficient matrix for spatially varying post-processing. In training stage, we compute a coefficient matrix for each feature class (representing different geometric distances to the image center in the spatial domain) by

$$\min_{H_c}\left\|H_c y_{H,c} - y_c\right\|_2^2 \tag{8}$$

The close-form solution is given by

$$H_c = y_c y_{H,c}^T \left(y_{H,c} y_{H,c}^T\right)^{-1} \tag{9}$$

3.3   Testing Stage

At this stage, for the given test images (after the un-distortion process), we first extract LR features from them by the same extraction process in training stage. Then, the nearest dictionary atom $\mathbf{d}_{L,k}$ for each low-resolution feature $\mathbf{x}_{T,j}$ is determined. Meanwhile, we also obtain the corresponding projection matrix $\mathbf{P}_k$. Therefore, the high-resolution patch $\mathbf{y}_{T,j}$ of the first reconstruction procedure can be calculated by

$$\mathbf{y}_{T,j} = \boldsymbol{P}_k\boldsymbol{x}_{T,j} \tag{10}$$

Then, we utilize a post-processing procedure for the second reconstruction. The reconstructed HR patches $\mathbf{y}_T$ are divided into n_{group} classes based on the geometric distance to image center. For each class, we utilize the corresponding coefficient matrix to conduct the second reconstruction process. The equation we used to compute the reconstructed features $\mathbf{y}_{R,c}$ is given by

$$\mathbf{y}_{R,c} = \mathbf{H}_c\mathbf{y}_{T,c} \tag{11}$$

In the end, these HR patches $\mathbf{y}_{R,c}$ are aggregated to form the final high-resolution image, where the merging method is same as the decomposition process.

4   Experiments

In this section, we introduce experiment settings, explain the parameters, and analyze the results compared with other state-of-the-art approaches.

4.1   Experiment Settings

In this paper, the training dataset includes 91 nature images that were provided by Yang et al. [6]. We use two test datasets Set5 and Set14 proposed by Zeyde et al. [4]. They contain five and fourteen commonly used nature images, respectively. The evaluation metrics utilized in our paper are PSNR (Peak Signal-to-Noise Ratio) and SSIM (structural similarity) [12]. We compare our method with the bicubic interpolation method and the state-of-the-art super resolution methods, including K-SVD method [4], GR (Global Regression) method [5], ANR (Anchored Neighborhood Regression) method [5] and A+ method [3].

In our experiments, we set the up-scaling factor u equals to 2. Other parameter settings are the same as A+ method. The high-pass filter number R is 4, dictionary size K equals to 1024, λ is 0.1 and neighborhood size m is 2048. The class number n_{group} is set to 4 in post-processing procedure. The parameter a_2 that controls the distortion degree of image is set to 0.25 and 0.28 (a_1 is set to 1) throughout the experiments.

4.2   Experimental Results

In this section, we introduce the results of our method and other comparative methods [3–5]. Table 1 shows the average PSNR and SSIM results on Set5 and Set14 when

Table 1. Average PSNR (dB) and SSIM results on Set5 and Set14 when $a_2 = 0.25$.

Dataset	Bicubic	K-SVD	GR	ANR	A+	Ours
Set5	26.18	27.91	27.91	27.96	28.00	**28.23**
	0.7908	0.8318	0.8273	0.8326	0.8424	**0.8466**
Set14	24.36	25.93	25.94	25.97	25.89	**26.20**
	0.7547	0.8072	0.8092	0.8099	0.8097	**0.8180**

Table 2. Average PSNR (dB) and SSIM results on Set5 and Set14 when $a_2 = 0.28$.

Dataset	Bicubic	K-SVD	GR	ANR	A+	Ours
Set5	25.71	27.66	27.68	27.70	27.72	**28.02**
	0.7775	0.8259	0.8219	0.8266	0.8365	**0.8422**
Set14	24.25	25.69	25.70	25.73	25.70	**25.90**
	0.7515	0.7995	0.8012	0.8018	0.8035	**0.8097**

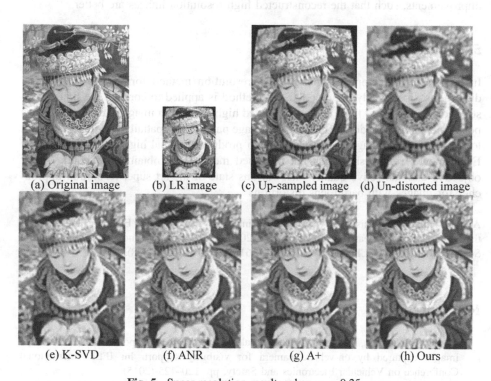

(a) Original image   (b) LR image   (c) Up-sampled image   (d) Un-distorted image

(e) K-SVD   (f) ANR   (g) A+   (h) Ours

Fig. 5. Super-resolution results when $a_2 = 0.25$.

distortion parameter $a_2 = 0.25$ while Table 2 shows the results when $a_2 = 0.28$. Figure 5 represents the resulting images of various methods for the parameter $a_2 = 0.25$.

In these two tables, the PSNR and SSIM results indicate that the performance of our method improves significantly over other state-of-the-art super-resolution methods for different values of a_2. Table 1 shows that our method provides higher PSNR values of Set5 and Set14 by 0.23 dB and 0.21 dB respectively. Meanwhile, the SSIM improvement is more than 0.0042 and 0.0083 on Set5 and Set14. Table 2 gives similar results, where our method has an advantage of 0.30 dB in PSNR on Set5 dataset.

Figure 5 shows the original images, low-resolution distorted images, bicubic interpolated images, un-distorted images, and resulting images of our approach and the several comparative approaches. The results indicate that our super-resolution method has successfully compensated more accurate high-frequency information, and improved the quality of reconstructed images. Specifically, the edges and the texture regions of the images reconstructed by the proposed method have better fidelity.

Experimental results show that our method is better than A+ method in terms of PSNR, SSIM values and visual evaluations. That is because for the given barrel distortion images, our approach uses both dictionary-based method and post-processing procedure, and utilizes different coefficient matrixes according to different spatially varying distortion characteristics of image. Therefore, our results show further improvements, such that the reconstructed high resolution images are better.

5   Conclusion

In this paper, we propose a novel super-resolution method for images with barrel distortions. Specifically, the popular A+ method is applied to conduct the first reconstruction procedure to form the reconstructed high-resolution image. After that, we use post-processing procedure to classify the image patches for spatially varying refinement to reconstruct the images again, in order to produce the final high-resolution images. Experimental results show that the proposed method can obtain significantly better objective and subjective quality than various state-of-the-art super-resolution approaches for images with barrel lens distortions.

Acknowledgments. This work was supported in part by the Shenzhen Emerging Industries of the Strategic Basic Research Project (No. JCYJ20160226191842793), and the National Natural Science Foundation of China (Nos. 61602312, 61602314, 61620106008).

References

1. Takano, T., Ono, S., Matsushita, Y., Kawasaki, H., Lkeuchi, K.: Super resolution of fisheye images captured by on-vehicle camera for visibility support. In: IEEE International Conference on Vehicular Electronics and Safety, pp. 120–125 (2015)
2. Hongzhi, W., Meijing, L., Liwei, Z.: The distortion correction of large view wide-angle lens for image mosaic based on OpenCV. In: International Conference on Mechatronic Science, Electric Engineering and Computer (MEC), pp. 1074–1077 (2011)
3. Timofte, R., De Smet, V., Van Gool, L.: A+: adjusted anchored neighborhood regression for fast super-resolution. In: Cremers, D., Reid, I., Saito, H., Yang, M.-H. (eds.) ACCV 2014. LNCS, vol. 9006, pp. 111–126. Springer, Cham (2015). doi:10.1007/978-3-319-16817-3_8

4. Zeyde, R., Elad, M., Protter, M.: On Single Image Scale-Up Using Sparse-Representations. In: Boissonnat, J.-D., Chenin, P., Cohen, A., Gout, C., Lyche, T., Mazure, M.-L., Schumaker, L. (eds.) Curves and Surfaces 2010. LNCS, vol. 6920, pp. 711–730. Springer, Heidelberg (2012). doi:10.1007/978-3-642-27413-8_47
5. Timofte, R., De Smet, V., Van Gool, L.: Anchored neighborhood regression for fast example-based super-resolution. ICCV **2013**, 1920–1927 (2013)
6. Yang, J., Wright, J., Huang, T.S., Ma, Y.: Image super-resolution via sparse representation. IEEE Trans. Image Process. **19**(11), 2861–2873 (2010)
7. Kun, L., Jingyu, Y., Jianmin, J.: Nonrigid structure from motion via sparse representation. IEEE Trans. Cybern. **45**(8), 1401–1403 (2015)
8. Tsai, R.: A versatile camera calibration technique for high-accuracy 3D machine vision metrology using off-the-shelf TV cameras and lenses. IEEE J. Robot. Autom. **3**(4), 323–344 (1987)
9. Ahmed, M., Farag, A.: Nonmetric calibration of camera lens distortion: differential methods and robust estimation. IEEE Trans. Image Process. **14**(8), 1215–1230 (2005)
10. Mallon, J., Whelan, P.F.: Precise radial un-distortion of images. In: Proceedings of the 17th International Conference on Pattern Recognition, pp. 18–21 (2004)
11. Mei, S., Sheng-hua, Z., Jianmin, J.: Transfer learning based on A+ for image super-resolution. In: KSEM 2016, pp. 325–336 (2016)
12. Wang, Z., Bovik, A.C., Sheikh, H.R., Simoncelli, E.P.: Image quality assessment: from error visibility to structural similarity. IEEE TIP **13**(4), 600–612 (2004)

Knowledge Retrieval

Mining Schema Knowledge from Linked Data on the Web

Razieh Mehri[1](✉) and Petko Valtchev[2]

[1] Concordia University, Montreal, Canada
r_mehrid@encs.concordia.ca
[2] Université du Québec à Montréal, Montreal, Canada
valtchev.petko@uqam.ca

Abstract. Datasets on the Web of Data (WoD) are often published without a precise schema which may discourage their reuse. Methods for schema acquisition from linked data have been proposed that mainly exploit the regularities in property and/or value distributions in resources to discover potentially useful classes as homogeneous clusters. Yet the crucial task of interpreting and naming the discovered classes is left to the human analyst. We prone a more holistic approach to schema discovery that, beside clustering, assists the analyst by suggesting plausible names for clusters. In doing that we: (1) rely on concept analysis for class discovery from linked data and (2) exploit known DBpedia types and shared properties to form candidate names. An evaluation of our approach with a dataset from the WoD showed it performs well.

Keywords: Linked data · Schema discovery · Formal concept analysis · DBpedia · Semantic web

1 Introduction

The Web of Data (WoD) is a rapidly expanding part of the general Semantic Web (SW) landscape where hundreds if not thousands Linked Data (LD) datasets are published every year. When publishing their dataset, authors often do not care to provide a dedicated schema but rather refer to openly available sources such as DBpedia, YAGO, GeoNames, etc. While these resources enable a rapid expansion of the WoD by reducing the publication effort, a more precise schema –tailored to the dataset content– would facilitate its further exploiting, e.g., querying or integration with other dataset [1].

The need to provide tightly-fitting schemas for already published datasets motivated the design of methods and tools for schema discovery from LD [2–4]. In its most general wording [3], the corresponding task is aimed at mining relevant schema elements, esp. classes (both identifying and producing definitions/descriptions thereof), subclass and subproperty links, rules, etc.

The overall schema discovery may be roughly split into the following subtasks: (1) class identification, (2) class hierarchy construction, (3) definition of

© Springer International Publishing AG 2017
G. Li et al. (Eds.): KSEM 2017, LNAI 10412, pp. 261–273, 2017.
DOI: 10.1007/978-3-319-63558-3_22

property domains and ranges, (4) class name assignment. The first two steps can be automated to a large extent: Many methods from the literature would address them as a single task and apply hierarchical clustering to 1 and 2 whereby the similarity measure is based on shared properties among resources. In particular, formal concept analysis (FCA) techniques have been frequently used to that end, due to the nice structural properties of its final result, the concept lattice. Compared to that, task 3 has often been implicitly addressed by the proposed methods.

In stark contrast to the clustering and domain/range assignment that exploit regularity in the distribution of the predicates or the (predicate, object) pairs across the triples of the LD dataset, class naming is inherently manual task. Indeed, it involves interpreting the resource cluster behind a new class and hence assessing its relevance. To the best of our knowledge, none of the methods from the literature have addressed the (partial) automation of the process, e.g., by providing plausible candidates for class names. Yet this could be a great relief for the human analyst, especially with approaches which tend to generate a large number of classes such as those exploiting FCA or affiliated pattern mining techniques.

In this paper, we address the problem of maximizing the automation of schema discovery from data. As the emphasis is on the aforementioned step 4, we choose a relatively unambitious variant of the first stage of the process: Our goal is to define the class hierarchy and feed in property domains and ranges (whenever applicable), which amounts to designing an RDF Schema for the initial dataset. More thorough class descriptions in expressive languages, e.g., OWL 2 EL, although relevant, are beyond our scope. Thus, our method applies FCA to jointly address the first three steps of the schema discovery task. For class naming, typing information from generally available external sources such as DBpedia is factored in whereas the properties of the FCA output, i.e., the concept lattice, are exploited to ensure produced names are both most specific and unique.

We conducted a preliminary evaluation of our method with a Russia-related dataset from the WoD. Its choice was dictated by the need to keep the FCA output reasonably-sized without applying strong filtering techniques so that the potential of our naming technique could be assessed in a fair manner. The output set of classes was assessed in terms of recall, precision and f-measure w.r.t. to a reference class set and the results were judged promising.

The remainder of the paper is as follows: Sect. 2 summarizes related work whereas Sect. 3 provides background to the study. Sections 4 and 5 describe our method and the experimental study, respectively. Finally, Sect. 6 concludes.

2   Related Work

The aforementioned schema discovery tasks are dealt with in the larger field of ontology learning [1] whereas individual methods address specific subsets thereof. Most methods only address the first two of them, i.e., the construction of the class hierarchy, a.k.a. the taxonomy. To that end, they apply a hierarchical clustering algorithm, either FCA-based or a statistical one, e.g., nested k-means. In [5], a comparison is drawn between three clustering paradigms, FCA and two statistical ones, with their respective merits for ontology learning. The authors conclude that FCA performs better and produces clusters that are easier to interpret yet there is a price to that as concept lattices tend to grow rapidly.

An interesting variant of the ontology learning from LD problem is addressed in [3]: An LD dataset where a schema, albeit present, is unexpressive (RDFS) and unmarked within the overall RDF graph is provided with a richer, OWL EL-level ontology. A related FCA application is presented in [4] where authors use FCA to detect incorrect or incomplete typing in LD datasets. Here FCA is a mere means to the discovery of strong associations between types to use as correcting templates for the RDF data.

Mining conceptual knowledge from RDF data dates back at least to the work of Delteil *et al.* [2]. The proposed method is FCA-based, yet the target concepts are not intended to become RDFS classes, as in our case, but rather fit the conceptual graph template. Thus, their intents comprise chains of triples of a fixed length (gradually increasing along an iterative construction process), rather than being set of attributes translating RDF properties as proposed here.

FCA has been applied to RDF data to acquire conceptual knowledge in a variety of other contexts: For instance, enhancing the functionalities of RDF repositories or the entire WoD w.r.t. navigating, browsing, visualizing, etc., motivated a large body of work [6–9]. Noteworthily, relational extensions of FCA to fit graph-shaped data are used in [6] and [8], called *logical concept analysis* (LCA) [10] and *relational concept analysis* (RCA) [11], respectively. Similarly, mining queries or query answer by means of FCA has been actively researched on [12–14].

Finally, assigning names to automatedly discovered concepts, a.k.a. concept labeling, has been addressed in the context of ontology learning from texts [15]. Yet, given the specific nature of the input data, the techniques used to that end are totally unrelated to the one we present here.

To sum up, no known and comparable method has jointly addressed the above four tasks.

3   Background

3.1   The RDF/S Data Models

As a generic data model, RDF[1] represents the information on the web in the form of `subject-predicate-object` triples. Each triple is a sentence describing a resource. A resource is an entity which can be a subject, predicate or object in an RDF triple. The subject or first part of an RDF triple is a resource which the statement describes. The predicate or second part of the triple is the property

---

[1] https://www.w3.org/TR/REC-rdf-syntax/.

or aspect which relates the resource to an object. Therefore, the object is the third part of the triple which could be another resource or a literal value defined as a string or a number, a date, etc.

On top of the RDF which does not provide significant semantics, RDFS[2] is an extensible knowledge representation language which adds vocabulary to RDF in order to express the information about classes and their relations including superclass/subclass relations and properties (predicate relationships between the classes).

3.2 DBpedia

DBpedia[3] is a project which aims at extracting structured information from the Wikipedia content. In addition to free text information, DBpedia also uses the different types of structured information from Wikipedia including the infobox templates[4], title, abstract, categorization information, images, geo information, and external url links. This open source data set is available on the web as linked data (RDF triples) [16].

3.3 Formal Concept Analysis

Formal Concept Analysis (FCA) is a lattice theory-based approach towards the discovery of conceptual knowledge from data [17]. Datasets are introduced as *formal contexts*, i.e., objects-to-attributes cross tables.

Definition 1 (Formal Context). A formal context is a triple $\mathcal{K}(G, M, I)$, where G and M are sets of objects and attributes, respectively, and $I \subseteq G \times M$ is an incidence relation.

With the above definition, $(g, m) \in I$ is interpreted as the object g having the attribute m. Figure 1 depicts on the left a sample context drawn from a Russia-related dataset found on the WoD (In this figure, labels are shown instead of full URIs). It is made of five resources as objects and six properties as attributes, whereas the incidence reflects the composition of the triples from the dataset.

Every incidence relation induces a *Galois connection* [18] that helps reveal hidden regularities in the distribution of attributes over the set of objects. The revealed conceptual structure of the context is called (formal) *concepts* in FCA:

Definition 2 (Concept). A pair $(A, B) \in \wp(G) \times \wp(M)$ is a (formal) concept of the context $\mathcal{K}(G, M, I)$ iff $A' = B$ and $B' = A$. A is called the *extent* and B the *intent* of the concept.

For instance, $(\{O2, O5\}, \{A1, A3, A5\})$ is a formal concept of the context in Fig. 1.

---

[2] https://www.w3.org/TR/rdf-schema/.
[3] http://dbpedia.org/.
[4] http://en.wikipedia.org/wiki/Help:Infobox.

An additional advantage of FCA as a tool for extracting conceptual knowledge comes from the hierarchical structure that is implicit in the concept set $\mathcal{C}_\mathcal{K}$. In fact, concepts are partially ordered by the set-theoretic inclusion of extents. Thus, a *subconcept-of* relation $\leq_\mathcal{K}$ is defined by:

$$(A_1, B_1) \leq_\mathcal{K} (A_2, B_2)\ iff\ A_1 \subseteq A_2 (\Leftrightarrow B_1 \supseteq B_2).$$

Moreover, following the *Central Theorem* of FCA [17], the partial order induces a *complete lattice* on the concept set.

property 1 (Concept Lattice). For a formal context $\mathcal{K}(G, M, I)$ with its concept set $\mathcal{C}_\mathcal{K}$, the partially ordered set $\mathcal{L}_\mathcal{K} = \langle \mathcal{C}_\mathcal{K}, \leq_\mathcal{K} \rangle$ is a complete lattice where the join ($\bigvee$) and meet ($\bigwedge$) operators over an arbitrary set of concepts $\{(A_i, B_i)\}_{i \in [1,k]}$ are defined as follows:

- $\bigvee_{i=1}^{k}(A_i, B_i) = ((\bigcup_{i=1}^{k} A_i)'', \bigcap_{i=1}^{k} B_i)$,
- $\bigwedge_{i=1}^{k}(A_i, B_i) = (\bigcap_{i=1}^{k} A_i, (\bigcup_{i=1}^{k} B_i)'')$.

In the above property, $''$ stands for both compound mappings $^{\triangle\triangledown}$ and $^{\triangledown\triangle}$ which are known to be closure operators [18] in any Galois connection.

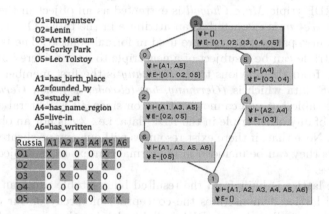

Fig. 1. Formal context drawn from a dataset about Russia (left) with its concept lattice (right)

The concept lattice is typically represented as the Hasse diagram where nodes representing concepts are annotated with respective intents and extents. Mathematically, the graph of that diagram follows the transitive restriction of the lattice order $\leq_\mathcal{K}$, i.e., the precedence $\prec$ relation. The lattice of our sample context is shown next to it in Fig. 1.

4   Schema Extraction

Our method comprises three steps: At step one, FCA algorithms are applied to an appropriate transformation of the RDF dataset to construct its concept lattice. At step two, the concepts are converted into RDFS classes and subconcept-of links from the lattice into subclass assertions. Moreover, property domains and ranges are set to the classes gathering the resources referred in the property triples. Finally, at step three tentative names are composed for the newly designed classes using labels recovered from DBpedia.

4.1   Mining Conceptual Knowledge from LD

In this step, the RDF data is converted to a formal context depicted as a binary table. As illustrated before, binary table has objects and attributes respectively in its rows and columns as well as binary values which indicate if an object has a specific attribute or not. Individuals that are not part of any RDF triple are considered noise in our data, i.e., they do not play any role in constructing our lattice. Each resource in RDF data is an object in the binary table and the predicates of that resource are attributes in the table. For example, to map the RDF triple *(Marc Chagall, live-in, Germany)* to formal context the first member of RDF triple *Marc Chagall* is extracted as an object in the table and its predicate *live- in* is extracted as an attribute in the table. One can see that only the two first parts of a triple are used in formal context. The last member of any RDF triple can be a subject of a new triple to be considered in the table. For example, from the previous triple *Germany* is the first member of another triple in RDF data which is *(Germany, has-tel-code, 49)*, i.e., *Germany* is an object in the table but one cannot continue on since *49* is a literal and cannot be a subject of any other triple in the RDF data, i.e., *49* is not an object in the binary table. Note that, if there exist resources without any predicate in dataset unlike literals they can be included in the formal context as the objects without any attributes.

Now, the lattice is built from the resulted binary table using an FCA tool. The resulted lattice demonstrates the conceptual hierarchy of our data. Each concept contains similar objects (RDF individuals); in other words, objects with similar attributes (predicates). Hierarchy structure also shows the superconcept and subconcept relations between the concepts.

4.2   Translating the Concept Lattice into RDFS

At this step, four basic rules are applied to turn the lattice into an RDFS schema:

Class Rule: All concepts, but the lattice top (G'', G') (extent comprises all dataset resources) and bottom become RDFS classes. Unless the extent of the lattice bottom (M', M'') is non void (at least one resource has triples for all properties from the dataset), it is ignored as well.

Subclass Rule: For each pair of concepts c_1 and c_2 such that $(c_1, c_2) \in \prec$, a `rdfs:subClassOf` assertion is created for their respective RDFS translations whenever those exist (see above).

Domain Rule: To establish domains of properties, we exploit a basic structural regularity of the concept lattice. In fact, for any attribute m, there is a unique maximal concept that has m in its intent. The concept $c = (m', m'')$, called the attribute concept of m, is such that for any other concept $\bar{c} = (A, B)$, $m \in B$ entails $\bar{c} \leq_{\mathcal{K}} c$. Now given an RDF property p, its domain is set to the class yielded by translation of the attribute concept of $m(p)$, the counterpart of p in the formal context. For example, the equivalent RDFS for the concept *Person* with three attributes *has-name* and *has-birthplace* and *live-in* is a class with the domain of the three predicates *has-name* and *has-birthplace* and *live-in*. Finally, in the generated RDFS, there is no need to draw the same properties for the subclasses of *Person* such as *Politician* (This also applies to the ranges).

Range Rule: Property ranges are a bit trickier to establish as our context does not reflect the (predicate, object) pairs in the RDF triples. Thus, there is no equivalent of the above attribute concept to rely on. However, another structural property of the lattice comes into play here: In fact, for any object subset $A \subseteq G$, there is a unique minimal concept whose extent contains A, i.e., (A'', A'). Now, for any property whose values are resources, we set its range to the class that translates the minimal concept comprising all property values (as objects) in its extent. For properties of literal values, ranges are not established. For example, both *has-birthplace* and *live-in* are related from one resource to another resource which are respectively the individuals of the concept *Person* and *Country*; but, *has-name* relates from one resource (individuals of the concept *Person*) to a literal. Therefore, the equivalent RDFS is a class *Country* (and *Person*) with range (and domain) of only two predicates *has-birthplace* and *live-in*.

4.3 Naming RDFS Classes

The final stage of our method generates candidate names for the discovered classes. The basic idea behind our method is to bring in typing knowledge from DBpedia and to exploit its ontological structure. We posit that for each class, an appropriate name must be designed which represents every class member.

Sources of Naming Information. For each dataset resource, we query DBpedia resources that match it. There are alternative typing sources in DBpedia, the most obvious being its ontology. Indeed the DBpedia classes shared by the members of an FCA-discovered concept can be a good starting point for naming. However, these classes might not be very specific and provide for a discriminant names in some cases. Another source is the `dc:description` property of DBpedia resources, yet it requires some light-weight NLP to be applied as explained below. Finally, discriminant names might be obtained by appending some dataset-specific information to overly general class names, e.g., the properties shared by class members (from the underlying concept intent). In the next paragraph, we describe four techniques to compose names from external as well as internal information.

Naming Techniques. We factor out shared DBpedia typing information in a much similar way FCA factors out the shared properties into concept intents. Thus, for a set of DBpedia resources that are recognized as matches for our dataset resources, we collect all the known types from the DBpedia ontology. Then, we intersect the resulting class lists and take the most specific type in the intersection. For instance, consider *Lake Onega* and *Neva River* as the resources to be grouped together and their concept should get a name based on the types they share in DBpedia. The types of any resource in DBpedia are retrieved from the `rdf:type` predicate. Figure 2 shows the DBpedia content for both *Lake Onega* and *Neva River* resources (`dbpedia-owl` prefix indicates the classes belong to the OWL ontology of DBpedia). The intersection list of the types in our example is [*Place*, *BodyOfWater*, *NaturalPlace*] whereby *BodyOfWater* is the most specific one.

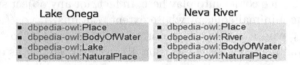

Fig. 2. Objects of the `rdf:type` predicate of *Lake Onega* and the *Neva River* in DBpedia

Resources with Several Known Names. The first challenge is to correctly spot the relevant resources inside DBpedia even though they are recognized by different names in a different context, e.g., *Neva* is used instead of the *Neva River* in some data. Luckily, DBpedia has a ready-made solution: For every such resource, the `dbpedia-owl:wikiPageRedirects` predicate lists known alternative names. Figure 3 shows the DBpedia page for the *Neva River* with different names.

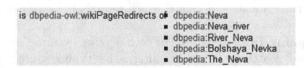

Fig. 3. Objects of the `dbpedia-owl:Wikipagesdirect` predicate for the *Neva River* in DBpedia

Beyond Shared DBpedia Types. Another challenge arises with overly general types retrieved by our basic naming technique. For instance, assume a resource represents a person and the `rdf:type` predicate only refers to the *Person* class.

Assume also two FCA-output classes whose members are in fact famous musicians and famous writers, respectively, but all we can retrieve from DBpedia is their common type *Person*. We have a clear naming conflict and a natural way out would the to retrieve the more specific information about their professions.

Such information is often comprised in the `dc:description` predicate of the DBpedia resources which represent people. The retrieval process is a bit more complex here as the content of the predicate is free text with few constraints, hence it requires some light-weight NLP. We illustrate it below through an example.

Assume *Rihanna* and *John Bottomley* are the only members of a newly discovered class. The `dc:description` for *Rihanna* contains the *"Singer, songwriter"* string while for *Facundo Cabral* it is *"Canadian singer and songwriter"*. Notice that *"and"* and *","* are phrase separators in the overall literal that are easy to spot. Moreover, in English multi-term phrases only the last word is of interest as it is the leading noun. Thus, with this heuristic rule, we can easily guess the types of the resource, in particular, the profession-related ones for representations of famous people. .

Past that step, in our example, *Rihanna* has a profession list comprising *"Singer"* and *"Songwriter"* and so has *John Bottomley* (capitalization is ours). The intersection of the lists is again *"Singer"* and *"Songwriter"*, hence, the assigned class name will be *"Singer Songwriter"*.

Appending Property Information. In a number of cases, the above techniques will output identical names for a number of classes. In such cases, we artificially produce discriminant names by appending the name of a property to the one retrieved by the previous techniques. The structural regularities of the lattice help again: due to the uniqueness of concept intents, there is at least one more attribute in an intent w.r.t. the intent of the parent concept. By choosing these differential properties, we ensure classes are named differently from their immediate super-classes, which, inductively, ensures the uniqueness of all names.

Finally, there might be a small number of classes for which no sensible name will be produced as they group resources of totally unrelated types. We keep their automated names and leave them to the analyst's judgment.

5 Experimental Study

We conducted an experimental study on the validity of our approach. A single dataset from the WoD was chosen to run on our schema discovery tool.

5.1 Dataset

We used an LD dataset about Russia which comprises 1159 triples. It covers data about Russia's culture (theaters, museum, galleries, etc.), nature (rivers, lakes, parks, etc.), famous Russian people, entertainments and other features. The dataset contains both instance and schema level triples. Since our goal is to assess the discovery of schema level by only looking at the instance level, we hid the former part from our method. Thus we used only the 539 triples about individual resources as input to the FCA-based tool. The remainder was used as a ground truth in the evaluation for the final schema.

5.2   Experimental Results

The formal context of the Russia dataset has 92 objects and 47 attributes. There are 256 (object, attribute) pairs in the context. The concept lattice of the Russia dataset has 69 concepts whereas the top and the bottom ones are trivial.

Figure 4 shows the RDFS graph extracted from the dataset. DBpedia is used to name the nodes in the graph. The version of DBpedia used for this research work is 3.8. The algorithm chooses the names for the classes according to the common information their objects share in DBpedia. Therefore, there is a reduction in the number of nodes compared to the concept lattice due to the same names applied to some nodes. In order to tune this reduction, the attribute concatenation technique (refer to Sect. 4.3 in the paper) has been applied for naming same nodes differently as much as possible.

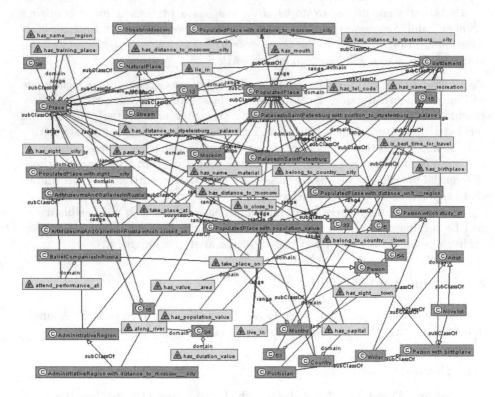

Fig. 4. Parts of the RDFS graph of the Russia dataset

The attribute concatenation is used for the concepts with the same name as their parents: The differential attribute name is appended to the temporal name.

For example, both concepts 7 and 23 got "*PopulatedPlace*". Since 7 is the parent of 23 a differential attribute, e.g., *distance_unit_region* is added, hence the name of 23 changes to "*PopulatedPlace with distance_unit_region*".

5.3 Evaluation

To evaluate our approach, we assess how well our method categorizes the dataset against the provided schema by the dataset. Notice that the naming process also indirectly affects our evaluation since applying the attribute concatenation technique which decreases the number of concepts with the same names (i.e., overall increasing in the number of concepts) can drastically improve the clustering performance.

In the following, the quantitative results of the evaluation are provided in terms of Precision, Recall and F-Measure.

To evaluate the approach against the provided schema by the dataset, only the relevant classes in the dataset are considered. Since in the provided schema there are also some classes without any instances (or instances with no properties), in the evaluation the relevant classes the provided schema stands for the classes which have instances and at least one of their instances is used to construct the concept lattice, i.e., the instances with properties.

Recall is a measure which calculates the number of the relevant classes extracted from the dataset:

$$Recall = \frac{relevant\ classes \cap discovered\ classes}{relevant\ classes} \qquad (1)$$

The number of relevant classes in the dataset is 35 and the number of classes which are extracted to construct the resulted RDFS graph is 36. Among all of the discovered classes, only 27 are relevant. Therefore, the value is 27/35.

Precision shows the number of extracted classes from the dataset which are relevant:

$$Precision = \frac{relevant\ classes \cap discovered\ classes}{discovered\ classes} \qquad (2)$$

The exact value is 27/36.

By applying the attribute concatenation technique the number of relevant classes increases from 17 to 27. The table below shows the improvement in the Precision, Recall, and F-Measure after applying that technique (Table 1). Figure 4 shows the final RDFS graph.

Table 1. Measurement table

	Precision	Recall	F-Measure
Without attribute concatenation	0.472	0.485	0.478
With attribute concatenation	0.75	0.771	0.763

5.4   Discussion

A comparison between the class names we produced and the original ones is provided below.

First, while the dataset is about Russia, our approach seems to distinguish better by considering locations. For example, the approach created both classes *Museum* and *Art Museums And Galleries In Russia* whereas the provided schema only contains *Museum*. Another example covers people with different professions. Our approach created both classes *Novelist* and *Writer* whereas initially we only had *Author*. It seems the automated approach provides more accurate names although there are also some deficiency compared to the original schema. Indeed, unnamed classes remain in our case: For example, the concept 13 comprises two objects *Russian_Winter_Folk_Festival* and *the_Festival_of_the_North*. Since no information is retrieved from DBpedia for those objects, no name can be created. Another example is concept 16 with *ice_skating*. The resource should be typed *sport* but the information is missing from DBpedia (search with *ice_skating* is fruitless). The corresponding classes in the original schema are *Festival* and *Sport*, respectively.

Finally, classes with only one instance might be considered as spurious. In a more general view, it would be reasonable to ignore the concepts with too few instances. Depending on the size of dataset one can decide about a threshold.

6   Conclusion

We presented here a method for extracting exploitable schema knowledge from datasets on the WoD. It reveals the implicit conceptual structure in RDF data by applying concept analysis to a straightforward representation of the linked data as a resource-to-property cross-table. The method then assigns names to discovered RDFS classes that reflect both known DBpedia types and shared properties of the member resources. The approach underwent a preliminary experimental evaluation whose results indicate it performs well.

Our next step is a comprehensive validation study of our method using datasets of variable sizes and provenance. In a future refinement of the approach, combining a range of external resources, e.g., WordNet or Wikipedia, for class naming will be investigated. Another avenue consists in feeding in property values into the conceptual analysis: For instance, links between resources could be mined to yield richer class descriptions, e.g., by means of dedicated mining frameworks [11].

References

1. Lehmann, J., Völker, J. (eds.): Perspectives on Ontology Learning, vol. 18. IOS Press, Amsterdam (2014)
2. Delteil, A., Faron, C., Dieng, R.: Building concept lattices by learning concepts from RDF graphs annotating web documents. In: Priss, U., Corbett, D., Angelova, G. (eds.) ICCS-ConceptStruct 2002. LNCS, vol. 2393, pp. 191–204. Springer, Heidelberg (2002). doi:10.1007/3-540-45483-7_15

3. Völker, J., Niepert, M.: Statistical schema induction. In: Antoniou, G., Grobelnik, M., Simperl, E., Parsia, B., Plexousakis, D., Leenheer, P., Pan, J. (eds.) ESWC 2011. LNCS, vol. 6643, pp. 124–138. Springer, Heidelberg (2011). doi:10.1007/978-3-642-21034-1_9
4. Alam, M., Buzmakov, A., Codocedo, V., Napoli, A.: Mining definitions from RDF annotations using formal concept analysis. In: IJCAI 2015, Buenos Aires, Argentina (2015)
5. Cimiano, P., Hotho, A., Staab, S.: Comparing conceptual, divisive and agglomerative clustering for learning taxonomies from text. ECAI **2004**, 435–439 (2004)
6. Ferré, S.: Conceptual navigation in RDF graphs with SPARQL-like queries. In: Kwuida, L., Sertkaya, B. (eds.) ICFCA 2010. LNCS (LNAI), vol. 5986, pp. 193–208. Springer, Heidelberg (2010). doi:10.1007/978-3-642-11928-6_14
7. Kirchberg, M., Leonardi, E., Tan, Y.S., Ko, R.K.L., Link, S., Lee, B.S.: Beyond rdf links-exploring the semantic web with the help of formal concepts. In: 9th Annual Semantic Web Challenge Conjunction with ISWC 2011 (2011)
8. Alam, M., Chekol, M.W., Coulet, A., Napoli, A., Smaïl-Tabbone, M.: Lattice based data access (LBDA): an approach for organizing and accessing linked open data in biology. In: d'Amato, C., Berka, P., Svátek, V., Wecel, K. (eds.) Data Mining on Linked Data Workshop (ECML/PKDD, 2013). DMoLD 2013, vol. 1082. Springer, Heidleberg (2013)
9. Reynaud, J., Toussaint, Y., Napoli, A.: Contribution to the classification of web of data based on formal concept analysis. In: What can FCA do for Artificial Intelligence (FCA4AI)(ECAI 2016) (2016)
10. Ferré, S., Ridoux, O., Sigonneau, B.: Arbitrary relations in formal concept analysis and logical information systems. In: Dau, F., Mugnier, M.-L., Stumme, G. (eds.) ICCS-ConceptStruct 2005. LNCS, vol. 3596, pp. 166–180. Springer, Heidelberg (2005). doi:10.1007/11524564_11
11. Rouane-Hacene, M., Huchard, M., Napoli, A., Valtchev, P.: Relational concept analysis: mining concept lattices from multi-relational data. Ann. Math. Artif. Intell. **67**(1), 81–108 (2013). doi:10.1007/s10472-012-9329-3
12. Rutledge, L., van Ossenbruggen, J., Hardman, L.: Making RDF presentable: integrated global and local semantic web browsing. WWW 2005, pp. 199–206, Chiba, Japan, ACM (2005). doi:10.1145/1060745.1060777
13. d'Aquin, M., Motta, E.: Extracting relevant questions to an rdf dataset using formal concept analysis. In: Proceedings of the Sixth International Conference on Knowledge Capture, pp. 121–128. ACM (2011)
14. Chekol, M.W., Napoli, A.: An FCA framework for knowledge discovery in sparql query answers. In: Proceedings of the 2013th International Conference on Posters & Demonstrations Track-Volume 1035, pp. 197–200. CEUR-WS. Org (2013)
15. Wong, W., Liu, W., Bennamoun, M.: Ontology learning from text: a look back and into the future. ACM Comput. Surv. **44**(4), 1–36 (2012)
16. Auer, S., Bizer, C., Kobilarov, G., Lehmann, J., Cyganiak, R., Ives, Z.: DBpedia: a nucleus for a web of open data. In: Aberer, K., Choi, K.-S., Noy, N., Allemang, D., Lee, K.-I., Nixon, L., Golbeck, J., Mika, P., Maynard, D., Mizoguchi, R., Schreiber, G., Cudré-Mauroux, P. (eds.) ASWC/ISWC -2007. LNCS, vol. 4825, pp. 722–735. Springer, Heidelberg (2007). doi:10.1007/978-3-540-76298-0_52
17. Ganter, B., Wille, R.: Formal Concept Analysis: Mathematical Foundations. Springer, Heidelberg (1999)
18. Denecke, K., Erné, M., Wismath, S.L.: Galois Connections and Applications, vol. 565. Springer, Heidelberg (2013)

Inferring User Profiles in Online Social Networks Based on Convolutional Neural Network

Xiaoxue Li[1,2], Yanan Cao[2], Yanmin Shang[2(✉)], Yanbing Liu[2],
Jianlong Tan[2], and Li Guo[2]

[1] School of Cyber Security, University of Chinese Academy of Sciences,
Beijing, China
[2] Institute of Information Engineering, Chinese Academy of Sciences,
Beijing, China
{lixiaoxue,caoyanan,shangyanmin,liuyanbing,
tanjianlong,guoliiie}@iie.ac.cn

Abstract. We propose a novel method to infer missing attributes (e.g., occupation, gender, and location) of online social network users, which is an important problem in social network analysis. Existing works generally utilize classification algorithms or label propagation methods to solve this problem. However, these works had to train a specific model for inferring one kind of missing attributes, which achieve limited precision rates in inferring multi-value attributes. To address above challenges, we proposed a convolutional neural network architecture to infer users' all missing attributes based on one trained model. And it's novel that we represent the input matrix using features of target user and his neighbors, including their explicit attributes and behaviors which are available in online social networks. In the experiments, we used a real-word large scale dataset with 220,000 users, and results demonstrated the effectiveness of our method and the importance of social links in attribute inference. Especially, our work achieved a 76.28% precision in the occupation inference task which improved upon the state of the art.

Keywords: User attributes mining · Deep neural network · Social network analysis

1 Introduction

Online social networks have become increasingly important platforms for users to interact with each other, process information, and diffuse social influence. A user in an online social network essentially has a list of social friends, a digital record of behaviors and a profile. To address users' privacy concerns, however, online social network provide users with fine-grained privacy settings. And users' private information is difficult to acquire. So, inferring social network users' profile get more and more attention in both industry and academia. In existing works, researchers aim to utilize social network user's explicit information (i.e. information which is registered or

© Springer International Publishing AG 2017
G. Li et al. (Eds.): KSEM 2017, LNAI 10412, pp. 274–286, 2017.
DOI: 10.1007/978-3-319-63558-3_23

easy to acquire) to infer their missing or incomplete attributes. To perform such privacy inference, they attempted to collect available data from online social networks and leverage several kinds of methods to analysis these explicit information.

Thus, existing attribute inference works can be roughly classified into two categories, classification-based method and label propagation-based one. Classification-based works are based on the intuition of you belong to what. Specially, they infer attributes for a user based on his neighbours' explicit attributes, and users' similarity is an important character for missing attributes classification. In general, SVM model obtains good performance in inferring gender, age, personality, etc. [17]. Other classification algorithms, such as Naïve Bayes, Logic Regression, Decision Tree also do well in similar tasks.

Label-propagation-based works propagate missing attribute values from label nodes to the unlabel nodes in a graph. The foundation of label-propagation-based work is homophily, which means that two linked users share similar attributes. The intuition behind label-propagation-based works is *what belongs to you*. For example, users are more likely to be friends with who has the same attributes. For instance, if more than half of the user's friends major in Computer Science in the certain university, the user might also major in Computer Science in the same university with a high probability. Motivated by this hypothesis, GSSL [11] attempted to infer users' university attributes.

However, these works generally train one specific model to infer one target attribute. That is to say, one trained model can only infer one kind of attribute and thus they need serval models to infer various attributes. Moreover, many classifiers are designed to candle binary classification problem, so the precision of inferring multi-valued attributes is much lower than inferring two-valued ones.

Our Work. We aim to precisely infer users' multi-valued attributes in one trained model. To this end, we first proposed a definition of ego-network to gracefully integrate social structures and user attributes in a unified framework. We used a matrix to represent this ego-network and its features.

Here, we design a convolutional neural network (CNN) architecture under the ego-network (UPE) to perform attribute inference. The information belong to a ego-network is used as the UPE input feature. Specially, CNN extract higher level features of users' public information, which performs better on inferring multi-valued attributes. Moreover, we could train our architecture once for inferring several attributes simultaneously.

In the experiments, we demonstrate the effectiveness of CNN architecture empirically. We used a large-scale dataset with 22,000 users collected from Zhihu, and compared UPE with several previous works for inferring gender and status. Experimental results showed that our work achieved a 76.28% precision in the occupation inference task which improved upon the state of the art which is 52.85%. Moreover, we observed that social structures are private in some social network, so we design another model to prove the importance of social structures in attributes inference. The result shows that social links have less influence when infer gender but have great influence when infer status.

In summary, our key contributions are as follows:

- We propose a ego-network to gracefully integrate social structures and users' attributes.
- We design a convolutional neural network architecture under ego-network (UPE) to infer users' missing attributes, which specially performs well on multi-valued attribute inference.

2   Related Work

Classify-Based Works. Lindamood et al. [9] transformed attributes inference to Naïve Bayes classification. They evaluated their method using Facebook social network with political attitudes—positive or negative. However, their approach is not applicable to user that share no attributes.

Kosinski et al. [2] studied various approaches to consider attributes might have various possible values. For instance, to infer a user's age, their method used logistic regression model. However, their approach need train a target model for an attribute. Thomas et al. [18] studied the inference of attributes such as gender, political views, and religious views. They used multi-label classification methods and leverage features from users' friends and wall posts. Moreover, they proposed the concept of multi-party privacy to defend against attributes inference.

Weinsberg et al. [19] investigated the inference of gender using the rating scores that users gave to different movies. In particular, they constructed a feature vector for each user; the ith entry of the feature vector is the rating score that the user gave to the ith movie if the user reviewed the ith movie, otherwise the ith entry is 0. They compared a few classifiers including Logistic Regression (LG), SVM, and Naïve Bayes, and they found that LG outperforms the other approaches.

Label-Propagation-Based Works. Li et al. [1] present a hidden factor in social connections—relationship type and propose a co-profile users' attributes and relationship types based on this development. Through iteratively profiles attributes by propagation via certain types of connections, and profiles types of connections based on attributes and the network structure, their algorithm profiles various attributes accurately. However, the weight matrix used large time to compute.

Ding et al. [7] design different strategies for computing the relational weights between users' attributes and social links and used a graph-based semi-supervised learning (GSSL) algorithm to infer attributes. Similar to the work performed by Mo et al. [11], they used a co-training method.

Dougnon et al. [3] note that several algorithms do not consider the rich information that is available on social networks, such as, group memberships. The authors proposed a new lazy algorithm named PGPI that can infer user profiles by using the rich information without training. Although this approach was successful, it requires large time to compute the scores in each iteration.

Yin et al. [14, 15] proposed a social-attribute network (SAN) to gracefully integrate social link and attribute information in a scalable way. They focused on a Random Walk with Restart (RWwR) algorithm to the SAN model to predict links as well as infer profile. Gong et al. [13] review this model and extend it to incorporate negative and mutex attributes. They generalize several leading supervised and unsupervised link and attribute-prediction algorithms in this improved model. Moreover, they make a novel observation that inferring attributes could help predict links. However, they did not make full use of this conclusion in their experimental section.

Other Approaches. Otterbacher et al. [21] studied the inference of gender using users' writing styles. Kosinski et al. [2] also construct an user-like martix. The dimensionalty of the user-like matrix was reduces using singular-value decomposition (SVD). Thus can be trained by its prediction model. Individual traits and attributes can be predicted to a high degree of accuracy based on records of users' likes.

Attributes inference using various mode could also be solved by a social recommeder system (e.g. [20]). However, such approach need train target model for different attributes and have a limitation when infer other attributes.

3 Problem Definition

As motivated in Sect. 1, we aim to study the problem of user profiling in both social network and ego networks and propose a novel model of inferring user's attributes and behaviors to suit the neural network. In this section, we will formally introduce some definition of the problem of inferring user profiles (Fig. 1).

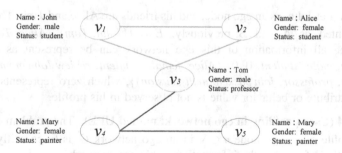

Fig. 1. An example of full social graph and ego-network. All of nodes and edges are belong to a full social graph and ego-network is represent as orange nodes and edges.

Definition 1 (full social graph). A social graph G is a quaternion $G = \{V, A, B, F\}$. V is the set of nodes in the social graph, and we assume the number of full social graph nodes is m. $A = \{A_1, A_2, \cdots A_i \cdots, A_m\}$ contains for each node V_i's attributes A_i and A_{ij} represents the value of node V_i's jth attribute. Meanwhile, $B = \{B_1, B_2, \cdots B_i \cdots, B_m\}$ is the set of each node's behaviors records. We define $F = \{F_1, F_2, \cdots F_i \cdots, F_m\}$, and $F_i = A_i \cup B_i$.

Example 1. Let be a social graph with five nodes $V = \{John, Alice, Tom, Mary, Alan\}$. Consider three attributes name, gender and status, and two behaviors review John's moments and take part in a basketball discussion. Therefore, the relation assigning attributes values to nodes are $A_1 = \{John, male, student\}$, $A_2 = \{Alice, female, student\}$, $A_3 = \{Tom, male, professor\}$, $A_4 = \{Mary, female, painter\}$ and $A_5 = \{Alan, male, student\}$. Thus the relation between the behavior's value and the node are $B_1 = \{John, null\}$, $B_2 = \{Alice, review John moment\}$, $B_3 = \{Tom, join basketball disscusion\}$, $B_4 = \{Mary, public picture in her moment\}$ and $B_5 = \{Alan, like Mary picture\}$. Then $F_1 = \{John, male, student, null\}$, $F_2 = \{Alice, female, student, review John moment\}$, $F_3 = \{Tom, male, professor, join basketball disscussion\}$, $A_4 = \{Mary, female, painter, pubic picture\}$ and $A_5 = \{Alan, male, student, like mary picture\}$.

Definition 2. (User profiling in a social graph by using attributes and behavior's information, called UPS). The problem of inferring the user profile of a node n in a social graph G is to correctly guess the missing attributes' value using other's attributes and behavior's information provided in the social graph. We address this problem using a CNN architecture and we will discuss the detail in next section.

Definition 3 (ego network). A user's ego network can be represented as a graph $G' = \{V', E', A', B', L\}$ where the A' and B' are defined as previously. V' is a set of the ego node's (e.g. v0) social friends. In addition, we define E' contains both the connections between the ego and his friends and the connections among their friends. L is a relation $L \in V' \times M$ (M is the sum of kinds of attributes and behavior records), which include the ego node and his friends' both attributes and behaviors records. Besides, a matrix L includes all information about an ego network.

Example 2. Let John, as an ego node, and his friends are Alice and Tom. The A' and B' are represented as A and B previously. $E' = \{(John, Tom), (Alice, Tom), (John, Alice)\}$. The all information of this ego network can be represent as the matrix $L = \{(John, male, student, 0, 0), (Alice, female, student, review John moment, 0), (Tom, male, professor, join basketball disscusion)\}$, which zero represents the corresponding attribute or behavior value is not observed in his profile.

Definition 4 (User profiling in ego network, named UPE). The problem of inferring the user profile of an ego node $n \in V$ in an ego network G' is to correctly guess the attributes value of n using the information in his ego network.

4   Proposed Architecture

In this section, we present our CNN architecture UPE and explain each layer of our architecture. UPE that designed for infer user attributes in the ego network using attributes, behaviors and links data. The foundation of our method is that CNN is a feedforward neural network, which could extract its topology from a matrix, and use back-propagation algorithm to optimize the network structure and compute the unknown parameters. Based on this, we could capture a relationship between these connections and missing attributes and compute the missing attributes value.

We observe that in some social network, such as Wechat and QQ, the social structures is private and hard to crawl. In order to address this problem we first propose neural network under public information (UPS) as our method's first version. UPS only used users' attributes. And the second version of our method is UPE. Then, we explain how these two versions infer users' missing attributes.

4.1 User Profiling in Social Network

Our proposed algorithm UPS for inferring profiles in social network by using attributes and behaviors information. This algorithm is inspired by the good performance of NN model in Natural Language Processing [12]. In the NLP, NN model changes each sentence as a feature vector. Similarly, we defined the set F, which includes the attributes and behaviors information of each user, as the feature vector. Then we will try to pro-process our features as numeric feature vectors and then fed to a multilayer convolutional neural network, trained in an end-to-end fashion. Contrasting with the traditional Attributes Infer approach, neural network (NN) architecture could achieve better performance when we fed it a larger dataset.

Our model is summarized in Fig. 2. The first layer extracts feature for each user. The second layer translates those features into the numeric feature vectors. In addition, the following layers are standard NN layers.

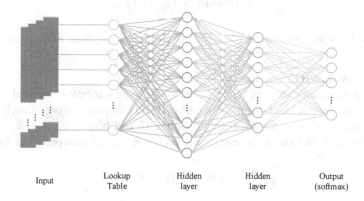

| Input | Lookup Table | Hidden layer | Hidden layer | Output (softmax) |

Fig. 2. UPS architecture

First, we given a Notation, we consider a neural network f(.). Any feed-forward neural network with P layers can be seen as a composition of functions f, corresponding to each layer p.

$$f_\theta(\cdot) = f_\theta^P\left(f_\theta^{P-1}\left(\cdots f_\theta^1(\cdot)\cdots\right)\right) \tag{1}$$

And initially, $f_\theta^0(\cdot) = F_i$ where $i \in (1, n)$. In the following, we will describe each layer we used in our architecture shown in Fig. 2.

Transforming User's F into Feature Vectors (Lookup Table Layer)

In NLP, researcher use word representing to transform each input word into a vector through looking up embedding. Similarity, in order to classify the attributes to each nodes in our attribute classification (we prefer consider our profile inferring problem as classification problem); we should first concentrate on learning discriminative word embedding, which carry less syntactic and semantic information. And, it usually takes a long time to train the word embedding. In addition, there are many trained word embedding that are freely available, for example, Google word2vec. Our experiments firstly utilize the trained embedding provided by Google. However, the size of those vector is abhorrent, so we take the KNN algorithms to cluster them into serval categories for each kind of attributes. Besides, some attributes are construct by serval words and we trained a small lookup table to represent those attributes perfectly.

More formally, for any node i' s word vector F_i, a numeric vector representation MF_i is given by the Lookup table layer LTF($\cdot$):

$$f_\theta^1(\cdot) = MF_i = LTF(F_i) \tag{2}$$

Hidden Layer

These two hidden layers are fully connected layers. The first hidden layer has n*n neuron, and dropout half of these neuron in the second hidden layer. We used ReLU as active function, which is represent by the following equation:

$$f_\theta^p(\cdot) = ReLU\left(f_\theta^{p-1}(\cdot)\right) \tag{3}$$

Besides, in order to compute the parameters quickly, we used SGD in our training steps.

Softmax Classifier (Output Layer)

To compute the finally value of the target user's missing attributes, we used a softmax classifier in our output layer. Softmax classifier could compute a score of each possible attribute value if we give the final weight matrix W and bias b.

Formally, the final output of each input vector F_i can be interpreted as follow:

$$f_\theta^{p-1} = softmax\left(Wf_\theta^{p-1}(F_i) + b\right) \tag{4}$$

4.2   User Profiling in Ego-Network

UPE use the CNN model to catch latent relations of users' attributes and social links and predict user attribute value using this relationship. This algorithm is inspired by the good performance of CNN in image Identification. In the process of image recognition, researchers catch the feature of the image through the CNN, and the input of CNN is an array of pixel values of the image. Similarly, when we add social links in our neural network, the input become a matrix L, as the feature matrix in image classification.

The pre-process is the same as UPS algorithm, but we construct a more fixed network to trained our dataset and predict the missing attribute (or classify the target user into the missing attribute possible values).

Our model is summarized in Fig. 3. The first layer is the same as the UPS layer. The second layer contains serval convolutional operation, and we take multiple convolutional kernels of different sizes to extract feature of input data. We would introduce our model layer by layer in the following parts.

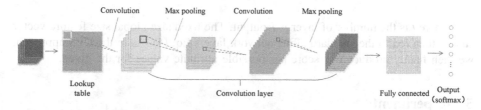

Fig. 3. UPE architecture

Convolutional Layer

The first two layer are same as UPS. That is to say, each word in L is first passed through the lookup table layer, producing a numeric vector of ML_i of the same size as L_i. This feature can be viewed as the initial input of the standard convolutional neural network. More formally, the initial input feature fed to the convolutional layer can be written as:

$$f_\theta^p(\cdot) = ML_i = LTF(F_i) \tag{5}$$

Convolutional can be seen as a linear operation between weight vector W and a numeric represented by elements in a matrix ML_i (in our lookup table (5)). Then the vector ML_i can be fed to one standard neural network layer which perform affine transformations over their inputs:

$$f_\theta^p(\cdot) = ReLU\left(Wf_\theta^{p-1}(\cdot) + b\right) \tag{6}$$

Where W is the weight matrix and b is the bias in this layer. We use ReLU as the active function. As for standard affine layers, convolutional layers often stacked to extract higher level features.

Max Pooling layer

The size of the output Eq. (6) depend on the number of ego nodes social friends in the ego-network fed to the network. Local feature vectors extracted by the convolutional layers have to be combined to obtain a global feature vector, with a fixed size independent of the L_i, in order to apply subsequent standard affine layers. Traditional convolutional network often apply an average or a max operation over the convolutional layer. The average operation does not make much sense in our case; as in general most friends attribute value in an ego-network do not have any influence on the attributes prediction. Instead, we used a max approach, which forces the network to capture the most useful local features produced by the convolutional layers. Given a matrix f_θ^{p-1} output by a convolutional layer $p-1$, the Max layer 1 output a vector f_θ^{p-1}

$$\left[f_\theta^p\right]_i = \max_t \left[f_\theta^{p-1}\right]_{i,t} \tag{7}$$

Where t is the number of layer $p-1$ output. The fixed size global size feature vector can be then fed to the standard affine network layers Eq. (4). As in the UPS approach, we then finally produce one score per possible attribute values for the given.

5   Experiment

We performed several experiments to access the accuracy of the proposed UPE and UPS algorithms for predicting different kinds of attribute values of nodes in our datasets, such as, gender, status and major. Contrast experiment were performed on a computer with a fourth generation 64-bit Core i5 processor running Ubuntu 14.5 and 16 GB of RAM.

We compared the performance of the proposed algorithm with four state-of-the-art algorithms: Linear Regression, Naïve Bayes classifiers [9], Graph Semi-Supervised Learning and Majority Voting. These four algorithms predict the value of target user's gender, status and major respectively.

Linear Regression (LR): we construct a linear function by using our training dataset, and predict the missing values using this function. Naïve Bayes (NB) classifiers: NB infer user profiles strictly based on correlation between attributes values which is as well as our UPS model.

Both the Linear Regression and Naïve Bayes classifiers have the state-of-art performance on binary classification. However, inferring user's missing profiles is a problem of multi-classification. Therefore, we extend those two algorithms to solve this problem by using One vs. Rest ways.

Graph Semi-Supervised Learning (GSSL) [11] and Majority Voting (MV) [6] infer user's profiles by using the social structures which is the same as our UPE model. For algorithms which need specific parameters, the best values have been empirically found to provide the best results.

5.1   Dataset

Experiments were carried on datasets are crawled from the real online social network websites: Zhihu. A series of data processing such as target attribute selection and data cleaning are conducted before running algorithms.

A series of data processing such as target attribute selection and data cleaning are conducted before running algorithms.

Target attribute selection: there are 15 features for each user, however, not all of them are needed. In fact, some features such as username provide little information for our classification problem. Besides, most people fill only a few of these features. Thus, we select 13 features for our algorithms. After excluding useless attributes, we finally choose gender, status and major as basic profile information of each target user.

Data cleaning: although we select 13 attributes which are most useful information for our problem, the number of missing value is still very large and noise information, like status values are love money, exist widespread in our datasets. For gender, 0.5 is used to represent missing value (1 represents male and 0 represents female). For status and major, 20 is represent the noise information and 0 is represent the missing value (we assume the possible value of status and major are 15 and 19 and represent by 1–15 and 1–19 respectively). And we also take the same method to clean other 10 attributes.

5.2 Experiment Result

Tables 1 and 2 give the results of experiments, from which various algorithms' performance can be evaluated. Figure 4 describes the accuracy of predict target users gender and status attributes via public information respectively.

Table 1. Result of experiment

Algorithms	UPS	UPE	NB	LR	GSSL	MV
Accuracy-gender	76.25%	79.11%	73.70%	72.80%	74.30%	52.50%
Accuracy-status	37.20%	76.28%	40.05%	41.60%	52.85%	32.90%
Time-gender	2 ms	3.2 ms	3.5 ms	4.2 ms	4.9 ms	5 ms
Time-status	2 ms	3.2 ms	3.5 ms	4.2 ms	5.2 ms	5.1 ms

Table 2. Result of UPE and UPS

Algorithms	UPS$_{NN}$	UPE$_{NN}$	UPS$_{CNN}$	UPE$_{CNN}$
Accuracy-gender	76.25%	77.41%	76..27%	79.11%
Accuracy-status	37.20%	74.62%	39.20%	76.28%
Time-gender	2 ms	2.1 ms	3.1 ms	3.2 ms
Time-status	2 ms	2.1 ms	3.1 ms	3.2 ms

Gender, the first version of Fig. 4 illustrates various algorithms' performance on gender prediction. It is clear to see in most cases the result of all methods are similarly. The reason leads to this result is that gender prediction is a binary classification. In specific, UPS and UPE methods perform better than other methods, and MV achieve the worst performance. This is because that the attribute of gender did not depend on the target user social friends.

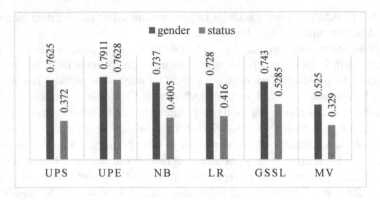

Fig. 4. The accuracy of predict target users gender and status attributes

Status, another version gives another experiment result. That is, UPE outperform other methods. In detail, the accuracy most of methods is between 30% and 40%. However, UPE algorithms obtain a rate of 76.28% of correctly predicting status, in which the accuracy is improved almost 40%. This result demonstrates that UPE have the great advantages when deal with multi classification problem.

Analysis Table 2, we can see two phenomena. The first one is UPE have same experiment result used NN and CNN architecture. So, we choice NN architecture in our UPS algorithms because it cost less time to predict the missing attributes and can achieve the same accuracy. Second, the accuracy of UPECNN is improved when used CNN architecture both in gender and status prediction. Compared UPE and UPS, the result improved significantly when we add the social links in the neural network, which is to see, UPE could easily learn the hidden relation between the ego node and its friends. And, this hidden relation, in our definition, is the nonlinear relation exist in our ego-network.

6   Conclusion

In contrast to machine learning algorithms, UPE predicts multi candidate attributes such as status in online social network more accurately. From our result, we find it is possible to learn hidden users' attributes based on relational information and profile similarity among users. However, we did not demonstrate the less time neural network used in our experiment. Thus, there is a need of big data in real online social network to test distinction between our algorithms and other methods. In addition, some attributes are mutex, such as gender. But we did not consider this relation in our algorithms. As our future work, we would like to add this mutex relation in our algorithms. Further, we want to evaluate our algorithms in a large dataset and apply it to different social networks.

Acknowledgement. This work was supported by the National Natural Science Foundation of China grants (NO. 61403369, NO. 61602466), the National Key Research and Development program (No. 2016YFB0801304).

References

1. Li, R., Wang, C., Chang, K.C.C.: User profiling in an ego network: co-profiling attributes and relationships. In: Proceedings of the 23rd International Conference on World Wide Web, pp. 819–830. ACM (2014)
2. Kosinski, M., Stillwell, D., Graepel, T.: Private traits and attributes are predictable from digital records of human behavior. Proc. Natl. Acad. Sci. U.S.A. **110**(15), 5802–5805 (2013)
3. Dougnon, R.Y., Fournier-Viger, P., Nkambou, R.: Inferring user profiles in online social networks using a partial social graph. In: Barbosa, D., Milios, E. (eds.) CANADIAN AI 2015. LNCS, vol. 9091, pp. 84–99. Springer, Cham (2015). doi:10.1007/978-3-319-18356-5_8
4. Ciot, M., Sonderegger, M., Ruths, D.: Gender inference of Twitter users in non-English contexts. In: EMNLP, pp. 1136–1145 (2013)
5. Petkos, G., Papadopoulos, S., Kompatsiaris, Y.: Social circle discovery in ego-networks by mining the latent structure of user connections and profile attributes. In: 2015 IEEE/ACM International Conference on Advances in Social Networks Analysis and Mining (ASONAM), pp. 880–887. IEEE (2015)
6. Jurgens, D.: That's what friends are for: inferring location in online social media platforms based on social relationships. ICWSM **13**, 273–282 (2013)
7. Dong, Y., Tang, J., Wu, S., et al.: Link prediction and recommendation across heterogeneous social networks. In: 2012 IEEE 12th International Conference on Data Mining (ICDM), pp. 181–190. IEEE (2012)
8. Mislove, A., Viswanath, B., Gummadi, K.P., et al.: You are who you know: inferring user profiles in online social networks. In: Proceedings of the third ACM International Conference on Web Search and Data Mining, pp. 251–260. ACM (2010)
9. Lindamood, J., Heatherly, R., Kantarcioglu, M., et al.: Inferring private information using social network data. In: Proceedings of the 18th International Conference on World Wide Web, pp. 1145–1146. ACM (2009)
10. Davis Jr., C.A., Pappa, G.L., de Oliveira, D.R.R., et al.: Inferring the location of twitter messages based on user relationships. Trans. GIS **15**(6), 735–751 (2011)
11. Mo, M., Wang, D., Li, B., et al.: Exploit of online social networks with semi-supervised learning. In: The 2010 International Joint Conference on Neural Networks (IJCNN), pp. 1–8. IEEE (2010)
12. Collobert, R., Weston, J., Bottou, L., et al.: Natural language processing (almost) from scratch. J. Mach. Learn. Res. **12**, 2493–2537 (2011)
13. Gong, N.Z., Talwalkar, A., Mackey, L., et al.: Joint link prediction and attribute inference using a social-attribute network. ACM Trans. Intell. Syst. Technol. (TIST) **5**(2), 27 (2014)
14. Yin, Z., Gupta, M., Weninger, T., et al.: A unified framework for link recommendation using random walks. In: 2010 International Conference on Advances in Social Networks Analysis and Mining (ASONAM), pp. 152–159. IEEE (2010)
15. Yin, Z., Gupta, M., Weninger, T., et al.: Linkrec: a unified framework for link recommendation with user attributes and graph structure. In: Proceedings of the 19th International Conference on World Wide Web. pp. 1211–1212. ACM (2010)
16. Gong, N.Z., Liu, B.: You are who you know and how you behave: attribute inference attacks via users' social friends and behaviors. arXiv preprint arXiv:1606.05893 (2016)

17. Xu, F.X., Liang Ying, H.-B.: A sock-puppet ralation detection method on social network. J. Chinese Inf. Process. **28**(6), 162–168 (2014)
18. Thomas, K., Grier, C., Nicol, David M.: unFriendly: multi-party privacy risks in social networks. In: Atallah, M.J., Hopper, N.J. (eds.) PETS 2010. LNCS, vol. 6205, pp. 236–252. Springer, Heidelberg (2010). doi:10.1007/978-3-642-14527-8_14
19. Weinsberg, U., Bhagat, S. Ioannidis, Taft, N.: Blurme: inferring and obfuscating user gender based on ratings. In: RecSys (2012)
20. Ye, M., Liu, X., Lee, W.-C.: Exploring social influence for recommendation a probabilistic generative model approach. In: SIGIR (2012)
21. Otterbacher, J.: Inferring gender of movie reviewers: exploiting writing style, content and metadata. In: CIKM (2010)

Co-saliency Detection Based on Superpixel Clustering

Guiqian Zhu[1], Yi Ji[1(✉)], Xianjin Jiang[1], Zenan Xu[1],
and Chunping Liu[1,2,3(✉)]

[1] School of Computer Science and Technology, Soochow University,
Suzhou 215006, China
zhuguiqian@163.com, {jiyi, cpliu}@suda.edu.cn,
{20154227026, 1527403027}@stu.suda.edu.cn
[2] Key Laboratory of Symbolic Computation and Knowledge Engineering
of Ministry of Education, Jilin University, Changchun 130012, China
[3] Collaborative Innovation Center of Novel Software Technology
and Industrialization, Nanjing 210046, China

Abstract. The exiting co-saliency detection methods achieve poor performance
in computation speed and accuracy. Therefore, we propose a superpixel clustering based co-saliency detection method. The proposed method consists of
three parts: multi-scale visual saliency map, weak co-saliency map and fusing
stage. Multi-scale visual saliency map is generated by multi-scale superpixel
pyramid with content-sensitive. Weak co-saliency map is computed by superpixel clustering feature space with RGB and CIELab color features as well as
Gabor texture feature in order to the representation of global correlation. Lastly,
a final strong co-saliency map is obtained by fusing the multi-scale visual saliency map and weak co-saliency map based on three kinds of metrics (contrast,
position and repetition). The experiment results in the public datasets show that
the proposed method improves the computation speed and the performance of
co-saliency detection. A better and less time-consuming co-saliency map is
obtained by comparing with other state-of-the-art co-saliency detection methods.

Keywords: Content-sensitive · Superpixel pyramid · Superpixel clustering ·
Co-saliency

1 Introduction

In recent years, co-saliency has become a research hotspot. Due to more information
can be obtained from multiple images shared common objects, co-saliency detection
can extract common foreground more accurately than saliency detection. Therefore,
it has a very wide range of applications, such as image retrieval [1], co-segmentation
[2, 3], video compression [4] and target tracking [5].

Co-saliency Detection is still a challenging problem. First, the change of illumination, occlusion, angle, etc. will impact the result of co-saliency detection seriously
under complex background. Second, it's difficult to obtain the co-saliency cues on
different scales. Third, co-saliency detection needs to process multiple images leading
to high computation complexity.

© Springer International Publishing AG 2017
G. Li et al. (Eds.): KSEM 2017, LNAI 10412, pp. 287–295, 2017.
DOI: 10.1007/978-3-319-63558-3_24

To solve these problems, the framework of our proposed co-saliency detection method is illustrated in Fig. 1. First, we generate a superpixel pyramid for each image in group, and obtain the saliency map by computing the saliency value for each superpixel. Then, we cluster all superpixels into several classes to capture global correlation. By using the clustering results to calculate the three cues, we obtain weakly co-saliency maps. Finally, the saliency maps and weak co-saliency maps are fused to form the final co-saliency maps.

In summary, our paper offers the following contributions:

(1) A superpixel clustering based co-saliency detection method is proposed, we compute the superpixel-level co-saliency instead of pixel-level co-saliency.
(2) Multi-scale visual saliency map is generated by multi-scale superpixel pyramid with content-sensitive.

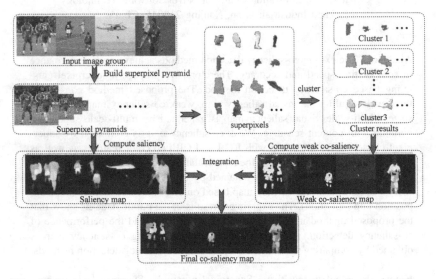

Fig. 1. Overview of the proposed superpixel clustering based co-saliency detection.

2   Related Works

Co-saliency detection involves two parts: saliency detection and common detection from two or more relevant images. In 1998, Itti et al. [6] proposed a saliency model based on neurobiology research firstly. Harel et al. [7] proposed a graph-based saliency model based on Itti model. This model used the markov random field instead of original linear combination to fuse multiple feature maps. In order to address the problem of noisy detection results and limited representations from bottom-up methods, Tong et al. [8] proposed a novel algorithm for salient object detection via bootstrap learning. Lee et al. [9] introduced an encoded low level distance map to denote low level saliency information, which assist high level semantic feature gained by deep framework gain detect salient regions. While saliency detection can find the regions with rich information, it cannot detect common foreground in multiple images. To obtain the common foreground, co-saliency detection was proposed.

The concept of co-saliency was first introduced by Jacobs et al. [10] in 2010. In [11, 12], the manually designed co-saliency cues are used to explore the co-saliency between images. It costs much time to cluster similar pixels. A serious problem these methods faced is how to choose proper features to describe co-saliency. Multiple saliency maps are fused to discover co-saliency information from the collection of multiple related images in [13, 14]. These methods compute multiple saliency maps, increase the computational complexity. We only compute one saliency map to guide the weak co-saliency map. Methods in [15, 16] transferred co-saliency detection problem to classification problem for each image region. These methods haven't been mature, and need much time to train the model. We use the bottom-up cues to detect the co-saliency can decrease the detection time.

3 Proposed Method

Our proposed co-saliency detection method is shown in Fig. 1. For the following brief description, our superpixel clustering based co-saliency detection method is named SCCD.

3.1 Build Superpixel Pyramid

To obtain content-sensitive superpixels on multi-scale, our idea is to build a superpixel pyramid for each image. Firstly, we build a Gaussian pyramid with three layers for the purpose of obtain saliency and co-saliency information on three scales in our experiments. Then we segment image on every scale by MSLIC [17], using MSLIC to segment image we can capture content-sensitive superpixels, i.e., small superpixels in content-dense regions and large superpixels in content-sparse regions.

3.2 Content-Sensitive Based Multi-scale Saliency (CSMS) Detection

After building the superpixel pyramid, we detect the salient object via bootstrap learning [8]. First, a weak saliency map is constructed based on three bottom-level features to generate training samples for a strong model. Second, a strong classifier based on multiple kernel boosting is learned to measure saliency where three features are extracted and four kernels are used to exploit rich feature representations, and then a strong saliency map is obtained. Third, in order to generate the final saliency map, we fuse the weak saliency maps and strong saliency maps on different scales by a weighted combination [8].

3.3 Superpixel Clustering

The clustering breaks the limit of an image by gathering the similar superpixels to same cluster. In our method, K-means clustering method is used [12]. We choose RGB,

CIELab and texture feature as the clustering features. We compute the texture feature by combining 8 orientations of Gabor filter. In order to obtain proper cluster number, we adopt a flexible equations to set the cluster number as

$$K = min(max(2 \times N_{img}, 10), 30) \tag{1}$$

where N_{img} denotes the superpixel numbers of image.

For notation, the i th superpixel is denoted by p_i^j in image I^j, where i range from 1 to N_j, N_j denotes the superpixel number of the j th image. Given M images $\{I^j\}_{j=1}^M$, we obtain K clusters $\{c^k\}_{k=1}^K$. The clusters are denoted by a set of vectors $\{u^k\}_{k=1}^K$, in which u^k denotes the cluster center.

3.4   WCS (Weak Co-saliency) Computation

Our method describes WCS by calculating the contrast cue, the repetition cue, and the position cue. The foreground regions often have high contrast, and the image center usually attract the first attention of human eyes, so we use contrast cue and position cue to find the salient regions. The contrast cue of cluster c_k is calculated by:

$$w_c(k) = \sum_{i=1, i \neq k}^{K} \frac{n_i}{N} \|u_k - u_i\| \tag{2}$$

where N denotes the sum of superpixels in all images, and n_i denotes the superpixel number of cluster c_i. And the position cue is expressed by:

$$w_p(k) = \frac{1}{n_k} \sum_{j=1}^M \sum_{i=1}^{N_j} [D(i,j)] \tag{3}$$

where $D(i,j)$ equals:

$$D(i,j) = \begin{cases} \|z_i^j - o^j\|^2, & if \ p_i^j \in c_k \\ 0, & otherwise \end{cases} \tag{4}$$

The repetition cue describes the frequency of which cluster appears in multiple images, and it is an important global attribute that reflects the co-saliency. It is easy to understand that the distribution of a cluster in all images is more uniform, its synergy is stronger. We employ the variances of clusters to roughly measure how widely is the cluster distributed among the multiple input images [12]. A M-bin histogram $\hat{q}^k = \{\hat{q}^k\}_j^M$ is adopted to describe the distribution of cluster c_k in M images:

$$\hat{q}_j^k = \frac{1}{n_k} \sum_{i=1}^{N_j} \delta[b(p_i^j) \cdot c_k], j = 1, \ldots, M \tag{5}$$

where $\delta(\cdot)$ is the Kronecker delta function, $b(p_i^j)$ returns the cluster index of superpixel p_i^j. Then, the repetition cue $w_r(k)$ is defined as:

$$w_r(k) = \frac{1}{var(\hat{q}^k) + 1} \tag{6}$$

where $var(\hat{q}^k)$ denotes the variance of histogram $\hat{q}^k$ of the cluster c_k. In order to avoid the denominator is 0, usually with 1. The cluster with the high repetition cue represents that the superpixels in this cluster evenly distribute in each image.

3.5 Integration

The process of co-saliency detection mainly contains three integrating operation. First, the weak co-saliency map on each scale is formed by pixel-wise multiplying three cues. Second, the WCS maps are generated by fusing weak co-saliency maps on different scales with the weight 2:3:5. Third, the final co-saliency map is obtained by linear summing and multiplying integration based on weak co-saliency map and saliency map.

4 Experimental Results

We evaluate our works on two aspects: the single image saliency detection, and the co-saliency detection on the multiple images. We compare our method with the state-of-the-art methods on a variety of benchmark datasets, more detail about datasets can be found in [18–21]. All the experiments are carried out using MATLAB R2013R on a desktop computer with an Intel i7-4790 CPU (3.60 GHz) and 12 G RAM.

4.1 Evaluation of CSMS Method

Firstly, we present some results of saliency maps generated by our CSMS method and other three methods (RARE [22], BL [8] and ELD [9]) on benchmark MSRA [18] and PASCAL-S [19] dataset in Fig. 2. Experiment shows CSMS method can get more complete and salient regions, and the residuals of the background are also less. It is obviously that the boundary and detail processing of CSMS is better, and the saliency maps are closer to the truth map.

Figure 3 shows the Precision and Recall (PR) curves of these four methods for quantitative evaluation, we can find CSMS's PR curve is close to the position of (0, 0), this indicates the performance of CSMS is better than other three methods.

4.2 Evaluation of SCCD Method

We present some results of co-saliency maps generated by our SCCD method and other three methods (CCD [12], CDRP [13] and MSG [14]) on benchmark iCoseg [20] and

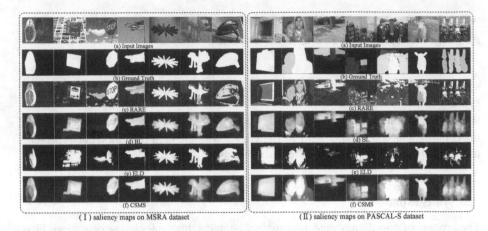

Fig. 2. Comparison of our saliency maps with three state-of-the-art methods on MSRA and PASCAL-S dataset. Maps in dashed box (I) and (II) are results on MSRA and PASCAL-S dataset respectively.

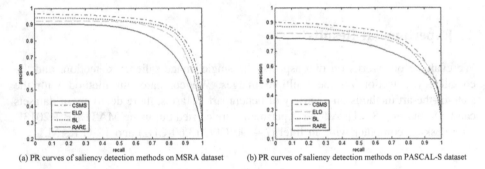

(a) PR curves of saliency detection methods on MSRA dataset   (b) PR curves of saliency detection methods on PASCAL-S dataset

Fig. 3. PR curves results on two datasets. CSMS is our result.

MSRC [21] dataset in Fig. 4. The SCCD method can obtain the meticulous co-salient object region by using the MSLIC, and get the correlation among the images by the clustering method. Therefore, the combination of the two can better detect the co-saliency among multiple images.

Figure 5. PR curves results on two datasets shows the PR curves of these four co-saliency methods on two datasets, we can find SCCD's PR curve is close to the position of (0, 0), this indicates the performance of SCCD is better than other three methods.

Furthermore, we compare the average running time of these four methods in Table 1. Our SCCD method adopts the bottom-up cues to measure the co-saliency. Comparing with the individual pixel based methods, it achieves an efficient and rapid computation. We can see that SCCD's computation is faster than other three methods'.

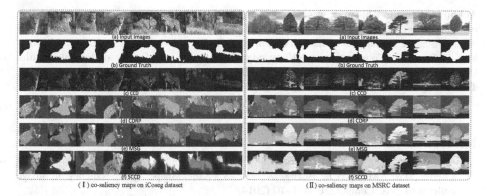

(I) co-saliency maps on iCoseg dataset   (II) co-saliency maps on MSRC dataset

Fig. 4. Comparison of our co-saliency maps with three state-of-the-art methods on iCoseg and MSRC dataset. Maps in dashed box (I) and (II) are results on iCoseg and MSRC dataset respectively.

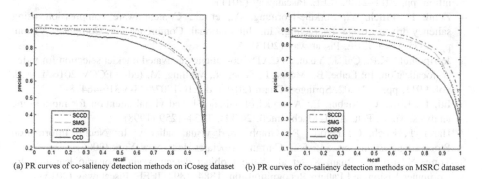

(a) PR curves of co-saliency detection methods on iCoseg dataset   (b) PR curves of co-saliency detection methods on MSRC dataset

Fig. 5. PR curves results on two datasets

Table 1. Average running time (seconds per image).

Methods	CCD	CDRP	SMG	SCCD
Time on iCoseg	7.35	30.04	8.65	5.43
Time on MSRC	13.20	70.56	17.96	10.31

5 Conclusion

In this paper, we propose a novel co-saliency detection method SCCD. SCCD uses superpixel clustering to break the limit of single image, makes similar superpixels to gather in one cluster. And using the superpixels instead of pixels to calculate the co-saliency shorts the computation time. Simultaneously, with the help of a content-sensitive superpixel segmentation method and superpixel pyramid, a saliency detection (CSMS) is proposed. CSMS can describe the boundary and internal detail of salient object accurately. Extensive experimental results demonstrate that the proposed approaches perform favorably.

Acknowledgment. This work was partially supported by National Natural Science Foundation of China (NSFC Grant Nos. 61170124, 61272258, 61301299, 61272005, 61572085), Provincial Natural Science Foundation of Jiangsu (Grant Nos. BK20151254, BK20151260), Key Laboratory of Symbolic Computation and Knowledge Engineering of Ministry of Education, Jilin University (Grant No. 93K172016K08), and Collaborative Innovation Center of Novel Software Technology and Industrialization.

References

1. Cheng, M.M., Mitra, N.J., Huang, X., Hu, S.: SalientShape: group saliency in image collections. Vis. Comput. **30**(4), 1–10 (2014)
2. Mukherjee, L., Singh, V., Peng, J.: Scale invariant cosegmentation for image groups. In: Computer Vision and Pattern Recognition, pp. 1881–1888. IEEE, Piscataway (2011)
3. Chang, K.Y., Liu, T.L., Lai, S.H.: From co-saliency to co-segmentation: an efficient and fully unsupervised energy minimization model. In: Computer Vision and Pattern Recognition, pp. 2011–2136. IEEE, Piscataway (2011)
4. Zund, F., Pritch, Y., Sorkine-Hornung, A., et al.: Content-aware compression using saliency-driven image retargeting. In: International Conference on Computer Vision, pp. 1845–1849. IEEE, Piscataway (2013)
5. Jerripothula, K.R., Cai, J., Yuan, J.: CATS: co-saliency activated tracklet selection for video co-localization. In: Leibe, B., Matas, J., Sebe, N., Welling, M. (eds.) ECCV 2016. LNCS, vol. 9911, pp. 187–202. Springer, Cham (2016). doi:10.1007/978-3-319-46478-7_12
6. Itti, L., Koch, C., Niebur, E.: A model of saliency-based visual attention for rapid scene analysis. Trans. Patt. Anal. Mach. Intell. **20**(11), 1254–1259 (1998)
7. Harel, J., Koch, C., Perona, P.: Graph-based visual saliency. In: Neural Information Processing Systems, pp. 545–552. Curran Associates, Inc., New York (2006)
8. Tong, N., Lu, H., Ruan, X., et al.: Salient object detection via bootstrap learning. In: Computer Vision and Pattern Recognition, pp. 1884–1892. IEEE, Piscataway (2015)
9. Lee, G., Tai, Y.W., Kim, J.: Deep saliency with encoded low level distance map and high level features. In: Computer Vision and Pattern Recognition, pp. 660–668. IEEE, Piscataway (2016)
10. Jacobs, D.E., Goldman, D.B., Shechtman, E.: Cosaliency: Where people look when comparing images. In: ACM Symposium on User Interface Software and Technology, pp. 219–228. ACM, New York (2010)
11. Li, H., Meng, F., Ngan, K.N.: Co-salient object detection from multiple images. Trans. Multimedia **15**(8), 1896–1909 (2013)
12. Fu, H., Cao, X., Tu, Z.: Cluster-based co-saliency detection. Trans. Image Process. **22**(10), 3766–3778 (2013)
13. Li, L., Liu, Z., Zou, W., et al.: Co-saliency detection based on region-level fusion and pixel-level refinement. In: International Conference on Multimedia and Expo, pp. 1–6. IEEE, Los Alamitos (2014)
14. Li, Y., Fu, K., Liu, Z., Yang, J.: Efficient saliency-model-guided visual co-saliency detection. Sig. Process. Lett. **22**(5), 588–592 (2015)
15. Zhang, D., Han, J., Li, C., et al.: Co-saliency detection via looking deep and wide. In: Computer Vision and Pattern Recognition, pp. 2994–3002. IEEE, Piscataway (2015)
16. Zhang, D., Meng, D., Li, C., et al.: A self-paced multiple-instance learning framework for co-saliency detection. In: International Conference on Computer Vision, pp. 594–602. IEEE, Piscataway (2015)

17. Liu, Y.J., Yu, C.C., Yu, M.J., et al.: Manifold SLIC: a fast method to compute content-sensitive superpixels. In: Computer Vision and Pattern Recognition, pp. 651–659. IEEE, Piscataway (2016)
18. Liu, T., Yuan, Z., Sun, J., Wang, J., Zheng, N., Tang, X., Shum, H.: Learning to detect a salient object. Trans. Patt. Anal. Mach. Intell. **33**(2), 353–367 (2011)
19. Li, Y., Hou, X., Koch, C., et al.: The secrets of salient object segmentation. In: Computer Vision and Pattern Recognition, pp. 280–287. IEEE, Piscataway (2014)
20. Batra, D., Kowdle, A., Parikh, D., et al.: iCoseg: Interactive co-segmentation with intelligent scribble guidance. In: Computer Vision and Pattern Recognition, pp. 3169–3176. IEEE, Piscataway (2010)
21. Winn, J., Criminisi, A., Minka, T.: Object categorization by learned universal visual dictionary. In: International Conference on Computer Vision Computer Vision, pp. 1800–1807. IEEE, Piscataway (2005)
22. Riche, N., Mancas, M., Duvinage, M., Mibulumukini, M., Gosselin, B., Dutoit, T.: Rare 2012: a multi-scale rarity-based saliency detection with its comparative statistical analysis. Sig. Process. Image Commun. **28**(6), 642–658 (2013)

ARMICA-Improved: A New Approach for Association Rule Mining

Shahpar Yakhchi[1]([✉]), Seyed Mohssen Ghafari[1], Christos Tjortjis[2], and Mahdi Fazeli[3]

[1] Computer Science Department, Azad University, Borujerd, Iran
computermsc.y@gmail.com, mohssenghafari@gmail.com
[2] School of Science and Technology Department,
International Hellenic University, Thermi, Greece
c.tjortjis@ihu.edu.gr
[3] Computing Department, Iran University of Science and Technology,
Tehran, Iran
m_fazeli@iust.ac.ir

Abstract. With increasing in amount of available data, researchers try to propose new approaches for extracting useful knowledge. Association Rule Mining (ARM) is one of the main approaches that became popular in this field. It can extract frequent rules and patterns from a database. Many approaches were proposed for mining frequent patterns; however, heuristic algorithms are one of the promising methods and many of ARM algorithms are based on these kinds of algorithms. In this paper, we improve our previous approach, ARMICA, and try to consider more parameters, like the number of database scans, the number of generated rules, and the quality of generated rules. We compare the proposed method with the Apriori, ARMICA, and FP-growth and the experimental results indicate that ARMICA-Improved is faster, produces less number of rules, generates rules with more quality, has less number of database scans, it is accurate, and finally, it is an automatic approach and does not need predefined minimum support and confidence values.

Keywords: Association rules mining · Data mining · Imperialist Competitive Algorithm (ICA)

1 Introduction

Association Rules Mining (ARM) is a data mining technique to extract frequent rules and patterns from a database. Many challenges are still remained in ARM techniques. Being a single dimension solution and focusing only on one aspect of ARM problems is the main drawback of these methods. For instance, many of them tend to be a fast approach and do not consider other parameters like the number of generated rules or accuracy; or only focus on accuracy of generated rules and do not investigate other factors like being a fast approach or having the least number of database scans. In conclusion, an efficient ARM approach should consider all these parameters at the same time.

© Springer International Publishing AG 2017
G. Li et al. (Eds.): KSEM 2017, LNAI 10412, pp. 296–306, 2017.
DOI: 10.1007/978-3-319-63558-3_25

One of the main parameters that have been frequently considered in many ARM approaches is having low execution time. Generating frequent and interesting rules in a short period of time is one of the primary goals of many ARM approaches. To be a fast approach, many researchers have worked on other parameters that may affect the execution time, like the number of database scans or the number for generated rules. In contrast, in many cases generating accurate rules or even rules with more quality, in this paper we assume that rules with more confidence are more qualify, is in higher priority compared to having low execution time. Finally, an automatic ARM method could be independent from user knowledge and can be applied on any databases. For this reason, some researchers have worked on making their approaches automatic.

In this paper, we focused on different aspects of ARM at the same time. We proposed a new ARM approach, named ARMICA-Improved, which extracts frequent rules accurately and in a short period of time. This approach is based on a heuristic algorithm called Imperialist Competitive Algorithm (ICA) [1]. ARMICA-Improved scans the database only once and generates less number of rules compared to the well-known ARM approaches. It does not consider infrequent items and only selects the most frequent items. In addition, it eliminates the transactions that do not contain any of these frequent items. Finally, our approach is an automatic approach and set the minimum support automatically and does not need minimum confidence to extract frequent rules. The experimental results indicate that ARMICA-Improved has lower execution time, less number of database scans, less number of generated rules, set the minimum support automatically and does not need minimum confidence value; also its generated rules are accurate, and generating more qualify rules compared to the Apriori [2, 3] and FP-growth [4].

The rest of this paper is organized as follows. Section 2 reviews the literature. Description of our proposed method and an example could be found in Sects. 3 and 4, respectively. Our experimental results would be in Sect. 5. Section 6 discusses the experimental results and there would be a conclusion statement in Sect. 7.

2 Literature Review

Apriori [2, 3] is the most famous ARM. Many algorithms tried to improve Apriori whilst others follow different approaches compared to Apriori. Apriori's mechanism is as: T = {t1, t2,..., tn} and I = {i1, i2,..., in} are the set of transactions and set of items that this dataset has, respectively. The algorithm tries to find all {X,Y} that both X and Y may contain at least one item. The extracted rule may be like: $X \rightarrow Y$

This rule means that if we find X in a transaction, then with a probability (Confidence) we also find Y in that transaction. The important thing is X and Y should not have any item in common: $X \cap Y = \Phi$

Most of ARM algorithms have two steps: first, finding the frequent itemsets. Frequent itemsets are sets of items that frequently occur together in the database. Secondly, generate frequent rules from the frequent itemsets. In Apriori, there is a parameter, named minimum support that items and itemsets with frequency of more than minimum support are frequent. Support of each item is the number of occurrence of that item in the database. In each level, Apriori generates candidate list of frequent items and itemsets. Then, it removes the items and itemsets with support of lower than minimum support. In this step, the algorithm employs a technique, named pruning to check that is there any itemsets, which has an item that was not on the candidate list in the previous levels; if it find one, so it removes this itemset. After pruning, it joins all the items and itemsets in the candidate list with each other and produces new candidate list. This process will continue and Apriori generates 2-length, 3-length, 4-length,… itemsets. It is worth to mention that, in this algorithm user should set the minimum support in advance and manually.

Finally, the last candidate list is the frequent list. At that point, Apriori extracts all non-empty subset of each item generates rules. In this step, Apriori needs another user defined parameter, named Minimum Confidence. Based on that, the algorithm removes the weak rules. Confidence of each rule could be calculated by:

$$Support\,(X \cup Y)/Support\,(X) \tag{1}$$

In the literature, many heuristic approaches have been proposed. One of heuristic approaches in ARM is [5]. Authors have proposed two new ARM algorithm based on GA, named IARMGA and Memetic algorithm, named IARMMA. They claim that most of bio-inspired-based algorithms have two main drawbacks: Generating false rules and considering some rules with low support and confidence as high qualify rules. They considered two parameters to evaluate their approaches and compared them with each other: Execution time and Accuracy of generated rules. Accuracy in this approached has been considered as value of their fitness function. For this reason, they propose a new method, named "delete and decomposition strategy" to have better accuracy. Finally, their experimental results indicate that IARMMA has higher execution time compared to IARMGA especially in a big dataset. However, IARMMA has better solution quality. In the end, they claimed that their approaches solved the problems of generating false and inaccurate rules. The main drawback of this work may be lack of comparison with other famous methods like Apriori.

Yan et al. have proposed a novel approach based on Genetic algorithms for ARM [6]. In their method, they did not use any fixed minimum support threshold. Instead, they employ relative minimum confidence term as fitness function to select only the best rules. At the beginning, they propose an algorithm, named ARMGA, which is designed to deal with Boolean association rule mining. However, since they also want to deal with quantitative association rule discovery, they propose another Genetic algorithm based method, named EARMGA which is an expansion of ARMGA. They also designed a FP-tree approach based on FP-growth for implementing EARMGA. Experimental results illustrate that their algorithms reduces computation costs and generates interesting association rules only.

In our previous work [7], we proposed an ARM approach, named ARMICA. Our main focus was on proposing a fast algorithm that extract frequent rules with small time consumption value. Hence, we employed Imperialist Competitive Algorithm (ICA) to extract frequent rules automatically. This approach did not required any predefined minimum support and confidence. Our experimental results illustrated that ARMICA is faster than Apriori. Moreover, ARMICA generates the same rules as Apriori, which can be the proof of its accuracy. However, what was the drawbacks of ARMICA? First, it requires a predefined parameter, named Number of Imperialists. Although defending a value for this parameter is not a complex task compared to the defining minimum support and confidence, it still relies on user to have this value. Secondly, ARMICA should be compared with other ARM approaches not only the Apriori. Finally, we should consider more parameters to improve the ARMICA, like number of database scans or number of generated rules.

3 Proposed Method

We propose new ARM approach based on ICA algorithm called ARMICA-Improved, which is a heuristic approach. This approach is an improved version of our previous method ARMICA [7]. ARMICA-Improved has some significant improvements compared to the ARMICA. One of the parameter that has a great impact on execution time is the number of database scans. Having higher number of database scans may increase the execution time. For this reason, ARMICA-Improved scans the database only once and at the same time calculate the frequency of each item in the database.

In this algorithm, we consider the frequency of each item as cost of that item in ICA and each item is a country. In addition, we assume that a country with more cost has more power. In the other worlds, a country (item) with high cost (frequency) is powerful. In the first step, the algorithm sends the countries' names and their cost (which were calculated in the database scan stage) to the ICA. ICA selects some of the most powerful countries as imperialists and divides other countries between them based on the power of each imperialist. More powerful imperialist can have more colonies. In this stage, the empires are built. At that point, ICA establishes competition between the empires. The more powerful empires try to steal the colonies of weaker empires. In the original ICA algorithm, the stolen colony become one the new colonies of the powerful empire, but like ARMICA, in ARMICA-Improved, this colony would be removed and added to a list, named Reserve List. Moreover, the stolen colony should be the weakest colony of the weaker empire. This process continues until there is only one colony left for the weaker empire. In this occasion, the colony and imperialist of the weaker empire become colonies of the powerful empire. This completion continues until there is only one empire left. At that point, the algorithm checks if there is any colony in the reserve list, which has more power than any colony in the final empire and exchange them. Finally, the members of the final empire are our frequent itemset. It is noticeable that since ICA selects the most powerful countries as imperialists and because we working on offline databases and the items have fix frequency, there is no chance for a colony to become more powerful than its imperialist; as a result, there is no need for Revolution process.

Next, the algorithm sorts the frequent rules based on their costs and calculates the median cost of them as the universal minimum support. Hence, ARMICA-Improved determines the minimum support automatically and it does not require any user knowledge to set this parameter in advance. One of the differences between ARMICA and ARMICA-Improved is that at this stage ARMICA-Improved removes each items that has frequency less than minimum support. Hence, when it extracts each combination of the remained frequent items, it is rarely to have infrequent itemsets (itemset, which their support value is less than minimum support). This could reduce the number of generated itemsets, significantly. At that point, the algorithm removes each transaction of the database that does not contain any of the frequent items. It also removes the columns of infrequent items. This process could dramatically decrease the size of the database.

Just like Apriori, FP-growth and ARMICA, ARMICA-Improved tries generate all the possible k-length frequent itemset that k is 1, 2, ..., n. However, in contrast to ARMICA, it in each stage, it stores all the generated frequent item sets along with their frequency to avoid any recalculation in the future. This could make the algorithm more efficient compared to the ARMICA. In the other worlds, one of the biggest difference between ARMICA and ARMICA-Improved is that in ARMICA-Improved we calculate the cost of frequent itemsets few times. We store all the costs of all itemsets in previous steps. This save us lots of time compared to ARMICA, which requires calculating costs of all frequent items and itemsets repeatedly.

4 Example

To familiarize the readers with ARMICA-Improved, here we made an example. Assume that we have transactional database like Table 1. This database has 17 items and 7 transactions. ARMICA-Improved scans this dataset and calculate the frequency of each items at the same time. At this stage, it sends the items and their frequency to the ICA. ICA selects some of them as imperialist. As it was mentioned before, the number of imperialists is a free parameter in the original ICA. As a result, since here we only have 17 countries, we cannot consider 10 percent of them as imperialist and we assume that we have 4 imperialist and distribute the rest of them between these imperialists based on their power. At this time, we the empires are built. Then the algorithm calculates the power of each empire based on power of their colonies and imperialists. The most powerful empire is empire 1 and the weakest one is empire 4. At this stage. Empire 1 tries to steal the I12 from the empire 4. The algorithm removes this colony from empire 4 and adds it to the reserve list. This process continues until there is only one empire left (Level n). The members of the last empire are the frequent items. ARMICA-Improved stores their names and costs in a list, named Save List for the future references.

Next, the algorithm tries to extract 2-lenth frequent itemsets. It generates them and calculate their costs, and stores them in the Save list. This process continues until all the possible frequent itemsets be produced. Then, ARMICA-Improved extract the frequent rules from these itemsets. Since the algorithms stores all the possible items and temsets and their costs in the Save list, in contrast to the ARMICA, there is no need to calculate the support of different parts of rules again. This would increase the execution time of the algorithm. Finally, ARMICA-Improved generate the frequent rules like other ARM approaches (Table 1).

Table 1. Example database

	I1	I2	I3	I4	I5	I6	I7	I8	I9	I10	I11	I12	I13	I14	I15	I16	I17
T1	t		t		t	t	t		t				t				t
T2	t			t		t		t	t	t					t	t	t
T3			t		t	t		t			t				t		
T4	t	t		t			t	t	t			t			t		
T5	t	t					t	t					t		t	t	
T6	t	t	t			t	t			t					t		
T7	t	t		t			t			t				t			T
Frequency	6	4	3	3	1	4	6	3	4	3	2	1	2	1	3	4	3

Level 1

Empire 1	Empire 2	Empire 3	Empire 4
Imperialist1: I1	Imperialist2: I7	Imperialist3: I2	Imperialist4: I6
I3 I9 I5 I10	I8 I15 I11	I4 I16 I13	I17 I14 I12
Power: 17	Power: 15	Power: 13	Power: 9

Level 2

Imperialist1: I1	Imperialist2: I7	Imperialist3: I2	Imperialist4: I6	Reserve List
I3 I9	I8 I15	I4 I16	I17 I14	
I10 I5	I11	I13		I12
Power: 17	Power: 15	Power: 13	Power: 8	

Level n

Imperialist1: I1										Reserve List					
I3	I1	I9	I5	I8	I17	I7	I	I16	I2	I12	I13	I14	I4	I15	I11

Final level

Imperialist1: I1										Reserve List					
I3	I9	I6	I15	I10	I2	I8	I17	I7	I16	I14	I12	I13	I4	I5	I11

Frequent Items	I3	I9	I1	I6
	I10	I15	I8	I2
	I16	I7		I17

5 Evaluation

We evaluated ARMICA-Improved using Java 1.7 in Netbeans IDE 7.2 and ran it on an Intel (R) Core (TM) i5 CPU at 2.40 GHz and 2 GB RAM. We also used the implementation of Apriori and FP-growth from Weka 3.6 [8] along with the Supermarket, Mushroom, Spect_Train, and Vote datasets from the UCI Machine Learning Repository [9] to benchmark our method. Moreover, to further study on ARMICA_Improved, we also employed LUCS-KDD ARM data generator [10] to generate different databases with different data density. In the other word, we wanted to investigate our approach under the different circumstances and see what is its characteristics in different databases with different data density. In addition, we considered four factors for this evaluation: the quality of generated rules (formula 2), the number of database scans, the number of generated rules, and execution time. Figure 1 indicates that in the Supermarket dataset ARMICA-Improved has the least number of generated rules with 347 rules. After that, FP-growth generates 350 and Apriori and ARMICA generate 372 rules.

Table 2. The characteristics of databases

Data set	Items	Transac-tions	Input Distribu-tion (Density) %	Algorithm	Predefined Min. Support	Predefined Min. Confidence
Supermarket	217	4627	-	Apriori	28.3	39
				ARMICA	—	—
				FP-growth	28.3	39
				ARMICA-Improved	—	—
Mushroom	119	8124	-	Apriori	36.18	47
				ARMICA	—	—
				FP-growth	36.18	47
				ARMICA-Improved	—	—
DataGenerator 1	110	1200	50	Apriori	1132.3	91
				ARMICA	—	—
				FP-growth	1132.3	91
				ARMICA-Improved	—	—
DataGenerator 2	110	1200	70	Apriori	94	95
				ARMICA	—	—
				FP-growth	94	95
				ARMICA-Improved	—	—
DataGenerator 3	110	1200	40	Apriori	75.8	78
				ARMICA	—	—
				FP-growth	75.8	78
				ARMICA-Improved	—	—
DataGenerator 4	140	2400	55	Apriori	91.43	92
				ARMICA	—	—
				FP-growth	91.43	92
				ARMICA-Improved	—	—
Spect_Train	46	267	-	Apriori	23.14	82
				ARMICA	—	—
				FP-growth	23.14	82
				ARMICA-Improved	—	—
Vote dataset	32	435	-	Apriori	52	86
				ARMICA	—	—
				FP-growth	52	86
				ARMICA-Improved	—	—

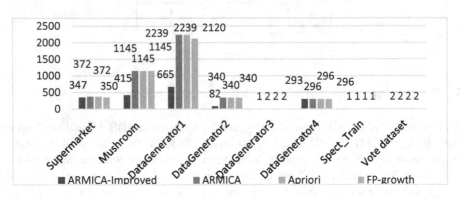

Fig. 1. The number of generated rules

$$The\ quality\ of\ generated\ rules = AVG(Generated\ Rules'\ Confidence) \qquad (2)$$

Figure 2 illustrates that, in the supermarket database the execution time of ARMICA-Improved is the lowest time compared to the other approaches. Its execution time is around 17 times less than Apriori. After ARMICA_Improved, FP-growth, ARMICA have the lowest execution times, respectively. The results in Fig. 3 indicate that, in the Supermarket database the average quality of generated rules, which we considered it as average confidence value of all generated rules, in ARMICA, Apriori and FP-growth is the same and is equal to 61.09. According to this figure, ARMICA-Improved with 62.2 has the highest average quality of rules compared to the other methods. It is worth to mention that Apriori needs many database scans to generate the frequent rules. After that ARMICA do few scans on the database for mining frequent rules. However, compared to many ARM approaches, FP-growth has one of the lowest number of database scans with 2 scans. This may have effect the execution time of this algorithm. Scanning the database is an I/O operation and having less I/O operation could make the approach faster. ARMICA-Improved with one database scan, scans the database even less than FP-growth.

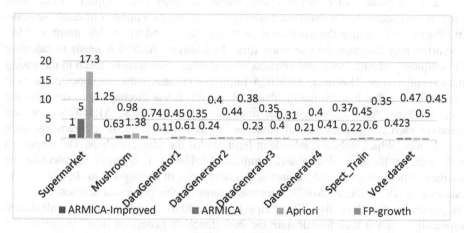

Fig. 2. Execution time

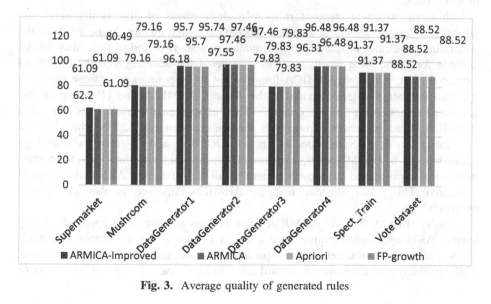

Fig. 3. Average quality of generated rules

6 Discussion

We believe that an ARM approach should optimize more than on parameter at the same time. This makes the approach more productive. At our previous research, ARMICA, we only focused on execution time and automatic procedure. However, in this paper, we considered other parameters like the number of generated rules, the number of database scans, execution time, and the quality of generated rules. Finally, like ARMICA, ARMICA-Improved also is an automatic approach.

Many researches have been done to propose fast ARM approaches. Extracting frequent rules from the database was also one of our primary goal in this paper. Consequently, ARMICA was improved in this paper. ARMICA-Improved tried to decrease the number of database scans, which may have some impact on execution time. It only scans the database once which is even less than number of database scans in FP-growth. Scanning the database is an I/O operation and having less number of I/O operation may decrease the execution time. In addition, ARMICA needs to calculate the frequency of each items and itemsets several times, which could result in increasing the execution time. However, ARMICA-Improved calculates the frequency of items at the same time that it scans the database. Moreover, it has mechanism to decrease the database size. After ICA algorithm finds the frequent items, ARMICA-Improved removes each items (columns) of database that is not frequent. It also removes each transactions, which does not include at least one of the frequent items. This approach will decrease the size of database, significantly. Hence, it is easier to calculate the frequency of each generated itemset in the next steps of the algorithm. In addition, in contrast to ARMICA, ARMICA-Improved stores all the generated itemsets and their frequency in the Save list. This helps the algorithm to prevent any recalculation especially when it tries to calculate the confidence of generated rules; the algorithm easily finds the support of each part of the rules in the save list and find the confidence of those rules.

Although extracting frequent rules in a short period of time is important, the generated rules also should be accurate and have the high quality. In the other words, generating false rules or rules with low quality in a short period of time, may not be suitable for users. As a result, we also considered the accuracy and the quality of extracted rules. Although ARMICA-Improved in the supermarket database produces the least number of rules compared to the Apriori, ARMICA, and FP-growth, it has generated rules with more quality. The experimental results indicate that ARMICA-Improved generates rules with average confidence of 62.2%, which is more than the quality of generated rules by Apriori, ARMICA, and FP-growth with average confidence of 61.09. Moreover, experimental results did not show any false rules generation. Apriori and FP-growth generated all the rules that ARMICA-Improved generated. This could illustrate that the ARMICA-Improved is also an accurate approach.

Finally, like ARMICA, ARMICA-Improved is an automatic approach. Many current ARM approaches require fixed and user defined minimum support and minimum confidence values. Setting these parameters before running the algorithm, especially in the bigger databases, is a hard task. In many case it needs a try and error approach to set these parameters. As a result, ARMICA-Improved tries to be independent from user knowledge about the database and set the minimum support automatically. This algorithm also does not need any minimum confidence. This feature makes ARMICA-Improved a database independent approach, which could be applied in any databases. However, it requires a predefine parameter like the number of imperialists. Although setting this parameter is not comparable with setting minimum support and confidence, like ARMICA we consider 10 percent of all countries as imperialist, it should be addressed in the future references; we should find a proper mechanism for setting this parameter.

After all, ARMICA-Improved showed some significant improvement in extracting frequent rules from the database. It is a fast approach, scans the database only once, generates less number of rules, generates rules with higher quality, does not generate false rules, removes unnecessary items and transactions from the database, decreases the database's size, and finally, it is an automatic approach and does not need any predefined minimum support and confidence values.

7 Conclusion

With the dramatic increase in amount of available data, applying data mining techniques to extracting useful knowledge from databases became more popular. One of these techniques is ARM. ARM approaches try to extract frequent patterns and rule form the databases. There have been proposed many ARM algorithms; however, many of them only focused on one aspect of the problem. This paper proposed a new ARM approach called ARMICA-Improved, which is an improved version of our previous research, ARMICA. The experimental results indicate that it is faster than Apriori, ARMICA, FP-growth. It scans the database only once; it decreases the size of database; it generates less number of rules and rules with higher quality compared to the other three mentioned approaches. Finally, it is an automatic approach and it does not need any predefined minimum support and confidence. For the future research, we should consider other parameters like interestingness and amount of memory usage. We also should try to test ARMICA-Improved on big data. Finally, ARMICA-Improved needs a proper mechanism to set the number of imperialists, which should be addressed in the future researches.

References

1. Atashpaz-Gargari, E., Lucas, C.: Imperialist competitive algorithm: An algorithm for optimization inspired by imperialistic competition. In: IEEE Congress on Evolutionary Computation, pp. 4661–4667 (2007)
2. Agrawal, R., Srikant, R.: Fast algorithms for mining association rules in large databases. In: VLDB Proceedings of the 20th International Conference on Very Large Data Bases, pp. 487–499 (1994)

3. Agrawal, R., Imielinski, T., Swami, A.: Mining association rules between sets of items in large databases. In: ACM SIGMOD Conference on Management of Data. ACM, New York (1993)
4. Han, J., Pei, J., Yin, Y., Mao, R.: Mining frequent patterns without candidate generation: a frequent-pattern tree approach. Data Min. Knowl. Disc. **8**, 53–87 (2004)
5. Drias, H.: Genetic algorithm versus memetic algorithm for association rules mining. In: Sixth World Congress on Nature and Biologically Inspired Computing (NaBIC). IEEE, Porto (2014)
6. Yan, X., Zhang, C., Zhang, S.: Genetic algorithm-based strategy for identifying association rules without specifying actual minimum support. Expert Syst. Appl. **36**(2), 3066–3076 (2009)
7. Ghafari, S.M., Tjortjis, C.: Association rules mining using the imperialism competitive algorithm (ARMICA). In: 12th IFIP International Conference on Artificial Intelligence Applications and Innovations (AIAI), Thessaloniki (2016)
8. Witten, I.H., Eibe, F., Hall, M.A.: Data Mining: Practical Machine Learning Tools and Techniques, 3rd edn. Morgan Kaufmann, Burlington (2011)
9. Bache, K., Lichman, M.: UCI machine learning repository (2013)
10. Coenen, F.: LUCS-KDD ARM data generator (2007)

Recommendation Algorithms and Systems

Collaborative Filtering via Different Preference Structures

Shaowu Liu[1(✉)], Na Pang[2], Guandong Xu[1], and Huan Liu[3]

[1] University of Technology Sydney, Sydney, Australia
{shaowu.liu,guandong.xu}@uts.edu.au
[2] School of Cyber Security, University of Chinese Academy of Sciences,
Beijing, China
pangna@iie.ac.cn
[3] Arizona State University, Tempe, USA
huan.liu@asu.edu

Abstract. Recently, social network websites start to provide third-parity sign-in options via the OAuth 2.0 protocol. For example, users can login *Netflix* website using their *Facebook* accounts. By using this service, accounts of the same user are linked together, and so does their information. This fact provides an opportunity of creating more complete profiles of users, leading to improved recommender systems. However, user opinions distributed over different platforms are in different preference structures, such as ratings, rankings, pairwise comparisons, voting, etc. As existing collaborative filtering techniques assume the homogeneity of preference structure, it remains a challenge task of how to learn from different preference structures simultaneously. In this paper, we propose a fuzzy preference relation-based approach to enable collaborative filtering via different preference structures. Experiment results on public datasets demonstrate that our approach can effectively learn from different preference structures, and show strong resistance to noises and biases introduced by cross-structure preference learning.

Keywords: Recommender system · Pairwise preference · Data mining

1 Introduction

Personalized recommendation is an important component of today's business. By observing user behaviors, recommender systems can identify potential users of a product, or products that could be interested by a targeted user. An important technique to make recommendations is collaborative filtering (CF). CF is based on the intuition that there exist shared patterns to transfer preferences across like-minded users. For example, whether a targeted user will like a movie can be inferred by other users who have similar taste to the targeted user. The taste of a user can be extracted from user preferences in different structures, such as ratings [7], rankings [8], pairwise comparisons [5], voting [11], text reviews, etc.

© Springer International Publishing AG 2017
G. Li et al. (Eds.): KSEM 2017, LNAI 10412, pp. 309–321, 2017.
DOI: 10.1007/978-3-319-63558-3_26

A common assumption made by CF is the homogeneity of preference structures, where only one type of preference structure is accepted at a time.

The last decade has seen a growing trend towards creating and managing more profiles in social network, such as *Facebook, LinkedIn, Netflix*, etc. Furthermore, the popularization of third-party sign-in via the OAuth 2.0 protocol has made it possible to link multiple profiles of the same user together. In light of this trend, it becomes possible to alleviate the *cold-start* problem by learning user preferences from multiple profiles, e.g., a new user of *Netflix* may have been used *Facebook* for a while. Nevertheless, user preferences collected from different platforms are often expressed in different preference structures. For example, 5-star rating is used by *Netflix*, but voting (thumbs up) is used by *Facebook*. Despite of explicit preferences, additional complexity is added if implicit preferences such as *page views* and *mouse clicks* are also taken into consideration.

Moreover, user preferences collected from different platforms may contain different noises and biases, as the user preferences not only reflect inherent quality of the product but also quality of the product providers. For example, a user may rate a movie 3 star on one platform, but 5 star for the same movie on another platform due to 3D support, which is called misattribution of memory [13] in psychology. Nevertheless, different preference structures need to be placed on the same scale for accurate discovering of shared patterns to achieve quality recommendations.

In this work, we propose a fuzzy preference relations-based approach to learn from different preference structures. Rather than trying to learn an independent model for each type of preference structure, we propose to *simultaneously learn user preferences in all structures in one model*. With the assistance of *PR*, user preferences in different forms can be fused seamlessly. For example, user preferences expressed as 5-star ratings, binary ratings, and *page views* can not be directly fused in general. However, all those user preferences can be deduced into the *PR* format by performing pairwise comparison on items. Once the user preferences are represented in *PR*, a direct merge can be performed. In fact, converting user preferences into *PR* not only provides a method to merge heterogeneous data but also reduces the biases that come with heterogeneity, i.e., the relative ordering of items is resistant to biases. *The main contribution of this work is proposing an approach to learn from multiple data sources with different preference structures such as ratings, page views, mouse clicks, reviews, etc.*

The rest of the paper is organized as follows. Section 2 introduces the basic concepts of CF and preference structures. Section 3 is devoted to describe the proposed method. In Sect. 4, the proposed method is applied to public datasets for top-N recommendation. Finally, conclusions are drawn in Sect. 5.

2   Preliminaries and Related Work

This section briefly summarizes necessary background related to the *heterogeneous sources* problem and the *preference relations* that form the basis of our solution.

2.1 Heterogeneous Sources

User preferences are usually assumed to come from a single *homogeneous source*. This assumption is becoming invalid given the rapid development of online social networks in which users maintain multiple profiles and the form of preferences diverges. We define two sources as heterogeneous if their preferences are (1) in different forms, e.g., *ratings* and *clicks*; (2) in different scales, e.g., *5-star* scale and *6-star* scale; (3) or biased differently due to factors irrelevant to the items' quality, e.g., quality of the service providers. Based on this definition, not only the physically separated sources are heterogeneous but a source changed significantly is also considered heterogeneous to itself.

In general, user preferences from heterogeneous sources cannot be merged directly as they may be in different forms. Even if their forms are the same, the scales could be different, where a force casting may change the meaning of preferences. In case that the scales are the same, biases are still introduced by the sources which make the recommendations inaccurate.

2.2 Preference Relation

Preference relation (PR) encodes user preferences in the form of *relative* ordering between items, which is a useful alternative representation to *absolute* ratings as suggested in recent works [3,5,9]. In fact, existing preferences such as ratings or other types of preferences can be easily represented as *PR* and then merged into a single dataset as shown in Fig. 1. This property is particularly useful for the *cold-start* problem but has been overlooked in literature.

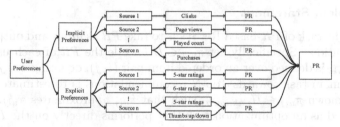

Fig. 1. Flow from user preferences to PR

We formally define the *PR* as follows. Let $\mathcal{U} = \{u\}^n$ and $\mathcal{I} = \{i\}^m$ denote the set of n users and m items, respectively. The *PR* of a user $u \in \mathcal{U}$ between items i and j is encoded as π_{uij}, which indicates the strength of user u's preference relation for the ordered item pair (i, j). A higher value of π_{uij} indicates a stronger preference to the first item over the second item.

The preference relation is defined as

$$\pi_{uij} = \begin{cases} (\frac{2}{3}, 1] & \text{if } i \succ j \,(\,u \text{ prefers } i \text{ over } j\,) \\ [\frac{1}{3}, \frac{2}{3}] & \text{if } i \simeq j \,(\,i \text{ and } j \text{ are equally preferable}) \\ [0, \frac{1}{3}) & \text{if } i \prec j \,(\,u \text{ prefers } j \text{ over } i\,) \end{cases} \qquad (1)$$

where $\pi_{uij} \in [0, 1]$ and $\pi_{uij} = 1 - \pi_{uji}$.

An interval is allocated for each preference category, i.e., *preferred, equally preferred*, and *less preferred*. Indeed, each preference category can be further break down into more intervals, though here in this paper we consider the minimal case of 3 intervals.

Similar to [3], the *PR* can be converted into *user-wise preferences* over items which encode the ranking of items evaluated by a particular user. The user-wise preference is defined as

$$p_{ui} = \frac{\sum_{j \in \mathcal{I}_u \setminus i} [\![\pi_{uij} > \frac{2}{3}]\!] - \sum_{j \in \mathcal{I}_u \setminus i} [\![\pi_{uij} < \frac{1}{3}]\!]}{|\Pi_{ui}|} \tag{2}$$

where $[\![\cdot]\!]$ gives 1 for *true* and 0 for *false*, and Π_{ui} is the set of user u's PR related to item i. The user-wise preference p_{ui} falls in the interval $[-1, 1]$, where 1 and -1 indicate that item i is the most and the least preferred item for u, respectively.

3 Preference Relation-Based Conditional Random Fields

In this section, we propose the *Preference Relation-based Conditional Random Fields* (PrefCRF) to model both the heterogeneous preferences and the side information. The rest of this section defines the *PR*-based RecSys problem, and introduces the concept of the *PrefNMF* [5] that forms our underlying model, followed by a detailed description of the *PrefCRF* and discussion on issues such as feature design, parameter estimation, and predictions.

3.1 Problem Statement

Generally, the task of *PR*-based RecSys is to take *PR* as input and output Top-N recommendations. Specifically, let $\pi_{uij} \in \Pi$ encode the *PR* of each user $u \in \mathcal{U}$, and each π_{uij} is defined over an ordered item pair (i, j), denoting $i \prec j$, $i \simeq j$, or $i \succ j$. The main task towards Top-N recommendations is to estimate the value of each unknown $\pi_{uij} \in \Pi_{unknown}$, such that $\hat{\pi}_{uij}$ approximates π_{uij}. This can be considered as an optimization task that performs directly on the *PR*

$$\hat{\pi}_{uij} = \arg \min_{\hat{\pi}_{uij} \in [0,1]} (\pi_{uij} - \hat{\pi}_{uij})^2 \tag{3}$$

However, it can be easier to estimate the $\hat{\pi}_{uij}$ by the difference between two user-wise preferences p_{ui} and p_{uj}, i.e., $\hat{\pi}_{uij} = \phi(\hat{p}_{ui} - \hat{p}_{uj})$, where $\phi(\cdot)$ is a function that bounds the value into $[0, 1]$ and ensures $\phi(0) = 0.5$. For example, the *inverse-logit* function $\phi(x) = \frac{e^x}{1+e^x}$ can be used when user-wise preferences involve large values. The objective of this paper is then to solve the following optimization problem

$$(\hat{p}_{ui}, \hat{p}_{uj}) = \arg \min_{\hat{p}_{ui}, \hat{p}_{uj}} (\pi_{uij} - \phi(\hat{p}_{ui} - \hat{p}_{uj}))^2 \tag{4}$$

which optimizes the user-wise preferences directly, and Top-N recommendations can be obtained by simply sorting the estimated user-wise preferences.

3.2   Preference Relation-Based Matrix Factorization

Matrix Factorization (MF) [7] is a popular *RecSys* approach that has mainly been applied to absolute ratings. Recently, the *PrefNMF* [5] model was proposed to accommodate *PR* input for *MF* models. Like traditional *MF* models, the *PrefNMF* model discovers the latent factor space shared between users and items, where the latent factors describe both the *taste* of users and the *characteristics* of items. The attractiveness of an item to a user is then measured by the inner product of their latent feature vectors.

Formally, each user u is associated with a latent feature vector $\mathbf{u}_u \in \mathbb{R}^k$, and each item i is associated with a latent feature vector $\mathbf{v}_i \in \mathbb{R}^k$, where k is the dimension of the latent factor space. The attractiveness of items i and j to user u are $\mathbf{u}_u^\top \mathbf{v}_i$ and $\mathbf{u}_u^\top \mathbf{v}_j$, respectively. When $\mathbf{u}_u^\top \mathbf{v}_i > \mathbf{u}_u^\top \mathbf{v}_j$, the item i is said to be more preferable to the user u than item j, i.e., $i \succ j$. The strength of this preference relation π_{uij} can be estimated by $\mathbf{u}_u^\top (\mathbf{v}_i - \mathbf{v}_j)$, and the *inverse-logit* function is applied to ensure $\hat{\pi}_{uij} \in [0,1]$: $\hat{\pi}_{uij} = \frac{e^{\mathbf{u}_u^\top (\mathbf{v}_i - \mathbf{v}_j)}}{1 + e^{\mathbf{u}_u^\top (\mathbf{v}_i - \mathbf{v}_j)}}$.

The latent feature vectors $\mathbf{u}_u$ and $\mathbf{v}_i$ are learned by minimizing regularized squared error with respect to the set of all known preference relations Π:

$$\min_{\mathbf{u}_u, \mathbf{v}_i \in \mathbb{R}^k} \sum_{\pi_{uij} \in \Pi \wedge (i < j)} (\pi_{uij} - \hat{\pi}_{uij})^2 + \lambda(\|\mathbf{u}_u\|^2 + \|\mathbf{v}_i\|^2) \tag{5}$$

where λ is the regularization coefficient.

3.3   Conditional Random Fields

Conditional Random Fields (CRF) [14] model a set of random variables having Markov property with respect to an undirected graph $\mathcal{G}$, and each random variable can be conditioned on a set of global observations $\mathbf{o}$. The undirected graph $\mathcal{G}$ consists of a set of vertexes $\mathcal{V}$ connected by a set of edges $\mathcal{E}$ without orientation, where two vertexes are neighboring to each other when connected. Each vertex in $\mathcal{V}$ encodes a random variable, and the Markov property implies that a variable is conditionally independent of others given its neighbors.

In this work, we use *CRF* to model interactions among user-wise preferences conditioned on side information with respect to a set of undirected graphs. Specifically for each user u, there is a graph $\mathcal{G}_u$ with a set of vertexes $\mathcal{V}_u$ and a set of edges $\mathcal{E}_u$. Each vertex in $\mathcal{V}_u$ represents a user-wise preference p_{ui} of user u on the item i. Each edge in $\mathcal{E}_u$ captures a relation between two preferences by the same user.

Each vertex is conditioned on a set of global observations $\mathbf{o}$, which is the *side information* in our context. Specifically, each user u is associated with a set of L attributes $\{\mathbf{o}_u\}^L$ such as *age*, *gender* and *occupation*. Similarly, each item i is associated with a set of M attributes $\{\mathbf{o}_i\}^M$ such as *genres* for movie. Those side information is encoded as the set of global observations $\mathbf{o} = \{\{\mathbf{o}_u\}^L, \{\mathbf{o}_i\}^M\}$.

Formally, let $\mathbf{p}_u = \{p_{ui} \mid i \in \mathcal{I}_u\}$ be the joint set of preferences expressed by user u, then we are interested in modeling the conditional distribution $P(\mathbf{p}_u \mid \mathbf{o})$ over the graph $\mathcal{G}_u$.

$$P(\mathbf{p}_u \mid \mathbf{o}) = \frac{1}{Z_u}\Psi_u(\mathbf{p}_u, \mathbf{o}) \tag{6}$$

$$\Psi_u(\mathbf{p}_u, \mathbf{o}) = \prod_{(ui)\in\mathcal{V}_u} \psi_{ui}(p_{ui}, \mathbf{o}) \prod_{(ui,uj)\in\mathcal{E}_u} \psi_{ij}(p_{ui}, p_{uj}) \tag{7}$$

where $Z_u(\mathbf{o})$ does normalization to ensure $\sum_{\mathbf{p}_u} P(\mathbf{p}_u \mid \mathbf{o}) = 1$, and $\psi(\cdot)$ is a positive function known as *potential*. The potential $\psi_{ui}(\cdot)$ captures the global observations associated to the user u and the item i, and the potential $\psi_{ij}(\cdot)$ captures the correlations between two preferences p_{ui} and p_{uj}

$$\psi_{ui}(p_{ui}, \mathbf{o}) = \exp\{\mathbf{w}_u^\top \mathbf{f}_u(p_{ui}, \mathbf{o}_i) + \mathbf{w}_i^\top \mathbf{f}_i(p_{ui}, \mathbf{o}_u))\} \tag{8}$$

$$\psi_{ij}(p_{ui}, p_{uj}) = \exp\{w_{ij} f_{ij}(p_{ui}, p_{uj})\} \tag{9}$$

where $\mathbf{f}_u$, $\mathbf{f}_i$, and f_{ij} are the features to be designed shortly in Sect. 3.4, and $\mathbf{w}_u$, $\mathbf{w}_i$, and w_{ij} are the corresponding weights realizing the importance of each feature. With the weights estimated from data, the unknown preference p_{ui} can be predicted as

$$\hat{p}_{ui} = \arg\max_{p_{ui}\in[-1,1]} P(p_{ui} \mid \mathbf{p}_u, \mathbf{o}) \tag{10}$$

where $P(p_{ui} \mid \mathbf{p}_u, \mathbf{o})$ measures the prediction confidence.

The *Ordinal Logistic Regression* [10] is then used to convert the user-wise preferences p_{ui} into ordinal values, which assumes that the preference p_{ui} is chosen based on the interval to which the latent utility belongs:

$$p_{ui} = l \text{ if } x_{ui} \in (\theta_{l-1}, \theta_l] \text{ and } p_{ui} = L \text{ if } x_{ui} > \theta_{L-1} \tag{11}$$

where L is the number of ordinal levels and θ_l are the threshold values of interest. The probability of receiving a preference l is therefore:

$$Q(p_{ui} = l \mid u, i) = \int_{\theta_{l-1}}^{\theta_l} P(x_{ui} \mid \theta)\,\mathrm{d}\theta = F(\theta_l) - F(\theta_{l-1}) \tag{12}$$

where $F(\theta_l)$ is the cumulative logistic distribution evaluated at θ_l.

3.4   PrefCRF: Unifying PrefNMF and CRF

The *CRF* provides a principled way of capturing both the side information and interactions among preferences. However, it employs the log-linear modeling as shown in Eq. 7, and therefore does not enable a simple treatment of *PR*. The *PrefNMF*, on the other hand, accepts *PR* but is weak in utilizing side information. The complementary between these two techniques calls for an unified *PrefCRF* model to take all the advantages.

Unification. Essentially, the proposed *PrefCRF* model captures the side information and promotes the agreement between the *PrefNMF* and the *CRF*. Specifically, the *PrefCRF* model combines the item-item correlations (Eq. 9) and the ordinal distributions $Q(p_{ui} \mid u, i)$ over user-wise preferences obtained from Eq. 12.

$$P(\mathbf{p}_u \mid \mathbf{o}) \propto \Psi_u(\mathbf{p}_u, \mathbf{o}) \prod_{p_{ui} \in \mathbf{p}_u} Q(p_{ui} \mid u, i) \tag{13}$$

where Ψ_u is the potential function capturing the side information and interaction among preferences related to user u. Though there is a separated graph for each user, the weights are optimized across all graphs.

Feature Design. A feature is essentially a function f of $n > 1$ arguments that maps the n-dimensional input into the unit interval $f : \mathbb{R}^n \to [0, 1]$. We design the following kinds of features:

Correlation Features. The item-item correlation is captured by the feature

$$f_{ij}(p_{ui}, p_{uj}) = g(|(p_{ui} - \bar{p}_i) - (p_{uj} - \bar{p}_j)|) \tag{14}$$

where $g(\alpha)$ normalizes feature values and α plays the role of deviation, and $\bar{p}_i$ and $\bar{p}_j$ are the average user-wise preference for items i and j, respectively.

Attribute Features. Each user u and item i has a set of attributes $\mathbf{o}_u$ and $\mathbf{o}_i$, respectively. These attributes are mapped to preferences by the following features

$$\begin{aligned} \mathbf{f}_i(p_{ui}) &= \mathbf{o}_u g(|(p_{ui} - \bar{p}_i)|) \\ \mathbf{f}_u(p_{ui}) &= \mathbf{o}_i g(|(p_{ui} - \bar{p}_u)|) \end{aligned} \tag{15}$$

where $\mathbf{f}_i$ models which users like the item i and $\mathbf{f}_u$ models which classes of items the user u likes.

Since one correlation feature exists for each pair of co-rated items, the number of correlation features can be large, and makes the estimation slow to converge and less robust. Therefore, we only keep strong correlation features $\mathbf{f}_{\text{strong}}$ extracted based on the *Pearson* correlation between items using a user-specified *minimum correlation threshold*.

Parameter Estimation. In general, *CRF* models cannot be determined by standard maximum likelihood estimations, instead, approximation techniques are used in practice. This study employs the pseudo-likelihood [1] to estimate parameters by maximizing the regularized sum of log local likelihoods:

$$log\mathcal{L}(\mathbf{w}) = \sum_{p_{ui} \in \Pi} \log P(p_{ui} \mid \mathbf{p}_u, \mathbf{o}) - \frac{1}{2\sigma^2}\mathbf{w}^\top \mathbf{w} \tag{16}$$

where $\mathbf{w}$ are the weights and $1/2\sigma^2$ controls the regularization. To optimize the parameters, we use the stochastic gradient ascent procedure.

Item Recommendation. The *PrefCRF* produces distributions over the user-wise preferences, which can be converted into point estimates by computing the expectation

$$\hat{p}_{ui} = \sum_{p_{ui}=l_{min}}^{l_{max}} p_{ui} P(p_{ui} \mid \mathbf{p}_u, \mathbf{o}) \tag{17}$$

where l refers to the intervals of user-wise preferences: from the least to the most preferred. Given the predicted user-wise preferences, the items can be sorted and ranked accordingly.

4   Experiment and Analysis

To study the performance of the proposed *PrefCRF* model, comparisons were done with the following representative algorithms: *KNN* [12], *NMF* [7], *PrefKNN* [3], and *PrefNMF* [5]. We employ two evaluation metrics *Normalized Cumulative Discounted Gain*@T (NDCG@T) [6] that is popular in academia, and *Mean Average Precision*@T (MAP@T) [4] that is common in contests.

4.1   Experimental Settings

Datasets and Experiment Design. Experiments are conducted on four public datasets: *MovieLens*-1M[1], *Amazon Movie Reviews*[2], *EachMovie*[3], and *MovieLens*-20M (see footnote 1). These datasets or their subsets are transformed to simulate four scenarios of heterogeneous data:

Side Information. The impact of side information is studied on the *Movie-Lens*-1M dataset which provides *gender*, *age*, and *occupation* information about users and *genres* of movies. The dataset contains more than 1 million ratings by 6, 040 users on 3, 900 movies. For a reliable comparison, the dataset is split into training and test sets with different sparsities.

Different Forms. *Amazon Movie Reviews* dataset contains two forms of preferences: *textual reviews* and 5-star *ratings*. We extracted a dense subset by randomly selecting 5141 items with at least 60 reviews for each, and 2000 users with at least 60 movies reviews for each, and this results in 271K ratings. For each user, 50 random reviews are selected for training, and the rest are put aside for testing. The training set is further split into half ratings and half textual reviews. Rating-based models are trained on the ratings only, where *PR*-based models utilize textual reviews as well.

Different Scales. *EachMovie* dataset contains ratings in 6-star scale that can be easily converted into binary scale, i.e., ratings 1–3 and 4–6 are mapped to 0 and 1 respectively. We extract a subset by randomly selecting 3000 users who have rated at least 70 items as a dense dataset is required for splitting.

---

[1] http://grouplens.org/datasets/movielens.
[2] http://snap.stanford.edu/data/web-Movies.html.
[3] http://grouplens.org/datasets/eachmovie.

The resultant dataset contains 120K ratings on 1495 items. For each user we randomly select 60 ratings for training and leave the rest for testing, and half of the ratings in the training set are mapped into binary scale. Rating-based models are trained on the 6-star ratings while *PR*-based models will exploit the binary ratings as well.

Different Biases. We study the impact of biases by adding biases into a stable dataset with minimal existing biases. To prepare such dataset we extract a stable subset from the latest *MovieLens*-20M released on April-2015. Specifically, 258K ratings by 2020 users on 4408 movies released between 2010 and 2015 are extracted, where each user has rated at least 60 ratings. Biases are then introduced by adding a different *Laplace noise* sampled from $Laplace(0, b)$ to each user and item.

For *PR*-based methods, the same conversion method as in [5] is used to converted ratings into *PR*. For example, 1, 0 and 0.5 are assigned to the preference relation π_{uij} when $p_{ui} > p_{uj}$, $p_{ui} < p_{uj}$, and $p_{ui} = p_{uj}$, respectively.

Parameter Setting. For a fair comparison, we fix the number of latent factors to 50 for all algorithms. The number of neighbors for *KNN* algorithms is set to 50. We vary the minimum correlation threshold for the *PrefCRF* to examine the performance with different number of features. Different values of regularization coefficient are also tested.

4.2 Results and Analysis

Algorithms are compared on four heterogeneous scenarios: *side information, different forms, different scales* and *different biases*. The impact of sparsity levels and parameters is studied on the *MovieLens*-1M dataset, while these settings for other experiments are fixed. Each experiment is repeated ten times with different random seeds and we report the mean results with standard deviations. For each experiment, we also performed a paired *t*-test (two-tailed) with a significance level of 95% on the best and the second best results, and all *p*-values are less than 1×10^{-5}.

Fusing Side Information. Table 1 shows the *NDCG* and *MAP* metrics on Top-N recommendation tasks by compared algorithms. It can be observed that the proposed *PrefCRF*, which captures the side information, consistently outperforms others. To confirm the improvement, we plot the results in Fig. 2b by varying the position T. The figure shows that *PrefCRF* not only outperforms others but has a strong emphasize on top items, i.e., $T < 5$.

The impact of sparsity is investigated by plotting the results against sparsity levels as in Fig. 2a. We can observe that the performance of *PrefCRF* increases linearly given more training data, while its underlying *PrefNMF* model is less extensible to denser dataset.

Table 1. Mean results and standard deviation over ten runs on *MovieLens*-1M dataset.

Algorithm	Given 30				Given 40			
	NDCG@1	NDCG@10	MAP@1	MAP@10	NDCG@1	NDCG@10	MAP@1	MAP@10
UserKNN	0.4306 ± 0.0011	0.4081 ± 0.0029	0.3539 ± 0.0071	0.2744 ± 0.0025	0.3695 ± 0.0048	0.4252 ± 0.0036	0.3663 ± 0.0047	0.2877 ± 0.0034
NMF	0.5274 ± 0.0084	0.5195 ± 0.0040	0.5225 ± 0.0081	0.3549 ± 0.0037	0.5424 ± 0.0067	0.5291 ± 0.0034	0.5377 ± 0.0066	0.3631 ± 0.0035
PrefKNN	0.3462 ± 0.0073	0.4048 ± 0.0038	0.3430 ± 0.0072	0.2720 ± 0.0037	0.3651 ± 0.0065	0.4283 ± 0.0024	0.3620 ± 0.0063	0.2904 ± 0.0023
PrefNMF	0.5778 ± 0.0112	0.5680 ± 0.0041	0.5724 ± 0.0109	0.3992 ± 0.0033	0.5883 ± 0.0073	0.5732 ± 0.0028	0.5832 ± 0.0073	0.4019 ± 0.0032
PrefCRF	**0.6206 ± 0.0076**	**0.5856 ± 0.0028**	**0.6150 ± 0.0073**	**0.4195 ± 0.0028**	**0.6395 ± 0.0064**	**0.5990 ± 0.0023**	**0.6340 ± 0.0062**	**0.4294 ± 0.0021**

Algorithm	Given 50				Given 60			
	NDCG@1	NDCG@10	MAP@1	MAP@10	NDCG@1	NDCG@10	MAP@1	MAP@10
UserKNN	0.3831 ± 0.0063	0.4424 ± 0.0027	0.3803 ± 0.0060	0.3015 ± 0.0026	0.4035 ± 0.0090	0.4622 ± 0.0035	0.4002 ± 0.0085	0.3163 ± 0.0027
NMF	0.5430 ± 0.0083	0.5326 ± 0.0036	0.5390 ± 0.0082	0.3669 ± 0.0025	0.5547 ± 0.0109	0.5409 ± 0.0063	0.5504 ± 0.0113	0.3734 ± 0.0055
PrefKNN	0.3831 ± 0.0092	0.4483 ± 0.0030	0.3803 ± 0.0089	0.3070 ± 0.0022	0.3979 ± 0.0075	0.4689 ± 0.0039	0.3948 ± 0.0069	0.3223 ± 0.0033
PrefNMF	0.5873 ± 0.0096	0.5745 ± 0.0035	0.5830 ± 0.0098	0.4019 ± 0.0033	0.5854 ± 0.0145	0.5733 ± 0.0048	0.5808 ± 0.0142	0.4007 ± 0.0037
PrefCRF	**0.6548 ± 0.0055**	**0.6068 ± 0.0018**	**0.6499 ± 0.0059**	**0.4372 ± 0.0024**	**0.6677 ± 0.0074**	**0.6139 ± 0.0018**	**0.6625 ± 0.0072**	**0.4436 ± 0.0016**

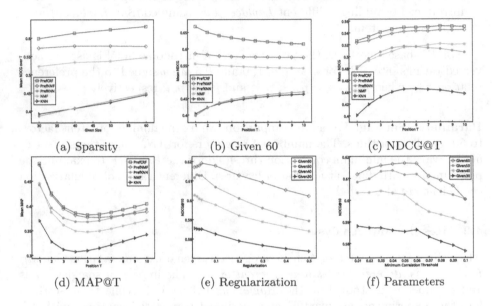

(a) Sparsity   (b) Given 60   (c) NDCG@T

(d) MAP@T   (e) Regularization   (f) Parameters

Fig. 2. Plots of experimental results

Fusing Preferences in Different Forms. In this experiment, we first converted textual reviews into negative (-1), neutral (0), and positive (1) values using the *NLTK* library [2], and then converted them into *PR*. We study how these additional information can assist *PR*-based methods, and results over ten runs are shown in Table 2. Surprisingly, the performance of all *PR*-based methods except *PrefCRF* have decreased by incorporating textual reviews. We suspect that this is due to the misclassification errors introduced by sentiment classification on text. However, in the next subsection we will see that an accurate conversion can actually improve the performance.

Fusing Preferences in Different Scales. In this experiment preferences in different scales are fused into *PR* to boost the performance of *PR*-based methods. The binary scale ratings are similar to the positive/negative textual reviews, however without incorrect values introduced by text classification.

Table 2. Results over ten runs on *Amazon* dataset.

Algorithm	Ratings		Ratings + Textual reviews	
	NDCG@10	MAP@10	NDCG@10	MAP@10
UserKNN	0.6244 ± 0.0040	0.4599 ± 0.0035	0.6244 ± 0.0037	0.4599 ± 0.0025
NMF	$\mathbf{0.8073 \pm 0.0040}$	$\mathbf{0.6689 \pm 0.0038}$	$\mathbf{0.8073 \pm 0.0041}$	$\mathbf{0.6689 \pm 0.0000}$
PrefKNN	0.6410 ± 0.0038	0.4690 ± 0.0029	0.5765 ± 0.0039	0.4083 ± 0.0029
PrefNMF	0.7495 ± 0.0040	0.5924 ± 0.0031	0.7377 ± 0.0030	0.5806 ± 0.0031
PrefCRF	$\mathbf{0.8223 \pm 0.0033}$	$\mathbf{0.6813 \pm 0.0027}$	$\mathbf{0.8259 \pm 0.0035}$	$\mathbf{0.6890 \pm 0.0026}$

Table 3. Results over ten runs on *EachMovie* dataset.

Algorithm	6-Star ratings		6-Star ratings + Binary ratings	
	NDCG@10	MAP@10	NDCG@10	MAP@10
UserKNN	0.4374 ± 0.0047	0.3418 ± 0.0029	0.4374 ± 0.0047	0.3418 ± 0.0029
NMF	0.5211 ± 0.0078	0.3710 ± 0.0034	0.5211 ± 0.0078	0.3710 ± 0.0034
PrefKNN	0.4908 ± 0.0070	0.3793 ± 0.0031	0.5074 ± 0.0061	0.3938 ± 0.0044
PrefNMF	0.5233 ± 0.0061	0.3820 ± 0.0033	0.5454 ± 0.0060	0.3881 ± 0.0032
PrefCRF	$\mathbf{0.5439 \pm 0.0056}$	$\mathbf{0.4006 \pm 0.0045}$	$\mathbf{0.5506 \pm 0.0053}$	$\mathbf{0.4038 \pm 0.0043}$

From Table 3, we can observe that the performance of all *PR*-based methods has increased by incorporating additional binary ratings, while the performance of rating-based methods remains the same.

Fusing Preferences with Different Biases. In this experiment we investigate the impact of different biases, particularly the user-wise and item-wise biases, which are sampled from $Laplace(0, b)$. From Table 4 we can see that the performance of rating-based methods has decreased while *PR*-based methods are unaffected by such biases.

Table 4. NDCG@10 on *MovieLens*-20M dataset.

Algorithm	Bias = None	User-bias = $Laplace(0, 2)$	Item-bias = $Laplace(0, 2)$
UserKNN	0.4465 ± 0.0033	0.3729 ± 0.0033	0.2914 ± 0.0017
NMF	0.4982 ± 0.0034	0.4566 ± 0.0032	0.3074 ± 0.0019
PrefKNN	$\mathbf{0.4683 \pm 0.0027}$	$\mathbf{0.4683 \pm 0.0027}$	0.3157 ± 0.0021
PrefNMF	$\mathbf{0.4950 \pm 0.0035}$	$\mathbf{0.4950 \pm 0.0035}$	0.3137 ± 0.0017
PrefCRF	$\mathbf{0.5288 \pm 0.0037}$	$\mathbf{0.5288 \pm 0.0037}$	0.3729 ± 0.0023

Impact of Regularization and Correlation Threshold. The proposed *PrefCRF* method has two user specified parameters: the *regularization coefficient* and a *minimum correlation threshold* that controls the number of correlation features. For the regularization, we can see from Fig. 2e that the performance gets better when a small regularization penalty applies. In other words, *PrefCRF* can generalize reasonable well without too much regularization. For the correlation threshold, Fig. 2f shows that a smaller threshold results better performance by including more correlation features, however, at the cost of more training time and more training data.

5   Conclusions and Future Works

In this paper we talcked the learning from different preference structures problem by the *PrefCRF* model, which takes advantages of both the representational power of the *CRF* and the ease of modeling *PR* by the *PrefNMF*. Experiment results on four public datasets demonstrate that different preference structures have been properly handled by *PrefCRF*, and significantly improved Top-N recommendation performance has been achieved. For future work, the computation efficiency of *PR*-based methods can be further improved given that the number of *PR* is usually much larger than ratings. Parallelization is feasible as each user has a separated set of *PR* that can be processed simultaneously.

Acknowledgment. This work was partially supported by the Guangxi Key Laboratory of Trusted Software (No. KX201528).

References

1. Besag, J.: Spatial interaction and the statistical analysis of lattice systems. J. Roy. Stat. Soc. **36**(2), 192–236 (1974)
2. Bird, S., Klein, E., Loper, E.: Natural Language Processing with Python. O'Reilly Media Inc., Newton (2009)
3. Brun, A., Hamad, A., Buffet, O., Boyer, A.: Towards preference relations in recommender systems. In: PL: ECML/PKDD (2010)
4. Chapelle, O., Metlzer, D., Zhang, Y., Grinspan, P.: Expected reciprocal rank for graded relevance. In: CIKM 2009, pp. 621–630. ACM (2009)
5. Desarkar, M.S., Saxena, R., Sarkar, S.: Preference relation based matrix factorization for recommender systems. In: Masthoff, J., Mobasher, B., Desmarais, M.C., Nkambou, R. (eds.) UMAP 2012. LNCS, vol. 7379, pp. 63–75. Springer, Heidelberg (2012). doi:10.1007/978-3-642-31454-4_6
6. Järvelin, K., Kekäläinen, J.: Cumulated gain-based evaluation of IR techniques. ACM TOIS **20**(4), 422–446 (2002)
7. Koren, Y., Bell, R., Volinsky, C.: Matrix factorization techniques for recommender systems. IEEE Comput. **42**(8), 30–37 (2009)
8. Liu, J., Caihua, W., Xiong, Y., Liu, W.: List-wise probabilistic matrix factorization for recommendation. Inf. Sci. **278**, 434–447 (2014)
9. Liu, S., Li, G., Tran, T., Jiang, Y.: Preference relation-based markov random fields for recommender systems. Mach. Learn. **106**(4), 523–546 (2017)

10. McCullagh, P.: Regression models for ordinal data. J. Roy. Stat. Soc. B **42**(2), 109–142 (1980)
11. Rendle, S., Freudenthaler, C., Gantner, Z., Schmidt-Thieme, L.: BPR: Bayesian personalized ranking from implicit feedback. In: UAI 2009, pp. 452–461. AUAI Press (2009)
12. Resnick, P., Iacovou, N., Suchak, M., Bergstrom, P., Riedl, J.: Grouplens: an open architecture for collaborative filtering of netnews. In: CSCW 1994, pp. 175–186. ACM (1994)
13. Schacter, D.L., Dodson, C.S.: Misattribution, false recognition, the sins of memory. Philos. Trans. Roy. Soc. Lond. B: Biol. Sci. **356**(1413), 1385–1393 (2001)
14. Tran, T., Phung, D.Q., Venkatesh, S.: Preference networks: probabilistic models for recommendation systems. In: AusDM 2007, pp. 195–202. ACS (2007)

A Multifaceted Model for Cross Domain Recommendation Systems

Jianxun Lian[1]($\boxtimes$), Fuzheng Zhang[2], Xing Xie[2], and Guangzhong Sun[1]

[1] University of Science and Technology of China, Hefei, Anhui, China
`lianjx@mail.ustc.edu.cn, gzsun@ustc.edu.cn`
[2] Microsoft Research, Beijing, China
`{fuzzhang,xingx}@microsoft.com`

Abstract. Recommendation systems (RS) play an important role in directing customers to their favorite items. Data sparsity, which usually leads to overfitting, is a major bottleneck for making precise recommendations. Several cross-domain RSs have been proposed in the past decade in order to reduce the sparsity issues via transferring knowledge. However, existing works only focus on either nearest neighbor model or latent factor model for cross domain scenario. In this paper, we introduce a Multifaceted Cross-Domain Recommendation System (MCDRS) which incorporates two different types of collaborative filtering for cross domain RSs. The first part is a latent factor model. In order to utilize as much knowledge as possible, we propose a unified factorization framework to combine both CF and content-based filtering for cross domain learning. On the other hand, to overcome the potential inconsistency problem between different domains, we equip the neighbor model with a selective learning mechanism so that domain-independent items gain more weight in the transfer process. We conduct extensive experiments on two real-world datasets. The results demonstrate that our MCDRS model consistently outperforms several state-of-the-art models.

Keywords: Knowledge-boosted recommender systems · Collaborative filtering · Cross domain · Knowledge transferring

1 Introduction

With the boosting of online services, recommendation systems (RS) are playing an increasingly important role in filtering information for customers. Since most users are only connected to a small set of items, data sparsity becomes a major bottleneck for building an accurate RS. This is especially true for newly joined users/items, which is known as the cold start problem. To address this problem, researchers have introduced cross domain RSs which can transfer knowledge from some relatively dense source domains to the target domain. They assume that there are some consistent patterns across domains. Take book domain and movie domain for example, users with similar interests in the movie domain may

© Springer International Publishing AG 2017
G. Li et al. (Eds.): KSEM 2017, LNAI 10412, pp. 322–333, 2017.
DOI: 10.1007/978-3-319-63558-3_27

also have similar interest in the book domain (rule of collaborative filtering); and users who like fantasy movies have a higher probability of liking fantasy books than users who do not (rule of content-based filtering). Therefore, when we do not have enough data on the book domain, leveraging the data from movie domain appropriately can improve the quality of book recommendations.

According to whether the items or users from different domains have overlapping or not, existing literatures for cross domain RSs can be roughly divided into two categories: with overlap and without overlap. CMF [26], CST [21], TCF [20], MV-DNN [8] and [9] assume the users or/and items from multiple domains are fully or partially aligned, and the knowledge is transferred through the known common user/item factors; On the contrary, CBT [12], RMGM [13] and [6] do not require any overlap of users/items between auxiliary domains and the target domain. They transfer useful knowledge through cluster-level rating patterns.

In this paper, we assume the users between multiple domains are overlapped. This can be the case when a company owns multiple products but only a few of them provide enough data, while the others' data are too sparse to build effective models; or when a company wants to promote new services or products to its customers, they can leverage customer data on existing services or products. The aforementioned cross domain methods only focus on transferring either content-based filtering or collaborative filtering knowledge. However, models which combine the two algorithms have been widely discussed in the single domain case. Thus we consider the first challenge: will it produce a large amount of improvement if we transfer both content-based filtering and CF across different domains simultaneously? To address the question, we propose a unified factorization model to transfer both CF and content-based filtering cross domains. The key idea is to learn content embeddings so that the content-based filtering has a same latent factor formulation as model-based CF. Actually we propagate the CF not only to user-item level but also to user-content level.

Meanwhile, we notice that the multifaceted model can outperform the pure latent factor model [10]. However, there may be some inconsistency between different domains so that the neighborhood built from the source domain may not hold for the target domain. Take news and movie recommendation for instance, suppose Bob and David come from the same city, they may be similar in the news domain because they both love to read news related to the city. However, the similarity between them in movie domain may be low, since their interest in movies are not quite related to where they live. Motivated by this, we propose a novel selective mechanism which can iteratively learn a specific weight for each item in the source domain. The goal is to transfer preference on domain-independent items rather than domain-dependent items.

The key contributions of this paper are summarized as follows:

- We introduce the multifaceted model for cross domain RSs, which includes collective learning (the latent factor model) and selective learning (the neighbor model) modules.
- We propose a novel selective neighbor model to automatically assign weights to items so that domain inconsistency problem can be mitigated.

- We propose a unified factorization model to collectively learn content-based filtering and CF for cross domain RSs. Although models for combining the CF and content have been widely studied in the single domain situation, we aim to figure out the improvement in the cross domain scenario.
- We conduct extensive experiments on two real-world datasets, and the experimental results reveal that our proposed model consistently outperform several baseline methods.

2   Content-Boosted CF for Cross Domain RS

Suppose we have two domains: the target domain and the source domain. For each domain the data are comprised of user-item ratings and item attributes. In the target domain, we denote the user-item ratings as a matrix $\mathbf{R}^{(1)} = [r_{ui}]_{N \times M^{(1)}}$ with $r_{ui} \in \{[r_{min}, r_{max}], ?\}$, where ? denotes a missing value, N and $M^{(1)}$ denote the number of users and items respectively. The item content is denoted as a matrix $\mathbf{A}^{(1)} = [a_{ik}]_{M^{(1)} \times T^{(1)}}$, where each row represents an item, $T^{(1)}$ is the number of attributes, $a_{ik} \in [0, 1]$ is a normalized weight on attribute k, and $a_{ik} = 0$ indicates item i does not have attribute k. Our goal is to predict the missing value in $\mathbf{R}^{(1)}$. For various reasons $\mathbf{R}^{(1)}$ is sparse, and we want to improve the model with the help of the auxiliary domain, whose corresponding data is denoted as $\mathbf{R}^{(2)} = [r_{ui}]_{N \times M^{(2)}}$ and $\mathbf{A}^{(2)} = [a_{ik}]_{M^{(2)} \times T^{(2)}}$. Note that the users are fully or partially overlapped, and we know the mapping of users between domains.

A widely used method in collaborative filtering is latent factor model. It factorizes the rating matrix $\mathbf{R}_{N \times M}$ into two low-rank matrices, $\mathbf{U}_{N \times D}$ and $\mathbf{V}_{M \times D}$, as latent factors for users and items. In probabilistic matrix factorization (**PMF**) model [24], these latent factors are assumed to be generated from zero-mean spherical Gaussian distributions, while each rating is generated from a uni-variate Gaussian. Actually, the PMF model only learns from the user-item rating pairs and makes recommendations through clustering similar rating patterns. However, auxiliary signals such as user demographics and item attributes can help improve the restaurant recommendation model. Motivated by [19], we try to embed the attributes into a shared latent space, and then augment the item latent vector V_j with the weighted average of the embedded representation of attributes.

Formally, for each attribute $k, k \in 1 \ldots T$, we denote its embedding representation by $B_k = \langle b_{k1}, b_{k2}, \ldots, b_{kL} \rangle$. So we have a attribute embedding matrix $\mathbf{B}$ with each row indicates an attribute. The augmented latent vector for item j is denoted by $\widetilde{V}_j = \{V_j, P_j\}$ where $P_j = A_j \mathbf{B}$. The user latent vector is simply extended as $\widetilde{U}_i = \{U_{ia}, U_{ib}\}$, where U_{ia} denotes the original CF part while U_{ib} denotes the content preference part. Then the mean rating of user i on item j is changed as follows:

$$r_{ij} = \widetilde{U}_i \cdot \widetilde{V}_j = U_{ia} \cdot V_j + U_{ib} \cdot (A_j \mathbf{B}) \tag{1}$$

and the generative procedure is updated as follows:

(i) For a user i, sample two vectors: $U_{ia} \sim \mathcal{N}(0, \sigma_{ua}\mathbf{I})$, $U_{ib} \sim \mathcal{N}(0, \sigma_{ub}\mathbf{I})$.
(ii) For an item j, sample a vector: $V_j \sim \mathcal{N}(0, \sigma_v\mathbf{I})$.
(iii) For an attribute k, sample a vector: $B_k \sim \mathcal{N}(0, \sigma_b\mathbf{I})$.
(iv) For each user-item entry, generate the rating $r_{ij} \sim \mathcal{N}(U_{ia} \cdot V_j + U_{ib} \cdot \sum_k a_{jk}B_k, \sigma_r)$.

We name the new model Probabilistic Preference Factorization (**PPF**).

Cross Domain Collective Learning. When $\mathbf{R}^{(1)}$ is too sparse, the learning of latent factors may be inadequate and thus lead to overfitting on the training set. Next we study how to make use of the data from the auxiliary domain. We adapt the Collective Matrix Factorization (**CMF**) [26] model to our situation. The key idea of CMF is to factorize multiple matrices jointly through a multi-task learning framework. Since we have one-one mapping on the user side between the target domain and source domain, we simply assume the users share the same latent factors across these domains: $\mathbf{U} = \mathbf{U}^{(1)} = \mathbf{U}^{(2)}$, just like the approach in [17]. We also apply the aforementioned **PPF** generative process to the source domain. This model can be easily extended by introducing priors on the hyperparamenters and applying fully Bayesian methods such as MCMC [23]. However, to avoid high computational cost, here we regard $\sigma_{u*}, \sigma_v, \sigma_r$ as constant hyperparameters and use grid search to find their best values. The parameter set are reduced to $\Theta = \{\mathbf{U_a}, \mathbf{U_b}, \mathbf{V}^{(1)}, \mathbf{V}^{(2)}, \mathbf{B}^{(1)}, \mathbf{B}^{(2)}\}$ and we want to maximize the following Bayesian posterior formulation:

$$p(\Theta|\mathbf{R}^1, \mathbf{R}^2) \propto p(\mathbf{R}^1, \mathbf{R}^2|\Theta)p(\Theta) \qquad (2)$$

Maximizing Eq. (2) is equivalent to minimizing the following loss function:

$$\mathcal{L} = \sum_{i,j}(r_{ij}^{(1)} - U_{ia} \cdot V_j^{(1)} - U_{ib} \cdot \sum_k a_{jk}^{(1)}B_k^{(1)})^2$$
$$+ \lambda_1 \sum_{i,j}(r_{ij}^{(2)} - U_{ia} \cdot V_j^{(2)} - U_{ib} \cdot \sum_k a_{jk}^{(2)}B_k^{(2)})^2 \qquad (3)$$
$$+ \lambda_2 \sum_i \left\|V_i^{(*)}\right\|^2 + \lambda_3 \sum_k \left\|B_k^{(*)}\right\|^2 + \lambda_4 \sum_i \left\|U_{ia}^{(*)}\right\|^2 + \lambda_5 \sum_i \left\|U_{ib}^{(*)}\right\|^2$$

where we replace σ_* with λ_* after eliminating some constants for notation simplicity.

3   The Multifaceted Model

Koren [10] introduces a multifaceted model to smoothly merge the factor and neighborhood models. Following this idea, we plan to equip the **PPF** model with a neighborhood module. However, as we have discussed before, in the cross domain situation, the neighborhoods of the same user under different domains may be different. Thus it is questionable to compute the user similarity scores directly through items from the source domain. To address this problem, we propose a selective learning algorithm to assign weight for each item in the source domain. Our goal is to select those domain-free items to build an accurate neighborhood for the target domain. The new prediction rule becomes:

$$\hat{r}_{ij} = b_{ij} + \frac{w(i)}{\sqrt{|N^m(i,j)|}} \sum_{k \in N^k(i,j)} (r_{kj} - b_{ij})s(i,k) \tag{4}$$

Here, b_{ij} represents the prediction of the aforementioned latent factor model. $w(i) = 0.2e^{-0.5|R^{(1)}_{i*}|}$ is a tradeoff function controlling the weight of the neighborhood model, and it decreases when the number of rating records of user i in the target domain increases. $N^m(i,j)$ represents user i's top m nearest neighbors who have rated item j. $s(i,k)$ denotes the similarity value between user i and k. Due to the sparsity of target domain, we have to compute $s(i,k)$ using the data from source domain. In the traditional neighbor-based CF model, the similarity is static and calculated from their common rated items. Now we assign each item in the source domain with some parameters, which determine the weight of the item for the similarity calculation:

$$s(i,k) = e^{-\gamma_1 \frac{\sum_{t \in V(i,k)} D(i,k,t)*\beta(t)}{\sum_{t \in V(i,k)} \beta(t)}} \tag{5}$$

$V(i,k)$ indicates the common items rated by user i and k. $D(i,k,t)$ measures the difference of ratings on item t from user i and user k. We assign a small value, δ, to it when r_{it} equals to r_{kt}:

$$D(i,k,t) = max\{(r_{it} - r_{kt})^2, \delta\} \tag{6}$$

$\beta(t) \in (0,1)$ represents the weight of item t. We assume that the weight is determined via item id and item's content information. So for each item t, there is a corresponding parameter η_t; similarly, for each item attribute k, its corresponding parameter is denoted by α_k. We apply a logistic regression process to learn the optimal parameters:

$$\beta(t) = \frac{1}{1 + e^{-(\eta_t + \alpha_0 + \sum_k \alpha_k A_{tk})}} \tag{7}$$

Now, with Eq. (4) we introduce a Multifaceted model for Cross Domain Recommendation Systems (**MCDRS**). Its first part is a collective latent factor model, in which we embed both collaborative filtering and content-based filtering in order to exploit as much knowledge as possible. And its second part is a novel neighborhood model whose main purpose is to learn residuals based on the latent factor model.

Learning. There are several groups of parameters in our multifaceted model, i.e. latent factors for CF, latent factors for content, and weights in the neighbor model. It is hard for SGD to learn a good solution due to the fact that it does not discriminate between infrequent and frequent parameters. So we use AdaDelta [30] to automatically adapt the learning rate to the parameters, leading to the update rule:

$$\Delta x_t = -\frac{RMS[\Delta x]_{t-1}}{RMS[g]_t} g_t \tag{8}$$

where RMS represents the root mean squared criterion with exponentially decaying:

$$RMS[g]_t = \sqrt{E[g^2]_t + \epsilon} \tag{9}$$
$$E[g^2]_t = \rho E[g^2]_{t-1} + (1 - \rho)g_t^2 \tag{10}$$

In the experiments we set $\rho = 0.9$ and $\epsilon = 1e - 8$.

4 Experiments

We evaluate the proposed model on two rating datasets: Douban[1] and MovieLens 20M[2].

Douban is a leading Chinese online community which allows users to record information related to multiple domains such as movies, books and music. Users can rate movies/books/songs on a scale from 1 to 5, and each movie/book/song has a list of tags (tag_id,count) indicating how many users have rated the tag_id on this item, and the tags can be used as the item's attributes. Thus Douban is an ideal source for our cross domain experiments. We build a distributed crawler to fetch the item information and user-item rating records from Douban. After filtering out users who appear in only one domain or have less than 20 ratings, and movies/books/songs which have fewer than 10 ratings, we randomly sample 100k users, 50k books, 30k movies and 30k music for experiments. Via switching the choice of source domain and target domain, we report the results of three cross domain tasks, i.e., $\langle movie \rightarrow music \rangle$, $\langle movie \rightarrow book \rangle$, and $\langle book \rightarrow music \rangle$. In each task we split the target domain into training/valid/test set by 70%/15%/15%. In order to study the performance lift under different sparsity levels, we keep the validation and test set unchanged and randomly sample a subset from the training set with different sparsity levels of 1%, 2%, 5%, 8%, 10% correspondingly.

Besides the Douban dataset, we also evaluate the proposed algorithm on a widely used benchmark dataset, MovieLens 20M. It is currently the latest stable benchmark dataset from GroupLens Research for new research, and it contains more than 20 million ratings of 27,000 movies by 128,000 users, with rating score from 0.5 to 5.0. Because our proposed model is to study the combination of CF and content-based methods, we filter out the movies which have no tags, and then use tags as the movie attributes.

Since the MovieLens dataset has only one domain, we split the movies into two disjointed parts to simulate two different domains. This approach has also

---

[1] http://www.douban.com.
[2] http://grouplens.org/datasets/movielens/.

been used in existing works such as [20,21]. Specifically, denote the full rating matrix as $\mathbf{R}_{1\sim N,1\sim M}$. We take the first half sub-matrix $\mathbf{R}_{1\sim N,1\sim \frac{M}{2}}$ as the target domain, while the other half $\mathbf{R}_{1\sim N,\frac{M}{2}+1\sim M}$ is the auxiliary domain. Again we split the target domain into training/valid/test set by 70%/15%/15% and extract several subsets from the training set according to different sparsity levels.

4.1   Baselines and Evaluation Metrics

We compare our model with the following methods:

Bias Matrix Factorization (BMF) [11]. This is a standard SVD matrix decomposition with user and item bias: $\widehat{r}_{uv} = \mu + b_u + b_v + U_u \cdot V_v$, where b_u, b_v indicate the deviations of user u and item i respectively, and U_u, V_v are latent factor vectors.

SVDFeature [5]. This is a famous recommendation toolkit for feature-based CF. The authors of the toolkit have used it to win the KDD Cup for two consecutive years. It can include attributes into CF process and is one of the state-of-the-art recommendation methods for single domain.

STLCF [17]. This is a selective knowledge transfer method and can outperform some cross domain RSs such as CMF. It applies the AdaBoosting framework in order to capture the consistency across domain in CF settings, and it does not consider the rich content information of items.

MV-DNN [8]. This is a content-based multiple domain recommendation. It uses deep learning to match rich user features to item features. However, it does not exploit collaborative filtering.

MCDRS-. This is a variant of our proposed method which removes the neighbor model submodule.

We adopt Root Mean Square Error (RMSE) to evaluate the performance of different methods, where $\mathfrak{R}$ indicates the test set:

$$RMSE = \sqrt{\frac{1}{|\mathfrak{R}|} \sum_{(u,v,r_{uv})\in\mathfrak{R}} (r_{uv} - \widehat{r}_{uv})^2}$$

We use grid search to find the best parameters for each method. We have tried λ_* from $\{0.001, 0.005, 0.01, 0.05, 0.1, 0.5, 1.0\}$. To reduce computational cost, for all algorithms we use a same fixed latent feature dimension, i.e. we fix the dimensions of BMF, SVDFeature and STLCF to be 16, and fix the size of the last layer in MV-DNN model to be 16 while changing the sizes of previous hidden layers in $\{32, 64, 128\}$. For our proposed model, we set the CF dimension to be 8 and the content-related dimension to be 8, so that the total size of the latent factors is 16. After running these groups of parameters, we pick the best parameter set for each model according to the validation set. We re-run the experiment pipeline five times for each model by fixing the best parameter set, and report the corresponding average $RMSE$ on the test set. For our model, the best setting is $\{\lambda_1 = 0.8, \lambda_3 = \lambda_5 = 0.005, \lambda_4 = \lambda_2 = 0.1\}$.

4.2 Results

Figure 1 shows the RMSE results of different algorithms. Generally speaking, due to consistency problem across domains, a small volume of new data in the target domain matches up to a large volume of data from the source domain. And our MCDRS is designed to consider as much knowledge as possible in the target domain, and at the same time transfer knowledge from the source domain. So it is expected to observe the fact that our proposed model consistently outperforms the other models under various tasks.

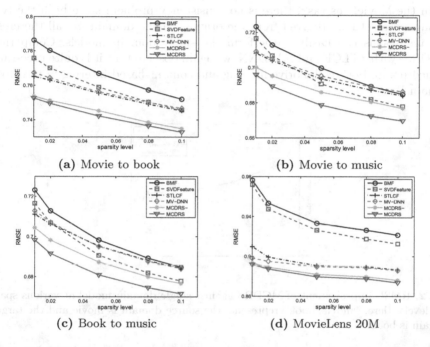

Fig. 1. RMSE evaluation on Douban dataset (a–c) and MovieLens 20M dataset (d). Here, "movie to book" represents the source domain is movie and target domain is book.

Figure 1a reports the performance comparison when we use movie domain as the source domain and take book domain as the target domain. The BMF model is the weakest one because it is a single domain CF algorithm, and it does not consider the rich content information. SVDFeature is the single domain model which combines both CF and content information. It's no surprise that SVDFeature significantly outperforms BMF. STLCF and MV-DNN work better than SVDFeature only under the low sparsity levels, and the superiority trend to disappear when the data become denser. MCDRS- significantly outperforms STLCF and MV-DNN due to its utilization of both collaborative filtering and content-based filtering in the cross domain situation. At last, MCDRS shows a further improvement, which demonstrates the necessary of selective learning part.

We observe some difference in RMSE trends when comparing Fig. 1b, c with 1a. Movie and book are two close domains and they share a lot of topics, such as adventure, history fiction, and science fiction. Some movies are even adaptations of famous books. So it will be easier to build cross domain RSs between movie and book domain. However, music domain is relatively not so close with movie or book domain. Thus in Fig. 1b and c, STLCF and MV-DNN are worse than SVDFeature when the sparsity level is above 5%. MCDRS model works well in the two tasks, which verifies that our designed selective learning module can mitigate the inconsistency problem between domains.

In the MovieLens task, there is no consistency problem since both the two domains are actually derived from the original movie domain. So all the cross domain models can easily outperform the single domain models. Comparing MCDRS- with STLCF and MV-DNN, we again observe that it is quite necessary to incorporate both collaborive filtering and content-based filtering in the cross domain RSs.

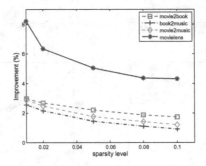

Fig. 2. RMSE improvement of MCDRS against single domain model at various sparsity levels. Here, "movie2book" represents the source domain is movie and the target domain is book.

Figure 2 summarizes the performance gain when comparing MCDRS against the SVDFeature. There is a clear pattern that the cross domain RS works well when data in the target domain is severely sparse, and the degree of improvement decreases when the data of target domain becomes denser. Relatively speaking, the MovieLens task shows the greatest improvement as expected. Results of the task *movie2book* is better than the *movie2music* because that the inconsistency on the movie-book domain is smaller than that on the movie-music domain.

5   Related Works

Single Domain Recommendation Systems. In general, recommendation systems can be divided into content-based model, collaborative filtering (CF) models and hybrid models [4]. Content-based models represent user and item under a same feature space and make recommendation according to their similarity [15]. CF models make predictions about the interests of a user by collecting preferences or taste information form many users. Among CF models the latent factor models are the most widely studied methods [7, 23, 24]. Hybrid models try to integrate content-based method and CF method into one unified intelligent system [2, 16, 18, 28]. In the recent years, the great success of deep learning models on computer vision and natural language processing has motivated researchers to develop deep learning models for recommender systems [1, 25, 27, 29, 31].

Cross Domain Methods with User/Item Aligned. [3] studies some earliest CDCF models including the cross domain neighbor model. CMF [26] simultaneously factorizes several matrices while sharing latent factors when an entity participates in multiple relations. CST [21] is an adaptation style tri-factorization algorithm which transfers knowledge from the auxiliary domain to improve performance in the target domain. TCF [20] uses the tri-factorization method to collectively learn target and auxiliary matrices. In their case both users and items are aligned between the two matrices. The major advantage of the TCF approach is that through tri-factorization it is able to capture the data-dependent effect when sharing data independent knowledge. [17] proposes a criterion which can be embedded into a boosting framework to perform selective knowledge transfer. [22] suggests a cross domain CF based on CCA to share information between domains. [8, 14] use multi-view deep learning models to jointly latent features for users and items from multiple domains. Our paper belongs to this category of cross domain RSs. While the existing works focus on transfer knowledge through collaborative filtering or content-based filtering, we focus on incorporating the advantages of both the two models and further explore how to reduce the inconsistency problem through the rich information.

Cross Domain Methods with Latent Clusters. CBT [12] studies the cases in which neither users nor items in the two domains overlap. It assumes that both auxiliary and target domains share the cluster-level rating patterns. So it compresses the auxiliary rating matrix into a compact cluster-level rating pattern representation referred to as a codebook, and reconstructs the target rating matrix by expanding the codebook. RMGM [13] extends CBT to a generative model and they share the same assumption. [6] exploits tensor factorization to model high dimensional data and transfers knowledge from the auxiliary domain through a shared cluster-level tensor. In this paper, we discuss the scenario in which users of the two domains are well aligned, and thus is different from these hidden cluster-linking approaches.

6   Conclusion

In this paper, we introduce a multifaceted model for cross domain RSs. It includes two parts which are delicately designed for cross domain situation. First we propose a latent factor model to incorporate both collaborative filtering and content-based filtering, and different domains are bridged through the aligned latent user features. Second, we propose a selective neighbor model to reduce the inconsistency problem across domains. Experimental results demonstrate that our proposed model consistently outperforms several state-of-the-art methods on both Douban and MovieLens datasets. For future works, we will explore how to exploit deep learning techniques to learn better representation from the rich content information for cross domain RSs.

References

1. Bansal, T., Belanger, D., McCallum, A.: Ask the GRU: multi-task learning for deep text recommendations. In: Proceedings of the 10th ACM Conference on Recommender Systems, pp. 107–114. ACM (2016)
2. Basilico, J., Hofmann, T.: Unifying collaborative and content-based filtering. In: Proceedings of the Twenty-First International Conference on Machine Learning, p. 9. ACM (2004)
3. Berkovsky, S., Kuflik, T., Ricci, F.: Cross-domain mediation in collaborative filtering. In: Conati, C., McCoy, K., Paliouras, G. (eds.) UM 2007. LNCS, vol. 4511, pp. 355–359. Springer, Heidelberg (2007). doi:10.1007/978-3-540-73078-1_44
4. Bobadilla, J., Ortega, F., Hernando, A., Gutirrez, A.: Recommender systems survey. Knowl.-Based Syst. **46**, 109–132 (2013)
5. Chen, T., Zhang, W., Lu, Q., Chen, K., Zheng, Z., Yu, Y.: SVDFeature: a toolkit for feature-based collaborative filtering. J. Mach. Learn. Res. **13**(1), 3619–3622 (2012)
6. Chen, W., Hsu, W., Lee, M.L.: Making recommendations from multiple domains. In: Proceedings of the 19th ACM SIGKDD International Conference on Knowledge Discovery and Data Mining, pp. 892–900. ACM, New York (2013)
7. Ekstrand, M.D., Riedl, J.T., Konstan, J.A.: Collaborative filtering recommender systems. Found. Trends Hum.-Comput. Interact. **4**(2), 81–173 (2011)
8. Elkahky, A.M., Song, Y., He, X.: A multi-view deep learning approach for cross domain user modeling in recommendation systems. In: Proceedings of the 24th International Conference on World Wide Web, WWW 2015, Florence, Italy, pp. 278–288 (2015)
9. Jiang, M., Cui, P., Yuan, N.J., Xie, X., Yang, S.: Little is much: bridging cross-platform behaviors through overlapped crowds. In: AAAI, pp. 13–19. AAAI Press (2016)
10. Koren, Y.: Factorization meets the neighborhood: a multifaceted collaborative filtering model. In: Proceedings of the 14th ACM SIGKDD International Conference on Knowledge Discovery and Data Mining, pp. 426–434. ACM (2008)
11. Koren, Y., Bell, R., Volinsky, C.: Matrix factorization techniques for recommender systems. Computer **42**(8), 30–37 (2009)
12. Li, B., Yang, Q., Xue, X.: Can movies and books collaborate?: cross-domain collaborative filtering for sparsity reduction. In: IJCAI 2009, pp. 2052–2057. Morgan Kaufmann Publishers Inc., San Francisco (2009)
13. Li, B., Yang, Q., Xue, X.: Transfer learning for collaborative filtering via a rating-matrix generative model. In: ICML 2009, pp. 617–624. ACM, New York (2009)
14. Lian, J., Zhang, F., Xie, X., Sun, G.: CCCFNet: a content-boosted collaborative filtering neural network for cross domain recommender systems. In: WWW 2017 Companion, pp. 817–818 (2017)
15. Lops, P., De Gemmis, M., Semeraro, G.: Content-based recommender systems: state of the art and trends. In: Ricci, F., Rokach, L., Shapira, B., Kantor, P.B. (eds.) Recommender Systems Handbook, pp. 73–105. Springer, New York (2011). doi:10.1007/978-0-387-85820-3_3

16. Lu, Z., Dou, Z., Lian, J., Xie, X., Yang, Q.: Content-based collaborative filtering for news topic recommendation. In: AAAI, pp. 217–223. Citeseer (2015)
17. Lu, Z., Zhong, E., Zhao, L., Xiang, E.W., Pan, W., Yang, Q.: Selective transfer learning for cross domain recommendation. In: Proceedings of the 2013 SIAM International Conference on Data Mining, pp. 641–649. SIAM (2013)
18. Melville, P., Mooney, R.J., Nagarajan, R.: Content-boosted collaborative filtering for improved recommendations. In: AAAI/IAAI, pp. 187–192 (2002)
19. Mikolov, T., Dean, J.: Distributed representations of words and phrases and their compositionality. In: Advances in Neural Information Processing Systems (2013)
20. Pan, W., Liu, N.N., Xiang, E.W., Yang, Q.: Transfer learning to predict missing ratings via heterogeneous user feedbacks. In: IJCAI 2011, pp. 2318–2323. AAAI Press (2011)
21. Pan, W., Xiang, E.W., Liu, N.N., Yang, Q.: Transfer learning in collaborative filtering for sparsity reduction. In: AAAI 2010, pp. 230–235. AAAI Press (2010)
22. Sahebi, S., Brusilovsky, P.: It takes two to tango: an exploration of domain pairs for cross-domain collaborative filtering. In: RecSys 2015, pp. 131–138. ACM, New York (2015)
23. Salakhutdinov, R., Mnih, A.: Bayesian probabilistic matrix factorization using Markov chain Monte Carlo. In: Proceedings of the 25th international conference on Machine learning, pp. 880–887. ACM (2008)
24. Salakhutdinov, R., Mnih, A.: Probabilistic matrix factorization. In: NIPS, vol. 20, pp. 1–8 (2011)
25. Salakhutdinov, R., Mnih, A., Hinton, G.: Restricted Boltzmann machines for collaborative filtering. In: Proceedings of the 24th International Conference on Machine Learning, pp. 791–798. ACM (2007)
26. Singh, A.P., Gordon, G.J.: Relational learning via collective matrix factorization. In: KDD 2008, pp. 650–658. ACM, New York (2008)
27. Van den Oord, A., Dieleman, S., Schrauwen, B.: Deep content-based music recommendation. In: Advances in Neural Information Processing Systems, pp. 2643–2651 (2013)
28. Wang, C., Blei, D.M.: Collaborative topic modeling for recommending scientific articles. In: Proceedings of the 17th ACM SIGKDD International Conference on Knowledge Discovery and Data Mining, pp. 448–456. ACM (2011)
29. Wang, H., Wang, N., Yeung, D.-Y.: Collaborative deep learning for recommender systems. In: Proceedings of the 21th ACM SIGKDD International Conference on Knowledge Discovery and Data Mining, pp. 1235–1244. ACM (2015)
30. Zeiler, M.D.: ADADELTA: an adaptive learning rate method. CoRR, abs/1212.5701 (2012)
31. Zhang, F., Yuan, N.J., Lian, D., Xie, X., Ma, W.-Y.: Collaborative knowledge base embedding for recommender systems. In: Proceedings of the 22nd ACM SIGKDD International Conference on Knowledge Discovery and Data Mining, pp. 353–362. ACM (2016)

Cross Domain Collaborative Filtering by Integrating User Latent Vectors of Auxiliary Domains

Xu Yu[1]([✉]), Feng Jiang[1], Miao Yu[2], and Ying Guo[1]

[1] School of Information Science and Technology,
Qingdao University of Science and Technology, Qingdao 266061, China
yuxu0532@163.com
[2] The College of Textiles and Fashion,
Qingdao University, Qingdao 266071, China

Abstract. Cross-Domain Collaborative Filtering solves the sparsity problem by transferring rating knowledge across multiple domains. However, how to transfer knowledge from auxiliary domains is nontrivial. In this paper, we propose a model-based CDCF algorithm by Integrating User Latent Vectors of auxiliary domains (CDCFIULV) from the perspective of classification. For a user-item interaction in the target domain, we first use the trivial location information as the feature vector, and use the rating information as the label. Thus we can convert the recommendation problem into a classification problem. However, such a two-dimensional feature vector is not sufficient to discriminate the different rating classes. Hence, we require some other features for the classification problem with the help of the rating information from the auxiliary domains.

In this paper, we assume the auxiliary domains contain dense rating data and share the same aligned users with the target domain. In this scenario, we employ UV decomposition model to obtain the user latent vectors from the auxiliary domains. We expand the trivial location feature vector in the target domain with the obtained user latent vectors from all the auxiliary domains. Thus we can effectively add features for the classification problem. Finally, we can train a classifier for this classification problem and predict the missing ratings for the recommender system. Hence the hidden knowledge in the auxiliary domains can be transferred to the target domain effectively via the user latent vectors. A major advantage of the CDCFIULV model over previous collective matrix factorization or tensor factorization models is that our model can adaptively select significant features during the training process. However, the previous collective matrix factorization or tensor factorization models need to adjust the weights of the auxiliary domains according to the similarities between the auxiliary domains and the target domain. We conduct extensive experiments to show that the proposed algorithm is more effective than many state-of-the-art single domain and cross domain CF methods.

Keywords: Cross domain collaborative filtering · User latent vectors · UV decomposition · Classification problem

© Springer International Publishing AG 2017
G. Li et al. (Eds.): KSEM 2017, LNAI 10412, pp. 334–345, 2017.
DOI: 10.1007/978-3-319-63558-3_28

1   Introduction

The rapid growth of the information on the Internet demands intelligent information agent that can sift through all the available information and find out the most valuable to us. In recent years, recommender systems [1, 2] are widely used in e-commerce sites and online social media and the majority of them offer recommendations for items belonging to a single domain. Collaborative Filtering (CF) algorithm [3] is the most widely used method for recommender systems and it can be boiled down to analyzing the tabular data, i.e., the user-item rating matrix.

However, in real-world recommender systems, users usually dislike rating items and the items rated are very limited. Thus the rating matrix is very sparse. The sparsity problem has become a major bottleneck for most CF methods. To alleviate this difficulty, recently a number of Cross-Domain Collaborative Filtering (CDCF) methods have been proposed [4]. CDCF methods exploit knowledge from auxiliary domains (e.g. movies) containing additional user preference data to improve recommendation on a target domain (e.g. books). They can effectively relieve the sparsity problem in the target domain. Currently, CDCF methods can be categorized into two classes. One class assumes shared users or items [5–8]. This assumption is an important and commonly appeared case in many large-scale websites. For instance, Amazon website contains different domains, including books, music CDs, DVDs and video tapes. They share the identical users though their items are totally different. This is also the scenario studied in this paper. The other class do not require shared users or items [9, 10].

Among the previous works of the first class, Berkovsky et al. [5] mention an early neighborhood based CDCF (N-CDCF), which can be considered as the cross-domain version of a memory-based method, i.e., N-CF [11]. Likewise, the traditional matrix factorization (MF) model can also be developed to solve the CDCF problems. The cross-domain version of a matrix factorization method is denoted by MF-CDCF [6], in which an augmented rating matrix is constructed by horizontally concatenating all the rating matrices in different domains. With this matrix in hand, any classical MF algorithm, such as UV Decomposition (UVD) model [12], can be exploited to construct user factor matrix and item factor matrix. These factor matrices are used for prediction. Both N-CDCF and MF-CDCF accommodates items from all domains into a single matrix so as to employ single-domain CF methods. However, single domain models assume the homogeneity of items. Obviously, items in different domains may quite heterogeneous, and the above two models fail to take this fact into account.

Pan et al. [7] explore how to take advantage of knowledge in the form of binary ratings (like and dislike) to alleviate the sparsity problem in numerical ratings. They propose a novel transfer learning framework, Transfer by Collective Factorization (TCF). However, they require users and items of the target rating matrix and the auxiliary like/dislike matrix be both aligned besides the heterogeneity of the two matrices. In addition, they can only deal with the scenario of one auxiliary domain.

Singh and Gordon propose a Collective Matrix Factorization (CMF) [8] model, which is a multi-task learning (MTL) [13] version of UV decomposition model. CMF couples rating matrices for all domains on the *User* dimension so as to transfer knowledge through the common user-factor matrix. Hu et al. [6] propose a generalized Cross Domain

Triadic Factorization (CDTF) model over the triadic relation user-item-domain. Since not all the auxiliary domains are equally correlated with the target domain (for example, the task of predicting user preferences on books should be more related to predicting user preferences on movies than predicting user preferences on food), CMF and CDTF consider the different degrees of relatedness between each auxiliary domain and the target domain. This is an advantage of them over N-CDCF and MF-CDCF. However, CMF does not provide a mechanism to find an optimal weights assignment for the auxiliary domains. Though CDTF assigns the weights based on genetic algorithm (GA), the performance is susceptible to the setting of the initial population.

Among the previous works of the second class, Li et al. [9] propose a codebook-based knowledge transfer (CBT) for recommender systems. CBT achieves knowledge transfer with the assumption that both auxiliary and target data share the cluster-level rating patterns (codebook). Further, Li et al. [10] propose a rating-matrix generative model (RMGM). RMGM can be considered as a MTL version of CBT with the same assumption. Both CBT and RMGM require two rating matrices to share the cluster-level rating patterns. In addition, CBT and RMGM cannot make use of user side shared information, and only take a general explicit rating matrix as its auxiliary input. Hence, both CBT and RMGM are not applicable to the problem studied in this paper.

In this paper, from the perspective of classification, we propose a model-based CDCF algorithm by Integrating User Latent Vectors of auxiliary domains (CDCFIULV). We first extract the location information of a user-item interaction in the target domain as the feature vector, and use the corresponding rating information as the label. Thus we can convert the recommendation problem into a classification problem. However, the two-dimensional feature vector is not sufficient to discriminate the different rating classes, $\{1, 2, 3, 4, 5\}$. In our experimental tests, the classification model cannot obtain a satisfactory result. Hence we require more features in the classification model.

Note that the first dimension of the feature vector is the location information of the user, representing the user feature in the recommender system, and the second dimension of the feature vector is the location information of the item, representing the item feature in the recommender system. Inspired by this observation, we try to expand the feature vector in the target domain with some other significant features of users or items. Since we assume that the auxiliary domains contain dense rating data and share the same aligned users with the target domain, it is possible to extract user features from the auxiliary domains. In this scenario, we use the user latent vector obtained by UV decomposition model from the auxiliary domains as the extra features and add them to the feature vector in the target domain. As a result, we can effectively expand the feature vector for the classification problem. Our classification-based model has three advantages, (a) the hidden knowledge in the auxiliary domains can be transferred to the target domain effectively via the user latent vectors; (b) the importance of the features will be adaptively adjusted during the training process; (c) it can easily deal with the scenario of multi auxiliary domains.

The remainder of this paper is organized as follows: Sect. 2 reviews the related works on CDCF methods. Section 3 proposes our CDCFIULV model. In Sect. 4, we conduct extensive experiments to test the performance of the proposed algorithm. We conclude the paper and give future works in Sect. 5.

2 Related Works

Currently, CDCF methods can be categorized into two classes. One class assumes shared users or items, and the other class requires shared users and items in different domains. In the first class, Berkovsky et al. [5] mention a neighborhood based CDCF (N-CDCF). As neighborhood based CF (N-CF) computes similarity between users or items, which can be sub-divided into two types: user-based nearest neighbor (N-CF-U) and item-based nearest neighbor (N-CF-I), the N-CDCF algorithm can also be divided into two types: a user-based neighborhood CDCF model (N-CDCF-U) and an item-based neighborhood CDCF model (N-CDCF-I). For simplicity, we only give a detail review on N-CDCF-U, and the detail method of N-CDCF-I is in the same manner.

Let $\mathbf{D} = \{D_0, D_1, \cdots, D_m\}$ denote all the domains for modeling, $U = \{u_1, u_2, \cdots, u_n\}$ denote the users in $\mathbf{D}$ and $I_k = \{i_1^k, i_2^k, \cdots, i_{n(k)}^k\}$ denote items belonging to the domain D_k ($0 \leq k \leq m$), where $n(k)$ denotes the item set size of D_k. For a user-based CDCF algorithm, we first calculate the similarity, $s_{u,v}$, between the users u and v who have co-rated the same set of items. The similarity can be measured by the Pearson correlation:

$$s_{u,v} = \frac{\sum_{i \in i_{u,v}} (r_{u,i} - \bar{r}_u)(r_{v,i} - \bar{r}_v)}{\sqrt{\sum_{i \in i_{u,v}} (r_{u,i} - \bar{r}_u)^2 \sum_{i \in i_{u,v}} (r_{v,i} - \bar{r}_v)^2}} \tag{1}$$

where $i_{u,v} = i_u \cap i_v$ ($i_u = \bigcup_{d \in D} i_u^d$, $i_v = \bigcup_{d \in D} i_v^d$) denotes the items over all domains $\mathbf{D}$ co-rated by u and v; $r_{u,i}$ and $r_{v,i}$ are the ratings on item i given by users u and v respectively; $\bar{r}_u$ and $\bar{r}_v$ are the average ratings of user u and v for all the items rated respectively. Then the predicted rating of an item p for user u can be calculated by a weighted average strategy [11]:

$$\hat{r}_{u,p} = \bar{r}_u + \frac{\sum_{v \in U_{u,p}^k} s_{u,v}(r_{v,p} - \bar{r}_v)}{\sum_{v \in U_{u,p}^k} |s_{u,v}|} \tag{2}$$

where $U_{u,p}^k$ denotes the set of top k users (k neighbors) that are most similar to user u who rated item p.

Since there are more co-rated items in all the domains, the total similarity over all the domains may be more accurate. However it is not very reasonable to replace the local similarity in the target domain with this total similarity, as the user preference will vary on different domains. Thus the performance of N-CDCF-U is not always satisfactory.

In addition to the above model, the traditional MF model can also be developed to solve the CDCF problems straightforward. We can pour all the items from different domains together and then an augmented rating matrix, M_D, can be built by horizontally concatenating all matrices. Thus we can use MF model to obtain the latent user factors and latent item factors. These latent factors are used for prediction. In this paper, the MF model on CDCF problems is denoted as **MF-CDCF**. MF-CDCF accommodates items from all domains into a single matrix so as to employ single-domain MF. However, single domain model assumes the homogeneity of items. Obviously, items in different domains may quite heterogeneous, so N-CDCF-U and MF-CDCF fail to take this fact into account.

Pan et al. [7] present a novel transfer learning framework, Transfer by Collective Factorization (**TCF**), to transfer knowledge from auxiliary data of explicit binary ratings (like and dislike), which alleviates the data sparsity problem in numerical ratings. TCF collectively factorizes a 5-star numerical target data **R** and a binary like/dislike auxiliary data, and assumes that both user-specific and item-specific latent feature matrices are the same. Besides the shared latent features, TCF uses two inner matrices to capture the data-dependent information, which is different from the inner matrix used in CBT [9] and RMGM [10]. TCF requires users and items of the target rating matrix and the auxiliary like/dislike matrix be both aligned. Also, they can only deal with the scenario of one auxiliary domain. Hence, it is not applicable to the problem studied in this paper.

Singh and Gordon [8] propose the Collective Matrix Factorization (CMF) model. CMF jointly factorizes multiple matrices with correspondences between rows and columns while sharing latent features of matching rows and columns in different matrices. Hu et al. [6] propose the CDTF model, in which they consider the full triadic relation user-item-domain to effectively exploit user preferences on items within different domains. They represent the user-item-domain interaction with a tensor of order three and adopt a tensor factorization model to factorize users, items and domains into latent feature vectors. The rating of a user for an item in a domain is calculated by element-wise product of user, item and domain latent factors. A major problem of tensor factorization however, is that the time complexity of this approach is exponential as it is $O(k^m)$ where k is the number of factors and m is the number of domains. Both CMF and CDTF need to adjust the weights of the auxiliary domains according to the similarities between the auxiliary domains and the target domain.

In the second class, Li et al. [9] propose a CBT model. They first compress the ratings in the auxiliary rating matrix into an informative and yet compact cluster-level rating pattern representation referred to as a codebook. Then, they reconstruct the target rating matrix by expanding the codebook. CBT achieves knowledge transfer with the assumption that both auxiliary and target data share the cluster-level rating patterns (codebook). Further, Li et al. [10] propose a RMGM model. In this model, the knowledge is shared in the form of a latent cluster-level rating model. Each rating matrix can thus be viewed as drawing a set of users and items from the user-item joint mixture model as well as drawing the corresponding ratings from the cluster-level rating model. RMGM is a MTL version of CBT with the same assumption. Both CBT and RMGM require two rating matrices to share the cluster-level rating patterns. They assume that the items in an auxiliary data source (e.g. books) is related to the target data (e.g. movies). Hence they are not applicable to the scenario studied in this paper.

3 Our Model

3.1 Convert the Recommendation Problem into a Classification Problem

Assume D_1 is the target domain, and U_1 and I_1 are the sets of users and items in domain D_1. We model the standard recommendation problem in the target domain D_1 by a target function $y : U_1 \times I_1 \rightarrow R$. We represent each user-item interaction $(u, i, r) \in U_1 \times I_1 \times \{1, 2, 3, 4, 5\}$ with a feature vector (L_u, L_i) and a class label r, where L_u and L_i denote the location information of user u and item i respectively. Thus we can represent each user-item interaction as a training sample, and the original rating problem can be converted into a classification problem. We use Fig. 1 to illustrate our method.

Fig. 1. The rating matrix

In Fig. 1, u_1, u_2, u_3, and u_4 are four users, and i_1, i_2, and i_3 are three items. Firstly, we use 1, 2, 3, and 4 to denote the location of u_1, u_2, u_3, and u_4, and use 1, 2, and 3 to denote the location of i_1, i_2, and i_3. Then we can use (1, 1) to denote the features of the user-item interaction $(u, i_1, 2)$, and use 2 to denote the corresponding class label. In the same manner, we can also represent other user-item interactions as training samples. Thus the rating matrix can be converted into a training set and we can convert the recommendation problem into a classification problem.

3.2 Feature Vector Expansion

The training samples of the converted classification problem have only two location features. However, the two-dimensional feature vector is not sufficient to discriminate the different rating classes, $\{1, 2, 3, 4, 5\}$. In our experimental tests, the classification model with two-dimensional feature vector cannot obtain a satisfactory result. Hence we require more features in the classification model. Note that the first dimension of the feature vector represents the user feature in the recommender system, and the second dimension of the feature vector represents the item feature. Inspired by this observation, we try to expand the feature vector in the target domain with some other significant features of users or items.

Though different domains may not share the same item set, they share the same user set. In addition, we also assume that the auxiliary domains contain dense rating data. Thus it is possible to extract the user features from the auxiliary domains and transfer them to the target domain. In our model, we first use the UV decomposition model to obtain the user latent vectors from the auxiliary domains. We will use the user latent vectors to expand the user feature vector in the target domain.

In the UV decomposition model, each item i is associated with a latent vector $q_i \in R^f$, and each user u is associated with a latent vector $p_u \in R^f$. q_i measures the distribution of item i on those latent factors, and p_u measure the interest distribution of user u on those latent factors. The resulting dot product, $q_i^T p_u$, captures the interaction between user u and item i. This approximates user u's rating on item i, which is denoted by $\hat{r}_{ui}$ in the following form

$$\hat{r}_{ui} = q_i^T p_u \tag{3}$$

To learn the latent vectors (p_u and q_i), the UV decomposition model minimizes the regularized squared error on the set of known ratings

$$\min_{q^*, p^*} \sum_{(u,i) \in \kappa} (r_{ui} - q_i^T p_u)^2 + \lambda(\|q_i\|^2 + \|p_u\|^2) \tag{4}$$

Here, κ is the set of the (u, i) pairs for which r_{ui} is known. The constant λ controls the extent of regularization to avoid over-fitting and is usually determined by cross-validation [14]. An effective approach to minimize optimization problem (4) is stochastic gradient descent, which loops through all ratings in the training set. For each given training case, the system predicts r_{ui} and computes the associated prediction error

$$e_{ui} \stackrel{def}{=} r_{ui} - q_i^T p_u \tag{5}$$

Then it modifies the parameters by a magnitude proportional to γ (i.e., the learning rate) in the opposite direction of the gradient, yielding:

$$\begin{aligned} q_i &\leftarrow q_i + \gamma(e_{ui} p_u - \lambda q_i) \\ p_u &\leftarrow p_u + \gamma(e_{ui} q_i - \lambda p_u) \end{aligned} \tag{6}$$

Next, we will give the reason why we select the user latent vectors as the user features in the auxiliary domains. Firstly, since we assume the auxiliary domains contain sufficient rating data, the UV decomposition model can obtain relatively accurate user latent vectors. In other words, the accuracy of the user latent vectors can be guaranteed. Secondly, as the UV decomposition model can find a low-rank approximation for the rating matrix [12], so the user latent vectors are a good low-dimensional representation of the user features, which will make the classification model more efficient.

Then we expand the two-dimensional feature vectors in the target domain with them. Given a user-item interaction $(u, i, r) \in U_1 \times I_1 \times \{1, 2, 3, 4, 5\}$ in the target domain, we can expand the location feature vector (L_u, L_i) of the target domain with the latent vectors of user u from the auxiliary domains. Thus the user-item interaction can be represented by $(L_u, L_i, p_u^1, \ldots, p_u^m)$, where p_u^i $(i = 1, \ldots, m)$ represents the latent vector of user u in the i-th auxiliary domains.

3.3   Classification Problem Solving and the Proposed Algorithm

Many traditional classification algorithms can be employed to solve the converted classification problem. In this part, we employ Support Vector Machines (SVMs) [15] to solve the problem. Since the converted classification problem in this paper is a multi-class classification problem, we select one-against-all (1–v–r) approach [16], which transforms a c-class problem into c two-class problems. Also we select the following radial basis function (RBF) as the kernel function [15],

$$K(\mathbf{x}, \mathbf{y}) = \exp(-\frac{\|\mathbf{x} - \mathbf{y}\|^2}{2\sigma^2})\tag{7}$$

where σ is a width parameter, $\mathbf{x}$ and $\mathbf{y}$ are n-dimensional vectors in the original feature space. Note that SVMs can implement feature selection [17] by adjusting the kernel functions. Hence our model can adaptively select significant features during the training process.

We summarize our algorithm, a model-based CDCF algorithm by Integrating User Latent Vectors of auxiliary domains (CDCFIULV) in Algorithm 1.

Algorithm 1: the CDCFIULV algorithm

Input: the rating matrix M in the target domain; the rating matrix $M_1, \ldots, M_s$ in the auxiliary domain

Output: the missing ratings in the target domain

Method:

(1) Convert the recommendation problem in the target domain into a classification problem;

(2) Perform UV decomposition in each auxiliary domain to obtain the user latent vectors;

(3) Expand the feature vectors with the user latent vectors;

(4) Train a classifier on the obtained training set;

(5) Predict the missing ratings.

---

4   Experiments

In this section, we conduct extensive experiments to test the performance of the proposed algorithm. We compare our algorithm to six state-of-the-art algorithms, namely, N-CF-U, UVD, N-CDCF-U, MF-CDCF, CMF, and CDTF, where N-CF-U and UVD are two single domain CF algorithms, and N-CDCF-U, MF-CDCF, CMF and CDTF are four cross domain algorithms. All experiments are run on 2.20 GHz, Intel (R) Core

(TM) i5-5200U CPU with 8 GB main memory under windows 7. All algorithms are implemented with Matlab 2015B on top of two open source libraries for recommender systems: MyMediaLite [18] which implements most common CF approaches including Matrix Factorization, and LibSVM [19] which implements SVMs learning algorithms.

4.1 Data Sets

We conducted our experiments on Amazon dataset [20] which consists of rating information of users in 4 different domains: books, music CDs, DVDs and video tapes. The dataset contains 7,593,243 ratings on the scale 1–5 provided by 1,555,170 users over 548,552 different products including 393,561 books, 103,144 music CDs, 19,828 DVDs and 26,132 VHS video tapes.

We build the training and testing set in two different ways similar to [6] to be able to compare our approach with them. In detail, we selected Book and Music CD as the target domain to evaluate respectively. We filtered out users who have rated at least 50 books or 30 music CDs so that there are enough observations to be split in various proportions of training and testing data for our evaluation. Finally, 2,505 users were selected, and in addition we retrieved all items rated by these users in these four domains and set aside top K rated items for each domain respectively. Table 1 shows the statistics of the data for evaluation. Then, we constructed rating matrices over filtered out data for each domain.

Table 1. Statistics of amazon data for evaluation

Domain	Items	Avg. # ratings for each user	Density
Book	6000	57	0.0097
Music	5000	30	0.0062
DVD	3000	37	0.0124
VHS	3000	35	0.0117

To simulate the sparse data problem, we constructed two sparse training sets, TR_{20} and TR_{75}, by respectively holding out 80% and 25% data from the target domain Book, i.e. the remaining data of target domain for training is 20% and 75%. The hold-out data server as ground truth for testing. Likewise, we also construct two other training sets TR_{20} and TR_{75} when choosing Music as the target domain.

4.2 The Setting of the Compared Methods

(1) N-CF-U: A user-based neighborhood CF model. In this experiment, we use $k = 10$ closest users.
(2) UVD (the UV decomposition model): Map both users and items to a joint latent factor space of dimensionality f. In the empirical tests, we observed that the performance is rather stable when f is in the range of [30, 70]. Here we simply set

$f = 50$ from the range. The weight of the regularization terms λ is determined by cross-validation. The learning rate γ is a constant typically having a value between 0.0 and 1.0. If the learning rate is too small, then learning will occur at a very slow pace. If the learning rate is too large, then oscillation between inadequate solutions may occur. In this paper, for simplicity, we set $\gamma = 0.3$.

(3) N-CDCF-U: A cross domain version of N-CF-U. In this experiment, we use $k = 10$ closest users.

(4) MF-CDCF: A cross domain version of UVD. Here we also set $f = 50$ and $\gamma = 0.3$.

(5) CMF: Collective matrix factorization, which couples rating matrices for all domains on the *User* dimension so as to transfer knowledge through the common user-factor matrix.

(6) CDTF: Cross Domain Triadic Factorization, which takes one of the above domains as target domain to perform prediction and others as auxiliary domains to borrow knowledge. We use the same setting as in [6].

(7) CDCFIULV: The proposed method, which first converts the recommendation problem into a classification problem, then expands the feature vectors with user latent vectors from the auxiliary domains, and finally employs SVMs to solve the classification problem. In this experiment, we also set $f = 50$ and $\gamma = 0.3$. The parameters in the SVMs are determined by cross-validation.

4.3 Evaluation Protocol

We used mean absolute error (MAE) as evaluation metrics in our experiments. MAE is defined as

$$(\sum_{i \in T} |r_i - \tilde{r}_i|)/|T| \tag{8}$$

where T denotes the set of test ratings, r_i is the ground truth and $\tilde{r}_i$ is the predicted rating. A smaller value of MAE means a better performance.

4.4 Results

The comparison results are reported in Table 2.

As shown in Table 2, the five CDCF models (N-CDCF-U, MF-CDCF, CMF, CDTF and CDCFIULV) all perform better than UVD and N-CF-U, because UVD and N-CF-U are single domain CF algorithms which cannot deal with the sparsity problem effectively. The performances of N-CDCF-U and MF-CDCF are roughly equal. Since both N-CDCF-U and MF-CDCF fail to consider the differences among domains, as expected they perform worse than the three models (CMF, CDTF and CDCFIULV) which take this differences into account. Our model CDCFIULV perform much better than CMF and CDTF. This is because the classification-based model can adaptively implement feature selection from the whole feature vector according to the performance of the classification. However, CMF does not provide a mechanism to find an optimal weights assignment for the auxiliary domains. Though CDTF assigns the weights based on genetic algorithm (GA), the performance is susceptible to the setting of the initial population.

Table 2. MAE scores

Methods	Target domain: book		Target domain: music	
	TR_{75}	TR_{20}	TR_{75}	TR_{20}
N-CF-U	0.672	0.903	0.779	0.973
UVD	0.622	0.886	0.753	0.951
N-CDCF-U	0.475	0.796	0.703	0.931
MF-CDCF	0.492	0.789	0.718	0.925
CMF	0.461	0.763	0.701	0.812
CDTF	0.335	0.675	0.677	0.769
CDCFIULV	0.306	0.659	0.501	0.738

It is also worth noting that from TR_{20} to TR_{75}, our method possesses the largest performance improvements. Because with the number of training ratings increasing, the training set size of the converted classification problem is increased greatly. Thus the classification model can effectively avoid over-fitting, and the performance can be improved. N-CDCF-U also achieves a not bad performance when the data is relative dense, i.e. TR_{75}, but the performance decreases very fast when the data becomes sparser. Because when the data are sparse, the total similarity used in N-CDCF-U cannot represent the local similarity in the target domain well. However, according to Eq. (1), with the number of training ratings increasing, the total similarity can represent the local similarity in the target domain better.

5   Conclusion

In this paper, from the perspective of classification, we propose a model-based CDCF algorithm by Integrating User Latent Vectors of auxiliary domains (CDCFIULV). We expand the feature vectors in the target domain with user latent vectors obtained from auxiliary domains. Hence the hidden knowledge in the auxiliary domains can be transferred to the target domain via the user latent vectors. We use SVMs to train classifiers for the converted classification problems. Since SVMs can implement feature selection by adjusting the kernel functions, so our model can adaptively select significant features during the training process. The experiment results have shown that CDCFIULV significantly outperform all other state-of-the art baseline algorithms at various sparsity levels. Since we have only discussed how to expand the feature vector in the target domain with some other significant features of users, in the future we will also explore how to expand the feature vector with some other significant features of items.

Acknowledgments. This work is sponsored by the National Natural Science Foundation of China (No. 61402246), a Project of Shandong Province Higher Educational Science and Technology Program (No. J15LN38), Qingdao indigenous innovation program (No. 15-9-1-47-jch), and the Natural Science Foundation of Shandong Province (ZR2016FQ10).

References

1. Bobadilla, J., Ortega, F., Hernando, A., et al.: Recommender systems survey. Knowl.-Based Syst. **46**(1), 109–132 (2013)
2. Goldberg, D., Nichols, D., Oki, B.M., et al.: Using collaborative filtering to weave an information tapestry. Commun. ACM **35**(12), 61–70 (1992)
3. Liu, H., Hu, Z., Mian, A., et al.: A new user similarity model to improve the accuracy of collaborative filtering. Knowl.-Based Syst. **56**(3), 156–166 (2014)
4. Cremonesi, P., Tripodi, A., Turrin, R.: Cross-domain recommender systems. In: Data Mining Workshops, pp. 496–503 (2011)
5. Berkovsky, S., Kuflik, T., Ricci, F.: Cross-domain mediation in collaborative filtering. In: User Modeling 2007, Proceedings of the International Conference, Um 2007, Corfu, Greece, June 25–29 2007, pp. 355–359 (2007)
6. Hu, L., Cao, J., Xu, G., Cao, L., Gu, Z., Zhu, C.: Personalized recommendation via cross-domain triadic factorization. In: WWW 2013, Rio de Janiero, Brazil, pp. 595–606 (2013)
7. Pan, W., Liu, N.N., Xiang, E.W., et al.: Transfer learning to predict missing ratings via heterogeneous user feedbacks. In: IJCAI 2011, Proceedings of the, International Joint Conference on Artificial Intelligence, Barcelona, Catalonia, Spain, July 2011
8. Singh, A.P., Kumar, G., Gupta, R.: Relational learning via collective matrix factorization. In: ACM SIGKDD International Conference on Knowledge Discovery and Data Mining, Las Vegas, Nevada, USA, pp. 650–658. DBLP, August 2008
9. Li, B., Yang, Q., Xue, X.: Can movies and books collaborate? Cross-domain collaborative filtering for sparsity reduction. In: IJCAI 2009, Proceedings of the, International Joint Conference on Artificial Intelligence, Pasadena, California, USA, pp. 2052–2057, July 2009
10. Li, B., Yang, Q., Xue, X.: Transfer learning for collaborative filtering via a rating-matrix generative model. In: International Conference on Machine Learning, ICML 2009, Montreal, Quebec, Canada, pp. 617–624, June 2009
11. Resnick, P., Iacovou, N., Suchak, M., et al.: GroupLens: an open architecture for collaborative filtering of netnews. In: Proceedings of the 1994 ACM Conference on Computer Supported Cooperative Work, pp. 175–186. ACM (1994)
12. Koren, Y., Bell, R., Volinsky, C.: Matrix factorization techniques for recommender systems. Computer **42**(8), 30–37 (2009)
13. Caruana, R.: Multitask learning. Mach. Learn. **28**(1), 41–75 (1997)
14. Kohavi, R.: A study of cross-validation and bootstrap for accuracy estimation and model selection. In: International Joint Conference on Artificial Intelligence, pp. 1137–1143 (2001)
15. Vapnik, V.N.: The Nature of Statistical Learning Theory. Springer, New York (1995)
16. Hsu, C.W., Lin, C.J.: A comparison on methods for multi-class support vector machines. IEEE Trans. Neural Netw. **13**(2), 415–425 (2001)
17. Bishop, C.M.: Pattern Recognition and Machine Learning (Information Science and Statistics). Springer, New York (2006)
18. Gantner, Z., Rendle, S., Freudenthaler, C., et al.: MyMediaLite: a free recommender system library. In: ACM Conference on Recommender Systems, Recsys 2011, Chicago, IL, USA, pp. 305–308, October 2011
19. Chang, C.C., Lin, C.J.: LIBSVM: A library for support vector machines. ACM Trans. Intell. Syst. Technol. **2**(3), 27 (2011)
20. Leskovec, J., Adamic, L.A., Huberman, B.A.: The dynamics of viral marketing. ACM Trans. Web **1**(1), 5 (2005)

Collaborative Filtering Based on Pairwise User-Item Blocking Structure (PBCF): A General Framework and Its Implementation

Fengjuan Zhang[1,2]([✉]), Jianjun Wu[1,2], Jianzhao Qin[1,2], Xing Liu[1,2], and Yongqiang Wang[1,2]

[1] GF Securities, Guangzhou, China
{zhangfengjuan,JianzhaoQin,gfliuxing,wangyongqiang}@gf.com.cn
[2] University of Science and Technology of China, Hefei, China
wjianjun@mail.ustc.edu.cn
http://en.gf.com.cn

Abstract. To our knowledge, all existing collaborative filtering techniques need to find neighbouring relationship between users or items by using some kind of similarity measurement in the feature space. However, a hypothesis hidden behind most existing works is that the similar relationship between users remains static over the whole item sets, which is not true in reality. Users who share similar opinions on some items may have totally different opinions on other items. Users can form many clusters in terms of their opinions on a set of items, However, these clusters may collapse and a new cluster structure will be built in terms their opinions on the new item sets. Analogously, clusters of items formed based on their popularity among a group of users would be disintegrated when encounter a new group of users. In a nutshell, user cluster structure varies across item sets, and vice versa, item cluster structure also varies across user sets.

To deal with this collapse problem, we strive to find block structures embedded in the rating matrix in this paper. Block structure is used to characterize the interaction between users and items. This paper proposes a general framework of collaborative filtering based on pairwise user-item blocking structure and its implementation. At last, existing collaborative filtering algorithms are used to learn the latent factor at the global and block level and further make prediction on the unknown rating in the rating matrix. Experiment evidences show that the recommendation performance can be improved with utilization of these block structures.

Keywords: Collaborative filtering · Dynamically similar · Matrix blocking

1 Introduction

With the explosive growth amount of content available on the web, customers are suffering from the problem of information overload, which has motivated the

---

This work is done when Fengjuan Zhang works at GF Securities.

G. Li et al. (Eds.): KSEM 2017, LNAI 10412, pp. 346–358, 2017.
DOI: 10.1007/978-3-319-63558-3_29

development of recommendation system. Recommendation system aims at recommending ones from a tremendous number of items to users that might interest them according to their historical behaviors and item attributes or some other information. Collaborative Filtering (CF), as the most successful recommendation technique, has been very popular in both academia and industry because it is easy-to-implement and highly effective.

As we all know, most of existing collaborative filtering methods need to measure similarity between users and items in some feature space. We find that similarity is a basic problem in collaborative filtering. Similarity can be divided into two kinds, independent similarity and dependent similarity. Independent similarity means that there is no interaction between user and item similarity. Independent similarity plays an important role in the early development of collaborative filtering. However, user similarity only exists over some items because there can not be two users hold the same opinion over all items in the system. It is possible that user similarity over partial item sets may be covered by user similarity computed over all the items. Item similarity should be computed over a certain user set in the same idea. So the relationship between user similarity and item similarity should be close and mutually equal. Users and items can not be simply divided separately as existing works.

However, all the existed dependent similarity methods assume that the user similarity remains static when faced with different item sets. In reality, although two users may hold similar opinions on some items, they may hold totally different opinions on other items. Users can be grouped into several clusters when faced with some items, however, these clusters may collapse when faced with new item sets. At the same, item clusters are constantly changing towards different users.

To our delight, we are the first one to propose the idea that user similarity and item similarity are dynamic dependent on each other mutually. Block structure is used to describe user and item clusters and characterize their interdependence. We put forward a general framework of collaborative filtering algorithm based on pairwise user-item block structure. This framework includes extracting the unified feature of users and items, user alignment based on items, improved spectral clustering based on items and finally some existing collaborative filtering techniques are used to predict unknown ratings. Simulation results show that our method can effectively improve the performance of recommendation algorithm.

2   Related Work

We have mentioned that the basic of collaborative filtering algorithm is similarity. In this section, we present some previous works based on independent similarity and dependent similarity. We first introduce some famous CF algorithms to show the foundation status of independent similarity. Our own new understanding of dependent similarity comes later, because we find that user similarity and item similarity are interacting with each other in reality.

The common feature of all CF algorithms based on independent similarity is that they all use independent similarity between users or items. User-based and item-based CF algorithms [8,14] first measure the similarity of each pair users or items, and then predict unknown ratings based on the similarity. Pearson correlation coefficient(PCC) is usually used to measure similarity [8]. However, this kind of similarity have false high correlation problem. In reality, there are seldom two users rate totally the same item sets, in this condition the value of $|U_i \bigcap U_j|$ and $|I_u \bigcap I_v|$ is very small. [17] proposes a kind of similarity fusion algorithm. The core of this algorithm is that each known rating can be the estimate value of the unknown ratings. Recently, matrix factorization has been widely used in recommendation algorithms [10]. It mainly uses low-dimensional parameters to characterize the interaction of users and items. SVD is the best low-dimensional matrix approximation method if the approximate standard is Frobenius norm.

There exist some works in which user similarity has relation with item similarity. Co-clustering used in recommendation algorithm was first proposed in [6]. [6] partitions users and items simultaneously, then combine each user cluster and item cluster and form many small rating matrixes. Finally, each unknown rating was predicted according to the known ratings in each small matrix. [1,3] both use this kind of similarity. Although user similarity and item similarity are dependent on each other, their common deficiency is that user similarity is static when faced with different item sets, which means that two users always have the same opinions on all items. Another shortcoming in [6] is that the combination of user cluster and item cluster has no clear reason and meaning. It doesn't explain the meaning of the combination of user and item cluster though they are generated in turn. Although [1] doesn't encounter this combination problem, the cluster structure is just used to reduce the number of comparisons when finding neighbors. Its objective is to linear combine user-based and item-based methods by generating the two kinds of cluster structure simultaneously. However it dose not fully use the obtained user and item clusters and neglect the global information.

3   The Framework of Pairwise User-Item Blocking

Most existing CF algorithms assume user similarity and item similarity are independent. Although some works have realized user similarity and item similarity are mutually influenced, they think user similarity is static when faced with different items. In reality it is not the case. Users can form clusters when faced with a item set. This cluster structure may collapse and form a new cluster structure when faced with a new item set (Fig. 1).

We present a general framework in the following to extract our expected block structure from the rating matrix. The framework is divided into 6 steps.

1. Extract a unified low-dimensional feature representation of users and items. This dimensionality reduction compression step helps avoid computing a large number of users and items. Also we do not need to distinguish users and items, if it is this case, we can apply typical clustering algorithms for users and items simultaneously.

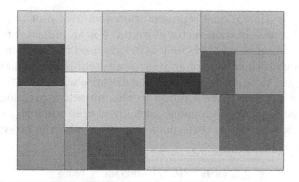

Fig. 1. Expected block structure.

2. Divide users into groups based on each item. For each item, we divide users into groups depending on their opinions on this item. A comparable way should be used to represent the division results based on different items.
3. Cluster items depending on user division result based on each item. At this step, each item is already represented in a standardized format which can be used to clustering directly. However, it is a difficult problem because we need to search for sparse cluster structure in high-dimensional sparse space. After this step, we can extract the block structures hidden in the rating matrix.
4. Feature learning. There are two factors which may influence users' opinions on items. One is users themselves and items' attributes, this factor is stable and not easy to change, we call it global factor. The other factor is influences caused by neighboring items and neighboring users, this factor is easy to change so we call it local factor. For instance, it is a user's inherent properties if he loves comedy films. He may love comedy films more if many of his friends also love it. In this case, the degree of his loving for comedy may be affected by his neighbors. We will learn these two factors in this step.
5. Predication. In this step, the combination of global factor and local feature is used to predict missing ratings in the rating matrix.

4   Implementation of Pairwise User-Item Block Learning

4.1   Unified Feature Extraction

In order to extract a unified low-dimensional representation of users and items we need to reexpress users and items in a same space. The higher rating of an user to an item, the closer they are in the feature space. On the contrary, the lower rating of an user to an item, the farther they are in the feature space. Enlightened by Laplacean Eigenmaps and Locality Preserving Projections(LPP), we define our objective function:

$$\frac{\sum_{u,i} \|\mathbf{q}_u - \mathbf{p}_i\|_f^2 R_{ui} \delta(R_{ui} < r_\theta)}{\sum_{u,i} \|\mathbf{q}_u - \mathbf{p}_i\|_f^2 R_{ui} \delta(R_{ui} \geq r_\theta)} \tag{1}$$

$\mathbf{q}_u \in \mathbb{R}^k$ is the low-dimensional representation vector of user u. $\mathbf{p}_i \in \mathbb{R}^k$ is the low-dimensional representation vector of item i. δ is an indicator function, if the condition is true, it equals 1 otherwise it is 0. r_θ is used to determine the distance of users and items. If the rating of an user to an item is larger than r_θ, we hope their features are as close as possible, and the distance with the rating are inverse. Missing values won't affect the solution to this objective function because the value of the objective function remains unchanged even when $R_{ui} = 0$. However, Eq. 1 has poor generalization performance, so we revise the object function as follows:

$$\frac{\sum_{u,i} \|\mathbf{q}_u - \mathbf{p}_i\|_f^2 R_{ui}\delta(R_{ui} < r_\theta)}{\sum_{u,i} \|\mathbf{q}_u - \mathbf{p}_i\|_f^2 R_{ui}\delta(R_{ui} \geq r_\theta) + \lambda(\sum_u \|\mathbf{q}_u\|_f^2 + \sum_i \|\mathbf{p}_i\|_f^2)} \tag{2}$$

λ is used to control the degree of regularization.

$$\sum_{u,i} \|\mathbf{q}_u - \mathbf{p}_i\|_f^2 R_{ui}\delta(R_{ui} < r_\theta)$$

$$= \sum_{u=1}^m \sum_{i=1}^n \|\mathbf{q}_u - \mathbf{p}_i\|_f^2 [R_2]_{ui} \quad ([R_2]_{ui} = R_{ui}\delta(R_{ui} < r_\theta))$$

$$= \sum_{u=1}^m \sum_{i=1}^n (\mathbf{q}_u^T \mathbf{q}_u [R_2]_{ui} + \mathbf{p}_i^T \mathbf{p}_i [R_2]_{ui}) - \sum_{u=1}^m \sum_{i=1}^n 2\mathbf{q}_u^T \mathbf{p}_i [R_2]_{ui}$$

$$= \sum_{u=1}^m \mathbf{q}_u^T \mathbf{q}_u [R_2^{row}]_{uu} + \sum_{i=1}^n \mathbf{p}_i^T \mathbf{p}_i [R_2^{col}]_{ii} - 2\sum_{u=1}^m \sum_{i=1}^n \mathbf{q}_u^T \mathbf{p}_i [R_2]_{ui}$$

$$= tr(Q^T R_2^{row} Q) + tr(P^T R_2^{col} P) - 2tr(Q^T R_2 P)$$

$$= tr([Q^T \quad P^T] \begin{bmatrix} R_2^{row} & -R_2 \\ -R_2^T & R_2^{col} \end{bmatrix} \begin{bmatrix} Q \\ P \end{bmatrix}$$

We also know $tr(Q^T R_2 P) = tr(P^T R_2^T Q)$. Equation 2 can be transformed into the following formulation:

$$\frac{tr\left(X^T M_2 X\right)}{tr\left(X^T (M_1 + \lambda I) X\right)} \tag{3}$$

where,

$$X = \begin{bmatrix} Q \\ P \end{bmatrix}, Q^T = [\mathbf{q}_1, \mathbf{q}_2, \ldots, \mathbf{q}_m], P^T = [\mathbf{p}_1, \mathbf{p}_2, \ldots, \mathbf{p}_n]$$

$$M_1 = \begin{bmatrix} R_1^{row} & -R_1 \\ -R_1^T & R_1^{col} \end{bmatrix}, M_2 = \begin{bmatrix} R_2^{row} & -R_2 \\ -R_2^T & R_2^{col} \end{bmatrix}$$

$[R_1]_{ui} = R_{ui}\delta(R_{ui} \geq r_\theta)$, R_1^{row} is a diagonal matrix. The u-th element on the main diagonal of R_1^{row} is $\sum_{i=1}^n [R_1]_{ui}$. R_1^{col} is also a diagonal matrix, the i-th element in the main diagonal is $\sum_{u=1}^m [R_1]_{ui}$. Similarly, R_2^{row} is a diagonal matrix, each element in the main diagonal is the sum of elements in the corresponding rows. R_2^{col} is also a diagonal matrix, each element in the main diagonal in the sum of elements in the corresponding columns. I is an unit matrix. $X \in \mathbb{R}^{(m+n) \times k}$, the former m rows represents user features and the latter n rows represents item features.

Any multiple of the optimal solution X^* to Eq. 3 is also an optimal solution. We adopt column orthogonal constraint to get a unique to avoid the situation when $X = 0$. So our formal formulation for the unified feature extraction objective function is:

$$\begin{cases} \max & \dfrac{tr\left(X^T M_2 X\right)}{tr\left(X^T (M_1 + \lambda I) X\right)} \\ \text{s.t} & X^T X = I \end{cases} \qquad (4)$$

Equation 4 is a standard form of the Trace Ratio optimization problem. This problem often occurs in the lower dimension of literatures [16,18]. Recently [12] proposes a method to solve Trace Ratio problem directly, which is more accurate and efficient than other works. The matrix $M_2 - \rho(M_1 + \lambda I)$ is very large and highly sparse, EIGFP [11] is used to get the largest eigenvalues of the matrix.

4.2   Users Align Base on Items

After getting an unified representation of users and items, we cluster users into groups based on each item. The distance between users and items shows the degree of users' love for items. For each item, users are grouped into clusters and then we examine the similarity of the cluster results on each item and combine similar items into one cluster. What's unique in this paper is that we use a standard form of representation to express the cluster results which are useful for us to compare results. The standard representation consists of values on all the standard units. In this paper, the standard unit is a real interval. We divide interval $[0,1]$ into n_b intervals, each interval is called a bucket. The distribution of users on buckets to represents the cluster results relative to each item. Users belong to buckets based on the following condition:

$$bucket[u][i] = \lfloor f(\mathbf{q}_u, \mathbf{p}_i) n_b \rfloor$$

$bucket[u][i]$ represents the bucket number of user u towards item i. The distance between users and items may vary in a wide range, so we need to map the distance into the standard interval $[0,1]$. Function f is the standardization function in this paper which is used to measure the closeness degree between user u and item i. In this paper, we have:

$$f(\mathbf{q}_u, \mathbf{p}_i) = e^{-\alpha(\|\mathbf{q}_u - \mathbf{p}_i\|)} \qquad (5)$$

f is called bucket mapping function, α is the bucket resolution parameter. If we divide a movie into three buckets, corresponding to like so much, okay and dislike respectively. Then each item can be represented by the distribution of users among the buckets as shown in Fig. 2.

like so much   okay   dislike

Fig. 2. Aligned users based on items.

4.3   Compute Item Similarity by MinHashing

Each item can be represented by users' clustering results. Item i can be represented by:

$$I_i = \{B_{i1}, B_{i2}, \ldots, B_{in_b}\}$$

B_{il} represents user sets in the bucket l towards item i. If item i and j have similar user sets in the bucket B_l, then item i and item j will be combined into one cluster. Jaccard similarity is used to measure the similarity between sets.

We have to explore efficient method to compute similarity between items. In this paper, we use Min-Hashing to compute the Jaccard similarity between user sets on each bucket. Hash function is used to simulate random permutation. Algorithm 1 illustrates the detail of the similarity computing process. In this algorithm, P is the smallest prime number larger than m. $No(u)$ is the original serial number of user u. The final similarity matrix is high-dimensional and sparse. So sparse format is used to store the matrix.

Algorithm 1. Compute Jaccard similarity on the bucket l by Min-Hashing

---

1: Number all the users, $\{1, 2, 3, \ldots\}$. Let $k = 1$
   Determine the signature length K_h. Input user number m.
2: **while** $k \le K_h$ **do**
3:    Randomly generate an integer ξ ranges in $[1, m]$.
4:    Generate the k-th signature of each item i $h_l(i, k)$:
      $h_l(i, k) = min_{u \in B_{il}}\{(No(u) + \xi)\%P\}$
5:    k++.
6: **end while**
7: **for** each item i **do**
8:    **for** each item j **do**
9:       $\widehat{Jaccard}_l(i, j) = (\sum_k \delta(h_l(i, k) = h_l(j, k)))/K_h$.
10:   **end for**
11: **end for**

4.4 Improved Spectral Clustering Based on Items

Items can be grouped into clusters after getting the similarity between items. Although there are many excellent cluster algorithms, spectral clustering is used in this paper. Spectral clustering maps objects to points in an undirected graph, similarity between objects is equivalent to the edge weight. Then some suitable graph partitioning algorithms are designed. The most famous spectral cluster algorithms are ratio-cut [7] and normalized-cut [15].

We improve spectral clustering algorithm in this paper to make it more suitable for collaborative filtering. Graph $G = (V, W)$ is divided into $\mathcal{V}_1, \mathcal{V}_2$, $\mathcal{V}_1 \bigcup \mathcal{V}_2 = V$, $\mathcal{V}_1 \bigcap \mathcal{V}_2 = \emptyset$. The division cost is:

$$cut(\mathcal{V}_1, \mathcal{V}_2) = \sum_{u \in \mathcal{V}_1, v \in \mathcal{V}_2} w(u, v)$$

An item is more influential and important if it owns more known ratings. So our objective is to minimize the following function:

$$\frac{cut(\mathcal{V}_1, \mathcal{V}_2)}{(1 - \rho)nas(\mathcal{V}_1, V) + \rho R(\mathcal{V}_1)} + \frac{cut(\mathcal{V}_1, \mathcal{V}_2)}{(1 - \rho)nas(\mathcal{V}_2, V) + \rho R(\mathcal{V}_2)} \quad (6)$$

$$nas(\mathcal{V}_i, V) = \frac{assoc(\mathcal{V}_i, V)}{assoc(V, V)} \quad (7)$$

$assoc(\mathcal{V}_1, V) = \sum_{u \in \mathcal{V}_1, v \in V} w(u, v)$. $R(\mathcal{V}_1) = \sum_{i \in \mathcal{V}_1} T_i / T$. T_i is the number of existing ratings owned by item i. T is the number of all existing ratings. ρ is an adjustable parameter. The edge is standardized so that they all range in $[0, 1]$, which has no effect on the final optimal solution. In other words, the following formulation defines the weight of each vertex:

$$weight(i) = (1 - \rho) \frac{\sum_{j \in V} W(i, j)}{\sum_{i, j \in V} W(i, j)} + \rho T_i / T$$

Suppose

$$\mathbf{W_g} = \begin{bmatrix} weight(1) & 0 & \cdots & 0 \\ 0 & weight(2) & & 0 \\ \vdots & & \ddots & \vdots \\ 0 & 0 & \cdots & weight(|V|) \end{bmatrix}$$

The object function in [4] is:

$$\mathcal{Q}(\mathcal{V}_1, \mathcal{V}_2) = \frac{cut(\mathcal{V}_1, \mathcal{V}_2)}{weight(\mathcal{V}_1)} + \frac{cut(\mathcal{V}_1, \mathcal{V}_2)}{weight(\mathcal{V}_2)} \quad (8)$$

$weight(\mathcal{V}_i) = \sum_{v \in \mathcal{V}_i} weight(v)$, it is a kind of method to define cluster size. Suppose $\mathbf{q}$ represents a kind of division, then we can revise our objective Eq. 8 as:

$$\mathcal{Q}(\mathcal{V}_1, \mathcal{V}_2) = \frac{\mathbf{q}^T \mathbf{L} \mathbf{q}}{\mathbf{q}^T \mathbf{W_g} \mathbf{q}}$$

$\mathbf{L}$ is the Laplacian matrix of the graph, $\mathbf{L} = \mathbf{D} - \mathbf{W}$, $\mathbf{D}$ is a diagonal matrix. The i-th element in the main diagonal is the sum of elements in i-th row in $\mathbf{W}$. Our objective is to minimize $\mathcal{Q}(\mathcal{V}_1, \mathcal{V}_2)$. So our objective is:

$$\begin{cases} \min & \dfrac{\mathbf{q}^T \mathbf{L} \mathbf{q}}{\mathbf{q}^T \mathbf{W_g} \mathbf{q}} \\ \text{s.t} & \mathbf{q} \neq 0, \mathbf{q}^T \mathbf{W_g} \mathbf{e} = 0 \end{cases}$$

The solution is generalized eigenvalues:

$$\mathbf{L}\mathbf{x} = \lambda \mathbf{W_g} \mathbf{x} \tag{9}$$

If the number of defined clusters is k_l, we only need to find the top 9 minimum eigenvalues. Then we can get the final cluster results by doing k-means clustering on these eigenvalues.

5 Prediction Based on the Block Structure

There are two factors which affect users' opinions on items. One is users' own interests and items' own attributes, which we call global feature. The other is the influence caused by neighbors, which we call local feature.

5.1 Global Feature Learning

Matrix factorization has a good performance on extracting global feature. Similar to RSVD, we learn a k-dimensional feature $\mathbf{p}_u$ for each user and a k-dimensional feature $\mathbf{q}_i$ for each item. In reality, people's ratings on items are always not in the same scale, so we adopt Improved Regularized SVD [13]. The prediction formulation is:

$$\widehat{r}_{u,i} = \mu + b_u + b_i + \mathbf{p}_u^T \mathbf{q}_i$$

b_u represents the rating scale of user u, b_i represents the scoring scale of item i, μ is the mean value of all known ratings. Our objective function is:

$$\begin{aligned} \mathcal{L} = &\sum_{(u,i)|r_{u,i}\text{known}} (r_{u,i} - \mu - b_u - b_i - \mathbf{p}_u^T \mathbf{q}_i)^2 \\ &+ \sum_{(u,i)|r_{u,i}\text{known}} \lambda(\|\mathbf{p}_u\|^2 + \|\mathbf{q}_i\|^2 + b_u^2 + b_i^2) \end{aligned} \tag{10}$$

Follow the discipline of stochastic gradient descent, we have the following iterative formulas: The following iterative formula can be deduced if we let $r_{u,i} - \widehat{r}_{u,i} = e_{ui}$ and load each known rating successively rather than load them all in a time. $lrate$ is learning rate. We first use all known ratings to train the first dimensional feature, the second dimensional feature comes next. So the iteration number k represents the dimension.

$$
\begin{cases}
b_u = b_u + 2lrate * (e_{ui} - \lambda b_u) \\
b_i = b_i + 2lrate * (e_{ui} - \lambda b_i) \\
\mathbf{p}_{uk} = \mathbf{p}_{uk} + 2lrate * (e_{ui}\mathbf{q}_{ik} - \lambda \mathbf{p}_{uk}) \\
\mathbf{q}_{ik} = \mathbf{q}_{ik} + 2lrate * (e_{ui}\mathbf{p}_{uk} - \lambda \mathbf{q}_{ik})
\end{cases}
\tag{11}
$$

5.2   Local Block Feature Learing

We decide to use RSVD to learn local block feature. After we learn the global feature, we need to continue analyze the residual ratings. The final rating prediction formulation is:

$$
\widehat{r}_{u,i} = \overbrace{\mu + b_u + b_i + \mathbf{p}_u^T \mathbf{q}_i}^{global} + \overbrace{(\mathbf{p}_u^b)^T \mathbf{q}_i^b}^{block}
\tag{12}
$$

$b = block(u, i)$ is the block number which the user and item are in. $\mathbf{p}_u^b \in \mathbb{R}^k$, $\mathbf{q}_i^b \in \mathbb{R}^k$ is computed by learning the residual ratings in the block which (u, i) in. At the first step, we make predictions according to the global features learned from all the known ratings. If both the training ratings and testing ratings minus the global prediction ratings, we can get the residual ratings. Each residual rating associates with a block. Then we can treat each block as a small matrix, local block features are learned by doing RSVD on the residual rating in each block.

6   Simulation

6.1   Experiment Setup and Evaluation Methodology

We use MovieLens ml-100k as our experimental data set. This data set consists of 100,000 ratings by 943 users on 1682 movies. These ratings take value from 1(lowest rating) to 5(highest rating). Each user rated at least 20 movies. We adopt 5-fold cross validation, which means the data set is divided into five disjoint splits. One split is used for test and the rest are used for training, in turn. We always report the average results, respectively in terms of several evaluation metrics discussed in the following.

There are many metrics to evaluate the performance of a recommendation system, which can be divided into three classes: predictive accuracy metrics, classification accuracy metrics, and rank accuracy metrics [9]. Cacheda et al. [2] takes predictive accuracy metrics MAE and classification accuracy metrics Precision and Recall together into account and obtain two new metrics, GIM and GPIM. GIM computes the absolute errors committed by the prediction on the relevance of the user-item pairs that are really relevant and GPIM computes that are thought to be relevant according to the prediction of recommendation system. Analogously, in this paper, we compute ordinary RMSE, GPIR and GIR. GPIR computes errors committed by the prediction for the pairs which are relevant thought by recommendation system. GIR computes errors committed by the prediction for truly relevant user-item pairs.

6.2   Experiments Comparisons

In this section, we compare our method with typical collaborative filtering algorithms RSVD [5] and IRSVD [13]. We randomly divide the data set MovieLens ml-100k two times and conduct experiments respectively. 80% of the records are used as training data, the rest is used as test data. Comparison results are shown in Tables 1 and 2.

Table 1. Compare results (The first random division)

	RMSE	GPIR	GIR	MAE	GPIM	GIM
RSVD	0.9419	0.8723	0.8440	0.7374	0.6679	0.6678
IRSVD	0.9282	0.8605	0.8039	0.7259	0.6572	0.6367
PBCF-nc	0.9258	0.8626	0.7880	0.7229	0.6575	0.6232
PBCF-inc	0.9337	0.8719	0.8010	0.7229	0.6638	0.6333

Table 2. Compare results (The second random division)

	RMSE	GPIR	GIR	MAE	GPIM	GIM
RSVD	0.9319	0.8523	0.8460	0.7290	0.6539	0.6678
IRSVD	0.9282	0.8554	0.8256	0.7258	0.6573	0.6513
PBCF-nc	0.9254	0.8560	0.8076	0.7223	0.6565	0.6360
PBCF-inc	0.9265	0.8590	0.8092	0.7237	0.6586	0.6381

PBCF-nc in Tables 1 and 2 represents using normalized-cut algorithm to cluster items in our pairwise user-item block collaborative filtering algorithm (PBCF, Collaborative Filtering Based on User-item Pairwise Blocking). PBCF-inc represents using improved normalized-cut algorithm to cluster items. RSVD and IRSVD both set $K = 30$, $lrate = 0.005$, $\lambda = 0.02$. In PBCF-nc, $\alpha = 3000000$, $\rho = 0$. In PBCF-inc, $\alpha = 3000000$, $\rho = 0.01$.

We can find that the performance of PBCF-inc is greatly improved than other algorithms. In the two random divisions, PBCF-inc has slight improvements or unchanged compared with RSVD and IRSVD over the measurements (RMSE, MAE, GPIR, GPIM). However, PBCF-inc has significant improvement compared with RSVD and IRSVD over the measurements GIR and GIM. GIR and GIM specially measure the prediction accuracy that items which users really love. In other words, our method can predict what users really love more accurate.

7   Conclusions

In this paper, we think that the user cluster structure is always changing when faced with different items. At the same time, item cluster structure is also changing when faced with different users. So we propose that user similarity and item similarity should be dynamic dependent on each other mutually. We try to discover the hidden block structure in the rating matrix. A block structure is used to characterize the interaction between uses and items. Each block consists of many users and items. Users in each block hold similar opinions on the items in that block.

We propose a general framework of pairwise user-item block structure algorithm. This framework includes the unified feature extraction of users and items. Divide users into groups based on each item. Cluster items according to user division results and then extract our expected block structure. At last we use existing collaborative filtering algorithms to learn global and local within block feature and then predict unknown ratings.

Experiments shows that our method can efficiently improve the performance of a recommendation system. After extracting blocks from the original rating matrix, we can greatly improve the performance of CF algorithms. Especially we can predict what users really love in reality accurately, which can promote users' satisfaction with our recommendation system. What's more, our method can deal with changing data efficiently. When new ratings come, we only need to retrain the feature within the block which the new ratings located. We don't need to retrain their global feature.

References

1. Banerjee, A., Dhillon, I., Ghosh, J., Merugu, S., Modha, D.S.: A generalized maximum entropy approach to bregman co-clustering and matrix approximation. In: Proceedings of the Tenth ACM SIGKDD International Conference on Knowledge Discovery and Data Mining, pp. 509–514. ACM (2004)
2. Cacheda, F., Carneiro, V., Fernández, D., Formoso, V.: Comparison of collaborative filtering algorithms: limitations of current techniques and proposals for scalable, high-performance recommender systems. ACM Trans. Web (TWEB) 5(1), 2 (2011)
3. Chen, G., Wang, F., Zhang, C.: Collaborative filtering using orthogonal nonnegative matrix tri-factorization. Inf. Process. Manag. 45(3), 368–379 (2009)
4. Dhillon, I.S.: Co-clustering documents and words using bipartite spectral graph partitioning. In: Proceedings of the Seventh ACM SIGKDD International Conference on Knowledge Discovery and Data Mining, pp. 269–274. ACM (2001)
5. Funk, S.: Netflix update: try this at home. http://sifter.org/simon/journal/20061211.html (2006)
6. George, T., Merugu, S.: A scalable collaborative filtering framework based on co-clustering. In: Fifth IEEE International Conference on Data Mining, p. 4. IEEE (2005)
7. Hagen, L., Kahng, A.B.: New spectral methods for ratio cut partitioning and clustering. IEEE Trans. Comput.-Aided Des. Integr. Circuits Syst. 11(9), 1074–1085 (1992)
8. Herlocker, J.L., Konstan, J.A., Borchers, A., Riedl, J.: An algorithmic framework for performing collaborative filtering. In: Proceedings of the 22nd Annual International ACM SIGIR Conference on Research and Development in Information Retrieval, pp. 230–237. ACM (1999)

9. Herlocker, J.L., Konstan, J.A., Terveen, L.G., Riedl, J.T.: Evaluating collaborative filtering recommender systems. ACM Trans. Inf. Syst. (TOIS) **22**(1), 5–53 (2004)
10. Ma, H., Zhou, D., Liu, C., Lyu, M.R., King, I.: Recommender systems with social regularization. In: Proceedings of the Fourth ACM International Conference on Web Search and Data Mining, pp. 287–296. ACM (2011)
11. Money, J.H., Ye, Q.: Algorithm 845: EIGIFP: a MATLAB program for solving large symmetric generalized eigenvalue problems. ACM Trans. Math. Softw. (TOMS) **31**(2), 270–279 (2005)
12. Ngo, T.T., Bellalij, M., Saad, Y.: The trace ratio optimization problem. SIAM Rev. **54**(3), 545–569 (2012)
13. Paterek, A.: Improving regularized singular value decomposition for collaborative filtering. In: Proceedings of KDD cup and workshop, vol. 2007, pp. 5–8 (2007)
14. Sarwar, B., Karypis, G., Konstan, J., Riedl, J.: Item-based collaborative filtering recommendation algorithms. In: Proceedings of the 10th International Conference on World Wide Web, pp. 285–295. ACM (2001)
15. Shi, J., Malik, J.: Normalized cuts and image segmentation. IEEE Trans. Pattern Anal. Mach. Intell. **22**(8), 888–905 (2000)
16. Wang, H., Yan, S., Xu, D., Tang, X., Huang, T.: Trace ratio vs. ratio trace for dimensionality reduction. In: 2007 IEEE Conference on Computer Vision and Pattern Recognition, CVPR 2007, pp. 1–8. IEEE (2007)
17. Wang, J., De Vries, A.P., Reinders, M.J.: Unifying user-based and item-based collaborative filtering approaches by similarity fusion. In: Proceedings of the 29th Annual International ACM SIGIR Conference on Research and Development in Information Retrieval, pp. 501–508. ACM (2006)
18. Yan, S., Xu, D., Zhang, B., Zhang, H.J.: Graph embedding: A general framework for dimensionality reduction. In: IEEE 2005 Computer Society Conference on Computer Vision and Pattern Recognition, CVPR 2005, vol. 2, pp. 830–837. IEEE (2005)

Beyond the Aggregation of Its Members— A Novel Group Recommender System from the Perspective of Preference Distribution

Zhiwei Guo[1], Chaowei Tang[1]([✉]), Wenjia Niu[2]([✉]), Yunqing Fu[4], Haiyang Xia[3], and Hui Tang[1]

[1] College of Communication Engineering,
Chongqing University, Chongqing 400044, China
zwguo@gmail.com, cwtang@cqu.edu.com, tangh@net-east.com
[2] Beijing Key Laboratory of Security and Privacy in Intelligent Transportation,
Beijing Jiaotong University, Beijing 100044, China
nwj6688@gmail.com
[3] School of Computer Science, Xi'an ShiYou University, Xi'an 710065, China
haiyangxia15@gmail.com
[4] College of Software, Chongqing University, Chongqing 400044, China
yqfu@cqu.edu.com

Abstract. This paper focuses on recommending items to group of users rather than individual users. To model group profile, existing researches almost aggregate preferences of members into a single value, and thus cannot reflect actual group profile of groups with conflicting characteristics. Therefore, we propose a novel group recommender system mechanism. It views group profile as preference distribution, and then models item recommendation process as a multi-criteria decision making process, in order to obtain better recommendation results. Finally, experiments are conducted to verify the proposed approach.

Keywords: Group recommender system · Aggregation method · Preference distribution · Multi-criteria decision making

1 Introduction

Recommending items to a group of users rather than individuals has become a new problem in field of recommender systems. *Group Recommender Systems* (GRS) suffer from a main barrier: how to formulate group profile because preferences of members are conflicting.

This paper focuses on GRS targeting ephemeral groups whose members gather occasionally or randomly, not because of social relation or anything in common (e.g. passengers occasionally taking the same airplane). Existing relevant researches primarily concentrate on aggregation of group members and can be divided into two categories: (1) aggregation of individual ratings [6–8,13,18]; (2) aggregation of individual recommendation lists [1,2,10,15,17]. The former

© Springer International Publishing AG 2017
G. Li et al. (Eds.): KSEM 2017, LNAI 10412, pp. 359–370, 2017.
DOI: 10.1007/978-3-319-63558-3_30

aggregate members' ratings towards candidate items into a single value as group profile by simple aggregation functions like average. The latter merge members' recommended lists into a singe list for the group by similar aggregation functions. Nevertheless, single aggregations fail to distinguish conflicting characteristics of groups, and thus cannot well reflect actual group profile. Figure 1 gives an typical example of such case. Let mean value μ denote aggregated group profile and variance value σ^2 denote conflicting degree of the group. Though μ_A and μ_B are close, Group B is more heterogeneous than Group A.

$$\mu_A = 6.25 \quad \sigma_A^2 = 2.44$$

Member	A1	A2	A3	A4	A5	A6	A7	A8
Rating	7	5	8	6	8	7	6	3

$$\mu_B = 6.00 \quad \sigma_B^2 = 6.75$$

Member	A1	A2	A3	A4	A5	A6	A7	A8
Rating	3	8	9	5	8	3	9	3

Fig. 1. Even with similar average value, the actual group profiles of the two groups are remarkably different

Instead of simple aggregations, this paper regards group profile as a distribution of preference (members' ratings), which is able to represent comprehensive preference characteristics of a group. Thus, a novel **G**roup **Re**commender System From the Perspective of **P**reference Distribu**tion** (**Greption**) is proposed. First, idea of *Label Distribution Learning* (LDL) [4] is introduced to finely model group profile. Then, by proposing a modified VIKOR method, we transform the process of selecting items for a group into a multi-criteria decision making process (MCDM).

The rest of the paper is organized as follows. In Sect. 2, we summarizes related work. Section 3 describes framework and mathematic modeling of the Greption approach. Section 4 presents the experimental results and analysis. And we conclude this paper in Sect. 5.

2   Related Work

In this section, we present the two general categories of approaches to generating group recommendations, and review both types of approaches. Also, we give a discussion of limitations of existing methods.

In [18], Wang et al. proposed a group recommender system which assigns each member a member contribution score and aggregates members' profiles into the group profile. In [6], Kagita et al. took all members' transitive precedence of items into consideration, and constructed a virtual user to represent group. In [7], Kim et al. presented a recommender system which adopts graph-based approach to model relation between users and items and then aggregates members into a group. In [8], Lin et al. merged members' historical records as a group's historical records, and then proposed a NLMS method to suggest group recommendations. In [13], Ortega et al. aggregated members by average method, and performed group recommendations using Matrix Factorization based Collaborative Filtering.

In [17], Skowron et al. assumed that each user had some intrinsic utility on each item, and that items are ranked accordingly. Then, rankings are aggregated. In [1], Baltrunas et al. aggregated individual recommendation lists produced by collaborative filtering to realize group recommendation. In [10], Meena and Bharadwaj developed a novel approach based on Kemeny optimal aggregation using genetic algorithm to aggregate rankings. In [15], Salehi and Boutilier firstly predicted unknown individual preferences with form of ranking through probabilistic inference, and then aggregated the individual preferences into a whole by simple aggregation.

In all, principle rules of almost all of existing methods are simple aggregation of group members. Some aggregate individual ratings, while some aggregate individual rankings. However, these methods cannot comprehensively model group profile in terms of conflicting characteristics.

3   Mathematic Modeling of Group Recommendation Mechanism

In this section, we first introduce framework of the Greption Mechanism. Then, its mathematical modeling is described.

3.1   Framework

As is shown in Fig. 2, suppose that a number of users make up a group G, and that a set of candidate items I_l $(l = 1, 2, \ldots, M)$ will be recommended to the group. Firstly, LDL is introduced to formulate a model to predict group profile concerning candidate items I_l $(l = 1, 2, \ldots, M)$. To estimate model parameters, user registry that contains historical items' ratings given by group members will be input as training set. The obtained group profile is a type of preference distribution, and is represented as $d_{I_l}^y$. Then, recommendation results are suggested on the basis of predicted group profile. To achieve this, we transform the process into an MCDM and propose a modified VIKOR method. After that, top-m recommendation results for the group, represented as $L_G = \{L_1, L_2, \ldots, L_m\}$, are obtained.

3.2   Modeling of Prediction of Group Profile

LDL is a general learning framework that views each instance as a distribution of labels rather than a single label or multiple labels. In a label distribution, a real number $P_y \in [0, 1]$ is assigned to each label y, representing the degree that

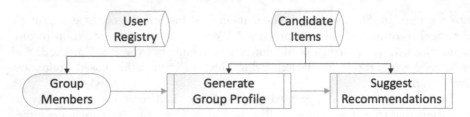

Fig. 2. Greption framework

the corresponding label describes the instance. The numbers for all the labels sum to 1, meaning full description of the instance.

We follow the idea proposed in [3] that group members' rating distribution for an item is viewed as a label distribution. We assume that training set contains a number of users constituting a group G as well as n items rated by users, and that the ratings given by users comprise c levels. And not all the items are rated by each user of the group. Let x_i $(i = 1, 2, \ldots, n)$ denote items of training set, y_j $(j = 1, 2, \ldots, c)$ denote all the labels, and $P_i(y_j)$ denote label distribution of x_i. The input space of LDL is obtained as $S = \{(x_1, P_1(y_j)), (x_2, P_2(y_j)), \ldots, (x_n, P_n(y_j))\}$. Given above, the goal of LDL is to learn a conditional p.d.f. $d \in \mathbb{R}^c$ with parameters:

$$d_x^y = p(y|x; \theta) \tag{1}$$

where θ is the vector of parameters. Due to the fact that degree that each label describes an item is in range of $[0, 1]$, the description degree of labels of x is modeled utilizing logistic function:

$$d = f(x) = \frac{1}{1 + \exp[-(W\varphi(x) + b)]} \tag{2}$$

where $\varphi(x)$ is a nonlinear transformation of x to a higher feature space $\mathbb{R}^{\mathcal{H}}$, $W \in \mathbb{R}^c$ and $b \in \mathbb{R}^c$ are the model parameters.

Next, in order to output distribution of the c labels, multi-dimensional regression analysis can be adopted to solve it. Logistic regression cannot well deal with operations with high computational complexity, thus is not suitable for learning from label distribution. Therefore, Multi-dimensional support vector regression (MSVR) [16] is instead employed to solve the problem, leading to solving the following optimization problem:

$$\min \frac{1}{2} \sum_{j=1}^{c} \|w^j\|^2 + C \sum_{i=1}^{n} L(u_i) \tag{3}$$

In MSVR, L_1-norm loss function will be calculated for each dimension independently, which will make solution complexity grow with the number of dimensions [16]. Instead, loss function is expected to be formulated as L_2-norm:

$$L(u_i) = \begin{cases} 0, & u_i < \varepsilon \\ (u_i - \varepsilon)^2, & u_i \geq \varepsilon \end{cases} \tag{4}$$

$$u_i = \|e_i\| = \sqrt{e_i^T e_i}, \tag{5}$$

$$e_i = d_i - f(x_i) \tag{6}$$

In order to predict group preference distribution, it is expected to learn model parameters by finding solution of Eq. (3). Unfortunately, substituting Eqs. (2), (4), (5), (6) into Eq. (3) cannot lead to a convex quadratic form. Therefore, it is hard to find the optimum as well as to apply the kernel trick. In [3], Gent et al. proposed an alternative loss function which can reform the Eq. (3) to a convex quadratic programming process:

$$u_i' = \|e_i'\| = \sqrt{(e_i')^T e_i'} \tag{7}$$

$$e_i' = -\log\left(\frac{1}{d_i} - 1\right) - (W\varphi(x_i) + b) \tag{8}$$

$$u_i' \geq 4u_i \tag{9}$$

Replacing u_i in Eq. (3) with $u_i'/4$, the goal of LDL can be transformed into minimizing the following target function:

$$T(W,b) = \frac{1}{2}\sum_{j=1}^{c}\|w^j\|^2 + C\sum_{i=1}^{n} L\left(\frac{u_i'}{4}\right) \tag{10}$$

It is still hard to obtain exact solution of the transformed problem. Instead, an iterative quasi-Newton method called Iterative Re-Weighted Least Square (IRWLS) [14] is utilized to search for approximate solution. Due to the space limit, we leave out the detailed process of derivation of optimal solution finding. Finally, the optimal solutions w^j and b^j can be obtained and then substituted into Eq. (2) to calculate predicted the label distributions of candidate items I_l ($l = 1, 2, \ldots, M$) who are to be recommended. So far, group profile $d_{I_l}^y$ is obtained.

3.3   Decision of Recommendation Results

Having modeled group profile, the scope of this stage is to select appropriate items for the group. However, preference distribution is highly ambiguous, and is composed by some components. Here, we propose a modified VIKOR method that incorporates fuzzy set theory, and transform the item selection problem into a MCDM.

VIKOR, proposed by Opricovic and Tzeng [12], focuses on ranking and selecting from a set of alternatives under uncertainty with conflicting criteria. To construct MCDM, following assumptions are established:

1. Let I_l $(l = 1, 2, \ldots, M)$ be the set of items to be recommended, that is, alternatives.
2. For items I_l $(l = 1, 2, \ldots, M)$, they have group preference distribution $d_{I_l}^y = p(y_1, y_2, \ldots, y_c | I_l)$ concerning c-level rating. Let all the y_j $(j = 1, 2, \ldots, c)$ denote the criteria for decision.
3. A set of preference distribution values of I_l $(l = 1, 2, \ldots, M)$ corresponding to criteria $y_j (j = 1, 2, \ldots, c)$ are called $V = \{V_{lj}; l = 1, 2, \ldots, M, j = 1, 2, \ldots, c\}$.

Given items I_l and predicted group profile $d_{I_l}^y$, the objective of the modified VIKOR method is to find optimal ranking order for items I_l $(l = 1, 2, \ldots, M)$ under c criteria.

Criteria Assessment: As c criteria possess different meanings, we firstly adopt fuzzy set theory to assess the nature of criteria. Here, the c criteria are divided into three clusters C_1, C_2, and C_3 with the aid of membership function:

$$y_j \in \begin{cases} C_1, & 0 < \tilde{\mu}_j \leq 0.3 \\ C_2, & 0.3 < \tilde{\mu}_j \leq 0.6 \\ C_3, & 0.6 < \tilde{\mu}_j \leq 1 \end{cases} \tag{11}$$

where

$$\tilde{\mu}_j = \frac{1}{2} + \frac{1}{2} \sin \frac{\pi}{c} \left(j - \frac{c}{2} \right), j = 1, 2, \ldots, c \tag{12}$$

Decision Matrix Generation: A pseudo decision maker is simulated to assess criteria values of items I_l $(j = 1, 2, \ldots, M)$ and to give decision scores for each criterion of items I_l $(j = 1, 2, \ldots, M)$. As for criteria of C_3, higher values signify utility. And for criteria of C_1, higher values imply dissatisfaction. Their decision scores are assumed to range from 0 to 100. Yet criteria of C_2 are relatively neutral, their decision scores are not expected to be too high or too low, and are assumed to range between two positive integers a_1 and a_2. Therefore, decision score of criteria can be calculated as follows:

$$Z_{lj} = \begin{cases} 50 \cdot \left[\frac{2}{\pi} \arcsin (2V_{lj} - 1) + 1 \right], & y_j \in C_3 \\ \frac{b - V_{lj}}{a_2 - a_1}, & y_j \in C_2 \\ 50 \cdot \left[\frac{2}{\pi} \arcsin (1 - 2V_{lj}) + 1 \right], & y_j \in C_1 \end{cases} \tag{13}$$

where Z_{lj} denotes the decision score of jth criteria of lth item. Thus, decision matrix D of the pseudo decision maker can be constructed:

$$D = \begin{bmatrix} Z_{11} & Z_{12} & \cdots & Z_{1c} \\ Z_{21} & Z_{22} & \cdots & Z_{2c} \\ \vdots & \vdots & \ddots & \vdots \\ Z_{M1} & Z_{M2} & \cdots & Z_{Mc} \end{bmatrix} \tag{14}$$

Item Ranking: For all criteria, we denote the best value by f_j^* and the worst value by f_j^-. Due to different nature of three categories of criteria, best value and worst value can be determined as follows:

$$f_j^* = \begin{cases} \max Z_{lj}; y_j \in C_3 \\ \min Z_{lj}; y_j \in (C_1 \cup C_2) \end{cases} ; f_j^- = \begin{cases} \min Z_{lj}; y_j \in C_3 \\ \max Z_{lj}; y_j \in (C_1 \cup C_2) \end{cases} \tag{15}$$

We denote the aggregated value of lth item with a maximum group utility by S_l and compute as following formula:

$$S_l = \sum_{j=1}^{c} \bar{w}_j \left(f_j^* - Z_{lj} \right) / \left(f_j^* - f_j^- \right) \tag{16}$$

We denote the aggregated value of lth item with a minimum individual regret of "opponent" by R_l and compute it as following formula:

$$R_l = \max_{j=1,\dots,c} \left[\bar{w}_j \left(f_j^* - Z_{lj} \right) / \left(f_j^* - f_j^- \right) \right] \tag{17}$$

where $\bar{w}_j$ is the weight of criterion j, and is initially defined as:

$$\bar{w}_j = \sum_{I=1}^{M} V_{lj} \tag{18}$$

We denote value of benefit ratio for lth item by Q_l and compute as following formula:

$$Q_l = v \left(S_l - S^* \right) / \left(S^- - S^* \right) + \left(1 - v \right) \left(R_l - R^* \right) / \left(R^- - R^* \right)$$
$$S^* = \min_{l=1,\dots,M} S_l, S^- = \max_{l=1,\dots,M} S_l$$
$$R^* = \min_{l=1,\dots,M} R_l, R^- = \max_{l=1,\dots,M} R_l \tag{19}$$

where v is a weight for the strategy of maximum group utility, and $1 - v$ is the weight of individual regret.

So far, all the items are ranked by sorting the values S_l, R_l, and Q_l respectively in ascending order, obtaining three ranking lists. If the following two conditions are satisfied simultaneously, then the scheme with minimum value of Q in ranking is considered the optimal solution. The two conditions are:

i. Acceptable advantage:

$$Q \left(I^\nabla \right) - Q \left(I^\Delta \right) \geq 1 / (M - 1) \tag{20}$$

where I^∇ is the item with second position in the ranking list by Q, I^Δ is the item with first position, and M is the number of items.

ii. Acceptable stability in decision making: The item I^Δ must be also the best ranked by S and/or R.

4   Experiments and Analysis

In this section, we present an empirical study of our approach on a real dataset, and analyze the results.

4.1   Experimental Setup

We employ the "MovieLens 10 M" dataset[1] which includes 10681 items and 10000054 ratings from 71567 different users. And the ratings range from 1 to 5. In experiments, items that have been rated by group members but will not be recommended to the group again. We randomly select 60% of the data for training and 40% for testing. Groups are formed by randomly selecting users. And the group size is set to 200, 300, and 400 respectively.

Table 1. Metadata included in the experimental dataset

Attribute	Type	ϕ	Attribute	Type
Genre	Categorical	0	Year	Numerical
Director	Categorical	5	Running time	Numerical
1st actor	Categorical	5	Budget	Numerical
2nd actor	Categorical	5		
Country	Categorical	10		
Language	Categorical	10		
Production company	Categorical	20		

To extract feature, metadata of the experimental dataset is obtained from IMDb[2]. Table 1 lists all the metadata included in the dataset. Note that all the attributes in Table 1 are either numerical attributes or categorical attributes. Thus, we set a threshold ϕ and re-assign a new value 'other' to those who appear less times in the dataset than the threshold, in order to simplify the complexity of data. The categorical attributes are then transformed into numerical ones by numbering each value of all the categorical attributes. Finally, all the numbers are normalized to the same scale through the min-max normalization.

The penalty parameter C in Eq. (3) is set to 1, and the insensitivity parameter ε in Eq. (4) is set to 0.1. a_1 and a_2 in Eq. (13) are set to 15 and 85 respectively, and the weight v in Eq. (19) is set to 0.5. As for the "MovieLens 10 M" dataset, the number of labels is 5.

---

[1] http://www.grouplens.org/.
[2] http://www.imdb.com/.

4.2   Metrics and Baselines

To evaluate the proposed recommender system based on recommendation lists, we adopt several metrics that have been utilized in other relevant researches: normalized discounted cumulative gain (nDCG) [18], F measure [13,18], precision@5 (P@5), precision@10 (P@10)[13], mean average precision@5 (MAP@5), mean average precision@10 (MAP@10), mean reciprocal rank@5 (MRR@5), mean reciprocal rank@10 (MRR@10) [5]. Detailed explanations of these measures are described in references and are left out here.

To measure the improvement of our Greption approach, we implement several popular group recommendation approaches as baselines: LM [11], AVG [11], AM [9], MCS [18].

4.3   Results and Analysis

Figure 3 shows the nDCG results obtained by LM, AVG, AM, MCS, and our approach with the change of m and group size. And it has three subfigures, denoting results with different group size values respectively: 200, 300, and 400. Obviously, recommendation lists obtained by Greption method are more ideal than other methods, because the Greption approach can model group profile more finely. In the meanwhile, different from previous methods, the Greption method is designed to serve groups with large scale. Therefore, when group size grows, our Greption method has better performance.

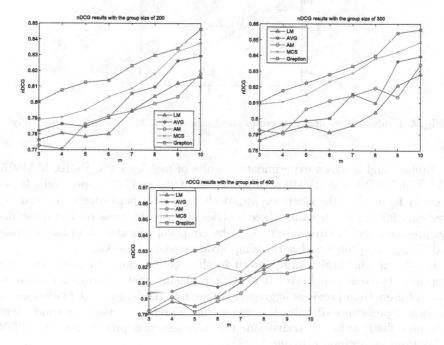

Fig. 3. nDCG results with group size value of 200, 300, and 400 respectively

Figure 4 shows the F measure results obtained by baselines and our approach with the change of m and group size. And it has three subfigures, denoting results with different group size values respectively: 200, 300 and 400. Obviously, the Greption approach outperforms other methods under any group size in the experiment. The MCS method has better performance than AVG, LM, AM in most cases. However, when group size is increasing, performance of the MCS method decreases distinctly. But the proposed Greption framework can work well in the experiment.

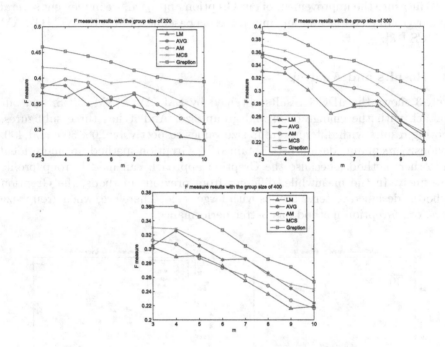

Fig. 4. F measure results with group size value of 200, 300, and 400 respectively

Tables 2 and 3 shows experimental results of metrics P@5, P@10, MAP@5, MAP@10, MRR@5, MRR@10 with group size of 250 and 350 respectively. It can be easily found that the Greption approach has better performance in terms of these metrics. This is because these metrics are used to measure recommendation precision and ranking effect. And the Greption contains two phases: finely modeling group profile and optimizing recommendation ranking.

In all, experimental results shown in this section can well prove that the Greption approach can better deal with problems that recommend items to group of users than previous aggregation-based methods. First, it models group profile as a preference distribution rather than a preference value. Second, given preference distribution, it transforms the item selection process into a MCDM and obtains an optimal ranking.

Table 2. Experimental results of Greption and baselines concerning metrics: P@5, P@10, MAP@5, MAP@10, MRR@5, MRR@10 (Group size: 250)

	P@5	P@10	MAP@5	MAP@10	MRR@5	MRR@10
Greption	0.7547	0.7291	0.7826	0.7419	0.6582	0.6948
LM	0.4651	0.4065	0.5082	0.4844	0.5526	0.5903
AVG	0.5023	0.4840	0.5327	0.5296	0.4739	0.5134
AM	0.4201	0.4018	0.4827	0.4416	0.4713	0.5228
MCS	0.7176	0.6854	0.7343	0.7110	0.6072	0.6328

Table 3. Experimental results of Greption and baselines concerning metrics: P@5, P@10, MAP@5, MAP@10, MRR@5, MRR@10 (Group size: 350)

	P@5	P@10	MAP@5	MAP@10	MRR@5	MRR@10
Greption	0.7378	0.6984	0.7562	0.7253	0.6674	0.7032
LM	0.4472	0.4283	0.4784	0.4639	0.5693	0.5889
AVG	0.5367	0.4971	0.5576	0.5363	0.4571	0.4966
AM	0.4158	0.3935	0.4425	0.4309	0.4467	0.4767
MCS	0.6946	0.6549	0.7207	0.7008	0.6076	0.6274

5 Conclusion

In order to tackle with problems that recommend items to group of users rather than individual users, We propose Greption, a novel group recommender system from the perspective of preference distribution. Specifically, framework of Greption mechanism is proposed and its mathematical modeling is described. Besides, we present a series of experiments as the demonstration of efficiency of our approach.

Acknowledgments. The authors would like to thank the guidance of Professor Wenjia Niu, Professor Chaowei Tang, and Professor Hui Tang. Meanwhile this research is supported by the National Natural Science Foundation of China (No. 61672091).

References

1. Baltrunas, L., Makcinskas, T., Ricci, F.: Group recommendations with rank aggregation and collaborative filtering. In: Proceedings of the Fourth ACM Conference on Recommender Systems, pp. 119–126. ACM (2010)
2. Crossen, A., Budzik, J., Hammond, K.J.: Flytrap: intelligent group music recommendation. In: Proceedings of the 7th International Conference on Intelligent User Interfaces, pp. 184–185. ACM (2002)
3. Geng, X., Hou, P.: Pre-release prediction of crowd opinion on movies by label distribution learning. In: Twenty-Fourth International Joint Conference on Artificial Intelligence (2015)

4. Geng, X., Ji, R.: Label distribution learning. In: Proceedings of the 2013 IEEE 13th International Conference on Data Mining Workshops, pp. 377–383. IEEE Computer Society (2013)
5. Jhamb, Y., Fang, Y.: A dual-perspective latent factor model for group-aware social event recommendation. Inf. Process. Manag. **53**(3), 559–576 (2017)
6. Kagita, V.R., Pujari, A.K., Padmanabhan, V.: Virtual user approach for group recommender systems using precedence relations. Inf. Sci. **294**, 15–30 (2015)
7. Kim, H., Bloess, M., El Saddik, A.: Folkommender: a group recommender system based on a graph-based ranking algorithm. Multimedia Syst. **19**(6), 509–525 (2013)
8. Lin, K., Shiue, D., Chiu, Y., Tsai, W., Jang, F., Chen, J.: Design and implementation of face recognition-aided IPTV adaptive group recommendation system based on NLMS algorithm. In: 2012 International Symposium on Communications and Information Technologies (ISCIT), pp. 626–631. IEEE (2012)
9. McCarthy, J.F., Anagnost, T.D.: MusicFX: an arbiter of group preferences for computer supported collaborative workouts. In: Proceedings of the 1998 ACM Conference on Computer Supported Cooperative Work, pp. 363–372. ACM (1998)
10. Meena, R., Bharadwaj, K.K.: Group recommender system based on rank aggregation – an evolutionary approach. In: Prasath, R., Kathirvalavakumar, T. (eds.) MIKE 2013. LNCS, vol. 8284, pp. 663–676. Springer, Cham (2013). doi:10.1007/978-3-319-03844-5_65
11. Oconnor, M., Cosley, D., Konstan, J.A., Riedl, J.: Polylens: a recommender system for groups of users. In: Prinz, W., Jarke, M., Rogers, Y., Schmidt, K., Wulf, V. (eds.) ECSCW 2001, pp. 199–218. Springer, Heidelberg (2001)
12. Opricovic, S., Tzeng, G.: Multicriteria planning of post-earthquake sustainable reconstruction. Comput.-Aided Civil Infrastruct. Eng. **17**(3), 211–220 (2002)
13. Ortega, F., Hernando, A., Bobadilla, J., Kang, J.H.: Recommending items to group of users using matrix factorization based collaborative filtering. Inf. Sci. **345**, 313–324 (2016)
14. Pérez-Cruz, F., Navia-Vázquez, A., Alarcón-Diana, P.L., Artes-Rodriguez, A.: An IRWLS procedure for SVR. In: 2000 10th European Signal Processing Conference, pp. 1–4. IEEE (2000)
15. Salehi-Abari, A., Boutilier, C.: Preference-oriented social networks: group recommendation and inference. In: Proceedings of the 9th ACM Conference on Recommender Systems, pp. 35–42. ACM (2015)
16. Sánchez-Fernández, M., de Prado-Cumplido, M., Arenas-García, J., Pérez-Cruz, F.: SVM multiregression for nonlinear channel estimation in multiple-input multiple-output systems. IEEE Trans. Sig. Process. **52**(8), 2298–2307 (2004)
17. Skowron, P.K., Faliszewski, P., Lang, J.: Finding a collective set of items: from proportional multirepresentation to group recommendation. In: Twenty-Ninth AAAI Conference on Artificial Intelligence (2015)
18. Wang, W., Zhang, G., Lu, J.: Member contribution-based group recommender system. Decis. Support Syst. **87**, 80–93 (2016)

Exploring Latent Bundles from Social Behaviors for Personalized Ranking

Wenli Yu[1], Li Li[1(✉)], Fan Li[1], Jinjing Zhang[1], and Fei Hu[1,2]

[1] Faculty of Computer and Information Science, Southwest University,
Chongqing, China
m13101332539@163.com, lily@swu.edu.cn
[2] Network Centre, Chongqing University of Education,
Chongqing 400065, China

Abstract. Users in social networks usually have different interpersonal relationships and various social roles. It is common that a user will synthesize all of his/her roles before taking any action. Understanding how products relate to each other is crucial in Recommender Systems (RSs). Predicting personalized sequential behaviors which are influenced by users' various social roles and product bundle relationships is one of key tasks for the success of RSs. In this paper, a novel method combining social roles and sequential patterns is proposed to explore the latent bundle dimensions from the perspective of user's sequential pattern and his/her social roles as well. The extracted vector represents the most distinctive features of interpersonal relationships for users. The proposed method tries to explore the latent bundle relationship by learning personal dynamics which is influenced by the user's social roles. The method is evaluated on Amazon datasets and demonstrates our framework outperforms alternative baseline by providing top k recommendations.

Keywords: Social roles · Sequential patterns · Latent bundle relationships

1 Introduction

Generally, the user has different roles in the society. And users with different roles may have different conformity behaviors, which is missed and disregarded in RSs. And, the types of relationships between products in which we are interested can be mined from behavioral data and social roles, such as browsing, common friends and co-purchasing logs. However the latent bundle information reflected by user-level sequential patterns and social roles remains unutilized. Therefore, there is a clear need to formally define the social roles, explicitly model role effect to personalized sequential behaviors and latent bundle relationships. More over, in order to predict user actions what is the next product to purchase, movie to watch, or place to visit comprehensively considering her his roles in social

© Springer International Publishing AG 2017
G. Li et al. (Eds.): KSEM 2017, LNAI 10412, pp. 371–379, 2017.
DOI: 10.1007/978-3-319-63558-3_31

network, it is challenging because putting user-level sequential patterns, social roles and latent bundle relationships, them together need careful consideration of both personal roles and sequential transitions.

In sequential recommendation domains, Markov chains have been studied by several earlier works, from investigating their strength at uncovering sequential patterns (e.g. [1,2]), and there is a line of work that employs (personalized) probabilistic Markov embeddings to model sequential data like play lists and POIs [3–5]. In a more recent paper [6] proposed to combine the strength of Markov chains at modeling the smoothness of subsequent actions and the power of Matrix Factorization at modeling personal preferences for sequential recommendation, which captures the sequential behaviors with which the next product to purchase. In addition, there has been a large body of work that models social networks for mitigating cold-start issues in recommender systems [7–11]. Trust-aware recommender systems have been widely studied, given that social trust provides an alternative view of user preferences other than item ratings [12–14]. However, users are not just two roles in reality and users with different roles often have different conformity tendency in a real life, and we introduce a role vector to control user behavior tendency.

In the paper, we develop the framework Personal Social Sequential Behaviors for Personalized Ranking (SSB-ranking) combining social influence from social network and personal sequential behaviors, which incorporates latent bundling features in social networking service environment for the task of personalized ranking on feedback datasets. By learning the social dimensions people consider when selecting products, the framework is able to alleviate cold start issues, help explain recommendations in terms of social signals, and produce personalized rankings that are more consistent with users preferences. In terms of latent bundle relationships intertwining with each product and the strength of sequential behaviors in the sparse real-world datasets, the framework naturally integrates sequential behaviors and hidden bundle relationships between products by learning a personalized weighting scheme over the sequence of items to characterize users.

2   Problem Formulation

Formally, let $U = \{u_1, u_2, \ldots, u_{|U|}\}$ represent the set of users and $I = \{i_1, i_2, \ldots, i_{|I|}\}$ the set of items. Each user u is associated with a sequence of actions S^u (e.g. items purchased by u, or places u has checked in): $S^u = (S_1^u, S_2^u, \ldots, S_{|S^u|}^u)$, where $S_k^u \in I$. The action history of all users is denoted by $S = (S^{u_1}, S^{u_2}, \ldots, S^{u_{|U|}})$. Users may connect to others in a social network. We define $T \in R^{U \times U}$ to indicate user-user social relations, where $T_{i,j} = 1$ means user u_i has a relation to u_j and zero otherwise. Notation is summarized in Table 1. The users with different roles in the social network can show different rating behaviors.

Table 1. Notations.

Notion	Description
U, I, u, i, k	User set, item set, a specific user, item, time step, positive item set of user u
S_t^u, I_u^+	The item user u interacted with at time step t
S^u	Action sequence of user u, $S^u = (S_1^u, S_2^u, \ldots, S_{|S^u|}^u)$
N_u	The neighborhoods of user u in social network
K, D, L	Dimension of latent factors, number of roles, order of Markov Chains
$p_u(j|i)$	Probability that user u chooses item j after item i
$\hat{p}_{u,i,t}$	Prediction that user u chooses item i at time step t
η^u	Personalized weighting vector, $\eta^u = (\eta_1^u, \eta_2^u, \ldots, \eta_L^u)$
γ_u, r_u, r_i	User u's bias, item i's bias, latent factors of user u, item i($K \times 1$)
θ_u	Social role of user u($D \times 1$)

Definition 1. *Role Vectors for Users:* We use role vectors to describe users profile. A user can play different roles while facing with different items. Here, we allocate a vector $\theta_u \in R^{D \times 1}$ to each user u, which θ_u^q represents the probability that user u belongs to role q. And D is the number of roles in the model ($\sum_{q=1}^{D} \theta_u^q = 1$).

3   The Framework

3.1   Modeling Markov Chains Incorporating Social Roles

While the Factorizing Personalized Markov Chains(FPMC) [6] only makes use of the collaborative data, without being aware of the underlying content of the items themselves. Such a formulation may suffer from cold item issues where there aren't enough historical observations to learn accurate representations of each item. Modeling the content of the items can provide auxiliary signals in cold-start settings and alleviate such issues.

In this section, we are interested in modeling personalized actions with role features. The role vector θ_u for user u adopted from multinomial distribution can been incorporated into the SSB-Ranking. Given the previous action sequence S^u of user u and the latent features r_{S^u} corresponding to items in sequence S^u of user u, the probability that user u, as role q in her/his social surroundings, will interact with item i in the next action is $p_u^q(i|S^u)$. And when taking into account various roles in social network, the final rating that user u will interact with item i in the next action is $p_u(i|S^u)$, where i is the choice u made at time step t and $|S^u| = t - 1$. Following this intuition, we propose to factorize the personalized Markov chain with the following formulation:

$$\hat{p}_u^q\left(i|S^u\right) = \gamma_u + \gamma_u + \theta_u^q\left\langle r_u + \sum_{l=1}^{L} r_{S_{t-l}^u}, r_i \right\rangle + (1-\theta_u^q)\frac{1}{|N_u|}\sum_{f\in N_u}\langle r_f, r_i\rangle$$

$$\hat{p}_u\left(i|S^u\right) = \sum_{q\in\theta_u}\hat{p}_u^q \tag{1}$$

where r_u and r_i are employed to capture user u's latent preferences and item i's latent properties respectively, N_u is the neighborhood of user u in the social information, and θ_u^q is the weight to balance the users own preference and conformity tendency corresponded with role q. $\left\langle \sum_{l=1}^{L} r_{S_{t-l}^u}, r_i \right\rangle$ brings together the personalized set Markov chains with the factorized transition cube results in the factorized personalized Markov chain.

3.2 Latent Bundle Relationships

While most recommender systems focus on analyzing personalized patterns influenced by social roles to provide personalized recommendations, another important problem is to understand relationships between products, in order to surface recommendations that are relevant to a given social context. In this section, we want to model the relationships that the transition of user u from sequential actions S^u to item i (at time step t) can be explained: in social environment, there are latent bundling relationships between item i and the items in sequence S^u of user u, and users' behaviors are influenced by the latent bundles. In other words, users will choice items that most relevant to the previous action sequence S^u according to her/his social role. We propose that each user is associated with a vector $\eta^u = \left(\eta_1^u, \eta_2^u, \ldots, \eta_{|S^u|}^u\right)$. The rationale behind this idea is that each of the previous L actions should contribute with different weights to the high-order smoothness. The transition of user u from item S_{t-1}^u at time step $t-1$ to item i at time step t in Eq.1 capture the long-term preferences of user u and temporary interest of user u. Following this intuition, we want the personalized weighting factor as:

$$\hat{p}_u^q\left(i|S^u\right) = \gamma_u + \gamma_u + \theta_u^q\left\langle r_u + \sum_{l=1}^{L} \eta_l^u r_{S_{t-l}^u}, r_i \right\rangle + (1-\theta_u^q)\frac{1}{|N_u|}\sum_{f\in N_u}\langle r_f, r_i\rangle$$

$$\hat{p}_u\left(i|S^u\right) = \sum_{q\in\theta_u}\hat{p}_u^q \tag{2}$$

$$where \quad \eta_l^u = \left\langle \varphi_u, \psi(r_i, r_{S_{t-l}^u})\right\rangle$$

$$\psi(r_i, r_{S_{t-l}^u}) = (r_{i,1} + r_{S_{t-l}^u,1}, r_{i,2} + r_{S_{t-l}^u,2}, \ldots, r_{i,k} + r_{S_{t-l}^u,k})$$

where $\langle\varphi_u, \psi\left(r_i, r_j\right)\rangle$ takes a positive value if i is related with $j \in S^u$. We want the latent factors associated with each product to be related for logistic regression in the sense that we are able to learn a logistic regressor parametrized by φ_u that predicts. by defining our features to be the element wise product between $r_{S_{t-l}^u}$ and r_i, we are saying that products are likely to be linked. The logistic vector φ_u for user u then determines which u's personal preference for different bundles.

3.3 Model Learning

For each user u and for each time step t, a sigmoid function $\sigma\left(\hat{p}_{u,t,S_t^u} - \hat{p}_{u,t,i}\right)$ ($\hat{p}_{u,t,\cdot}$ is a shorthand for the prediction in Eq. 2) is employed to characterize the probability that ground-truth item S_t^u is ranked higher than a negative item j given the model parameters Θ, $p(S_t^u >_{u,t} i|\Theta)$. Assuming independence of users and time steps, model parameters Θ are inferred by optimizing the following Maximum A Posteriori (MAP) estimation:

$$\arg\max_{\Theta} = \sum_{u \in U} \sum_{t=2}^{|S^u|} \sum_{i \neq S_t^u} \ln \sigma\left(\hat{p}_{u,t,S_t^u} - \hat{p}_{u,t,i}\right) + \ln p(\Theta) \qquad (3)$$

$p(\Theta)$ is a Gaussian prior over the model parameters. We adopt Stochastic Gradient Descent (SGD) to train parameters. First, it uniformly samples a user u from U as well as a time step t from $(S_1^u, S_2^u, \ldots, S_{|S^u|}^u)$. Next, a negative item $i \in I$ and $i \notin (S_1^u, S_2^u, \ldots, S_{|S^u|}^u)$ is uniformly sampled, which forms a training triple (u, t, i). Finally, the optimization procedure updates parameters in the following fashion:

$$\Theta \leftarrow \Theta + \tau(\sigma\left(\hat{p}_{u,t,S_t^u} - \hat{p}_{u,t,j}\right) \frac{\partial\left(\hat{p}_{u,t,S_t^u} - \hat{p}_{u,t,i}\right)}{\partial\Theta} - \lambda_\Theta\Theta) \qquad (4)$$

where τ is the learning rate and λ_Θ is a regularization hyperparameter.

4 Experiments

4.1 Dataset and Evaluation Metric

To evaluate the ability and applicability of SSB-ranking to handle different real-world scenarios, we include a spectrum of large datasets from different domains in order to predict actions ranging from the next product to purchase, next movie to watch, to next review to write, and next place to check-in. Note that these datasets also vary significantly in terms of user and item density (i.e., number of actions per user/item).

The group of large datasets is obtained obtained from Amazon.com[1] recently introduced by [15]. This is one of the largest datasets available that includes social information and time stamps spanning from May 1996 to July 2014. Each top-level category of products on Amazon.com has been constructed as an independent dataset by [15]. In this paper, we take four large categories including Movies, Books, clothing, and Electronics, shown in Table.2.

Evaluation Methodology: For each dataset, we use the two most widely used metric AUC (Area Under the ROC curve) the precision recall(@k).

Experimental Setting: To obtain ground-truth for pairs of latent bundling products we select two types behaviors from Amazon: 'Users who bought x also

---

[1] https://www.amazon.com/.

Table 2. Datasets statistics on Amazon.

Dataset	Items	Users	Action	Relationships *alse-brought*	Avg. *action-user*	Avg. *action-item*	Social relation
Movies	208 K	211 K	6.17 M	5.3 M	9.41	18.25	2.58 M
Electronics	2498 K	4.25 M	11.4 M	8.22 M	7.3	12.3	1.2 M
Books	2.73 M	8.2 M	25.9 M	16.5 M	5.88	6.41	7.85
Women's Clothing	838 K	1.82 M	14.5 M	10.61	5.66	7.23	7.58M

bought y and 'Users frequently bought x and y together', shown in Table 2. In addition to the obtained bundle relationships (assuming there are N pairs), we randomly generated a balanced dataset by sampling the same size of dataset without any bundling relationships. As a result, we have 2*N pairs of items with different subcategories. Note that and they share the same distribution over the items.

4.2 Baselines

- **BPR-MF:** [16] combines the strengths of the BPR framework, which implicitly treats non-observed interactions as negative feedback, and the efficiency of MF, and forms the basis of many state-of-the-art personalized methods.
- **FPMC:** [6] is a method that uses a Personalized Markov Chain for the sequential prediction task we are interested in.
- **FMC(Factorized Markov Chains):** factorizes the item-to-item transition matrix ($|I| \times |I|$) to capture the likelihood that an arbitrary user transitions from one item to another. Here we use a first-order Markov Chain as higher orders to avoid a state-space explosion.
- **SocTru:** [17] proposed a social CF model, which is a truster and trustee model on a twofold influence of trust propagation.

4.3 Performance and Quantitative Analysis

Error rates on both of all items and cold start sets, for all experiments are reported in Table 3 (Lower is better). For experiments on also bought

Table 3. Test errors $(1 - AUC)$ of the predictions on all items or cold start set using products on datasets of the Amazon.

Dataset	Setting	SocTru(%)	FMC(%)	FPMC(%)	BPR(%)	SSB Ranking	% impr. *SSB vs FPMC*	% impr. *SSB vs SocTru*
Movies	All items	24.4	24.5	23.02	26.69	**21.01**	8.73	16.13
	Cold start	28.51	29.2	28.21	31.81	**25.84**	8.4	8.57
Electronis	All items	22.47	23.24	21.92	24.01	**21.01**	4.2	6.54
	Cold start	27.5	28.98	27.12	29.55	**24.56**	9.43	11.05
Women's Clothes	All items	21.58	20.81	19.55	22.41	**18.23**	6.75	15.52
	Cold start	28.54	28.8	27.03	29.5	**24.15**	10.65	15.38
Books	All items	22.58	23.45	20.82	24.71	**18.87**	9.36	13
	Cold start	28.57	28.14	27.25	28.83	**23.25**	14.67	18.6

relationships, SSB-Ranking uses $K = 30$ dimensions for latent, $D = 5$ dimensions for role factors and $L = 5$ order of Markov Chains, and BPR, FPMC and FMC also use $K = 30$, $L = 5$. We make a few observations to explain and understand our findings as follows:

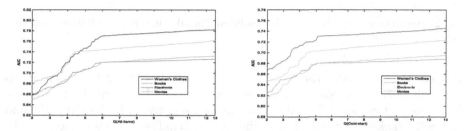

Fig. 1. AUC with the role number increasing for four datasets in terms of all items and cold start.

(1) **BPR** *vs* **SocTru** *vs* **SSB-Ranking:** BPR makes the worse than other methods whether for all items. This confirms our conjecture that only raw similarity of products and relations of users are inappropriate for our task to tackle sparse real-world datasets, and in order to learn the relationship dimensions among products apart from similarity and diverse roles of users, some sort of expressive transforms are needed for manipulating the raw features.

(2) **SocTru** *vs* **FPMC** *vs* **SSB-Ranking:** SocTru considers the influence of roles lost the information between products. FPMC is an inherent combination of BPR-MF and FMC, which makes it strong on modeling long-term and short-term dynamics respectively. SSB-Ranking is able to capture the latent bundle relationships by modeling users' 'also-brought' behaviors as well as gains from a fully personalized Markov chains with higher orders. As such, it beats FPMC in all settings significantly especially in related item recommending scenarios to capture personal style, and makes the better accurate than BPR and SocTru. SSB-Ranking outperforms BPR, SocTru, FMC and FPMC in all datasets where pure SSB-Ranking extracts meaningful factors.

Users' Roles in Social Environment: The learned parameter θ_u^q by SSB-Ranking represents the roles of user u in social network. Our model learns the mean value of each attribute for a role. Shown in Fig. 1, by using the users' roles for rating the related items, we are able to learn the social information for a person performed by the user's conversion between different roles, and we find SSB-Ranking not only learns the hidden bundles among products, but also discovers users will have a lot of roles in life and the impacts of social context for a user is a combination of multiple roles.

Latent Bundle Relationships: We want to examine whether the substitutes in ranking for the query are complementary to user's preference and needing.

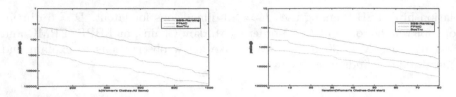

Fig. 2. Recall rates for Women's Clothes latent bundles in terms of all items and cold start.

A direct way to exhibit these dimensions extracted by Eq. 2 is to rank items highly that achieve maximal values for each user, seen Fig. 2 (results for movies, electronicsand women's clothes are similar and suppressed for brevity). We find that SSB-Ranking not only learns the hidden bundles, but also discovers the most relevant underlying social dimensions and maps items and users into the uncovered space.

5   Conclusion

Exploring the latent bundle relationships between items and social roles with users sequential patterns is challenging especially when it comes to large real-world datasets. In this paper, we tried to resolve the above issues by building sequential patterns to model correlative items and users' changing roles simultaneously. Empirically we evaluated our proposed framework on sequential patterns to test its ability to handle both roles of users and the hidden bundle relationships. Experimental results demonstrated the advantage of our approach over other baseline models.

Acknowledgments. This work is supported by NSFC(No.61170192) and the Fundamental Research Funds for the Central University for Student Program (XDJK2017D059 and XDJK2017D060).

References

1. Mobasher, B., Dai, H., Luo, T.: Using sequential and non-sequential patterns in predictive web usage mining tasks. In: ICDM, pp. 669–672. IEEE (2002)
2. Zimdars, A., Chickering, C., Meek, M.: Using temporal data for making recommendations. In: AI, pp. 580–588. MKPI (2001)
3. Chen, S., Moore, J.L., Turnbull, D., Joachims, T.: Playlist prediction via metric embedding. In: KDD, pp. 714–722 (2012)
4. Chen, S., Xu, J., Joachims, T.: Multi-space probabilistic sequence modeling. In: SIGKDD, pp. 865–873 (2013)
5. Feng, S., Li, X., Zeng, Y.: Personalized ranking metric embedding for next new POI recommendation. In: AI, pp. 2069–2075 (2015)
6. Rendle, S., Freudenthaler, C., Schmidt-Thieme, L.: Factorizing personalized Markov chains for next-basket recommendation. In: WWW, pp. 811–820. ACM (2010)

7. Chaney, A.J.B., Blei, D.M.: A probabilistic model for using social networks in personalized item recommendation. In: ACM Conference on Recommender Systems, pp. 43–50 (2015)
8. Guo, G., Zhang, J.: TrustSVD: Collaborative filtering with both the explicit and implicit influence of user trust and of item ratings (2015)
9. Ma, H., King, I.: Learning to recommend with social trust ensemble. In: International ACM SIGIR Conference on Research and Development in Information Retrieval, pp. 203–210 (2009)
10. Ma, H., Yang, H.: Sorec: social recommendation using probabilistic matrix factorization. In: ACM Conference on Information and Knowledge Management, CIKM 2008, Napa Valley, California, USA, October, pp. 931–940 (2008)
11. Zhao, T., Mcauley, J.: Leveraging social connections to improve personalized ranking for collaborative filtering. In: ACM International Conference on Information Knowledge Management, pp. 261–270 (2014)
12. Yao, J., He, W.: Modeling dual role preferences for trust-aware recommendation. In: International ACM SIGIR Conference on Research Development in Information Retrieval, pp. 975–978 (2014)
13. Fang, H., Bao, Y.: Leveraging decomposed trust in probabilistic matrix factorization for effective recommendation. In: Twenty-Eighth AAAI Conference on Artificial Intelligence, pp. 30–36 (2014)
14. Guo, G., Zhang, J.: Leveraging multiviews of trust and similarity to enhance clustering-based recommender systems. Knowl.-Based Syst. **74**(1), 14–27 (2015)
15. McAuley, J., Targett, C., Shi, Q.: Image-based recommendations on styles and substitutes. In: SIGIR, pp. 43–52. ACM (2015)
16. Rendle, S., Freudenthaler, C., Gantner, Z.: BPR: Bayesian personalized ranking from implicit feedback. In: AI, pp. 452–461. AUAI Press (2009)
17. Yang, B., Lei, Y.: Social collaborative filtering by trust. In: International Joint Conference on Artificial Intelligence, pp. 2747–2753 (2013)

Trust-Aware Recommendation in Social Networks

Yingyuan Xiao[1,2], Zhongjing Bu[2(✉)], Ching-Hsien Hsu[3],
Wenxin Zhu[2], and Yan Shen[2]

[1] Tianjin Key Laboratory of Intelligence Computing and Novel Software
Technology, Tianjin, China
[2] Tianjin University of Technology, Tianjin 300384, China
1635437949@qq.com
[3] Chung Hua University, Hsinchu 30012, Taiwan

Abstract. With the popularity of online social networks, social network information is becoming increasingly important to improve recommendation effectiveness of the existing recommender systems. In this paper, we propose an improved trust-aware recommendation approach, called *TRA*. *TRA* constructs a new social trust matrix based on users' trust relationships derived from online social networks to alleviate the problem of data sparsity, and meanwhile naturally fuses users' preferences and their trusted friends' favors together by means of probability matrix factorization. The experimental results show that *TRA* performs much better than the state-of-the art recommendation approaches.

Keywords: Recommender systems · Trust-Aware recommendation · Social networks · Social trust matrix · Probability matrix factorization

1 Introduction

Due to the exponential growth of information, it is difficult for people to find items they need. Nowadays, recommender systems are becoming more and more important for solving information overload problem. The traditional recommender systems usually leverage the sparse user-item rating matrix to predict users' preferences, while ignore the social relationships among users. However, the fact is the social relationships, such as trust relationship between users, play a very important role in influencing user selection. For example, people are typically used to accept an item recommended by their trusted friends. Online social networks are virtual communities, which allow people to connect and interact with each other, and build the corresponding social relationships to simulate the relationships of the real world. With the popularity of online social networks, social network information (e.g. trust relationship) is becoming increasingly important to improve recommendation effectiveness of the existing recommender systems.

In this paper, we give a fresh understanding to the relationships between users. We incorporate the user-user trust relationship with the user-user similarity to construct a new hybrid information matrix. Further, we employ a probabilistic framework to fuse users and their trusted friends' preferences together. The experimental results on a large

© Springer International Publishing AG 2017
G. Li et al. (Eds.): KSEM 2017, LNAI 10412, pp. 380–388, 2017.
DOI: 10.1007/978-3-319-63558-3_32

Epinions dataset show that our approach outperforms the state-of-the-art matrix factorization and trust relationship based recommendation methods.

2   Related Work

The collaborative filtering (CF) approach to recommenders has enjoyed much interest and progress [1]. The basic idea of CF is that if users shared the same interests in the past they will also have similar tastes in the future. CF methods utilize users' behaviors to infer a target user's preference for a particular item. The two main techniques of CF are the neighborhood approach [2–4] and matrix factorization technique [5, 6]. Neighborhood methods focus on relationships between items or users. Matrix factorization methods transform both items and users to the same latent factor space.

The traditional CF methods ignore the social trust relationships between users. To improve the recommendation accuracy and solve the data sparsity problem, users' social network should be taken into consideration. Liu and Hong [7] exploit to use social network information in a neighborhood approach. In [8], authors build factorization models that incorporate users' social network, action, tag, profile and items' taxonomy information. Although their method takes social networks into consideration, the authors' core concern is the relationship between user and object. Ma et al. [9] introduce social regularization to represent the social constraints on recommender systems and propose a matrix factorization framework with social regularization. Massa and Avesani [10, 11] exploit the trust-aware recommender systems. They replace the similarity finding process with the use of a trust metric. Ma et al. [12] propose a factor analysis approach based on probabilistic matrix factorization to solve the data sparsity and poor prediction accuracy problems by employing both users' social network information and rating records. The disadvantage of the approach is that the authors ignore the fact that the indirect trust relationship between users can also improve the accuracy of recommendation. In [13], authors propose a novel probabilistic factor analysis framework, called RSTE, which fuses the users' tastes and their trusted friends' favors together. Although RSTE employ the trusted friends' opinions in the social trust network to make recommendations for the users, it does not consider the possible propagation of trusts between various users.

3   Preliminary

We use U to denote the set of users and P to denote the set of items. Let u_i denote a user in U, p_j denote an item in P, and r_{ij} represent the rating of user u_i for item p_j. The trust-aware recommendation in this paper involves three central elements: user-item rating matrix, users' trust network, and probability matrix factorization (PMF) [14].

A user-item rating matrix R is a $|U| \times |P|$ matrix, and each entry r_{ij} of R represents the rating of user u_i for item p_j. A users' trust network is modeled as a directed edge-weighted graph $G = (E, U, W)$, where U is a set of vertexes, corresponding to the set of users of an online social network, E is a set of directed edges between two vertexes in U, and W is the set of weight values of all directed edges. If there is an edge

e_{ij} with weight w_{ij} from vertex u_i to vertex u_j, we say a direct trust relationship existing from u_i to u_j, denoted by $u_i \xrightarrow{w_{ij}} u_j$, where w_{ij} represents the direct trust degree from u_i to u_j, and is defined in the range of (0, 1]. The greater w_{ij} means u_i more trusts u_j. We use $Out(u_i)$ to denote the set of the users trusted by u_i, and $In(u_i)$ to represent the set of the users trusting u_i. We can infer that $|Out(u_i)|$ is equal to the out-degree of vertex u_i in graph G and $|In(u_i)|$ equal to the in-degree of vertex u_i in graph G by the definitions of $Out(u_i)$, $In(u_i)$ and graph G.

The probability matrix factorization (PMF) performs well on the large, sparse, and very imbalanced dataset. Suppose we have M items, N users, and integer rating values. Let column vectors U_i and P_j representing user-specific and item-specific latent feature vectors respectively. Since PMF performance is measured by computing the root mean squared error ($RMSE$) on the test dataset, a probability linear model with Gaussian observation noise is adopted. The conditional distribution over the observed ratings is defined as Eq. (1).

$$p\left(R|U, P, \sigma^2\right) = \prod_{i=1}^{N} \prod_{j=1}^{M} \left[\mathcal{N}\left(r_{ij}|g(U_i^T P_j), \sigma^2\right)\right]^{I_{ij}} \tag{1}$$

Where $\mathcal{N}(x|\mu, \sigma^2)$ is the probability density function of the Gaussian distribution with mean μ and variance σ^2, I_{ij} is the indicator function that is equal to 1 if user u_i rated item p_j and equal to 0 otherwise, and $g(x) = \frac{1}{1 + exp(-x)}$ is the logistic function, which makes it possible to bound the range of $U_i^T P_j$ within the range [0, 1].

4   The Trust-Aware Recommendation Approach

In this section, we first construct a novel social trust matrix by fusing direct and indirect trust relationships between users. Then, we propose an improved trust-aware recommendation approach, called *TRA*, on the basis of the existing RSTE [13].

4.1   Social Trust Matrix

Suppose we have a users' trust network graph $G = (E, U, W)$, where the vertex set $U = \{u_i\}_{i=1}^{N}$ represents all users in a social network, the edge set E denotes all direct trust relationships between users, and the weight set W represents all weights associated with the corresponding edges. In RSTE, authors define a user-user matrix $S = \{s_{ik}\}_{N \times N}$ as the social network matrix, where $s_{ik} = w_{ik}$ if existing an edge e_{ik} from u_i to u_k in graph $G = (E, U, W)$, $s_{ik} = 0$ otherwise. Obviously, the above-mentioned social network matrix is very sparse because it only considers the direct trust relationships between users. In the real world, however, trust propagation is ubiquitous. For example, if user A trusts user B and user B trusts user C, user A will also trust user C in general. That is, except for the explicit direct trust relationship, there are some latent trust relationships between non-adjacent users in graph $G = (E, U, W)$, which are important for improving the accuracy of recommendation methods. Motivated by this fact, we define three kinds of indirect trust relationships and introduce them into the social network matrix. On this basis, we define a novel social trust matrix.

Given a users' trust network graph $G = (E, U, W)$, let $u_i, u_k \in U$ and there is not the direct trust relationship between u_i and u_k. In the following, we formally define the three kinds of indirect trust relationships.

Definition 1. If $\exists u_j \in U \ s.t. \ u_i \xrightarrow{w_{ij}} u_j \wedge u_j \xrightarrow{w_{jk}} u_k$, we say there is the indirect trust relationship from u_i to u_k, denoted by $u_i \xrightarrow{c1_{ik}} u_k$, where $c1_{ik}$ represents the weight of the indirect trust relationship from u_i to u_k.

Definition 1 reflects the transitivity of trust relationships.

Definition 2. If $\exists u_j \in U \ s.t. \ u_i \xrightarrow{w_{ij}} u_j \wedge u_k \xrightarrow{w_{kj}} u_j$, we say there is the indirect trust relationship between u_i and u_k, denoted by $u_i \xrightarrow{c2_{ik}} u_k$ and $u_k \xrightarrow{c2_{ki}} u_i$, where $c2_{ik}$ and $c2_{ki}$ respectively represents the weights of the indirect trust relationships from u_i to u_k and from u_k to u_i.

Definition 3. If $\exists u_j \in U \ s.t. \ u_j \xrightarrow{w_{ji}} u_i \wedge u_j \xrightarrow{w_{jk}} u_k$, we say there is the indirect trust relationship between u_i and u_k, denoted by $u_i \xrightarrow{c3_{ik}} u_k$ and $u_k \xrightarrow{c3_{ki}} u_i$, where $c3_{ik}$ and $c3_{ki}$ respectively represents the weights of the indirect trust relationships from u_i to u_k and from u_k to u_i.

Definitions 2 and 3 are based on the reasonable assumption: two users, which trust or are trusted by a common user, often have a kind of latent trust relationship.

Further, we formally define the weights of these indirect trust relationships according to the theory of trust propagation in the real world. In the following definitions, we use $\bar{w}$ to represent the average weight of all direct trust relationships, i.e., $\bar{w} = \frac{1}{|E|} \sum w_{ij}$, where E is the edge set in the users' trust network graph G.

Definition 4. Let $u_i \xrightarrow{c1_{ik}} u_k$, then $c1_{ik}$ is defined by the equation: $c1_{ik} = \frac{|Out(u_i) \cap In(u_k)|}{\sqrt{|Out(u_i)| \times |In(u_k)|}} \times \bar{w}$.

Definition 5. Let $u_i \xrightarrow{c2_{ik}} u_k$, then $c2_{ik}$ is defined by the equation: $c2_{ik} = \frac{|Out(u_i) \cap Out(u_k)|}{\sqrt{|Out(u_i)| \times |Out(u_k)|}} \times \bar{w}$.

Definition 6. Let $u_i \xrightarrow{c3_{ik}} u_k$, then $c3_{ik}$ is defined by the equation: $c3_{ik} = \frac{|In(u_i) \cap In(u_k)|}{\sqrt{|In(u_i)| \times |In(u_k)|}} \times \bar{w}$.

Considering the fact that direct trust relationships often have a greater impact on users than indirect trust relationships, we use $\bar{w}$ as a slack factor to adjust the weight values of indirect trust relationships in the above Definitions 4–6.

Now, we give the definition of the social trust matrix.

Definition 7. Given a users' trust network graph $G = (E, U, W)$, the social trust matrix is a user-user trust relationship matrix, denoted by $Q = \{q_{ik}\}_{N \times N}$, where

$$q_{ik} = \begin{cases} w_{ik}, & \exists e_{ik} \in E \\ cx_{ik}, & \exists u_i \xrightarrow{cx_{ik}} u_k, \ cx_{ik} = c1_{ik}, c2_{ik}, c3_{ik}. \\ 0, & \text{otherwise} \end{cases}$$

4.2 The Trust-Aware Recommendation Approach

In the proposed trust-aware recommendation approach (*TRA*), we first build the user-user trust relationship matrix $Q = \{q_{ik}\}_{N \times N}$ and the user-user similarity matrix $S = \{s_{ik}\}_{N \times N}$ based on the users' trust network graph $G = (E, U, W)$ and the user-item rating matrix R, respectively. The above Definition 7 gives the way of building $Q = \{q_{ik}\}_{N \times N}$. We employ Pearson Correlation Coefficient [1] to compute the similarity between two users. Specifically, given two users u_i and u_k, the similarity between u_i and u_k, i.e., s_{ik}, is computed by Eq. (2).

$$s_{ik} = \frac{\sum_{j \in P} (r_{ij} - \overline{r_i})(r_{kj} - \overline{r_k})}{\sqrt{\sum_{j \in P} (r_{ij} - \overline{r_i})^2} \sqrt{\sum_{j \in P} (r_{kj} - \overline{r_k})^2}} \tag{2}$$

In Eq. (2), P represents the set of items, r_{ij} represents the rating of user u_i for item j, and $\overline{r_i}$ denote the average ratings of u_i.

Further, *TRA* fuses the user-user trust relationship matrix $Q = \{q_{ik}\}_{N \times N}$ with the user-user similarity matrix $S = \{s_{ik}\}_{N \times N}$ to construct a new hybrid information matrix $H = \{h_{ik}\}_{N \times N}$, where $h_{ik} = \frac{2s_{ik} \times q_{ik}}{s_{ik} + q_{ik}}$.

Finally, *TRA* leverages the modified PMF to calculate rating predictions based on $H = \{h_{ik}\}_{N \times N}$ by following RSTE. Specifically, similar to RSTE, *TRA* models the joint conditional distribution of latent user and item feature matrices U and P over the observed ratings and the hybrid information matrix H as:

$$p(U, P | R, H, \sigma^2, \sigma_U^2, \sigma_P^2)$$

$$= \prod_{i=1}^{N} \prod_{j=1}^{M} \left[\mathcal{N} \left(r_{ij} \middle| g \left(\alpha U_i^T P_j + (1 - \alpha) \sum_{k \in H(i)} h_{ik} U_k^T P_j \right), \sigma^2 \right) \right]^{I_{ij}^R}$$

$$\times \prod_{i=1}^{N} \mathcal{N}(U_i | 0, \sigma_U^2 I) \times \prod_{j=1}^{M} \mathcal{N}(P_j | 0, \sigma_P^2 I) \tag{3}$$

The log of the posterior distribution for the recommendation is given by the following Eq. (4).

$$\ln p\left(U, P | R, H, \sigma^2, \sigma_U^2, \sigma_P^2\right)$$

$$= -\frac{1}{2\sigma^2} \sum_{i=1}^{N} \sum_{j=1}^{M} I_{ij}^R \left(r_{ij} - g\left(\alpha U_i^T P_j + (1-\alpha) \sum_{k \in H(i)} h_{ik} U_k^T P_j\right)\right)^2$$

$$- \frac{1}{2\sigma_U^2} \sum_{i=1}^{N} U_i^T U_i - \frac{1}{2\sigma_V^2} \sum_{j=1}^{M} P_j^T P_j - \frac{1}{2}\left(\sum_{i=1}^{N} \sum_{j=1}^{M} I_{ij}^R\right) \ln \sigma^2$$

$$- \frac{1}{2}\left(Nl \ln \sigma_U^2 + Ml \ln \sigma_P^2\right) + C \tag{4}$$

Maximizing the log-posterior over two latent features is equivalent to minimizing the following sum-of-squared-errors objective functions:

$$L(R, H, U, P) = \frac{1}{2} \sum_{i=1}^{N} \sum_{j=1}^{M} I_{ij}^R \left(r_{ij} - g\left(\alpha U_i^T P_j + (1-\alpha) \sum_{k \in H(i)} h_{ik} U_k^T P_j\right)\right)^2$$

$$+ \frac{\sigma_R^2}{2\sigma_U^2} \times \sum_{i=1}^{N} \sum_{i=1}^{N} U_i^T U_i + \frac{\sigma_R^2}{2\sigma_P^2} \times \sum_{j=1}^{M} P_j^T P_j \tag{5}$$

In Eqs. (3)–(5), the difference of *TRA* and RSTE is that *TRA* uses hybrid information matrix $H = \{h_{ik}\}_{N \times N}$ to replace the social network matrix $S = \{s_{ik}\}_{N \times N}$ of RSTE.

We can find a minimum of the objective function $L(R, H, U, P)$ by performing stochastic gradient descent in U_i and P_j. According to stochastic gradient descent, first, $\frac{\partial L}{\partial U_i}$ and $\frac{\partial L}{\partial P_j}$ are calculated, respectively, and then, the following iterations are performed:

$$U_i = U_i - \eta \frac{\partial L}{\partial U_i} \tag{6}$$

$$P_j = P_j - \eta \frac{\partial L}{\partial P_j} \tag{7}$$

Where η is a step size (learning rate). Now, we can compute a rating prediction $\hat{r}_{ij}$ for user u_i giving to item p_j by the following Eq. (8).

$$\hat{r}_{ij} = \alpha U_i^T P_j + (1-\alpha) \sum_{k \in H(i)} h_{ik} U_k^T P_j \tag{8}$$

5   Experimental Evaluation

In this section, we evaluate the performance of *TRA* through extensive experiments. Similar to the literatures [12, 13], the performance metrics of concern in this paper are the *Mean Absolute Error* (*MAE*) and the *Root Mean Square Error* (*RMSE*). We choose Epinions as the data source for our experiments. We first evaluate the impact of parameter α and dimension d on the performance of *TRA*, and then we compare *TRA* with the state-of-the-art recommendation methods. For *TRA*, the parameter α makes a balance between the user-item rating matrix R and the hybrid information matrix H. α decides how much *TRA* should believe the user-item rating matrix R and the hybrid information matrix H. If $\alpha = 1$, *TRA* degrades into an ordinary probability matrix factorization.

Based on the past experimental experiences, the percentage of training set is set at 80% or 90% to calculate *MAE* and *RMSE* in our experiments. We assume the latent factor vector dimensionality $d = 20$ and the step size $\eta = 40$ in evaluating the impact of parameter α on the performance of *TRA*.

Figure 1 shows the impact of parameter α on *MAE* at $d = 20$ and $\eta = 40$. We can see from Fig. 1 that no matter using 80% or 90% training data *MAE* decreases at first as α increases, but when α is greater than a certain threshold (such as 0.4) *MAE* increases with α. This result is not surprising because α balances the influence from the user-item rating matrix R and the hybrid information matrix H. Obviously, too big or too small α is bad for fusing both R and H to make recommendations. Figure 2 shows the impact of parameter α on *RMSE* at $d = 20$ and $\eta = 40$. As we expect, no matter using 80% or 90% training data *RMSE* decreases at first as α increases, but when α is greater than a certain threshold (i.e., 0.4) *RMSE* increases with α. The reason is the same as the one for Fig. 1. Figures 3 and 4 depict the impact of dimensionality d on *MAE* and *RMSE* at $\alpha = 0.4$ and $\eta = 40$. As shown in Figs. 3 and 4, we can learn that no matter using 80% or 90% training data, both *MAE* and *RMSE* decrease (prediction accuracy increases) as d increases. We can also observe from Figs. 3 and 4 that the downward trend of *MAE* and *RMSE* becomes flat when d is greater than a certain threshold. In order to show the performance improvement of *TRA*, we compare *TRA* with the following four methods: (1) BS: this is a baseline method, which purely leverage direct trust relationships between users to make recommendations; (2) NMF: the method uses non-negative matrix factorization to make recommendations, only with the help of user-item matrix [6]; (3) PMF: i.e., probability matrix factorization [14] and (4) RSTE: the method introduces the direct trust relationships between users [13]. In the following experiments, the experimental parameter setting is as follows: $\alpha = 0.4$, $d = 60$ and $\eta = 40$. Similar to the previous experiments, we randomly select 80% or 90% of the ratings from Epinions dataset as the training data to predict the remaining ratings.

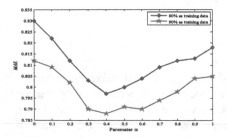

Fig. 1. Impact of α on *MAE*

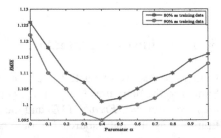

Fig. 2. Impact of α on *RMSE*

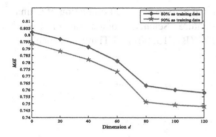

Fig. 3. Impact of d on *MAE*

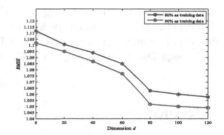

Fig. 4. Impact of d on *RMSE*

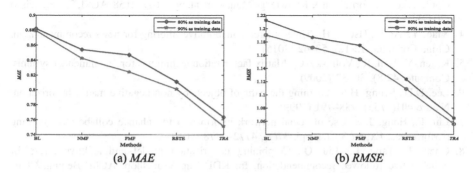

(a) *MAE*   (b) *RMSE*

Fig. 5. Performance comparison of different methods

Figure 5 shows the performance comparison of *TRA* with the above-mentioned four methods on *MAE* and *RMSE*. As we expect, *TRA* evidently outperforms the other four methods on both *MAE* and *RMSE*. This is because *TRA* makes a reasonable fusion between the user-item rating matrix and the hybrid information matrix for the recommendations. This result also demonstrates *TRA* and RSTE can provide a higher quality recommendation than BS, NMF and PMF, which only leverage the single matrix. In addition, the reason that RSTE is not as good as *TRA* is RSTE neglects the impact of indirect trust relationships on the recommendation accuracy.

6   Conclusion

In this paper, we introduce the trust propagation and diffusion mechanism and propose a new trust measure to extend the trust domain. On the basis of this, we propose an improved trust-aware recommendation approach, which consider both direct trust relationships and indirect trust relationships between users. The experimental results show that our approach outperforms the state-of-the-art matrix factorization and trust relationship based recommendation methods.

Acknowledgment. This work is supported by the National Nature Science Foundation of China (61170174), Major Research Project of National Nature Science Foundation of China (91646117), Natural Science Foundation of Tianjin (17JCYBJC15200) and Tianjin Science and Technology Correspondent Project (16JCTPJC53600).

References

1. Ricci, F., Rokach, L., Shapira, B., Kantor, P.B.: Recommender Systems Handbook. Springer, New York (2010)
2. Su, X., Khoshgoftaar, T.M.: A survey of collaborative filtering techniques. Adv. Artif. Intell. **2009**(12), 4 (2009)
3. Chatzis, S.: Nonparametric Bayesian multitask collaborative filtering. In: ACM International Conference on Information & Knowledge Management, pp. 2149–2158. ACM, San Francisco (2013)
4. Xiao, Y., Ai, P., Hsu, C.H.: Time-ordered collaborative filtering for news recommendation. China Commun. **12**(12), 53–62 (2015)
5. Koren, Y., Bell, R., Volinsky, C.: Matrix factorization techniques for recommender systems. Computer **42**(8), 30–37 (2009)
6. Lee, D.D., Seung, H.S.: Learning the parts of objects by non-negative matrix factorization. Nature **401**(6755), 788–791 (1999)
7. Liu, F., Hong, J.L.: Use of social network information to enhance collaborative filtering performance. Expert Syst. Appl. **37**(7), 4772–4778 (2010)
8. Chen, T., Tang, L., Liu, Q.: Combining factorization model and additive forest for collaborative followee recommendation. In: KDD-Cup Workshop. ACM, Beijing, China (2012)
9. Ma, H., Zhou, D., Liu, C.: Recommender systems with social regularization. In: Forth International Conference on Web Search and Web Data Mining, pp. 287–296, Hong Kong, China (2011)
10. Massa, P., Avesani, P.: Trust-aware collaborative filtering for recommender systems. In: Meersman, R., Tari, Z. (eds.) OTM 2004. LNCS, vol. 3290, pp. 492–508. Springer, Heidelberg (2004). doi:10.1007/978-3-540-30468-5_31
11. Massa, P., Avesani, P.: Trust-aware recommender systems. In: ACM Conference on Recommender Systems, pp. 17–24. ACM, Minnesota, USA (2007)
12. Ma, H., King, I., Lyu, M.R.: Learning to recommend with social trust ensemble. In: International ACM SIGIR Conference on Research and Development in Information Retrieval, pp. 203–210. ACM, Boston, MA, USA (2009)
13. Ma, H., Yang, H., Lyu, M.R.: SoRec: social recommendation using probabilistic matrix factorization. In: ACM Conference on Information and Knowledge Management, pp. 931–940. California, USA (2008)
14. Salakhutdinov, R., Mnih, A.: Probabilistic matrix factorization. In: International Conference on Machine Learning, pp. 880–887. ACM, Corvallis, Oregon, USA (2007)

Connecting Factorization and Distance Metric Learning for Social Recommendations

Junliang Yu[1,2], Min Gao[1,2]([✉]), Yuqi Song[1,2], Zehua Zhao[1,2], Wenge Rong[3],
and Qingyu Xiong[1,2]

[1] School of Software Engineering, Chongqing University, Chongqing 400044, China
{yu.jl,gaomin,songyq,zh.zhao,xiong03}@cqu.edu.cn
[2] Key Laboratory of Dependable Service Computing in Cyber Physical Society
(Chongqing University), Ministry of Education, Chongqing 400044, China
[3] School of Computer Science and Engineering, Beihang University,
Beijing 100191, China
w.rong@buaa.edu.cn

Abstract. Social relations can help to relieve the dilemmas called cold start and data sparsity in traditional recommender systems. Most of existing social recommendation methods are based on matrix factorization, which has been proven effective. In this paper, we introduce a novel social recommender based on the idea that distance reflects likability. It aims to make users in recommender systems be spatially close to their friends and items they like, and be far away from items they dislike by connecting factorization model and distance metric learning. In our method, the positions of users and items are decided by the ratings and social relations jointly, which can help to find appropriate locations for users who have few ratings. Finally, the learnt metric and locations are used to generate understandable and reliable recommendations. The experiments conducted on the real-world dataset have shown that, compared with methods only based on factorization, our method has advantages on both interpretability and accuracy.

Keywords: Social recommendations · Metric learning · Collaborative filtering · Matrix factorization

1 Introduction

Traditional Recommender systems usually suffer from the problem of cold-start. With the growing popularity of the online social platform, social recommender systems emerged [1]. These systems merge both rating information and social relations into recommendation, and consider that users can be affected by their friends in decision making. Therefore, the preference of users having few ratings can be inferred from that of their friends. It has been proven that social recommender systems can generate more accurate recommendations especially for so-called cold start users [2].

© Springer International Publishing AG 2017
G. Li et al. (Eds.): KSEM 2017, LNAI 10412, pp. 389–396, 2017.
DOI: 10.1007/978-3-319-63558-3_33

Most of existing recommender systems are based on matrix factorization [3], a basic model which has been extensively used for its scalability and efficiency. However, the meaning of latent factors are still ambiguous since the training process of matrix factorization is like a black box.

In this paper, we proposed a novel method having good interpretability, which is based on matrix factorization as well. The principle of the proposed method is that the distance reflects likability. However, in our model we use Mahalanobis distance to replace Euclidean distance to measure the gap since we consider that all dimensions in the low dimensional space are correlative. Therefore, distance metric is incorporated in our method. As an important topic, distance metric learning [4] has been applied in many domains, including information retrieval, supervised classification, and clustering. To the best of our knowledge, few work has combined distance metric learning and collaborative filtering. Different from the process of the general distance metric learning, our model needs to train the samples (latent factors) and the distance metric at the same time, which makes the model scalable. Finally, all users and items are embedded in a unified space. Users are spatially close to their friends and their liked items, and be far away from their disliked items. The distance between users and items can be used to generate recommendations. The experiments conducted on real-world dataset show that our method significantly improves the quality of social recommendations.

2 Social Recommender Connecting Factorization and Distance Metric Learning

In this section we put forward a **Social** recommendation model connecting **F**actorization and **D**istance metric learning called **SocialFD**.

2.1 Problem Definition

In social recommender systems, users can be defined as the set $\mathbf{U} = \{u_1, \ldots u_m\}$, and items can be defined as the set $\mathbf{I} = \{i_1, \ldots i_n\}$. Ratings given by users on items are marked with the matrix $\mathbf{R} = [r_{u,i}]_{m \times n}$, and $r_{u,i}$ denotes the rating from user u on item i. In a social network, each user has some friends denoted by the vector N_u. Regarding the edges in the social graph as trust statements which are real numbers in $[0, 1]$, we mark all trust statements from users with a adjacent matrix $\mathbf{T} = [t_{u,v}]_{m \times m}$. The task of a social recommender can be summarized as follows: given a user u and an item i, using the known information in $\mathbf{R}$ and $\mathbf{T}$ to predict the $r_{u,i}$.

2.2 Distance Metric Learning

Distance metric learning is crucial in real-world application. Generally, the training samples of distance metric learning will be cast into pairwise constraints. The target of distance metric learning is to learn a distance metric subjected to a given set of constraints in a global sense. Let the distance metric denoted by matrix $A \in \mathbb{R}^{k \times k}$, and the distance between any two data points x and y expressed by

$$d_A^2(x, y) = \|x - y\|_A^2 = (x - y)A(x - y)^T. \tag{1}$$

In Eq. 1, A has to be a positive semi-definite matrix to keep the distance non-negative and symmetric. The global optimization problem with constraints can be stated as

$$\min_{A \in \mathbb{R}^{k \times k}} \sum_{(x,y) \in S} \|x - y\|_A^2,$$

$$s.t. \quad A \succeq 0, \quad \sum_{(x,y) \in D} \|x - y\|_A^2 \geq \beta, \tag{2}$$

where S denotes the set of equivalent constraints in which x and y belong to the same class, and D denotes the set of inequivalent constraints in which x and y belong to different classes, and β is a constant to restrict the minimum distance between data points in different classes.

2.3 The SocialFD Model

The inspiration for SocialFD is that distance reflects likability. Different from those in the general distance metric learning, samples in our models are the latent factors which are not prepared at first, and labels are classified into two types (like and dislike) according to the rating expressed by users. Note that, we consider if a user rate an item with a higher score, it shows the item is positively rated; on the contrary, if a user rate an item with a lower score, it shows the item is negatively rated. And the sets of pairwise constraints are constructed as follows: given a user and an item, if the user positively rated the item, the pair will be distributed to the set of equivalent constraints; if the user negatively rated the item, they will be distributed to the set of inequivalent constraints.

To reach the goal of our model, we need to train the latent factors and the distance metric simultaneously. Firstly, we initialize two k-rank matrices filled with random values denoting the user latent matrix and the item latent matrix respectively, and defines a matrix $\in \mathbb{R}^{k \times k}$ to be the distance metric. The Mahalanobis distance between users and items can be calculated by the inner products of the differences of latent factors and the distance metric. During the training stage, constraints are imposed to guarantee that users should be spatially close to their friends and their liked items, and be far away from their disliked items. And the positions of users and items are decided by the ratings and social relations jointly. Therefore, if some users are lack of ratings, their social connections can help to find the best locations for them. Finally, the obtained latent factors can be interpreted as coordinates in the low dimensional space, and the distance calculated can be used to generate understandable recommendations.

According to the description above, in SocialFD, the predicted rating is defined as

$$\hat{r}_{ui} = \mu + b_u + b_i - (x_u - y_i)A(x_u - y_i)^T \tag{3}$$

where μ represents the overall average rating, b_u and b_i indicate the deviations of user u and item i, x_u and y_i denote the latent factors of user u and item i respectively, $A \in \mathbb{R}^{k \times k}$, and k is the same as the dimension of x and y, which is much less than the dimension of the primitive rating matrix. Instead of learning a positive semi-definite matrix A, $H \in \mathbb{R}^{k \times k}$ can be learnt with $A = HH^T$. However, H does not need to be positive semi-definite, which makes the problem can be solved with generic approaches. Hence, we can learn latent factors and the distance metric by solving the following optimization problem:

$$
\min_{H,x,y} \mathfrak{L} = \frac{1}{2} \sum_{r_{ui} \in \mathbf{R}} (r_{ui} - \mu - b_u - b_i + (x_u - y_i)HH^T(x_u - y_i)^T)^2
$$
$$
+ \frac{\alpha}{2} \sum_{(u,i) \in P} \|x_u - y_i\|_A^2 + \frac{\eta}{2} \sum_{u \in \mathbf{U}} \sum_{v \in N_u} \|x_u - x_v\|_A^2 \tag{4}
$$
$$
+ \frac{\alpha}{2} \sum_{(u,i) \in N} (\beta - min(\|x_u - y_i\|_A^2, \beta)) + \frac{\lambda}{2}(b_u^2 + b_i^2)
$$

where P is the set of pairs contains user u and his positively rated items, N is the set of pairs contains user u and his negatively rated items, the middle three terms are constraints used to adjust the distance into an appropriate range, λ controls the magnitudes of the biases, and α and η are algorithmic parameters to control the influence of the constraints.

A local minimum of the objective function given by Eq. 4 can be found by performing gradient descent in b_u, b_i, x_u, y_i, H,

$$
\frac{\partial \mathfrak{L}}{\partial b_u} = \lambda b_u - e_{ui} \qquad \frac{\partial \mathfrak{L}}{\partial b_i} = \lambda b_i - e_{ui}
$$
$$
\frac{\partial \mathfrak{L}}{\partial y_i} = -(e_{ui} \pm \alpha)(x_u - y_i)W
$$
$$
\frac{\partial \mathfrak{L}}{\partial x_u} = (e_{ui} \pm \alpha)(x_u - y_i)W - \eta \sum_{v \in N_u} (x_u - x_v)W \tag{5}
$$
$$
\frac{\partial \mathfrak{L}}{\partial H} = (e_{ui} \pm \alpha)H(x_u - y_i)^T(x_u - y_i) + \eta \sum_{v \in N_u} H(x_u - x_v)^T(x_u - x_v)
$$

where e_{ui} is the gap between the observed rating and the predicted rating, γ is the learning rate, $W = (HH^T + H^TH)$ and the exact operator of the plus-minus sign is determined by the sign of the term related to distance in Eq. 4.

3   Experimental Results

3.1   Experimental Setup

The real-world dataset Douban [5] is used in our experiments. This dataset contains 894,887 ratings (1–5 scale) rated by 2,848 users on 39,586 items, and 35,770 trust statements.

In our experiments, the Root Mean Square Error (RMSE) is chosen to measure the prediction error of all methods. Besides, we also use ranking-based metrics: Precision@50, Recall@50, and F1@50, to measure the quality of the recommendation list, which is more important to users. In order to show the performance improvement of SocialFD, we compare our method with the following methods: PMF [6], SoRec [7], SocialMF [8] and RSTE [9]. In all experiments, we set the parameters of these methods according to their best performance.

3.2 Performance for Predicting Missing Ratings

Predicting missing ratings in the rating matrix is the primary goal of most recommendation methods. In this section, we compare the performance of SocialFD with that of the baselines. We set the step size $\gamma = 0.05$, $\alpha = 0.2$, $\beta = 2$, and $\eta = 0.1$ for SocialFD. And the reduced dimension d is set at 20, and the regularization parameter λ is set at 0.001 for all the methods.

Figure 1 shows the change of the RMSE as the stochastic gradient descent algorithm proceeds. We can clearly observe that SocialFD outperforms the baselines. From Table 1, we can compare all methods involved in a more intuitive sight. In Table 1, we list the best RMSE of all methods, and the number in brackets is how many iterations past by the time the best performance was reached. Despite the fact that SocialFD is not fine tuned, it is obvious that SocialFD beat other methods by fairly large margins. Compared with PMF, SoRec, SocialMF, and RSTE, SocialFD reduces the RMSE by 3.40%, 3.42%, 5.21%, and 2.73% on 80% training data, and by 2.94%, 3.25%, 5.21%, and 2.47% on 90% training data, respectively.

In reality, the recommendations for users are usually presented as a recommendation list. Thus, we prefer the ranking metrics rather than the RMSE.

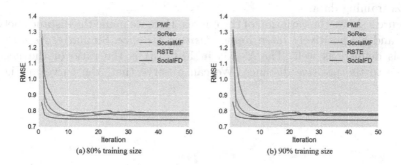

(a) 80% training size   (b) 90% training size

Fig. 1. RMSE of all methods

Table 1. Best RMSE of all methods

Training size	PMF	SoRec	SocialMF	RSTE	**SocialFD**
80%	0.7726 (9)	0.7728 (8)	0.7874 (25)	0.7674 (13)	**0.7463 (34)**
90%	0.7679 (8)	0.7704 (8)	0.7863 (23)	0.7642 (13)	**0.7453 (39)**

Table 2. The ranking quality of all methods

Training size	Metric	PMF	SoRec	SocialMF	RSTE	**SocialFD**
80%	F1@50	0.18%	0.81%	0.15%	1.89%	**1.96%**
	Recall@50	0.14%	0.79%	0.11%	**1.59%**	1.48%
	Precision@50	0.24%	0.83%	0.23%	2.32%	**2.91%**
90%	F1@50	0.09%	0.63%	0.08%	0.81%	**1.37%**
	Recall@50	0.08%	0.80%	0.08%	1.02%	**1.37%**
	Precision@50	0.09%	0.52%	0.08%	0.67%	**1.37%**

In the preprocessing stage, we binarize the explicit rating data by keeping the ratings of four or higher. In the prediction stage, we use the distance $\|x_u - y_i\|_A^2$ to generate ranking scores for users on all items. Sorting the ranking scores for all users, we can get the recommendation lists. Table 2 shows the lists measured by the ranking metrics. SocialFD transparently beats other methods on almost all the metrics except Recall@50. On 80% training data, RSTE wins by lesser superiority. However, it can not deny that using distance to reflect likability is a promising idea.

3.3 Change of the Distance

The goal of SocialFD is to make users be spatially close to their friends and their liked items, and be far away from their disliked items. In SocialFD, constraints for the distance are binding to help to reach the objective. In this part, we will confirm that whether the obtained distance is desired or not. Settings for SocialFD are the same with those in Sect. 3.2, and the experiment is conducted on 80% training data.

Figure 2 shows the variation of the curves. Here we name the distance between a user and his positively rated items *Positive distance*. Similarly, *Negative distance* is defined as well. The Y-axis in Fig. 2(a) is the ratio of the negative distance to the positive distance. We can observe that, the curve in Fig. 2(a)

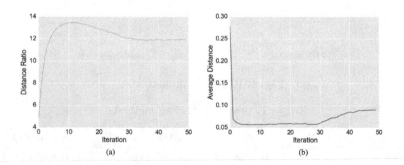

Fig. 2. Change of the distance

increases explosively at first, after 10 iterations, the curve gradually reaches a stable state. Finally, the ratio of the negative distance to positive distance converges at about 12. Figure 2(b) shows the change of the average distance among users with social connections. We can see the curve has a converse trend with that in Fig. 2(a). With the proceeding of the iteration, the average distance becomes shorter, and after 30 iterations, it starts to ascend. Eventually it reaches a flat state. So far, we can draw a conclusion that SocialFD do find appropriate locations for users and items.

4 Related Work

The most relevant research to ours are [8,10]. The authors of [8] proposed a method named SocialMF. The idea behind SocialMF is that a user's preference should be similar with that of his friends. Thus, in the training process, it concentrates on forcing a given user's latent factors approximate those of his friends. In [10], a method taking both trust and distrust into consideration was proposed. The authors deem that each user's latent factors should be close to those of his friends, and differ widely from those of distrusted users. It seems that the principle of the two methods are similar to that of SocialFD. However, these methods still regard latent factors as the representation of characteristics of users which are lack of explicit sense.

The first work integrates the distance into matrix factorization is [11]. The idea of the model is that all items and users can be embedded in a unified Euclidean space, and the learnt latent factors of users and items should reflect the negative correlation between the distance and the given rating. It means, the higher a given rating is, spatially the more closer the user and the item are. However, whether the Euclidean distance can measure the gap between users and items is questionable. In addition, no constraints are brought up to guarantee that the obtained distance is desired, and it all depends on the weak correlation with the ratings which may include some noises.

5 Conclusion

This paper is motivated by the assumption that distance can reflect likability. Based on the intuition, we proposed a novel recommendation method called SocialFD connecting distance metric learning and matrix factorization. The positions of users and items are decided by the ratings and social relations jointly, which can help to find appropriate locations for users who have few ratings. The distance eventually obtained are used to generate understandable and reliable recommendations. The experiments show that, SoicalFD outperforms the-state-of-art methods in recommendation accuracy.

In this paper, although we utilize the social connections to make recommendations for the users, we do not take the distrust into consideration. Generally speaking, negative opinions contains more information than positive opinions. Fusing the distrust into the distance-based social recommender can help to locate users more precisely. That is the potential direction we will explore in the future. In addition, we note that few work combines distance metric learning with the link prediction. We consider that the same idea of this paper can be extended to the area of the link prediction.

Acknowledgment. This research is supported by the Basic and Advanced Research Projects in Chongqing (cstc2015jcyjA40049), the National Key Basic Research Program of China (973) (2013CB328903), the National Natural Science Foundation of China (61472021), and the Fundamental Research Funds for the Central Universities (106112014 CDJZR 095502)

References

1. Massa, P., Avesani, P.: Trust-aware recommender systems. In: Proceedings of the 2007 ACM Conference on Recommender Systems, pp. 17–24. ACM (2007)
2. Tang, J., Xia, H., Liu, H.: Social recommendation: a review. Soc. Netw. Anal. Min. **3**(4), 1113–1133 (2013)
3. Koren, Y., Bell, R.M., Volinsky, C.: Matrix factorization techniques for recommender systems. IEEE Comput. **42**(8), 30–37 (2009)
4. Kulis, B.: Metric learning: a survey. Found. Trends Mach. Learn. **5**(4), 287–364 (2013)
5. Zhao, G., Qian, X., Xie, X.: User-service rating prediction by exploring social users' rating behaviors. IEEE Trans. Multimedia **18**(3), 496–506 (2016)
6. Salakhutdinov, R., Mnih, A.: Probabilistic matrix factorization, pp. 1257–1264 (2007)
7. Ma, H., Yang, H., Lyu, M.R., King, I.: Sorec: social recommendation using probabilistic matrix factorization. In: Proceedings of the 17th ACM conference on Information and knowledge management, pp. 931–940. ACM (2008)
8. Jamali, M., Ester, M.: A matrix factorization technique with trust propagation for recommendation in social networks, pp. 135–142 (2010)
9. Ma, H., King, I., Lyu, M.R.: Learning to recommend with social trust ensemble, pp. 203–210 (2009)
10. Ma, H., Lyu, M.R., King, I.: Learning to recommend with trust and distrust relationships, pp. 189–196 (2009)
11. Khoshneshin, M., Street, W.N.: Collaborative filtering via euclidean embedding, pp. 87–94 (2010)

Knowledge Engineering

Relevant Fact Selection for QA via Sequence Labeling

Yuzhi Liang[1], Jia Zhu[2], Yupeng Li[1], Min Yang[3(✉)], and Siu Ming Yiu[1]

[1] The University of Hong Kong, Pokfulam Road, Hong Kong, Hong Kong
[2] South China Normal University, Guangzhou, China
[3] SIAT, Chinese Academy of Sciences, Shenzhen, China
min.yang1129@gmail.com

Abstract. Question answering (QA) is a very important, but not yet completely resolved problem in artificial intelligence. Solving the QA problem consists of two major steps: relevant fact selection and answering the question. Existing methods usually combine the two steps to solve the problem. A major technique is to add a memory component to infer answers from the chaining facts. It is not very clear how irrelevant facts affect the effectiveness of these methods. In this paper, we propose to separate the two steps and only consider the problem of relevant fact selection. We used a graphical probabilistic model Conditional Random Field (CRF) to model the interdependent relationship among the chaining facts in order to select the relevant ones. In our experiments on a benchmark dataset, we are able to select correctly all relevant facts from 13 tasks out of 19 tasks (F-scores of the rest of the 6 tasks range from 0.8 to 0.97). We also show that using our selector to pre-select relevant facts can substantially improve the accuracies of existing QA systems (e.g. MemN2N (from 88% to 94%) and LSTM (from 66% to 91%) in 13 tasks with complete information).

Keywords: CRF · Question answering (QA) · Relevant fact selection

1 Introduction

Question answering (QA) is one of the key tasks in artificial intelligence and is not yet completely resolved. Given a query and a story containing several chaining facts, answering the query based on the story can be divided into two parts: identifying the relevant facts from the chaining facts and answering the question based on the relevant facts. Taking a sample from the bAbI dataset [14] as an instance (see below), to answer the question 'What is Emily afraid of?', we first identify Facts 7 and 5 from the chaining facts, then based on these two relevant facts, the answer 'wolf' is the output.

Fact 1: Wolves are afraid of mice.
Fact 2: Sheep are afraid of mice.
Fact 3: Winona is a sheep.

© Springer International Publishing AG 2017
G. Li et al. (Eds.): KSEM 2017, LNAI 10412, pp. 399–409, 2017.
DOI: 10.1007/978-3-319-63558-3_34

Fact 4: Mice are afraid of cats.
Fact 5: Cats are afraid of wolves.
Fact 6: Jessica is a mouse.
Fact 7: Emily is a cat.
Fact 8: Gertrude is a wolf.
Question: What is emily afraid of?
Supporting facts: 7, 5
Answer: wolf

Existing studies, mainly represented by Memory Network (MemNN) [15] and its variants, try to jointly solve these two tasks by adding a memory component, which contains fully distributed semantics, to a recurrent neural network [6]. However, the contents in the memory are hard to analyze, especially when using neural networks to decide how to write the memory. In other words, it is hard to trace the problem according to the contents in the memory when an error occurs. Although there are some works studying the neural network reasoning (e.g. [7]), the capability of reasoning and inferencing of neural network is still questionable.

Unlike previous works solving relevant fact selection and question answering jointly, we deal with the two tasks separately. This kind of separation makes the relevant fact selection and question answering evaluated independently. In this paper, we only focus on selecting relevant facts in chaining facts. To the best of our knowledge, this is the first work analyzing this problem. The goal of our research is to design a framework which can efficiently select relevant facts in chaining facts with limited amount of labelled data. The facts in a story can be interdependent such that reasoning and inferencing over the whole story is necessary during the process.

Conditional Random Field (CRF) [12] is used in our work to model the interdependent relationship among the chaining facts. CRF chooses the labels over the entire sequence at once which makes it utilize not only the local features but also the global information in defining the labels. In other words, CRF can define whether a fact is related to the given question based on inferencing over the whole story. In addition, as a discriminative model, CRF can model overlapping and non-independent features. Moreover, it is more traceable and requires less training data than neural networks based methods. The contributions of our work are as follow:

- To select relevant facts from chaining facts, we propose a framework called CRF fact selector which transforms this task to a sequence labeling problem.
- We evaluate the performance of CRF fact selector on the bAbI dataset. With limited amount of training data (merely 100 questions for each task, 10% of the overall training set), CRF fact selector can correctly select all the relevant facts in 13 tasks out of 19 tasks.
- We apply the trained CRF fact selector to all the training and testing data in the bAbI dataset, and use the selected facts as the input to some state-of-the-art QA systems, the accuracies of the answering part of the systems are significantly improved (e.g. MemN2N (from 88% to 94%) and LSTM (from 66% to 91%) in 13 tasks with complete information).

2   Related Work

One method of solving question answering problem is manually defining string matching rules or bag of words representations [5,9]. These methods are ineffective when the chaining facts in a story are interdependent. [2] tried to use recursive neural network (RNN) to model textual compositionality, but the vanish problem of RNN would hinder the answering accuracy when the story is long [3,18]. Very recently, memory network (MemNN)[15] is introduced to solve the memorization problem of applying RNN to QA task by adding a long-term memory component to the RNN. MemNN introduced a generalization process which updates old memories given the new input and an output feature map which produces a new output given the new input and the current memory state. In the generalization process, a function $H(\cdot)$ is responsible for selecting the slot. Theoretically, $H(\cdot)$ should organize the memory and discard the least useful fact when the memory is full. The output component reading from memory and performing inference over the facts. The MemNN is trained in a fully supervised setting: besides the question answer, the supporting facts are also provided in the training data (but not in the test data). End-to-end memory networks (MemN2N)[10] realizes end-to-end training (only the question answer is provided in the training data) by using a continuous representation in the memory component. Dynamic memory networks (DMN)[4] add an attention mechanism in memory networks. In particular, the attention mechanism calculates an attention score between the memory and the new input which indicates the importance of the new input, then the memory is updated according to the attention score. The attention score is computed using the Gated Recurrent Unit (GRU)[1]. MemNN and its related works solve the relevant fact selection jointly with question answering, the contents in the memory are hard to analyze, especially when using neural networks to decide how to write the memory. In addition, large amount of labelled data is required during the training process.

Information retrieval which detecting useful information from corpus (e.g. [13,17]) can be viewed as an inclusive problem of relevant fact selection, and community Question Answering(CQA) involves relevant information selection in QA. However, the main CQA task can be defined as: given a new question and a large collection of question-comment threads created by a user community, rank the comments according to the usefulness in answering the new question.

It is commonly solved as a ranking or clustering problem based on the similarities between the questions and existing answers (e.g. [8,16,19]). Reasoning and understanding the interdependency among the chaining facts are often not involved in the process. This is different from our problem setting.

3   Methodology

The relevant fact selection in chaining facts is a binary classification of deciding if a fact is relevant (thus, will be selected) or not. This process can be divided into 3 steps: fact reordering, fact to vector and fact selection by sequence labelling.

3.1 Fact Reordering

The facts of a story will be reordered to provide a more significant sequence pattern for the sequence labeling process. Three reordering methods are used in our experiment, namely ordering by subject, ordering by object and ordering by time. Ordering by subject is sorting the facts according to the subjects of the facts. Facts with the same subject will be put in a continuous sequence in the story. The other two types of ordering are similar. For a given task, we will choose a fact ordering, from the original ordering and the four reordering methods, which has the most significant sequence pattern. Specifically, when we treat the relevant fact selection as a binary classification, there are four patterns for the selection of a fact and its preceding fact:

- Both of the current fact and its preceding fact are not selected
- The current fact is selected while its preceding fact is not selected
- The current fact is not selected while its preceding fact is selected
- Both of the current fact and its preceding fact are selected.

The label of the four patterns can be represented as 00, 01, 10 and 11 respectively, and the set of patterns is $\mathcal{P} = \{00, 01, 10, 11\}$. Let T be the number of facts in a story and N be the amount of stories in the training set, the count of a pattern p is defined as Eq. 1

$$c_p = \sum_{i=1}^{N} \sum_{t=1}^{T} (y_{t-1}^{(i)} y_t^{(i)} = p) \tag{1}$$

Assume the set of count of the four patterns is $\mathcal{C} = \{c_{00}, c_{01}, c_{10}, c_{11}\}$, the standard deviation of $\mathcal{C}$ is used to indicates the dispersion of the sequence labels (2).

$$\sigma = \sqrt{\frac{1}{4} \sum_{p \in \mathcal{P}} (c_p - \mu)^2} \tag{2}$$

where μ is the mean of $\mathcal{C}$.

Large σ indicates the count of the patterns has high variance such that the sequence label pattern is significant, vice versa. For a given task, we will calculate the σ of different ordering methods and choose the one has the highest σ value as the ordering method of the task.

3.2 Fact to Vector

Given a story s containing T facts, each fact will be represented by a vector with a set of designed features. Let a_t denotes the tth fact, the set of features are functions of a_t, the stream history $\{a_1, \ldots, a_t\}$ and the whole story $s = \{a_1, \ldots, a_T\}$.

The corpus is preprocessed by stemming and replacing the demonstratives by their reference. We enumerate all the words that appear in the training set and the testing set, a word ID is assigned for each word[1]. The IDs of the specific words which appear in a fact serve as the most basic features of the fact. Then two kinds of features are created to evaluate the distance between current fact a_t and the question q, one is named as "semi-global" distance and the other is "global" distance. The semi-global distance is extracted from a_t and its stream history (i.e. a_1 to a_{t-1}). Let o_q be the query object of the question[2], o_t is a noun in a_t, the semi-global distance of o_t and o_q, denoted as d, is evaluated by the least steps they can be connected according to a_1 to a_{t-1}. We use $\mathcal{O}$ to represent the set of objects (nouns) appearing in story s and question q. o_t and o_q are an object in a_t and an object in q respectively. The process of semi-global distance calculation is listed in Algorithm 1. The semi-global distance is sequence sensitive, for the same o_t, the $d(o_t, o_q)$ might be different when it appears in different facts. On the other hand, the global distance is extracted from a_t and the whole story s. For a noun o_t from a_t and a query object o_q, the global distance is evaluated by the closest path from o_t to o_q based on the whole story. The calculation process is similar to that of semi-global distance, but the input is the whole story s instead of a_1 to a_t. The global distance is sequence insensitive, the global distance between a given o_t and o_q is consistent over the whole story. There might be more than one noun in a fact a_t, let $O_t = \{o_t\}$ denotes the set of nouns appear in a_t, $D_t = \{d(o_t, o_q), o_t \in O_t\}$ is the set of semi-global distances between o_q and the nouns in O_t. The minimum value in D_t serves as the the semi-global feature between a_t and o_q. Similarly, let $D'_t = \{d'(o_t, o_q), o_t \in O_t\}$ be the set of global distances between o_q and the nouns in O_t, the minimal value in D_t serves as the global feature between a_t and o_q.

3.3   Fact Selection

We select relevant facts in chaining facts by sequence labelling, Conditional Random Field (CRF) is used as the tagging algorithm. CRF is an undirected graphical model that encode a conditional probability distribution using a set of features. Selection of relevant facts can be interpreted as follows. Recall T is the total number of facts in a story, each fact in the story can be represented by several features. The features of the tth fact, denoted as x_t, serves as the observed variable at t. The label of the tth fact, denotes as y_t, is related to y_{t-1} according to the reordering process. The overall story $x = \{x_1, x_2, ..., x_t\}$ can be viewed as a chain of the facts and linear-chain CRF will be used to predict the labels. A linear-chain CRF can be described as a factor graph over x and y which defines the conditional probability of y given x. Figure 1 indicates the structure of the linear-chain CRF.

---

[1] In the bAbI dataset, the number of words is limited (less than 200 in one task) such that we just use an integer to represent the word ID in our experiment. Some complex word representation technics such as word embedding can be utilized when the number of words in the corpus increases.

[2] Any noun in q, normally, the meaning of sentences carried by the nouns in the sentences.

Algorithm 1. Semi-global distance calculation

Input : $a_1, \ldots, a_t, \mathcal{O}, o_t, o_q$ $(o_t, o_q \in \mathcal{O})$
Output: $d(o_t, o_q)$
1 Initialize an empty lookup table L;
2 **for** $o \in \mathcal{O}$ **do**
3 $\quad$ Initialize $d(o, o_q) = \infty$;
4 $\quad$ Add $o, d(o, o_q)$ to L;
5 **end**
6 Initialize $d(o_q, o_q) = 0$;
7 Add $o_q, d(o_q, o_q)$ to L;
8 **for** $i = 1$ *to* t **do**
9 $\quad$ Extract the set of objects O_i mentioned in a_i ;
10 $\quad$ $d_{min} \leftarrow \min\limits_{o \in O_i} d(o, o_q)$;
11 $\quad$ **for** $o \in O_i$ **do**
12 $\quad\quad$ **if** $d(o, o_q) > d_{min} + 1$ **then**
13 $\quad\quad\quad$ $d(o, o_q) \leftarrow d_{min} + 1$;
14 $\quad\quad\quad$ Update L;
15 $\quad\quad$ **end**
16 $\quad$ **end**
17 **end**
18 Find $d(o_t, o_q)$ in L;
19 Return $d(o_t, o_q)$;

Based on the set of features x, CRF defines the label of the sequence y using Eq. 3

$$p_\Lambda(y|x) = \frac{1}{Z(x)} \prod_{t=1}^{T} exp \left\{ \sum_{k=1}^{K} \lambda_k f_k(y_t, y_{t-1}, x_t) \right\} \tag{3}$$

The model parameters are a set of real weights $\Lambda = \{\lambda_k\}$, one weight for each feature, and $\{f_k(y, y', t)\}_{k=1}^{K}$ is a set of feature functions, each feature function is a transition function or a status feature function. K is the total number of feature functions. $Z(x)$ is an normalization factor over all state sequences for the sequence x.

$$Z(x) = \sum_{y} \prod_{t=1}^{T} exp \left\{ \sum_{k=1}^{K} \lambda_k f_k(y_t, y_{t-1}, x_t) \right\} \tag{4}$$

Given the model defined in Eq. 3, the most probable labeling sequence for an input x is Eq. 5

$$\hat{y} = \arg\max p_\Lambda(y|x) \tag{5}$$

To estimate the parameters, recall we used N to represent the number of labeled examples, $\{(x^{(i)}, y^{(i)})\}_{i=1}^{N}$, the log-likelihood of Eq. 3 is as Eq. 6.

$$\mathcal{L}(\lambda) = \sum_{i=1}^{N} \sum_{t=1}^{T} \sum_{k=1}^{K} \lambda_k f_k(y_t^{(i)}, y_{t-1}^{(i)}, x_t^{(i)}) - \sum_{i=1}^{N} log Z(x^{(i)}) \tag{6}$$

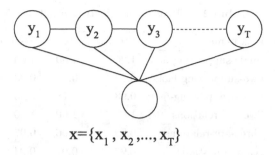

$$x=\{x_1, x_2, ..., x_T\}$$

Fig. 1. Linear-chain CRF

The Eq. 6 is concave, and can be efficiently maximized by second-order techniques such as conjugate gradient and L-BFGS.

4   Experiments

We use the bAbI dataset which contains 20 different types of questions to evaluate text understanding and reasoning. For each task, there are 1,000 questions

Table 1. bAbI dataset statics

ID	Story length				Proportion
	Max	Min	Avg	Dev	
1	10	2	6	2.83	16.67%
2	88	2	15.59	7.9	12.97%
3	228	4	50	34.5	5.99%
4	2	2	2	0	50%
5	94	2	19.89	13.46	5.26%
6	14	2	6.3	3.12	16.04%
7	34	2	8.9	5.09	25%
8	58	2	8.77	5.07	22.90%
9	10	2	6	2.83	16.67%
10	10	2	6	2.83	16.67%
11	10	2	6	2.83	33.33%
12	10	2	6	2.83	16.67%
13	10	2	6	2.83	33.33%
14	14	2	7.76	3.01	25.45%
15	8	8	8	0	25%
16	9	9	9	0	33.33%
17	2	2	2	0	100.00%
18	15	5	6.11	2.25	32.73%
19	5	5	5	0	40.00%
20	12	1	5.84	3.09	17.11%

Table 2. Relevant fact selection result

ID	Task name	Precision	Recall	F-score
1	Single-supporting-fact	1.00	1.00	1.00
2	Two-supporting-facts	0.96	0.90	0.93
3	Three-supporting-facts	0.93	0.78	0.85
4	Two-arg-relations	1.00	1.00	1.00
5	Three-arg-relations	0.99	0.96	0.97
6	Yes-no-questions	0.97	0.91	0.94
7	Counting	1.00	1.00	1.00
8	Lists-sets	1.00	1.00	1.00
9	Simple-negation	1.00	1.00	1.00
10	Indefinite-knowledge	1.00	1.00	1.00
11	Basic-coreference	1.00	1.00	1.00
12	Conjunction	1.00	1.00	1.00
13	Compound-coreference	1.00	1.00	1.00
14	Time-reasoning	0.92	0.86	0.89
15	Basic-deduction	1.00	1.00	1.00
16	Basic-induction	1.00	1.00	1.00
17	Positional-reasoning	/	/	/
18	Size-reasoning	0.82	0.78	0.80
19	Path-finding	1.00	1.00	1.00
20	Agents-motivations	1.00	1.00	1.00

for training and 1,000 questions for testing. A statistics overview of the dataset is listed in Table 1. We can notice that the lengths of the stories vary from 2 to 228 and the selected sentence proportion is less than 35% for most of the tasks, which means the classes of relevant facts and irrelevant facts are imbalanced. Task 17 is not involved in our experiment since all the facts in Task 17 are useful in answering the questions.

We evaluate the relevant fact selection result of CRF fact selector. We use 1/10 of the supporting fact labels provided by the bAbI training set. That is, we train the CRF fact selector based on only 100 stories. The selection result is evaluated by F-measure. From Table 2, CRF fact selector achieves a perfect result in 13 tasks out of 19 tasks. For three tasks, 3, 14, 18, the F-scores (0.80 to 0.89) are relatively lower. This might be because we simply used the linear-chain CRF in our experiment, some of the long-term dependencies in these tasks might need to be modeled by higher order CRF.

To better evaluate the effectiveness of CRF fact selector in QA systems, we divide the tasks into two categories, namely information complete tasks and information incomplete tasks. Information complete tasks are tasks that CRF fact selector can identifies all the relevant facts from the chaining facts, contrary to the information incomplete tasks. The information complete tasks including

Table 3. QA answering accuracy

ID	Without CRFfs		With CRFfs		/
	MemN2N	LSTM	MemN2N	LSTM	MemNN
1	**100**	**100**	**100**	**100**	**100**
2	90	50	74	71	**100**
3	60	20	54	23	**100**
4	94	63	**100**	**100**	**100**
5	89	62	96	94	**98**
6	92	51	93	92	**100**
7	81	79	92	**99**	85
8	90	77	96	**99**	91
9	87	61	**100**	61	**100**
10	87	48	**100**	96	98
11	99	75	**100**	**100**	**100**
12	**100**	76	**100**	**100**	**100**
13	**100**	94	**100**	95	**100**
14	98	35	77	71	**99**
15	**100**	32	**100**	**100**	**100**
16	**100**	51	**100**	**100**	**100**
17	/	/	/	/	/
18	85	91	88	91	**95**
19	31	10	**39**	36	36
20	79	94	**100**	**100**	**100**
C.I. Avg	88	66	**94**	91	93
I.I. Avg	86	51	80	74	**99**
All Avg	87	62	90	86	**95**

task 1, 4–7, 8–13, 15, 16, 19, 20 while the information incomplete tasks are task 2, 3, 5, 6, 14, 18 in the bAbI dataset. We compare the QA result of MemN2N[3] and LSTM [11][4] with and without CRF fact selector. We combine CRF fact selector and MemN2N/LSTM by pre-selecting relevant facts from the chaining facts using the trained CRF fact selector. Instead of using the whole stories, only the selected facts are used as the input. The result is shown in Table 3, the result of MemNN which is using a strongly supervised mode is also listed in the table for comparison. The CRFfs, C.I and I.I in the table are short of CRF fact selector, complete information and incomplete information.

---

[3] The experiments are based on the public source code https://github.com/facebook/MemNN.

[4] The experiments are based on the public source code https://github.com/fchollet/keras.

From Table 3, we can see the QA accuracies of MemN2N and LSTM improved from 88% to 94% and 66% to 91% by using CRF fact selector in the 13 complete information tasks. The QA accuracy of MemN2N with CRF fact selector is even slightly higher than that of MemNN which uses strong supervision. It shows that if irrelevant facts can be filtered correctly before inputting the facts to the QA systems, the accuracies of the answering part of the systems will be improved since there is less data to be analyzed by the system. On the other hand, CRF fact selector hinder the accuracy of MemN2N in 6 incomplete information tasks since some relevant facts are missing. An interesting phenomena is that the QA accuracy of LSTM improved by using CRF fact selector to pre-select the input even in incomplete information tasks. For the result of all the 20 tasks, the CRF fact selector can improve the average answering accuracy of MemN2N and LSTM from 87% to 90% and 62% to 86%, respectively. The accuracy of LSTM with CRF fact selector is comparable to that of MemN2N which means we have more flexibility in choosing the learning algorithm for QA by using CRF fact selector.

5   Conclusions

In this paper, we showed that we could separate the two major steps in solving the QA problem. In particular, we consider the first step of selecting relevant facts. We proposed a CRF (conditional random field) fact selector and showed that it can effectively identify relevant facts and can be used to improve the accuracy of existing QA systems by pre-selecting relevant facts as inputs to the systems. We plan to further investigate how to improve the effectiveness and accuracy of relevant fact selection and also work on how to combine the relevant fact selection with the question answering part to achieve a better overall performance result.

References

1. Chung, J., Gulcehre, C., Cho, K., Bengio, Y.: Empirical evaluation of gated recurrent neural networks on sequence modeling. arXiv preprint arXiv:1412.3555 (2014)
2. Iyyer, M., Boyd-Graber, J.L., Claudino, L.M.B., Socher, R., Daumé III, H.: A neural network for factoid question answering over paragraphs. In: EMNLP, pp. 633–644 (2014)
3. Jozefowicz, R., Zaremba, W., Sutskever, I.: An empirical exploration of recurrent network architectures. In: Proceedings of the 32nd International Conference on Machine Learning (ICML-15), pp. 2342–2350 (2015)
4. Kumar, A., Irsoy, O., Su, J., Bradbury, J., English, R., Pierce, B., Ondruska, P., Gulrajani, I., Socher, R.: Ask me anything: dynamic memory networks for natural language processing. CoRR, abs/1506.07285 (2015)
5. Lin, D., Pantel, P.: Discovery of inference rules for question-answering. Nat. Lang. Eng. 7(04), 343–360 (2001)
6. Mikolov, T., Karafiát, M., Burget, L., Cernockỳ, J., Khudanpur, S.: Recurrent neural network based language model. In: Interspeech, vol. 2, p. 3 (2010)

7. Peng, B., Lu, Z., Li, H., Wong, K.F.: Towards neural network-based reasoning. arXiv preprint arXiv:1508.05508 (2015)
8. Qiu, X., Huang, X.: Convolutional neural tensor network architecture for community-based question answering. In: IJCAI, pp. 1305–1311 (2015)
9. Ravichandran, D., Hovy, E.: Learning surface text patterns for a question answering system. In: Proceedings of the 40th Annual Meeting on Association for Computational Linguistics, pp. 41–47. Association for Computational Linguistics (2002)
10. Sukhbaatar, S., Weston, J., Fergus, R., et al.: End-to-end memory networks. In: Advances in neural information processing systems, pp. 2440–2448 (2015)
11. Sundermeyer, M., Schlüter, R., Ney, H.: LSTM neural networks for language modeling. In: Interspeech, pp. 194–197 (2012)
12. Sutton, C., McCallum, A.: An introduction to conditional random fields for relational learning. Introduction to statistical relational learning, pp. 93–128 (2006)
13. Tu, W., Cheung, D.W.L., Mamoulis, N., Yang, M., Lu, Z.: Real-time detection and sorting of news on microblogging platforms. In: PACLIC (2015)
14. Weston, J., Bordes, A., Chopra, S., Rush, A.M., van Merriënboer, B., Joulin, A., Mikolov, T.: Towards AI-complete question answering: a set of prerequisite toy tasks. arXiv preprint arXiv:1502.05698 (2015)
15. Weston, J., Chopra, S., Bordes, A.: Memory networks. arXiv preprint arXiv:1410.3916 (2014)
16. Wu, H., Wu, W., Zhou, M., Chen, E., Duan, L., Shum, H.Y.: Improving search relevance for short queries in community question answering. In: Proceedings of the 7th ACM International Conference on Web Search and Data Mining, pp. 43–52. ACM (2014)
17. Yang, M., Chow, K.-P.: An Information Extraction Framework for Digital Forensic Investigations. In: Peterson, G., Shenoi, S. (eds.) DigitalForensics 2015. IAICT, vol. 462, pp. 61–76. Springer, Cham (2015). doi:10.1007/978-3-319-24123-4_4
18. Yang, M., Tu, W., Yin, W., Lu, Z.: Deep Markov neural network for sequential data classification. In: The 53rd Annual Meeting of the Association for Computational Linguistics. ACL (2015)
19. Zhang, K., Wu, W., Wu, H., Li, Z., Zhou, M.: Question retrieval with high quality answers in community question answering. In: Proceedings of the 23rd ACM International Conference on Information and Knowledge Management, pp. 371–380. ACM (2014)

Community Outlier Based Fraudster Detection

Chenfei Sun, Qingzhong Li$^{(\boxtimes)}$, Hui Li, Shidong Zhang, and Yongqing Zheng

School of Computer Science and Technology, Shandong University, Jinan, China
sunchenfei@mail.sdu.edu.cn, {lqz,lih,zsd,l.liu}@sdu.edu.cn

Abstract. Healthcare fraud is causing billions of dollars in loss for public healthcare funds. In existing healthcare fraud cases, the convicted fraudsters are mostly physicians - the healthcare professionals who submit fraudulent bills. Fraudster detection can help us to find suspicious physicians and to combat healthcare fraud in advance. When it comes to the problem of fraudster detection, rule based fraud detection methods are not applicable because fraudsters will try everything to avoid detection rules. Meanwhile, outlier based fraud detection approaches primarily aim to find global outliers and can't find local outliers accurately. Therefore, we propose Community Outlier Based Fraudster Detection Approach - COBFDA in this paper. The proposed approach divides the physicians into different communities and looks for community outliers in each community. Extensive experiment results show that COBFDA outperforms the comparison approaches in terms of f-measure by over 20%.

1 Introduction

Healthcare fraud is a serious threat to the proper usage of public funds and exists in all countries around the world. Fraud has been difficult to detect owing to its various forms. Among all kinds of subjects in healthcare, not surprisingly, the largest number of convicted fraudsters are physicians - the healthcare professionals who submit fraudulent bills. Healthcare fraud detection systems are intended to help to detect suspicious fraudsters and then rely on human experts to manually identify the suspicious physicians. Existing rule-based methods often rely on rules designed by experts which can be used as a basis for identifying behaviors violating some of these rules [3], which are not applicable when it comes to the fraudsters detection problem. Nowadays, most fraudster detection methods are outlier based which aim to find the objects that different from most of the objects [7]. However, outlier based fraud detection approaches now available major in global outlier detection and can't find local outliers accurately.

Outlier detection is the main technique to conduct physician frauster detection at present. However, density-based, distribution-based and distance-based outlier detection method cannot detect local outlier objects as well as clustering-based outlier detection methods [3]. Meanwhile, it is difficult to cluster the physicians in the physician fraudster detection problem. Therefore, we model the healthcare claim as a physician-drug heterogeneous network and the physicians

G. Li et al. (Eds.): KSEM 2017, LNAI 10412, pp. 410–421, 2017.
DOI: 10.1007/978-3-319-63558-3_35

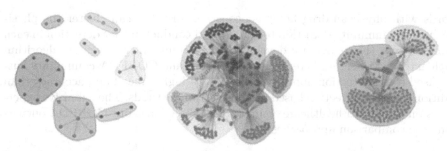

(a) disjoint   community (b) overlapping community (c) overlapping community
detection in homogeneous detection in homogeneous detection in heterogeneous
network   network   network

Fig. 1. Community detection

can be clustered by the structure of the network. Then these physician fraudsters can be detected by combining community detection and community outlier
detection in heterogeneous network. In this paper, we discuss the discovery of
physician communities and pick out the community outliers who conduct anomalous business practices. Within such a community, each physicians specialty
will match the kinds and quantity of the prescriptions they write under normal
circumstances. For instance, cardiologists will be divided into the same community with a high proportion of heart disease related medications whereas an
oncologist will tend to be in the same community compose of a high proportion
of chemotherapy drugs. If a cardiologist is divided into the oncologist community, which is called community outlier, then the cardiologist is most likely to
be a physician perpetrator. As shown in Fig. 1, the circle and the square nodes
refer to different drugs and physicians respectively. Physicians belonging to the
same community should have higher similarity with each other. Otherwise, the
physician which has low similarity with other physicians in the same community
is most likely to be fraudulent.

As shown in Fig. 1(a) and (b), traditional community detection method in
homogeneous network can be divided into disjoint community detection and
overlapping community detection [6]. Figure 1(c) shows ordinary overlapping
community detection in heterogeneous network, every node may belong to multiple communities. However, this case requires that circle nodes can be divided
into multiple communities, while the square nodes can be only belong to one
community. In other words, this is a specific community detection problem in
Heterogeneous Network–there is a constraint that each type of vertices have
different demands for community detection.

Because fraudster will try everything to avoid detection rules, rule based
fraud detection methods are not applicable. Meanwhile, outlier based fraud
detection approaches primarily aim to find global outliers and can't find global
outliers accurately. In brief, fraudster detection in healthcare cannot be solved
properly with Existing algorithm. Then we find that we can model the healthcare

records with physician-drug heterogeneous network. We can cluster the physicians using community detection technique and conduct outlier detection in each community. In this paper, we develop HSLPA community detection algorithm which extends SLPA into heterogeneous network and COBFDA-community outlier based fraud detection approach which has a higher recall and accuracy than traditional outlier detection based fraud detection methods. The extensive experiments on real-world healthcare claim dataset shows that our COBFDA outperforms the comparison approaches.

2   Related Work

Community detection problem is generally defined as clustering or identifying the community structure of the network. Despite the ambiguity in the definition of community, community detection has attracted strong attention in recent years. A wide variety of techniques have been used for community detection- identifying the community structure of networks, such as removal of high-betweenness edges [13], statistical inference [14] and so on. However, most of them handle disjoint community detection which assumes that a network is composed of a flat set of disjoint communities. This makes sense for many networks, for example, most papers are published just in a single conference. In the background of healthcare, drug/treatment can be used for multiple diseases are common situations. Therefore, we need to adopt overlapping community detection methods, which means that each vertex may appear in more than one community. This is more realistic in most cases. For example, the same drug/treatment may be included in prescriptions by physicians in different departments.

Overlapping community detection was previously proposed by Palla with the clique percolation algorithm (CPM) [15]. CPM assumes that a community is composed of fully connected graphs and detects overlapping by seeking adjacent cliques. Nevertheless, networks dont consist of fully connected graphs in most situations.

As network datasets become larger and larger, the speed of community detection algorithms becomes more and more important. For the time to come, algorithms will be capable of handling networks with millions of vertices within a reasonable time. As a consequence, any practical algorithm must have a very low time complexity. One of the fastest community detection algorithm to date is the Label Propagation Algorithm proposed by Raghavan et al. [10]. In LPA, each node holds only a single label and iteratively updates it to its neighborhood majority label. Despite its simple (no parameter) and near-linear time complexity (for sparse networks), it can only detect disjoint communities like most of other algorithms. SLPA is an extension of LPA for overlapping community detection [8]. However, our case requires that drug/treatment nodes can be divided into multiple communities, while the physician nodes can be only belong to one community. So we propose HSLPA algorithm which meet our demand that each type of nodes have different requirements for community detection.

3   Preliminaries

We start with a few fundamental concepts. Heterogeneous Information Network: A network with multiple types of vertex and multiple types of edges is called a Heterogeneous Network. In a Heterogeneous Information Network, each node represents an entity (e.g., physicians in a healthcare network) and each edge (e.g., tie) indicates a relationship between entities. Each node/edge may have attributes, labels (indicates the community label of the node), and weights. Link may carry rich semantic information.

Community: Consider a heterogeneous network with M different types of objects $\{Tp_1, Tp_2, \ldots, Tp_M\}$. A community is a probabilistic collection of similar objects, such that similarity between objects within the community is higher than the similarity between objects in different communities. For example, a therapeutic area is a community in a healthcare insurance network. For heterogeneous networks, one is often interested in identifying heterogeneous communities which contain objects of different types.

Community Outlier: An object o in a heterogeneous network G, is a community outlier if its attribute is far different from other objects of the same type in the same community. For example, an oncologist physician is a community outlier if other physicians in the same community are all cardiologists.

The community outlier based healthcare fraud detection problem can then be specified as follows.

Input: N prescriptions claims given by m physicians (with known departments), each prescription consist of a set of drug\treatment.

Output: Physician community outliers whose features are highly different from other physicians in the same community.

4   Community Outlier Based Fraudster Detection Approach

The COBFDA - Community Outlier Based Fraudster Detection Approach can be divided into three steps: (1) Constructing the heterogeneous network G composed of physician and drug/treatment. (2) Community detection in heterogeneous network G through the proposed HSLPA method. (3) Community outlier based fraudster detection.

In this section, we will present in detail how COBFDA works.

4.1   Network Construct

Construct the weighted heterogeneous network G = (V, E, W, T), where V, E, W and T represent the set of vertices, edges, weights of the graph and the type of the vertex, respectively. The vertices can be divided into two types: P for physician or

DT for drug\treatment. Vertex i of type P refers to a physician p_i, vertex j of type DT means drug/treatment dt_j.

There exists two kinds of edges in G: $e(p_i, dt_j)$ indicates that physician p_i give a prescription of drug\treatment dt_j; $e(dt_i, dt_j)$ denotes that drug\treatment dt_i and drug/treatment dt_j co-occurrence in the same prescription,dt_i and dt_j are used in chronological ascending order. The weight of edge $e(p_i, dt_j)$ between physician p_i and drug/treatment dt_j is defined as:

$$w_{(p_i dt_j)} = \frac{Num(p_i, dt_j)}{Num(p_i)} \tag{1}$$

$Num(p_i, dt_j)$ refers to the number of prescriptions consisting of drug/treatment dt_j which were written by physician p_i; $Num(p_i)$ indicates the number of prescriptions written by physician p_i. $w_{(p_i dt_j)}$ denotes the frequency that physician p_i uses drug/treatment dt_j in his prescriptions. The larger the $w_{(p_i dt_j)}$ is, the more often the physicians pi prescribes drug/treatment dt_j. Weight of edge $e(dt_i, dt_j)$ between drug/treatment dt_i and drug/treatment dt_j is calculated as the correlation of drugs:

$$w_{dt_i dt_j} = \sum_{\forall pre_n} \frac{1_{[distance(dt_i, dt_j, pre_n) < r]}}{N} \tag{2}$$

1 [condition] is a function which evaluates to 1 if [condition] is true. Otherwise, it evaluates to 0. The $1[distance(dt_i, dt_j, pre_n) < r] = 1$ indicates that drug/treatment dt_i and dt_j co-occurrence in prescription pre_n and their distance is smaller than the defined threshold r(decided by expert experiment). dt_i and dt_j occurrence in pre_n in chronological ascending order. The distance between dt_i and dt_j is calculated as follows.

$$distance(dt_i, dt_j, pre_n) = loc(dt_j, pre_n) - loc(dt_i, pre_n) \tag{3}$$

where $loc(dt_j, pre_n)$ and $loc(dt_i, pre_n)$ represent the corresponding locations of j and i in prescription pre_n. For instance, pre_1 and pre_2 are two different prescriptions composed of drug/treatment used in chronological ascending order and a, b, c, d, e indicate different drug/treatment. For prescription $pre_1 = (a \rightarrow b \rightarrow c \rightarrow d \rightarrow e)$ and $pre_2 = (b \rightarrow a \rightarrow d \rightarrow f \rightarrow g \rightarrow c)$:

In pre_1, distance(a, c, pre_1) = loc(c, pre_1)−loc(a, pre_1) = 3−1 = 2.

In pre_2, distance(a, c, pre_2) = loc(c, pre_2)−loc(a, pre_2) = 6−2 = 4.

The smaller the value of $distance(dt_i, dt_j, pre_n)$, the closer together dt_i and dt_j are in prescription pre_n.

N is the total number of prescriptions in history claims data. $w_{dt_i dt_j}$ shows the closeness between drug/treatment dt_i and dt_j. The larger the $w_{dt_i dt_j}$ is, the more often drug/treatment dt_i and dt_j co-occurrence in the same prescription, the closer dt_i and dt_j is.

Although there are different definitions of the weight of the two kinds of edges − $e(p_i, dt_j)$ and $e(dt_i, dt_j)$, they both indicate the closeness of the two vertices and they are both in the range of [0, 1]. So we can consider the network

G as a heterogeneous network. To find the suspicious physicians, we need to conduct community detection in G and proceed community outlier detection in each community.

4.2 Community Detection

Obviously, the fraudster detection problem is a local outlier detection problem. Existing clustering-based outlier detection methods are able to find local outliers. However, it is difficult to cluster the physicians because the demographic features of the physicians are not able to cluster the physicians properly in the problem of fraudster detection. So we construct physician-drug network and decide to cluster the physicians by community detection methods in G. The physicians can be divided into communities by the structure of the network.

This network G is different from a typical network in the following ways: (1) the network consists of both physicians and drug/treatment. Physician and drug/treatment are two types of entities. (2) the physician can only belong to one community while the drug/treatment can pertain to multiple communities. In the community detection of our healthcare heterogeneous network G, there is a specified constraint that different kinds of vertices have different demands for community detection. In other words, drug/treatment node can have more than one label and belong to multiple communities, which is called overlapping node. While physician node can only contain one label and pertain to one community, which is called disjoint node. To do community detection in heterogeneous network G, we propose heterogeneous speaker-listener based information propagation algorithm (HSLPA) which is an extension of speaker-listener based information propagation algorithm (SLPA) [8]. SLPA mimics human pairwise communication behavior. In each communication step, one node is a speaker (information provider), and the other is a listener (information consumer). However, SLPA didnt consider this specific community detection in heterogeneous network.

In the dynamic label propagation process, we need to determine (1) how to spread ones label to other nodes (2) how to process the label received from other nodes (3) how to satisfy the constraint to labels of different type of vertices. These critical problems all point to the question that how information should be maintained. HSLPA follows the idea of mimicking human communication behavior as SLPA but applies different dynamics to adapt to the needs of the heterogeneous network.

In HSLPA, each node can be a speaker or a listener. The roles are switched depending on whether a node serves as an information provider or information consumer. For overlapping nodes, a node can hold as many labels as it likes, which is up to what it has experienced during the propagation procedure driven by the underlying network structure. For disjoint node, a node can maintain the one label which has highest score. A node accumulates knowledge of repeatedly observed labels. Moreover, the more a node observes a label, the more likely it will spread this label to other nodes (mimicking peoples preference of spreading most frequently discussed opinions).

HSLPA is done in three main stages: (1) Firstly, the record of each node is initialized with this nodes id (i.e., with a unique label) (2) Secondly, the following steps are repeated until the stop criterion requirement is satisfied. a. One node is selected as a listener. b. Each neighbor of the selected node sends out labels following certain speaking rule, such as selecting the K label with highest probability from its record.c. The listener accepts labels from the collection of labels received from neighbors following certain listening rule, such as selecting the most K popular label from what it observed in the current step. d. If the listener is an overlapping node, keep the labels received from the neighbors satisfying specified condition. If the listener is a disjoint node, keep the one label with the highest score while erasing all other labels. (3) Finally, the post-processing based on the labels in the records of nodes is applied to output the communities.

A. Propagation Rules

Speaking Rule: For overlapping node v with multiple labels, it will send out the label i to its neighbors according to its record with certain probability.

$$P(v, label_i, t) = \frac{P(v, label_i, t-1)}{\sum_{l=1}^{n} P(v, label_l, t-1)} \tag{4}$$

t is the count of iteration. listening rule: The score of labels for the listener node u can be calculated by its neighbor speakers:

$$Score_{label_i}(u) = \sum_{\forall E(u,v) \in G} P(v, label_i) * w_{uv} \tag{5}$$

w_{uv} indicates the weight of edge e(u,v) in G, the $P(v, label_i)$ denotes the probability that neighbor v transform label i to u. The higher $Score - (label_i)(u)$ is, the more possible the node u contain the label i. If node u is a disjoint node, it will accept the label with highest score; If node u is a overlapping node, it will accept the k labels with highest score (k is a predefined threshold according to domain knowledge).

B. Stop Criterion

In the original SLPA algorithms, SLPA simply stops when the predefined maximum prideT is reached. In general, SLPA produces relatively stable outputs, independent of network size or structure, when T is greater than 20. Because of our strong applicable requirement for healthcare insurance domain, we set the stop criterion to that when the drug/treatment can be divided into communities which are corresponding to domain knowledge such as pharmacopeia.

C. Post-processing and Community Detection

SHLPA algorithm collect only label information which reflects the structure of the underlying network during the evolution. The community detection is performed when the stored label propagation information is post-processing. Convert the record of each node into a probability distribution of labels. For example, for a given node u, we can obtain the label distribution of u as follows.

$label_1, label_2, \cdots, label_n$ indicate different label, $p_1, p_2, \cdots, p_n$ denotes the probability that node u contain the corresponding label. This probability distribution defines the strength of association to communities to which the node belongs. The distribution can be used for fuzzy community detection [7]. However, for a given community, we would like to produce explicit answer if a node is a membership of this community. In other words, the answer should be binary, i.e., either a node is in this community or not. As a consequence, we perform a simple discretization procedure. Given a threshold $\gamma \in [0, 1]$, if the probability of seeing a particular label during the whole process if less than r, this label is deleted from a node's record. After this discretization, connected nodes having the same label are grouped together and form a community. If a node contains multiple labels, it belongs to more than one community which is called overlapping node. We remove the nested communities so that the final communities are maximal. Consequently, we obtain the community detection result $C = c_1, c_2, \ldots, c_m$, m is the number of detected community (Table 1).

Table 1. Node record

Label	$label_1$	$label_2$	$\cdots$	$label_i$	$\cdots$	$label_n$
Probability	p_1	p_2	$\cdots$	p_i	$\cdots$	p_n

4.3 Community Outlier Based Fraudster Detection

To search for physicians who perform fraud prescriptions, we need to find physician community outliers from the detected communities. The community outliers are physicians who have low feature similarity from other physicians in the same community.

Feature Selection: Select features which have significant influence on fraud detection. Denote the feature set as $F = (f_1, f_2, \ldots, f_n)$, where $f_1, f_2, \ldots, f_n$ represent the selected features. Each physician can be denoted by a vector composed of these features. The features can be selected through statistical analysis of historical data. For a given feature f_i, we compare the fraud distribution P(x) of the entire dataset and the fraud distribution Q(x) in the subset of fraud data with feature f_i. The Kullback-Leibler divergence [16] between P(x) and Q(x) can be calculated as:

$$D_{KL}(P\|Q) = \sum ln\frac{p(x)}{q(x)} * p(x) \tag{6}$$

If $D_{KL}(P\|Q)$ of f i is greater than a predefined threshold β (which can be determined empirically based a given application), f_i will be selected as a feature. Divide the features into nominal type and quantitative type For nominal type features (i.e., department), when the value of this feature of two physicians is same, the distance is 0, otherwise, the distance is 1. For quantitative type features (i.e., average cost of prescription), the distance of this feature between two physicians is calculated as Euclidean Distance.

$$Distance(f_{it}, f_{jt}) = \begin{cases} g(x) & f_t \quad is \quad nominal \quad feature \\ \frac{|f_{it}-f_{jt}|}{Max(f_t)-Min(f_t)} & f_t \quad is \quad quantitative \quad feature \end{cases} \quad (7)$$

$$g(x) = \begin{cases} 0 & f_{it} = f_{jt} \\ 1 & f_{it} \neq f_{jt} \end{cases} \quad (8)$$

For each detected community c, we calculate the dissimilarity between each physician pi and other physicians in the same community.

$$DisSim(p_i, p_j) = \sum_{t=1}^{n} \frac{|Distance(f_{it}, f_{jt}) - \mu(f_t)|}{\sigma(f_t)} \quad (9)$$

p_j denotes other physicians in community c except p_i, f_{it} is the feature ft of physician p_i, the same is with f_{jt}. $\mu(f_t)$ indicates the mean of f_t, while $\sigma(f_t)$ is the standard deviation of f_t. For each physician p_i, we can measure his possibility of fraud according to his similarity to the physicians who are in the same community with p_i. To simply the calculation, we select the top-K largest similarity physician pairs and the outlier probability of p_i is calculated as

$$outlierScore(p_i) = \log_{10} \frac{\sum_{j=1}^{K} DisSim(p_i, p_j)}{K} \quad (10)$$

The log function is applied for the sake of normalization. The larger outlierScore(p_i) is, the more probability that p_i is a community outlier in community c, the more likely that p_i is a healthcare insurance fraudster (Fig. 2).

Fig. 2. An example of original healthcare claim records

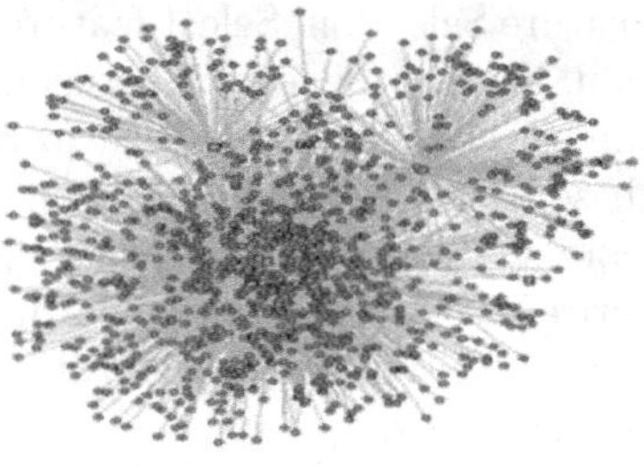

Fig. 3. Example of physician-drug network

According to the outlierScore of each physician p_i, we can sort the physicians. Then the physicians with highest outlierScore will be flagged as suspicious and sent to manual inspection. Domain experts will analysis the detailed healthcare claims related to these suspicious physicians and find out the fraud claims.

5   Experiments

In this section, we evaluate the performance of COBFDA using a real-world dataset. The dataset used in this experiment is collected from the Healthcare Insurance Claim System which is currently used in reality. The dataset contains more than 10 million records of medical insurance claim activities from over 500 physicians. The performance of the COBFDA approach is compared against four state-of-the-art approaches.

5.1   Experimental Settings

In the experiment, we select claim records from the same hospital which is the largest local hospital. The database records are then preprocessed heterogeneous network G, in which there are two types of nodes: physician and drug/treatment. The edges between nodes of different types indicate the relation between physician and drug, while the edges between nodes of the same type show the closeness between drug/treatment and drug/treatment.

We conduct community detection on network G using proposed HSLPA. Then in the community outlier process, 12 features are selected according to the Eq. 5, such as department, average cost of prescription and so on. Divide the 12 features into nominal type and quantitative type.

We adopt commonly used metrics including precision, recall, and f-measure to evaluate how effectively each approach identifies fraudulent physicians. Precision $= \frac{tp}{tp+fp}$ is the fraction of physicians identified as fraudulent which are indeed fraudulent. $Recall = \frac{tp}{tp+fn}$ is the fraction of all fraudulent physicians that have been correctly identified. f-measure $= \frac{2*precision*recall}{Precision+Recall}$ is the weighted harmonic mean of precision and recall. Here, tp (true positive) is the number of physicians correctly classified as fraudulent, fp (false positive) is the number of physicians incorrectly classified as fraudulent, and fn (false negative) is the number of physicians incorrectly classified as non-fraudulent.

5.2   Results and Analysis

Firstly, COBFDA construct physician-drug heterogeneous network G (shown as Fig. 3) according to the original healthcare claim records. Then we compare the result of our HSLPA and original SLPA in overlapping community detection of G. Our HSLPA can divide the network into communities which satisfy the rule that physician belong to only one community while drugs can pertain multiple communities. However, SLPA divide both physicians and drugs into multiple communities. HSLPA is more effective than SLPA in community detection in heterogeneous network.

Figure 4(a) shows the performance of COBFDA against four other outlier detection approaches [3] (distance based, LOF, distribution based and clustering based) in terms of fraudsters detection precision. It can be observed that

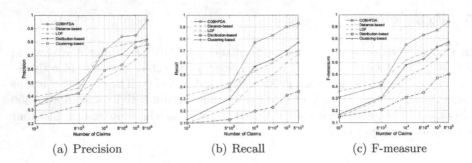

(a) Precision	(b) Recall	(c) F-measure

Fig. 4. Experiment results

COBFDA, which combines community detection and community outlier detection in heterogeneous network, achieves significantly better performance than other comparison approaches when number of claims exceeds 5,0000.

Figure 4(b) shows the performance of COBFDA against four other approaches in terms of fraud recall. COBFDA outperforms comparison approaches significantly across different N sizes. The recall values of the studied methods increase with increasing N sizes. COBFDA is able to correctly identify over 95% of the fraudsters from the datasets and has significantly higher recall than the other methods.

As a result of high precision and high recall, when these two metrics are combined together to form the f-measure (Fig. 4(c)), COBFDA consistently beats the comparison approaches in the experiments. On average, COBFDA outperforms the comparison approaches in terms of f-measure by over 20%.

6 Conclusion

In this paper, we propose a community outlier based healthcare fraud detection approach - COBFDA - to detect physician fraudsters who commit healthcare insurance fraud. COBFDA is evaluated against four state-of-the-art approaches using a large-scale real-world dataset from Healthcare Insurance System. The dataset contains more than 10 million medical claim activity records from more than 500 physicians. Extensive experiments show that COBFDA is over 20% more effective than the existing approaches. Currently, COBFDA has been implemented as a recommendation agent to provide decisions support for approval officers in a medical insurance claim system to assess the probability of fraud for physicians. In subsequent research, we will focus on human-agent interaction techniques to help the COBFDA agent build trust with the users.

References

1. Liao, K., Liu, G.: An efficient content based video copy detection using the sample based hierarchical adaptive k-means clustering. J. Intell. Inf. Syst. **44**(1), 133–158 (2015)

2. Thorpe, N., Deslich, S., Sikula, Sr., A., et al.: Combating medicare fraud: a struggling work in progress (2012)

3. Aggarwal, C.C.: Outlier analysis. In: Aggarwal, C.C. (ed.) Data Mining, pp. 237–263. Springer International Publishing, Heidelberg (2015)

4. Gupta, M., Gao, J., Sun, Y., et al.: Integrating community matching and outlier detection for mining evolutionary community outliers. In: Proceedings of the 18th ACM SIGKDD International Conference on Knowledge Discovery and Data Mining, pp. 859–867. ACM (2012)

5. Gao, J., Liang, F., Fan, W., et al.: On community outliers and their efficient detection in information networks. In: Proceedings of the 16th ACM SIGKDD International Conference on Knowledge Discovery and Data Mining. ACM, 2010: 813–822

6. Chen, D., Shang, M., Lv, Z., et al.: Detecting overlapping communities of weighted networks via a local algorithm. Physica A: Stat. Mech. Appl. **389**(19), 4177–4187 (2010)

7. van Capelleveen, G., Poel, M., Mueller, R.M., et al.: Outlier detection in healthcare fraud: a case study in the medicaid dental domain. Int. J. Account. Inf. Syst. **21**, 18–31 (2016)

8. Xie, J., Szymanski, B.K., Liu, X.: SLPA: Uncovering overlapping communities in social networks via a speaker-listener interaction dynamic process. In: 2011 IEEE 11th International Conference on Data Mining Workshops (ICDMW), pp. 344–349. IEEE (2011)

9. Gregory, S.: Finding overlapping communities in networks by label propagation. New J. Phys. **12**(10), 103018 (2010)

10. Raghavan, U.N., Albert, R., Kumara, S.: Near linear time algorithm to detect community structures in large-scale networks. Phys. Rev. E **76**(3), 036106 (2007)

11. Psorakis, I., Roberts, S., Ebden, M., et al.: Overlapping community detection using Bayesian non-negative matrix factorization. Phys. Rev. E **83**(6), 066114 (2011)

12. Xie, J., Szymanski, B.K.: Towards linear time overlapping community detection in social networks. In: Tan, P.-N., Chawla, S., Ho, C.K., Bailey, J. (eds.) PAKDD 2012. LNCS, vol. 7302, pp. 25–36. Springer, Heidelberg (2012). doi:10.1007/978-3-642-30220-6_3

13. Girvan, M., Newman, M.E.J.: Community structure in social and biological networks. Proc. Nat. Acad. Sci. **99**(12), 7821–7826 (2002)

14. Hofman, J.M., Wiggins, C.H.: Bayesian approach to network modularity. Phys. Rev. Lett. **100**(25), 258701 (2008)

15. Palla, G., Dernyi, I., Farkas, I., et al.: Uncovering the overlapping community structure of complex networks in nature and society. Nature **435**(7043), 814–818 (2005)

16. Hershey, J.R., Olsen, P.A.: Approximating the Kullback Leibler divergence between Gaussian mixture models. In: IEEE International Conference on Acoustics, Speech and Signal Processing. ICASSP 2007, vol. 4, pp. IV-317–IV-320. IEEE (2007)

An Efficient Three-Dimensional Reconstruction Approach for Pose-Invariant Face Recognition Based on a Single View

Minghua Zhao[1]($\boxtimes$) (iD), Ruiyang Mo[1], Yonggang Zhao[2],
Zhenghao Shi[1], and Feifei Zhang[1]

[1] Faculty of Computer Science and Engineering,
Xi'an University of Technology, Xi'an 710048, China
mh_zhao@126.com
[2] Faculty of Earth Science and Engineering,
Xi'an Shiyou University, Xi'an 710065, China

Abstract. A three-dimensional (3D) reconstruction approach based on a single view is proposed to solve the problem of lack of training samples while addressing multi-pose face recognition. First, a planar template is defined based on the geometric information of the segmented faces. Second, 3D faces are resampled according to the geometric relationship between the planar template and original 3D faces, and a normalized 3D face database is obtained. Third, a 3D sparse morphable model is established based on the normalized 3D face database, and a new 3D face can be reconstructed from a single face image. Lastly, virtual multi-pose face images can be obtained by texture mapping, rotation, and projection of the established 3D face, and training samples are enriched. Experimental results obtained using BJUT-3D and CAS-PEAL-R1 face databases show that recognition rate of the proposed method is 91%, which is better than other methods for pose-invariant face recognition based on a single view. This is primarily because the training samples are enriched using the proposed 3D sparse morphable model based on a new dense correspondence method.

Keywords: Pose-invariant face recognition · 3D face reconstruction · Single view · 3D morphable model · Dense correspondence

1 Introduction

Currently, face recognition technologies exhibit good performance under normal conditions [1–3]. However, face recognition with pose variations is still a major challenge because an object with different poses may appear considerably different owing to nonlinear deformation and self-occlusion [4]. Multi-pose face recognition based on a single view has high research value because it can be used in a wide range of practical applications. Researches on multi-pose face recognition technology based on a single view can be divided into two primary categories, i.e., pose correction methods and virtual multi-pose image synthesizing methods. Virtual multi-pose image synthesizing methods can improve the accuracy of face recognition because they generate appropriate virtual multi-pose face images to enrich training samples.

G. Li et al. (Eds.): KSEM 2017, LNAI 10412, pp. 422–431, 2017.
DOI: 10.1007/978-3-319-63558-3_36

Among virtual multi-pose image synthesizing methods, three-dimensional (3D) face morphable models are an important area of research. The establishment of this model is based on a set of 3D faces [5]. However, original 3D face data cannot be used for linear calculation because the numbers of vertices and patches are different for each 3D face. It is necessary for all vertices and patches of the 3D faces to have one-to-one correspondence. A large number of methods have been proposed for dense correspondence of 3D faces. Blanz et al. proposed an optical flow algorithm and a bootstrapping algorithm to solve the dense correspondence problem of 3D faces [6, 7]. The method is valid when the 3D face is similar to the reference face. Yongli Hu et al. proposed a dense correspondence method based on a grid resampling method, and the result was more accurate [8]. However, the process was very complicated because considerable manual interaction was required. Gu et al. proposed a uniform grid resampling algorithm [9]. Gong Xun et al. proposed a grid resampling method based on a planar template [10]. The method located features based on a combination of two-dimensional (2D) and 3D texture information, and divided the surface of a 3D face using these features.

To simplify the construction of a 3D face model, an improved dense correspondence method based on planar template resampling with geometric information is proposed in this paper. A 3D face is reconstructed using a 3D sparse morphable model, and multi-pose face images are generated using texture mapping, model rotation, and projection. The rest of the paper is organized as follows: In Sect. 2, an improved dense correspondence method based on planar template resampling with geometric information is proposed. In Sect. 3, 3D face reconstruction based on the 3D face morphable model is presented. In Sect. 4, the experiments performed using BJUT-3D and CAS-PEAL-R1 databases are described, and the proposed method is compared to other methods. Finally, conclusions are stated in Sect. 5.

2 An Improved Dense Correspondence Method Based on Planar Template Resampling with Geometric Information

2.1 Definition of Planar Template

After face segmentation on the cylindrical expansion of the original 3D faces, segmented 2D texture maps are aligned using the nose tip positions and overlapped. Among the overlapped points, the pixels that belong to more than half of the images are used to construct a planar template, which is shown in Fig. 1(a).

The corresponding vertices of the selected pixels are scattered, and meshes are built using these vertices to construct a normalized 3D face model. The topological structure is constructed through the following steps: First, vertices are connected according to the connection rules of vertices on the vertical and horizontal iso-lines. Then, each small rectangular area is divided into two triangles by the diagonal between the bottom left corner and the top right corner of this area. Lastly, the triangles are numbered from top to bottom and left to right, which is shown in Fig. 1(b). The template can be used to generate the normalized 3D face model.

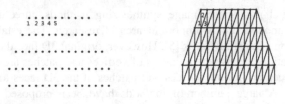

(a) Scattered points and numbers   (b) Topological structure

Fig. 1. Definition of planar template

2.2 Resampling of Original 3D Faces

An original 3D human face is not consistent with the definition of the planar template, and it is necessary to resample it.

(a) Vertical sampling

For an original 3D face, f_i $(1 \leq i \leq m)$, the vertex on the kth line can be calculated using interpolation. pre_{ver} and fol_{ver} shown in Eqs. (1) and (2) are used to represent the line numbers before and after the kth line, respectively.

$$pre_{ver} = floor(ratio_{ver} \times k) \tag{1}$$

$$fol_{ver} = ceil(ratio_{ver} \times k) \tag{2}$$

In Eqs. (1) and (2), *floor* is a rounded down function, *ceil* is a rounded up function, and $ratio_{ver}$ is the ratio between the height of the segmented 2D texture image, $f_i(h)$, and the height of the planar template, $Tmpt(h)$.

$$ratio_{ver} = f_i(h)/Tmpt(h) \tag{3}$$

The resampling data (including the shape and texture data) of the vertex, $v_k(vert)$, on the kth line can be calculated using interpolation between the vertex, $v_{pre}(ori)$, on the *pre* line and the vertex, $v_{fol}(ori)$, on the *fol* line.

$$v_k(vert) = v_{pre}(ori) \times (1 - \Delta r) + v_{fol}(ori) \times \Delta r$$
$$\Leftrightarrow \begin{cases} s_k(vert) = s_{pre}(ori) \times (1 - \Delta r) + s_{fol}(ori) \times \Delta r \\ t_k(vert) = t_{pre}(ori) \times (1 - \Delta r) + t_{fol}(ori) \times \Delta r \end{cases} \tag{4}$$

where $\Delta r = ratio \times k - pre$.

(b) Horizontal sampling

The interpolation operation that was used in vertical sampling is used to calculate the lth data on the kth line. pre_{hor} and fol_{hor} shown in Eqs. (5) and (6) are used to represent the row numbers before and after the lth row, respectively.

$$pre_{hor} = floor(ratio_{hor} \times l) \qquad (5)$$

$$fol_{hor} = ceil(ratio_{hor} \times l) \qquad (6)$$

where $ratio_{hor}$ is the ratio between the width of the segmented 2D texture image, $f_i(w)$, and the width of the planar template, $Tmpt(w)$.

$$ratio_{hor} = f_i(w)/Tmpt(w) \qquad (7)$$

The lth resampling data on the kth line (including the shape data and texture data) of the vertex, $v_{k,l}(final)$, can be obtained using interpolation between the vertex, $v_{k,pre}(vert)$, on the pre row and the vertex, $v_{k,fol}(vert)$, on the fol row.

$$v_{k,l}(final) = v_{k,pre}(vert) \times (1 - \Delta r) + v_{k,fol}(vert) \times \Delta r$$
$$\Leftrightarrow \begin{cases} s_{k,l}(final) = s_{k,pre}(vert) \times (1 - \Delta r) + s_{k,fol}(vert) \times \Delta r \\ t_{k,l}(final) = t_{k,pre}(vert) \times (1 - \Delta r) + t_{k,fol}(vert) \times \Delta r \end{cases} \qquad (8)$$

where $\Delta r = ratio \times l - pre$.

After arranging the resampled data according to the topology structure of the planar template, each 3D face can be represented as two vectors.

$$\begin{cases} s_i = (x_1 y_1, z_1, \ldots, x_j, y_j, z_j, \ldots, x_n, y_n, z_n)^T \in R^{3n} \\ t_i = (r_1, g_1, b_1, \ldots, r_j, g_j, b_j, \ldots, r_n, g_n, b_n)^T \in R^{3n} \end{cases}, 1 \le i \le N \qquad (9)$$

In Eq. (9), s_i is the shape vector composed of three coordinates of the ith 3D face, t_i is its corresponding texture vector composed of R value, G value, and B value, N is the number of 3D faces, and n is the number of facial points on the 3D face. Values with the same subscript represent the same facial feature points on different face vectors.

3   3D Face Reconstruction Based on Sparse Morphable Model

3D face reconstruction based on a sparse morphable model consists of two steps, i.e., model construction and face shape reconstruction. The first step involves acquisition and normalization of prototype 3D face data and establishment of a 3D morphable model. The second step is to match a target face image with the morphable model and accomplish the reconstruction of the face.

3.1   Construction of 3D Morphable Model

Normalized 3D faces have an equal number of vertices and patches. They can be regarded as a linear space, where each element can be expressed linearly using other elements. The shape of a 3D face can be expressed as Eq. (10).

$$s_i = (x_1, y_1, z_1, \ldots, x_k, y_k, z_k, \ldots, x_n, y_n, z_n)^T \in R^{3n}, i = 1, \ldots, m \qquad (10)$$

where (x_k, y_k, z_k) is the coordinate of the k th vertex, v_k, n is the number of vertices, and m is the number of 3D faces. The linear space constructed using m 3D faces can be expressed as Eq. (11).

$$S = (s_1, \ldots, s_m) \in R^{3n \times m} \tag{11}$$

The shape of a new face can be expressed as Eq. (12).

$$s_{new} = S \cdot a, a = (a_1, \ldots, a_i, \ldots, a_m) \tag{12}$$

where $a_i \in [0, 1]$ and $\sum_{i=1}^{m} a_i = 1$.

The $m'(m' \leq m - 1)$ vectors corresponding to the $m'(m' \leq m - 1)$ largest eigenvalues can be used to construct the feature matrix, $Q = (q_1, \ldots, q_{m'})$, and Eq. (12) can be rewritten as Eq. (13).

$$s_{new} = \bar{s} + Q \cdot \beta = \bar{s} + \Delta s \tag{13}$$

In Eq. (13), $\bar{s} = \frac{1}{m} \sum_{i=1}^{m} s_i$ and $\beta = (\beta_1, \ldots, \beta_i, \ldots, \beta_{m'})^T \in R^{m'}$. This implies that a specific face can be obtained using a deformation on the average face.

3.2   Face Shape Reconstruction Based on Sparse Morphable Model

Face shape reconstruction based on the sparse morphable model can be described as the analysis of global shape deformation, Δs, according to the shape deformation, Δs^f.

$$\Delta s^f = L(s - \bar{s}) = L(Q \cdot \beta) = Q^f \cdot \beta \tag{14}$$

where L represents the selected feature points.

The computation of β can be transformed into determining the optimal solution of the objective function, as shown in Eq. (15).

$$E(\beta) = \left\| Q^f \cdot \beta - \Delta s^f \right\|^2 + \eta \cdot \|\beta\|^2 \tag{15}$$

where the first term represents reconstruction error, the second term represents random fluctuation, which improves the robustness of the model to noise, and $\eta \geq 0$ is an adjusting parameter.

According to singular value decomposition, Eq. (16) is obtained by taking the derivative of β with respect to the objective function.

$$\arg \min \|E(\beta)\| = V \cdot \left(\frac{\lambda_i}{\lambda_i^2 + \eta} \right) \cdot U^T \cdot \Delta s^L \tag{16}$$

where $U \in R^{l \times l}$ and $V \in R^{m' \times m'}$.

Combining Eq. (16) with Eq. (13), the final reconstruction result of the given face can be described as Eq. (17).

$$s_{new} = \bar{s} + Q \cdot V \cdot \left(\frac{\lambda_i}{\lambda_i^2 + \eta} \right) \cdot U^T \cdot \Delta s \qquad (17)$$

4   Experiments and Analysis

4.1   3D Face Database and the Segmentation Performance

The BJUT-3D database [11] is used to evaluate the performance of the proposed method. There are approximately 200000 vertices and 400000 triangular faces for one 3D face model. An example of original 3D face data is shown in Fig. 2. Face segmentation results based on geometric information are shown in Fig. 3.

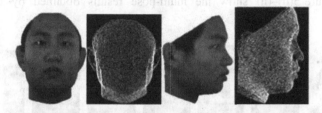

Fig. 2. Geometry and texture information of one person in BJUT-3D database

Fig. 3. Face segmentation results based on geometric information

4.2   Dense Correspondence Performance of the Proposed Method

The data of 100 3D faces from the BJUT-3D database, consisting of 50 females and 50 males, are used to evaluate the dense correspondence performance of the proposed method. A comparison between the original 3D point cloud data and the dense correspondence results is shown in Fig. 4. It can be observed from Fig. 4 that the 3D face database can be normalized using the proposed method and is suitable for linear operations to be performed later.

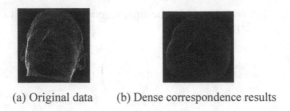

(a) Original data   (b) Dense correspondence results

Fig. 4. Dense correspondence results of the proposed method

4.3   3D Reconstruction Results of the Proposed Method

A hundred face images with a natural expression and no light variation are used from the CAS-PEAL-R1 database to evaluate the reconstruction performance of the proposed method. Two original face images are shown in Fig. 5(a). The images are normalized to a size of 164 × 146 pixels. The 3D reconstruction results obtained using the proposed method are shown in Fig. 5(b). The texture mapping results are shown in Fig. 5(c). Figure 5(d)–(n) show the multi-pose results obtained by rotation and

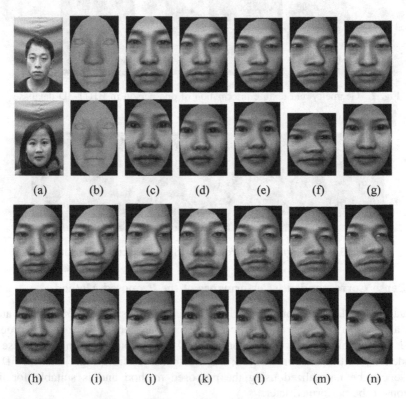

Fig. 5. 3D face reconstruction results (a) Input images (b) Reconstructed results (c) Texture mapping results (d)–(n) Multi-pose mapping results of the 3D face

projection of the 3D faces. Figures 5(d), (e), and (f) show the projected results of 15°, 30°, and 45° rotation, for the pose of looking in front horizontally. Figures 5(g), (h), (i), and (j) show the projected results of 0°, 15°, 30°, and 45° rotation, for the pose of looking down. Figures 5(k), (l), (m), and (n) show the projected results of 0°, 15°, 30°, and 45° rotation, for the pose of looking up.

4.4 Recognition Results of the Proposed Method

A hundred subjects with a natural expression and no light variation are used from the CAS-PEAL-R1 database to evaluate the performance of the proposed method for face recognition. Twelve multi-pose images for each subject obtained by rotation and projection of the 3D sparse morphable model, as described in Sect. 4.3, are used as training samples. Five real multi-pose face images for each subject are used as testing samples. The performance of the proposed method is compared to that of conventional methods, and the results are shown in Table 1.

Table 1. Recognition performance comparison of the proposed method and other methods

Methods	Recognition rate
Enhanced $(PC)^2A$ [12]	0.71
PAFE [13]	0.88
TPSM [14]	0.90
Proposed method	0.91

For the enhanced projection-combined principal component analysis $((PC)^2A)$ method, the recognition rate is relatively low. This is primarily because the new images are dependent and are highly correlated with the original image. The recognition rates of the pose adaptive feature extraction (PAFE) method and the triplet pose sparse matrix (TPSM) method are relatively high because a priori information about 3D face data is used. For the proposed method, training images are enriched by rotation and projection of the reconstructed 3D face. It can be observed from Table 1 that the recognition rate of the proposed method is 91%, which is superior to state-of-the-art methods for pose-invariant face recognition.

5 Conclusions and Discussions

To synthesize virtual multi-pose faces, a 3D face reconstruction method based on a single view is proposed in this paper. This method includes three aspects. The first is planar template establishment based on geometric information. The second is vertical and horizontal resampling of 3D face data based on the geometric relationship between the planar template and original 3D face data, based on which normalized 3D face data is obtained. The third is that a 3D sparse morphable model is constructed based on the normalized 3D face data, and multi-pose face images are generated by texture mapping, rotation, and projection of the established 3D face. The proposed method has two contributions. First, the dense correspondence process is performed automatically and the model construction process does not involve manual interaction, which makes it superior to other methods. Second, virtual multi-pose face images are synthesized using the proposed method and training samples are enriched, because of which recognition performance improves.

A pose-invariant face recognition problem without illumination variations is considered in this study. Face recognition with pose and illumination variations is more difficult. Future research will focus on a pose-invariant face recognition problem with illumination variations based on a single view.

Acknowledgments. This work was partially supported by a grant from the National Natural Science Foundation of China (No. 61401355), a grant from the Key Laboratory Foundation of Shaanxi Education Department, China (No. 14JS072) and a grant from Science and Technology Project Foundation of Beilin District, Xi'an City, China (No. GX1621). The authors also thank anonymous reviewers for their valuable comments.

References

1. Wright, J., Yang, A.Y., Ganesh, A., Sastry, S.: Robust face recognition via sparse representation. IEEE Trans. Pattern Anal. Mach. Intell. **31**(2), 210–227 (2009)
2. Ding, C., Choi, J., Tao, D., Davis, L.S.: Multi-directional multi-level dual-cross patterns for robust face recognition. IEEE Trans. Pattern Anal. Mach. Intell. **38**(3), 518–531 (2016)
3. Ahonen, T., Hadid, A., Pietikainen, M.: Face description with local binary patterns: application to face recognition. IEEE Trans. Pattern Anal. Mach. Intell. **28**(12), 2037–2041 (2006)
4. Sharma, R., Patterh, M.S.: A new pose invariant face recognition system using PCA and ANFIS. Optik-Int. J. Light Electron Opt. **126**(23), 3483–3487 (2015)
5. Blanz, V., Vetter, T.: A morphable model for the synthesis of 3D faces. In: Computer Graphics Proceedings SINGRAPH 1999, pp. 187–194 (1999)
6. Blanz, V., Vetter, T.: Face recognition based on fitting a 3D morphable model. IEEE Trans. Pattern Anal. Mach. Intell. **25**(9), 1063–1074 (2003)
7. Blanz, V.: Face recognition based on a 3D morphable model. In: Proceedings of the 7th International Conference on Automatic Face and Gesture Recognition, vol. 25(9), pp. 617–624. IEEE Computer Society (2006)
8. Hu, Y., Yin, B., Gu, C., Cheng, S.: 3D face reconstruction based on the improved morphable model. Chin. J. Comput. **28**(10), 1671–1679 (2005)
9. Gu, C., Yin, B., Hu, Y., Cheng, S.: Resampling based method for pixel-wise correspondence between 3D faces. In: The Proceedings of the International Conference on Information Technology: Coding and Computing, pp. 614–619 (2004)
10. Gong, X., Wang, G.: 3D face deformable model based on feature points. J. Softw. **20**(3), 724–733 (2009)
11. Hu, Y., Zhang, Z., Xu, X., Fu, Y., Huang, T.S.: Building large scale 3D face database for face analysis. In: Sebe, N., Liu, Y., Zhuang, Y., Huang, T.S. (eds.) MCAM 2007. LNCS, vol. 4577, pp. 343–350. Springer, Heidelberg (2007). doi:10.1007/978-3-540-73417-8_42
12. Songcan, C., Daoqiang, Z., Zhihua, Z.: Enhanced (PC)^2 A for face recognition with one training image per person. Pattern Recognit. Lett. **25**(10), 1173–1181 (2004)

13. Yi, D., Lei, Z., Li, S.Z.: Towards pose robust face recognition. In: Proceedings of IEEE Conference on Computer Vision and Pattern Recognition, vol. 9(4), pp. 3539–3545. IEEE Computer Society (2013)
14. Moeini, A., Moeini, H., Faez, K.: Real-time pose-invariant face recognition by triplet pose sparse matrix from only a single image. In: 2014 Proceedings of 22nd International Conference on Pattern Recognition, pp. 465–470 (2014)

MIAC: A Mobility Intention Auto-Completion Model for Location Prediction

Feng Yi[1,2], Zhi Li[1(✉)], Hongtao Wang[1,2], Weimin Zheng[1], and Limin Sun[1]

[1] Beijing Key Laboratory of IOT Information Security,
Institute of Information Engineering, CAS, Beijing 100093, China
lizhi@iie.ac.cn
[2] School of Cyber Security, University of Chinese Academy of Sciences,
19 A Yuquan Rd, Shijingshan District,
Beijing 100049, People's Republic of China

Abstract. Location prediction is essential to many proactive applications and many research works show that human mobility is highly predictable. However, existing works are reported with limited improvements in using generalized spatio-temporal features and unsatisfactory accuracy in complex human mobility. To address these challenges, a *Mobility Intention and Auto-Completion* (MIAC) model is proposed. We extract mobility patterns to capture common spatio-temporal features of all users, and use mobility intentions to characterize these mobility patterns. A new predicting algorithm based on auto-completion is then proposed. The experimental results on real-world datasets demonstrate that the proposed MIAC model can properly capture the regularity in human mobility by simultaneously considering spatial and temporal features. The comparison results also indicate that MIAC model significantly outperforms state-of-the-art location prediction methods, and can also predict long range locations.

Keywords: Location prediction · Mobility intention · Auto-completion · Spatiotemporal data

1 Introduction

Along with the popularity of smart devices with sensing technology, the past decade has witnessed a tremendous increase in the availability of mobility data, ranging from early cellular tower data of *Personal Communication Systems* (PCS) [1], GPS trajectories [2] to the check-in data of various location-based services (LBSs) [3]. In addition, many systems originally designed for fare purpose are naturally enriched with mobility data, for example, the *Smart Card Data* (SCD) of public transport [4]. Those available massive mobility data boom many appealing research works in human mobility, including location prediction, social strength inference and privacy preservation [5] etc. Location prediction is considered as the core function of various proactive systems, such as mobile marketing, emergency response [6] and public security [7].

© Springer International Publishing AG 2017
G. Li et al. (Eds.): KSEM 2017, LNAI 10412, pp. 432–444, 2017.
DOI: 10.1007/978-3-319-63558-3_37

For its potential value, location prediction has attracted significant research efforts in the past few years, and various location prediction models have been proposed to explore the regularity embedded in human mobility data. Typically two essential sub tasks are involved in those models: *how to discover and represent the mobility regularity*; and *how to utilize the mobility regularity for prediction*. Traditional location prediction models usually represent the mobility regularity as the common locations or the spatio-temporal features, such as Markov model [8]. By assuming that human mobility patterns are uniform and arise consecutively, many prediction algorithms can be designed based on the Markov property.

Although theoretical study shows that the predictability of human mobility has a boundary of 93% [1], the performance of state-of-the-art location prediction models is far below this limit. Accordingly, there are two significant challenges to reach this theoretic limit of location prediction: first, how to effectively characterize those regularities embedded in human mobility data; second, how to properly utilize those mobility regularities for location prediction.

In this paper, we tackle above challenges by proposing a novel location prediction model called the *Mobility Intention Auto-Completion* (MIAC) model. Through a tensor decomposition method, we extract the mobility intentions from mobility data and then characterize regular patterns by mobility intentions. Here, the *mobility intention* refers to the common cause that prompts a user to transit from the current location to the next location, such as commuting, entertainment and recreation. As will be shown in Sect. 4.4, more regularities are occurring in the mobility patterns characterized by mobility intentions. After that, we propose a location prediction algorithm based on the *Query Auto-Completion* (QAC) mechanism which predicts a user's intended query in information retrieval [9]. As showing in Fig. 1, if considering the partial mobility intention as a prefix and the mobility pattern composed by intentions as a query, the location prediction task in our context is very similar to the QAC prediction in information retrieval: when more mobility intentions are available, it is easier to predict future mobility intentions, which can then be used to predict future locations.

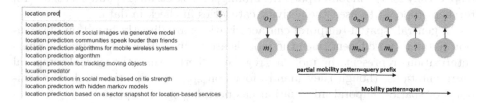

Fig. 1. An analogy between mobility intent prediction and query autocompletion.

The main contributions of this paper are summarized as follows:

- We propose a new representation to characterize the regularities in mobility pattern through mobility intentions, which simultaneously captures the diversity and the similarity of human movement.
- We develop a new predicting algorithm based on auto-completion mechanism, which can make long range location prediction in context of multiple mobility patterns.

The rest of this paper is organized as below. In Sect. 2, we review related work. Section 3 provides detail description of MIAC model. In Sect. 4, the performance of MIAC model is evaluated and analyzed. Finally, we summarize the conclusions and envisage the future work in Sect. 5.

2 Related Work

2.1 Mobility Regularity Characterization

Spatio-Temporal Characterization. Most pioneering works have been mostly focused on the mobility regularity in locations [2,10,11], while some also resorted to temporal mobility patterns. For example, Gao et. al. [12] proposed *HPY Prior Hourly Daily* model (HPHD) and extracted the temporal regularities in hours and days by Gaussian distribution. Salvatore et. al. [13] focused on the regularity on the epoch and dwell time of a location. Lian et. al. and Wang et. al. studied the location regularity constrained by times [7,14]. Some researchers also considered the spatial-temporal regularity on the mobility dataset. For example, [3] proposed the *periodic mobility model* (PMM) which considers both temporal and spatial regularity between home and workplaces.
Semantic Characterization. Semantic characterization captures the mobility regularity as hidden states which characterize the *semantic* information such as location category, activity category or user relationship. Likhyani et. al. [15] illustrated that the accuracy can be improved by using the location categories. YING et. al. [16] generated the geographic, temporal and semantic intentions respectively for the mobility patterns mining, and Ye et. al. [17] and Yu et. al. [18] further proposed to used the activity categories in check-in data.

In general, spatial-temporal characterizations are directly based on observation dataset and efficient once the models are trained. However, semantic characterizations represent the mobility regularity better than the spatial-temporal representations, though they mainly focus on spatial semantic, while the temporal or spatio-temporal information has been largely unexplored.

2.2 Prediction Algorithm

Various algorithms have been proposed for location prediction. Petzold et. al. [19] evaluated five machine learning techniques and found that Markov predictors usually achieve better accuracy with lower resource requirements. Since then, variants of Markov predictors have been widely used in location prediction models [8,11,18], such as the high order Markov [15] and the *Variable Order Markov* (VOM) [10]. However, those models mainly focus on the simple mobility pattern between the next location and a fixed number of past observations. In fact, human mobility is complex and dynamic, this calls for more flexible location prediction algorithm that can capture and utilize multiple mobility patterns which may contain different number of elements.

3 MIAC Model

3.1 Problem Statement

Let $\mathcal{L} = \{\ell_1, \ell_2, \cdots, \ell_M\}$ be a set of M locations, and $\mathcal{U} = \{u_1, u_2, \cdots, u_N\}$ be a set of N users. Each observation o is defined as a tuple (ℓ, t) which represents that user $u_j \in \mathcal{U}$ visited location $\ell \in \mathcal{L}$ at time t. The history observations of user u_j are represented by $O_j = \{o_{j,1}, o_{j,2}, \cdots, o_{j,n_j}\}$ with length n_j, and n_j may vary with different users. We use $\mathcal{D} = \{O_1, O_2, \cdots, O_N\}$ to denote the mobility dataset.

Definition 1 (Location Prediction). *Given a mobility dataset $\mathcal{D}$, the location prediction aims to predict a user u_j's future locations $\ell_{j,n_j+1}, \cdots$ at different time.*

By leveraging the mobility intention and the QAC mechanism, we propose a novel location prediction model called *Mobility Intention and Auto-Completion* (MIAC) model. We decompose the location prediction into the following three sub-problems:

Problem 1 (Mobility Intention Extraction and Transformation). Given a mobility dataset $\mathcal{D}$, how to extract those mobility intentions from $\mathcal{D}$ and map $O_j (1 \leq j \leq N)$ into the mobility intention sequence $s_j (1 \leq j \leq N)$.

Problem 2 (Mobility Intention Prediction). Given s_j and O_j, how to design a prediction algorithm for u_j's future mobility intentions $m_{j,n_j+1}, \ldots$.

Problem 3 (Future Location Prediction). Given $\mathcal{D}$, $S = \{s_j | 1 \leq j \leq N\}$ and $m_{j,n_j+k} (1 \leq k \leq k_{max})$, how to predict the future locations $\ell_{j,n+k} (1 \leq k \leq k_{max})$.

3.2 Mobility Intention Extraction and Transformation

In general, the mobility of a creature is driven by various intentions such as mating needs and food resources [20]. Similarly, the human mobility is fundamentally driven by diverse intentions, ranging from job commuting, family entertainment to routine and social activities. Mobility intention is the key factor for human mobility, while mobility patterns characterized by mobility intention are more regular than other characterizations, as will be shown in experiments. Hence, we use the mobility intention to characterize the mobility patterns.

Non-negative Tensor Factorization (NTF) is effective in analyzing the interrelationship between spatial and temporal features in mobility dataset and *CP* decomposition is a widely used NTF algorithm [21]. Through the *Non-negative Tensor Factorization (NTF)*, we can extract basic mobility pattern which can explain why a user visited a place at a given time and day from mobility dataset $\mathcal{D}$ [22]. These basic mobility patterns can be considered as mobility intentions. Here, the mobility intention extraction aims to construct a three-dimensional tensor $Y \in \mathbb{R}^{M \times H \times D}$ composed by location-hour-day. The element y_{r_i,t_j,d_k} of three-way tensor Y is computed as

$$y_{r_i,t_j,d_k} = \frac{Count(r_i, t_j, d_k)}{\sum_{q=1}^{M} Count(r_q, t_j, d_k)} \tag{1}$$

where r_i, t_j, and d_k are index of *location, time bins* and *day* respectively, and $Count(r_i, t_j, d_k)$ is the number of users who visited location r_i at time t_j on the d_k-th day. Then we factorize the tensor into a linear combination of rank-one tensors by CP [21] which can be represented as follows:

$$\mathbf{Y} \approx \sum_{r=1}^{R} \lambda_r Y_r \tag{2}$$

Every Y_r is the outer product of three vectors $\mathbf{h}_r$, $\mathbf{l}_r$ and $\mathbf{d}_r$:

$$Y_r = \mathbf{h}_r \circ \mathbf{l}_r \circ \mathbf{d}_r \tag{3}$$

If considering those vectors in the right hand side of Eq. (3) as inherent characteristics of mobility intention for *hour, location* and *day*, respectively. Y_k can be used to explain why people move to the location at certain time. Therefore, we consider every Y_k as a mobility intention, and use m_i for the i-th mobility intention and $\mathcal{M} = \{m_i | 1 \leq i \leq w\}$ for the set of mobility intentions.

Every mobility intention m in M can be considered as one class. Each history observation o corresponds to one mobility intention, in other words, belongs to a class. We choose the *support vector machine* (SVM) and adopt the $one - versus - the - rest$ strategy to construct multi-class SVM classifiers [23]. After finishing training, SVM classifier can map an observation sequence $O_j = \{o_{j,1}, o_{j,2}, \ldots, o_{j,n_j}\}$ to a mobility intention sequence $s_j = \{m_{j,1}, m_{j,2}, \ldots, m_{j,n_j}\}$. The regular mobility patterns can be mined from s_j to facilitate the location prediction.

3.3   Mobility Intention Prediction

Most existing location predicting algorithms assume that human mobility is composed by simple mobility patterns with a fixed number of elements. Schneider et. al. [24] further show that human mobility can be decomposed to multiple unique networks motifs with variable-length elements. Accordingly, human mobility can be considered as a set of mobility patterns rather than a single one.

In this work, we define the *mobility patterns* as an ordered subsequence composed by mobility intentions that repeatedly occur in the user's mobility intention sequence, as shown in Fig. 2. If the mobility pattern to which the current mobility intention belongs can be identified, future mobility intentions can then be predicted. Figure 2 shows two mobility patterns of one user: mobility pattern 1 is the weekdays pattern characterized by four mobility intentions: m_1 for buying fresh vegetables, two continuous m_2 for commuting between home and work place and the following m_3 for recreation; mobility pattern 2 is the weekend pattern characterized by three mobility intentions: m_4 for entertainment, m_2 for commuting back to home and m_5 for family dinners. If the last unknown mobility pattern is *mobility pattern 1*, it's easy to know that the next mobility intention is likely m_2. Furthermore, we can predict the second next mobility intention as m_3.

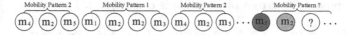

Fig. 2. Mobility patterns in mobility intentions sequence.

However, it is not easy to identify the mobility pattern from mobility intention sequence, because various mobility patterns may share the same intention subsequence. Though mobility pattern can hardly be solely determined by the current mobility intention, with more previous mobility intentions available, the mobility pattern gets more focused and clear. For example in Fig. 2, if only given the current mobility intention m_2, it is not sure whether the current mobility pattern is *mobility pattern 1* or *mobility pattern 2*. When m_1 is known to follow m_2, the mobility pattern can then be identified as *mobility pattern 1*.

This inference process is similar to QAC, the essential feature in modern search engines, such as *Google, Baidu* and *Bing*. QAC aims to predict users intention and formulate queries by providing completion suggestions as soon as the first few words are typed. When more query words are available, user's query intention get more and more clear. We adopt the QAC framework into our new predicting algorithm. Learning to QAC, in generally, is a supervised learning and has a two-step process [9]: *filtering by prefix-trees* and *learning to rank*.

The filtering by prefix-trees is often facilitated by using data structures such as prefix-trees (tries) that allow efficient lookups by prefix matching. We consider the arbitrary subsequence of s_j as one mobility pattern, and split it into prefix and corresponding candidate. we use **MITrie** which is similar to prefix-trees to store the pair of prefix and corresponding candidate. Once the prefix-tree is created, we can apply virtually any existing learning-to-rank algorithm for auto-completion ranker, such as *Most Popular Completion* (MPC) [9]. MPC aggregates the query frequencies over a query log, and uses these aggregated values to rank the QAC suggestions. When the prefix tree and the ranking strategy are decided, given a prefix $s_j^{w_j} = \{m_{j,n_j-w_j+1}, \ldots, m_{j,n_j}\}$ with length w_j, candidate mobility patterns are ranked by their past popularity.

It is noted that identified mobility pattern may include multiple future mobility intentions. Hence, the proposed algorithm has the ability of long-range mobility intention prediction. In the following sections, we use m_f to denote the next mobility intention. By now, given a user's mobility intention sequence s_i, we can obtain the future mobility intention m_f through the identified mobility patterns by QAC-based algorithm.

3.4   Location Prediction

Intuitively, given u_j's future mobility intention m_f, locations mapped to mobility intentions m_f in O_j are candidates for m_f. We use $C^*_{m_f}$ to denote the candidates set. The popularity of locations can be simply used as a ranking score. However in reality, human mobility behaviors are influenced by spatial-temporal context. Hence, candidate locations must satisfy spatial and temporal constraints in the context. Suppose u_j's current observation is $o_c = (\ell_c, t_c)$. Let

$$C_{m_f} = \{o_j | o_j \in C^*_{m_f} \wedge dist(\ell_j - \ell_c) < \delta \wedge t_j - t_c < \gamma\} \tag{4}$$

where δ, γ are thresholds for the distance and the time respectively. We consider the locations in set C_{m_f} as candidates of m_f. Then, the popularity of location is the probability of $p_{u_j}(\ell_k)$:

$$p_{u_j}(\ell_k) = Count(\ell_k)/r \qquad 1 \leq k \leq v, \ell_k \in L_{m_f} \tag{5}$$

where $Count(\ell_k)$ is number of ℓ_k's occurrence times in C_{m_f}, r is the cardinality of C_{m_f}, L_{m_f} is v distinct locations in C_{m_f}.

On the other hand, due to the personality trait of neophilia [14], which is associated with exploratory activity in response to novel stimulation, people also show their interests in exploring unvisited locations. A user seeks a new location for certain mobility intention m_f, the most possible choice is the preferred location by majority of other users with the same mobility intention m_f in the same context for social conformity. Following a similar analysis to those in one user's future location candidates, we obtain the location candidates $L^g_{m_f}$ with v_g locations come from the observation set

$$C^g_{m_f} = \{o_i | o_i \in C^{g^*}_{m_f} \wedge l_i \notin L_{m_f} \wedge dist(\ell_i - \ell_c) < \delta \wedge t_i - t_c < \gamma\} \tag{6}$$

where $C^{g^*}_{m_f}$ is observations of all users which is mapped to mobility intention m_f. The probability of locations in $C^g_{m_f}$ can be evaluated as:

$$p_g(\ell_k) = Count(\ell_k)/r_g \qquad 1 \leq k \leq r_g \tag{7}$$

where r_g is the cardinality of $C^g_{m_f}$.

The neophilia for a location is a personality trait related with mobility intention. We use $\lambda(0 \leq \lambda \leq 1)$ to measure the degree for a user with the mobility intention m_t to explore a novel location. If λ is close to 0, the future locations are mainly influenced by the user u_j's visited locations. Otherwise, the predicted location is mainly decided by other users with the same mobility intention. Hence, we utilize λ to combine $p_u(\ell_k)$ and $p_g(\ell_k)$, and the candidate's probability is evaluated as

$$p(l_k) = (1 - \lambda)p_u(\ell_k) + \lambda p_g(\ell_k) \tag{8}$$

Given a future mobility intention m_f, we use (8) to compute the probability of candidate locations and take the one with the max probability as the prediction. Or more generally, top-k ($k \geq 1$) results can be generated.

4   Experiment and Analysis

4.1   Data Set and Settings

In the experiment, we use three real-world datasets with decreasing sampling frequency: GeoLife, BBSC and Gowalla. Among them, Geolife dataset is a public available GPS trajectory mobility dataset [2]. We extract the stay point as the location to be predicted, following the method presented by Zheng et al. in [2]. The BBSC dataset collects prepaid smart card records for public transportation in Beijing, China. We obtained a dataset with $275,951,094$ bus transaction records about $16,161,460$ users in October of 2014, which contains more than 90% of Beijing urban public traffic lines. Gowalla dataset [3] is a publicly available check-in dataset with $6,442,890$ check-ins of $196,591$ users from February 2009 to October 2010. Mobility data of each user are split into training portion and testing portion in chronological order. We construct the prediction model from the training portion and test the performance of the constructed model on the test dataset.

Four baseline models are chosen for performance comparison: *Most Frequent Location* (MF), *First Order Markov Model* (Markov1), HPHD [12] and PMM [3]. Among them, MF uses the most popular location in user's history observation for prediction. Relationship between the current and the next location can be captured by Markov1 Model and then for location prediction. HPHD is a probabilistic mobility model based on spatial-temporal features. PMM adopts a Gaussian mixture model to learn user locations constrained by independently truncated Gaussian distribution temporal component.

For performance evaluation, we use the *accuracy* and *accuracy@top10*. The accuracy is the fraction of locations for which the predicted locations are exactly the true location among all predicted locations in the test dataset, while the *accuracy@top10* is the percentage of accurate predictions for a list of predictions with length 10.

4.2   Performance Comparison on the Next Location Prediction

The prediction accuracy of the MIAC model and baseline models on three datasets are shown in Fig. 3, which shows that the accuracy of MIAC is higher

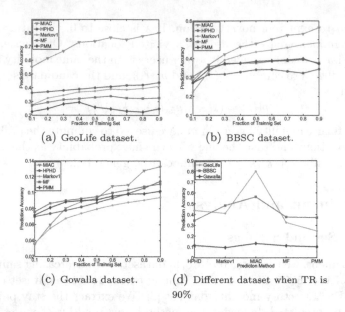

(a) GeoLife dataset.

(b) BBSC dataset.

(c) Gowalla dataset.

(d) Different dataset when TR is 90%

Fig. 3. The performance comparison of different predicting models on different dataset.

than that of baseline models in most cases. For the `GeoLife` dataset, MIAC's accuracy is nearly twice as those of baseline models. One possible reason is that `GeoLife` is with dense sampling and contains enough information of mobility regularity, and mobility patterns are more regular when characterized by mobility intention. `BBSC` dataset in Fig. 3(b) is much sparser than `GeoLife` dataset. Markov1 is the best baseline model on this data set, and this implies that frequent patterns exist in the observation. The accuracy of MIAC exceeds Markov1 by 15%, and this result implies that other forms of mobility patterns exist but can not be captured by Markov1.

Figure 3(c) shows that the performance of MIAC and baseline models on `Gowalla` dataset is not as good as on the other two datasets. The accuracy of all baseline models are less than 10%. This could be attributed to the sparsity and randomness in check-in data. Even so, MIAC exceeds other baseline models when training fraction (TR) is more than 0.5. As shown in Fig. 3(d), the performance of different prediction models on three datasets varies when TR is 90%. For example, HPHD achieves better performance than Markov1 on the `GeoLife` dataset, but it's not good as Markov1 on the `BBSC` dataset. However, MIAC is robust with different amount of mobility regularity available in three different dataset. Our experiments demonstrates that: the use of mobility intention representation and QAC-based prediction algorithm improve the location prediction performance.

4.3   Performance Comparison on the Future Location Prediction

The MIAC model predicts a number of future intentions, hence it can predict not only the immediately next but also future locations. Figure 4 shows *accuracy@top10* for future locations prediction on three mobility datasets. It shows that the *accuracy@top10* of the second next location is 0.56, 0.38 and 0.17 respectively, corresponding to GeoLife, BBSC and Gowalla dataset, respectively. Even for the fifth next location, the prediction *accuracy@top10* is 0.31 and 0.23 for GeoLife and BBSC dataset. It's indicates that the proposed MIAC model is promising in future locations prediction.

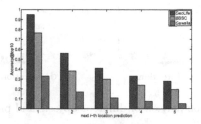

Fig. 4. The long range forecasting ability of MIAC.

4.4   Performance Comparison on Mobility Intention and QAC-Based Predicting Algorithms

In order to determine the effect of the characterization in mobility intention and the QAC-based predicting algorithm (see Sect. 3.3), We perform a set of comparisons on the BBSC dataset. we use location, location category [15] and mobility intention as the characterizations of mobility patterns, and compare the Markov1 and QAC-based predicting algorithm in experiments.

Both Fig. 5(a) and (b) show that the prediction accuracy increases remarkably using mobility intention to characterize mobility patterns. These results implies that mobility intentions can characterize the regularities in mobility patterns better than other features. Figure 5(c) and (d) show the comparison of Markov1 and QAC-based predicting algorithms. QAC-based prediction algorithm is 10% and 3% higher than Markov1 algorithm when characterization through mobility intention and locations, respectively. Figure 5(d) shows that there is limited improvement when mobility patterns is characterized by locations. On the other hand, Fig. 5(c) shows that there is almost 10% performance improvement by mobility intention, it also shows that complex patterns exist in the observation when using aggregate features to represent observation sequence. QAC-based algorithm has ability to exploit more regular mobility patterns and is more suitable for location predicting than Markov1.

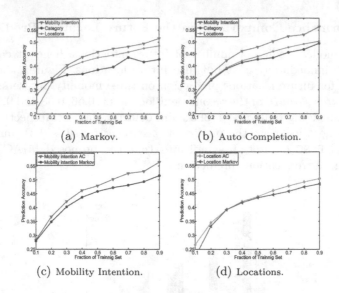

(a) Markov.   (b) Auto Completion.

(c) Mobility Intention.   (d) Locations.

Fig. 5. Performance comparison of different objects and predicting algorithms.

5   Conclusions

In this paper, we have proposed a new location prediction model called MIAC, which considers the mobility intention as hidden variable and adopts the auto completion mechanism into the location prediction. Extensive experiments indicate that the proposed model significantly outperforms existing location prediction models. In our future work, we plan to adopt more sophisticated learning-to-rank techniques, such as Lambda-MART or RankSVM, to further improve the performance of mobility intention prediction.

Acknowledgments. This work was supported by the National Natural Science Foundation of China (Grant No. 61472418), the Major R&D Plan (Grant No. 2016YFC 1202204).

References

1. Song, C., Zehui, Q., Blumm, N., Barabsi, A.-L.: Limits of predictability in human mobility. Science **327**(5968), 1018–1021 (2010)
2. Zheng, Y., Zhang, L., Xie, X., Ma, W.-Y.: Mining interesting locations and travel sequences from GPS trajectories. In: Proceedings of the 18th International Conference on World Wide Web, WWW 2009, pp. 791–800, New York, NY, USA. ACM (2009)
3. Cho, E., Myers, S.A., Leskovec, J.: Friendship and mobility: user movement in location-based social networks. In: Proceedings of the 17th ACM SIGKDD International Conference on Knowledge Discovery and Data Mining, San Diego, CA, USA, 21–24 August 2011, pp. 1082–1090 (2011)

4. Itoh, M., Yokoyama, D., Toyoda, M., Tomita, Y., Kawamura, S., Kitsuregawa, M.: Visual fusion of mega-city big data: an application to traffic and tweets data analysis of metro passengers. In: 2014 IEEE International Conference on Big Data (Big Data), pp. 431–440, October 2014
5. Li, H., Sun, L., Zhu, H., Lu, X., Cheng, X.: Achieving privacy preservation in wifi fingerprint-based localization. In: 2014 Proceedings IEEE INFOCOM, pp. 2337–2345. IEEE (2014)
6. Liu, W., Li, H., Chen, Y., Zhu, H., Sun, L.: Lares: latency-reduced neighbour discovery for contagious diseases prevention. Int. J. Ad Hoc Ubiquitous Comput. **16**(1), 3–13 (2014)
7. Wang, Y., Yuan, N.J., Lian, D., Xu, L., Xie, X., Chen, E., Rui, Y.: Regularity and conformity: location prediction using heterogeneous mobility data. In: Proceedings of the 21th ACM SIGKDD International Conference on Knowledge Discovery and Data Mining, KDD 2015, pp. 1275–1284. ACM (2015)
8. Ashbrook, D., Starner, T.: Using GPS to learn significant locations and predict movement across multiple users. Pers. Ubiquitous Comput. **7**(5), 275–286 (2003)
9. Bar-Yossef, Z., Kraus, N.: Context-sensitive query auto-completion. In: Proceedings of the 20th International Conference on World Wide Web, WWW 2011, Hyderabad, India, 28 March–1 April 2011, pp. 107–116 (2011)
10. Yang, J., Xu, J., Xu, M., Zheng, N., Chen, Y.: Predicting next location using a variable order markov model. In: Proceedings of the 5th ACM SIGSPATIAL International Workshop on GeoStreaming, IWGS 2014, pp. 37–42. ACM (2014)
11. Chen, M., Liu, Y., Yu, X.: Predicting next locations with object clustering and trajectory clustering. In: Cao, T., Lim, E.-P., Zhou, Z.-H., Ho, T.-B., Cheung, D., Motoda, H. (eds.) PAKDD 2015. LNCS (LNAI), vol. 9078, pp. 344–356. Springer, Cham (2015). doi:10.1007/978-3-319-18032-8_27
12. Gao, H., Tang, J., Liu, H.: Mobile location prediction in spatio-temporal context. In: Nokia Mobile Data Challenge Workshop, vol. 41, p. 44 (2012)
13. Scellato, S., Musolesi, M., Mascolo, C., Latora, V., Campbell, A.T.: NextPlace: a spatio-temporal prediction framework for pervasive systems. In: Lyons, K., Hightower, J., Huang, E.M. (eds.) Pervasive 2011. LNCS, vol. 6696, pp. 152–169. Springer, Heidelberg (2011). doi:10.1007/978-3-642-21726-5_10
14. Lian, D., Xie, X., Zheng, V.W., Yuan, N.J., Zhang, F., Chen, E.: Cepr: a collaborative exploration and periodically returning model for location prediction. ACM Trans. Intell. Syst. Technol. **6**(1), 8:1–8:27 (2015)
15. Likhyani, A., Padmanabhan, D., Bedathur, S.J., Mehta, S.: Inferring and exploiting categories for next location prediction. In: Proceedings of the 24th International Conference on World Wide Web Companion, WWW 2015, Florence, Italy, 18–22 May 2015 - Companion Volume, pp. 65–66 (2015)
16. Ying, J.J.-C., Lee, W.-C., Tseng, V.S.: Mining geographic-temporal-semantic patterns in trajectories for location prediction. ACM TIST **5**(1), 2 (2013)
17. Ye, J., Zhu, Z., Cheng, H.: What's your next move: user activity prediction in location-based social networks. In: Proceedings of the 2013 SIAM International Conference on Data Mining, pp. 171–179 (2013)
18. Yu, C., Liu, Y., Yao, D., Yang, L.T., Jin, H., Chen, H., Ding, Q.: Modeling user activity patterns for next-place prediction. Syst. J. IEEE, (99), 1–12 (2015)
19. Petzold, J., Bagci, F., Trumler, W., Ungerer, T.: Comparison of different methods for next location prediction. In: Nagel, W.E., Walter, W.V., Lehner, W. (eds.) Euro-Par 2006. LNCS, vol. 4128, pp. 909–918. Springer, Heidelberg (2006). doi:10.1007/11823285_96

20. Giannotti, F., Pappalardo, L., Pedreschi, D., Wang, D.: A complexity science perspective on human mobility. In: Mobility Data: Modeling, Management, and Understanding, pp. 297–314. Cambridge University Press (2013)
21. Kolda, T.G., Bader, B.W.: Tensor decompositions and applications. SIAM Rev. **51**, 455–500 (2009)
22. Fan, Z., Song, X., Shibasaki, R.: Cityspectrum: a non-negative tensor factorization approach. In: The 2014 ACM Conference on Ubiquitous Computing, UbiComp 2014, Seattle, WA, USA, 13–17 September 2014, pp. 213–223 (2014)
23. Bishop, C.M.: Pattern Recognition and Machine Learning. Springer, Heidelberg (2006)
24. Schneider, C.M., Belik, V., Couronné, T., Smoreda, Z., González, M.C.: Unravelling daily human mobility motifs. J. Royal Soc. Interface **10**(84) (2013)

Automatically Difficulty Grading Method Based on Knowledge Tree

Jin Zhang[1], Chengcheng Liu[1,3], Haoxiang Yang[1],
Fan Feng[1,2], and Xiaoli Gong[1,2(✉)]

[1] College of Computer and Control Engineering, Nankai University,
Tianjin 300-350, China
gongxiaoli@nankai.edu.cn
[2] State Key Laboratory of Computer Architecture,
Institute of Computing Technology, Chinese Academy of Sciences, Haidian, China
[3] China Academy of Information and Communications Technology, Beijing, China

Abstract. The aim of the current study is to propose a model, which can automatically grade difficulty for a question from "instruction system" question bank. The system mainly uses 4 attributes with 26 features based on principal component analysis, which are employed to be input of the Automatically Difficulty Grading Model (ADGM). A knowledge tree model and a machine learning algorithm are utilized as important parts for the classification module. The experimental dataset "instruction system" question bank is based on our built "Principles of Computer Organization" online education system, the accuracy result of difficulty classification could be 77.43% which is much higher than the accuracy of random guess 50%.

Keywords: Natural language processing · Difficulty classifier systems · Education · Knowledge tree · Automatically Difficulty Grading Method (ADGM)

1 Introduction

Textual entailment is an important issue in the study of natural language understanding. Due to early and well-developed English language analysis tools like WordNet or grammar parser, many approaches like [1] and so on for English textual entailment were proposed. In contrast, there are fewer applications for Chinese language analysis and the performance is not so good like that for English. Apparently, Chinese textual entailment is still quite a difficult issue.

With the development of Internet, online education is more and more popular, whose market size and user size are increasing year by year. So as to promote industrial development and innovative online education technology, our laboratory launched the "Principles of Computer Organization" online education system. There are many challenges and difficulties in achieving the functional modules of automation, intelligence and precision education. When achieving

G. Li et al. (Eds.): KSEM 2017, LNAI 10412, pp. 445–457, 2017.
DOI: 10.1007/978-3-319-63558-3_38

auto-generating test paper function module, the investigation contents and item difficulty for the questions of test paper need to be balanced and mastered. Then, how to automatically grade difficulty for the multiple-choice questions bank of the "instruction system" is a problem that need to be resolved.

The aim of the current study is to propose a model, which can solve the above problem. The model mainly uses about 26 features and a machine learning algorithm is utilized as a classifier for the model and features of "Instruction System" question bank are employed to be input for scoring process. The rest of this paper is organized as follows. Section 2 discusses previous related studies and indicates relations between our proposed method and previous methods. Section 3 introduces features and grading method developed by this study. Section 4 illustrates experimental result of adopting our proposed method for data set produced by "Principles of Computer Organization" online education system made by us own. Finally, this paper displays experimental result and some future work as well as conclusion.

The contributions of this paper are summarized as follows:

- In our research area, we are the first to investigate using questions' attributes for grading its difficulty.
- The knowledge tree system was established based on the proper nouns from the chapter of "instruction system" text of "Principles of Computer Organization" teaching material.
- Some pretreatments for question data of "instruction system", about 340 from 1409 question bank of "Principles of Computer Organization" online education system, including to gain an outcome of difficulty grading.
- Attributes system has been established, which acted in concert with the machine learning algorithm made the accuracy of difficulty classification to be 77.4336%, the result is much higher than the accuracy of random guess 50%.

2   Related Work

Many papers about classification are hot in today's research field (*e.g.* [2–4]). The classification of Chinese articles has been welcomed by many researchers. [5] utilized natural language processing to understand the context of Chinese questions. [6–8] used guide list method and word segmentation technology. Under normal circumstances, before word segmentation, pre-processing like removing punctuation, tables and graphics should have been done. Whether to retain English characters should rely on the specific needs of application. The current word segmentation algorithm is divided into four parts: string matching word segmentation, understanding word segmentation, statistical word segmentation and semantic segmentation. This study needs to get rid of redundant words referred to [9]. Word string matching calculation and other current algorithms are all based on semantic dictionary or corpus word semantic similarity algorithm.

It is the most important to complete the processes of pre-prepared data sets and attribute selection. In the process of setting up, data set generation was inspired by [10]. [11] seriously discussed four labels. [12] gave inspiration for attributes setting, which added external labels such as environment labels. In this paper, we combine these three methods and also referred to [13–15]. For assessing the feature of "Option similarity", Chinese scoring technique like [16] and so on are effective. In the meanwhile, [17] discusses the design and implementation of an optimization algorithm of automatic grading for subjective questions by using Chinese words segmentation, based on which, this paper filled in the appropriate parameters flexibly and optimize the mathematical formula.

Some papers can be referred for the questions attributes. [18] extracted the logical formula frame of simple mathematical problem. [19] consider a much more expressive class of logical forms, and show how to use dynamic programming to efficiently represent the complete set of consistent logical forms. However, their data is very simple to be understood.

Knowledge tree building referred to the research on structure like [20] and research on semantic knowledge base like [21–23] etc. Knowledge tree leaves also referred to [24, 25]. In order to make the difficulty classification be more credible, the final classification is still using the exact machine learning algorithm.

Though various classification models have been proposed, none of existing works has investigated or tried to grade Chinese questions' difficulty using 4 attributes with 26 features.

3 Methodology

3.1 Algorithm Architecture

Our Automatically Difficulty Grading Model (ADGM) is based on machine learning, but also includes a knowledge tree and a mechanism of attributes setting. As shown in Fig. 1, for the first, a knowledge search tree model should be built according to teaching material based on word segmentation module. Then,

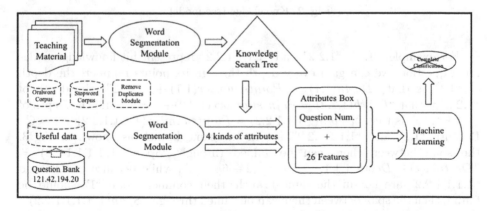

Fig. 1. An overview of architecture. Given a dataset, a high-precision model is induced.

a useful dataset should be prepared for training the model, which should remove the oral words, stop words and duplicate words. For this problem, there are four kinds of attributes should be extracted. Finally, necessary features sorted out by certain methods act as input for machine learning, and the difficulty grading is the output.

3.2   Knowledge Tree Model

Figure 2 shows the knowledge tree model based on the proper nouns from the chapter of "instruction system" text of "Principles of Computer Organization" teaching material. Each leaf node represents a knowledge point that was labeled by index with point and number. The index can be used to gain the depth of hit knowledge node and span between hit knowledge nodes. The relationship between knowledge nodes consists of the weight and the connection relation like equivalence, subclass or some other else.

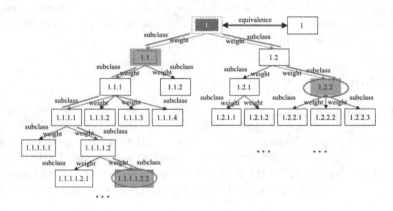

Fig. 2. Knowledge tree model

For example, "1.1", "1.2.2" and "1.1.1.1.2.2" represent hit knowledge nodes for a question, we can gain every depth through its points number, the depth of "1.1" is that "$Depth(1.1) = Pointsnumber(1.1) + 1 = 2$", the depth of "1.2.2" is that "$Depth(1.2.2) = Pointsnumber(1.2.2) + 1 = 3$", and the Depth of "1.1.1.1.2.2" is that "$Depth(1.1.1.1.2.2) = Pointsnumber(1.1.1.1.2.2) + 1 = 6$". Because "1.1" and "1.1.1.1.2.2" are in the same branch as shown in Fig. 3, the span between them can be gained through "$Span(1.1, 1.1.1.1.2.2) = \|Depth(1.1) - Depth(1.1.1.1.2.2)\| = \|2 - 6\| = 4$", while because "1.2.2" and "1.1.1.1.2.2" are not in the same branch, their common node "1" should be found, then the span between them can be gained through "$Span(1, 1.1.1.1.2.2) + Span(1, 1.2.2) = 5 + 2 = 7$", and so on. When necessary, multiply the weight between two near knowledge nodes.

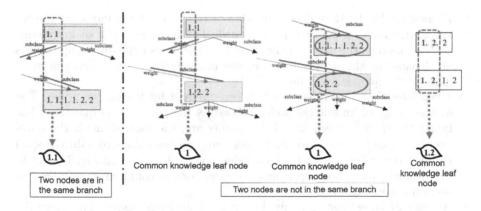

Fig. 3. The common node and branch of knowledge tree

3.3 Attributes Box

The method of designing the attributes box is brainstorming and expert collecting. From the actual situation, the objective reflects the difficulty of the examination. A question consists of conditions, questions, options, answers, and analysis, the treatment standard of the difficulty analysis for a question is shown as Table 1. The artificial thinking for judging the difficulty of a question generally includes four attributes (*e.g.* quantity, relationship, span range, environmental attributes and so on). From these four attributes, we choose the following specific attributes for each category, including the knowledge points, options, the question words, analysis words, the number of non-Chinese characters variables and basic conditions of the question writers, etc.

Table 1. Four attributes of question difficulty classification analysis

Knowledge tree	Quantity relationship	User portrait	Option similarity
Knowledge depth	Word number	Examination	ABCD similarity
Knowledge span	Analysis length	performance	Numeric options
Number of knowledge	Non-Chinese character no	of question	Text options
Word frequency	Number of unknown words	writers	Hybrid options
Number of relationship	/new words		

Attribute of Knowledge Tree. Knowledge tree attributes contains the knowledge depth, knowledge span, knowledge number, the word frequency in the textbook, and the number of the relationship between the near knowledge nodes for a question. The word segment of a question is the same as the leaf node of the knowledge tree is called hit knowledge node or hit knowledge point. Every attribute is as follows:

(1) *Knowledge Depth.* Base on the number of the point of the index for the hit knowledge node, a depth can be gained. The average depth of knowledge nodes is used as the depth of the question. The deeper the knowledge points are located in the knowledge tree, the more careful inspection, the more difficult the question will be.

(2) *Knowledge Span.* There is span between different hit knowledge nodes. The average span of knowledge nodes is used as the span of the question. The larger the span between knowledge points are, the more difficult the question is. As shown in Algorithm 1, there are two conditions to gain the span through the index for the hit knowledge nodes. As shown in Fig. 3, what in the dashed square box represents the index of the common knowledge leaf node between two hit nodes.

(3) *Number of knowledge.* The number of hit knowledge nodes. The more the number, the more species, the more difficult the question will be.

(4) *Knowledge Word Frequency.* The average frequency for the hit knowledge nodes. The higher the frequency, the more simple the question is.

(5) *Relationship Number.* Theoretically, there is connection relation like equivalence, subclass or some other else. The more the number of the kinds of the relationship between hit knowledge nodes, the more species, the more difficult the question is.

Algorithm 1. Span between near nodes

1: $L_n \leftarrow$ Every index of leaf node
2: **if** Two nodes are in the same branch **then**
3:     $Span = \|Depth(L_1) - Depth(L_2)\|$
4: **else**
5:     Find the common knowledge leaf node
6:     $L \leftarrow$ The index of the common node
7:     $Span = \|Depth(L_1) - Depth(L)\| + \|Depth(L_2) - Depth(L)\|$
8: **end if**
9: return Span

Attribute of Quantity Relationship. Quantity Relationship includes word number of the whole question, analysis length, the number of Non-Chinese character and the number of unknown words/new words. Word number of the whole question consists of condition word number and option word number, analysis length is counted separately, it is hard to understand non-Chinese character, so that it is necessary to count the number of non-Chinese characters such as letters, symbols and so on of conditions and options.

Attribute of User Portrait. The degree of difficulty is connected with the character of writers or some other external factors. Under normal circumstances, students with good performance, serious and challenging character are more probably going to deal with hard questions. Hence, external user portrait features with study performance can also be added to the attributes box.

Attribute of Option Similarity. Option similarity of a question consists of internal attributes with magnitude and knowledge coverage and external attributes with sentence length and grammatical morphology. Hybrid options contain numeric and text content. It is need to compare two options firstly. And with the increase of the number of options, the dimension of similarity features will increase. For example: if there are AB two options, the similarity of A and B is needed. If there are ABC three options, the similarities of AB, AC, BC are needed, and so on, in order to form a combination pattern.

The option similarity (OS) between option X_1 and option X_2 can be computed:

$$OS(X_1, X_2) = W * [SM, SK, SL, SC, SP]$$

where SM, SL, SC and SP are all functions with the input of (X_1, X_2), respectively represent StringMagnitude, StringKnowledgeSimila, StringLengthSimilarity, StringCutLengthSimilarity and StringPartSimilarity. W denotes the unit column vector.

The parameters represent option similarity features respectively. Specific labeling rules are as follows:

(1) *Internal Attributes*
   1. *The options involve the orders of magnitude for the number.* The more similar the orders of magnitude between different options is, the more similar the options will be. It focuses on numerical options. As shown in Algorithm 2, the value can ben computed.

Algorithm 2. StringMagnitude

1: **procedure** SM(*Option*1, *Option*2)
2:　　Op1=Option1+1
3:　　Op2=Option2+1
4:　　$magnitude = \| \lg(Op1)) - \lg(Op2)) \|$
5:　　**if** $magnitude > 1$ **then**
6:　　　　StringMagnitude=0
7:　　**else**
8:　　　　StringMagnitude=1
9:　　**end if**
10:　　return StringMagnitude
11: **end procedure**

---

   2. *Knowledge coverage computing.* The more similar the knowledge coverage between different options is, the more similar the options will be. It focuses on text options. As shown in Algorithm 3, Firstly, to gain the option segmentation vector through the segmentation function quecut(). Then, to gain the vector with average depth, average span, number, word frequency and number of unknown words/new words from the hit knowledge nodes through knowledge tree function knowledgetree(). Finally, to gain the Euclidean distance of two vectors. The smaller the distance, the closer the options, the more difficult the question is.

Algorithm 3. StringKnowledgeSimilary

1: option1cut=quecut(Option1), option2cut=quecut(Option2)
2: op1know=knowledgetree(option1cut), op2know=knowledgetree(option2cut)

$$StringKnowledgeSimila = EucDistance(op1Know, op2Know)$$

3: return StringKnowledgeSimila

(2) *External Attributes*

(1) *Length of option words (Word Number).* The smaller the length difference between different options is, the more similar the options will be. It focuses on text/mixed options. The SL can be gained from Algorithm 4.

(2) *Word segment length after segmenting option.* The more similar the word segment length between different options is, the more similar the options will be. It focuses on text/mixed options. The SC can be gained from Algorithm 4.

(3) *Part of speech of word segment.* The more similar the part of speech of word segment between different options is, the more similar the options will be. It focuses on text/mixed options. The SP can be gained from Algorithm 4.

Algorithm 4. Similarity value of SL, SC, SP

1: $Op1 \leftarrow Option1$; $Op2 \leftarrow Option2$
2: **if** len(Op1)==len(Op2)==0 **then**
3:     $StringLengthSimila = 1$
4: **else**

$$StringLengthSimila = 1 - \frac{\|(len(Op1) - len(Op2))\|}{(len(Op1) + len(Op2))}$$

$StringLengthSimila \in [0, 1]$
5: **end if**
6: **RETURN StringLengthSimila**

$$StringCutLengthSimila = 1 - \frac{\|CutNo(Op1) - CutNo(Op2)\|}{CutNo(Op1) + CutNo(Op2)}$$

$StringCutLengthSimila \in [0, 1]$
7: **RETURN StringCutLengthSimila**

$$StringPartSimila = \frac{GetPartSimila(Op1, Op2)}{MaxLength(CutNo(Op1), CutNo(Op2))}$$

$StringPartSimila \in [0, 1]$
8: **RETURN StringPartSimila**

4 Evaluation

4.1 Question Bank Data Preprocessing

This experiment data, 1409 reliable questions have been exacted from the database of the website of "121.42.194.20", *i.e.* "Principles of Computer Organization" online education system which is directed by our laboratory professor.[1] The unreliable questions with non-standard format, serious oral problems and so on have been removed, the reliable questions should be cleaned up and supplemented, and the superfluous spaces and symbols should be deleted. The extracted questions have been dissected as question number, condition, question writer number (student number or nickname), option A, option B, option C, option D, answer, analysis, chapter and so on. According to the classification of the subject category, the "instruction system" chapter has 341 questions, this 340 questions are classified by three levers including 125 easy questions, 114 medium questions and 101 hard questions.

4.2 "Instruction System" Ontology Knowledge Tree Building

Textbook "computer organization" is used to be knowledge corpus material. "Instruction System" knowledge tree has been built by Protégé. The method for knowledge tree model is constructed as follows: first of all, the content of the book is graded according to chapter. The chapter level is equivalent to the level of the leaf nodes in the knowledge tree. Secondly, we compare it with oralword (Speech corpus) and stopword (Punctuation corpus) to remove the colloquia words and the punctuation. Then we remove the replicate words and build a synonyms system, so the content of the book is cutted into different phrases. The rules for removing the replicate words are "First Cut First Keep". Finally, we complete the index number for every leaf node in the cleaned and supplemented knowledge tree from top to bottom, as shown in Fig. 4. A full knowledge tree is finished.

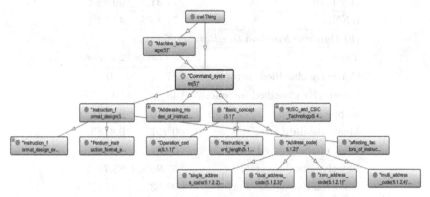

Fig. 4. Sketch map of treetrump of knowledge tree

---

[1] Data can be get from http://mobisys.cc/pages/resource.html.

4.3   Outcome of Classification Accuracy

This study has put the data of 26 (Attributes) $*$ 340 (Instances) into several machine learning algorithms classifiers, results of which are obtained in Table 2(a), among which, the "k" value for KNN is from 1 to 150, the average accuracy interval for 1000 times with 5 folds is [56.47%(k = 149),71.15%(k = 16)] for 2 classes, and [39.45%(k = 149), 47.14%(k = 21)] for 3 classes, meaning that the worst results of ADGM are better than random guesses. It can be seen that the random forest algorithm is the best choice for the classifier, based on which, the classification result of Table 4 can be get. As shown in Table 2(b), the correctly classified accuracy is 61.2069% by using 66% percentage split for training data and 34% left for testing data. The detailed accuracy is shown in Table 3. The correctly classified accuracy is 77.4336% by using cross-validation with 10 fold. And the detailed accuracy is shown in Table 3. From the results we can see

Table 2. Outcome summary

(a) Accuracy of different algorithms		
ML Algorithm	2 classes	3 classes
bayes.NaiveBayes	77.31%	59.48%
NaiveBayesSimple	77.31%	59.48%
bayes.BayesNet	77.31%	59.48%
tree.J48	48.28%	34.48%
tree.J48graft	48.28%	34.48%
tree.RandomForest	77.43%	61.21%
tree.RandomTree	53.45%	38.79%
rules.PART	48.28%	34.48%
lazy.LBR	77.31%	59.48%
KNN	71.15%	47.14%
(b) Outcome based on tree.RandomForest		
Type	3 classes	2 classes
Correctly classified instances	71	270
Incorrectly classified instances	45	70
Kappa statistic	0.4157	0.5882
Mean absolute error	0.257	0.4189
Root mean squared error	0.46	0.4329
Relative absolute error	57.96%	83.78%
Root relative squared error	98.53%	86.57%
Coverage of cases (0.95 level)	75%	
Mean rel. region size (0.95 level)	46.26%	
Total number of instances	116	340

Table 3. Detailed accuracy (Top lines for 3 classes and bottom lines for 2 classes)

TP rate	FP rate	Precision	Recall	F-measure	MCC	ROC area	PRC area	Class
0.7	0.184	0.667	0.7	0.683		0.818		1
0.512	0.28	0.5	0.512	0.506		0.677		2
0.629	0.123	0.688	0.629	0.657		0.857		3
0.612	0.2	0.614	0.612	0.612		0.78		W Avg.
0.976	0.475	0.718	0.976	0.827	0.577	0.935	0.942	1
0.525	0.024	0.946	0.525	0.675	0.577	0.935	0.928	3
0.774	0.274	0.820	0.774	0.759	0.577	0.935	0.935	W Avg.
0.848	0.509	0.646	0.848	0.734	0.365	0.767	0.798	1
0.491	0.152	0.747	0.491	0.593	0.365	0.767	0.704	2
0.678	0.339	0.694	0.678	0.666	0.365	0.767	0.753	W Avg.
0.807	0.545	0.626	0.807	0.705	0.282	0.756	0.789	2
0.455	0.193	0.676	0.455	0.544	0.282	0.756	0.695	3
0.642	0.379	0.650	0.642	0.630	0.282	0.756	0.745	W Avg.

Table 4. Confusion matrix for 3 classes and 2 classes

a	b	c	a	b	a	b	a	b
28	12	0	122	3	106	19	92	22
10	21	10	48	53	58	56	55	46
4	9	22						

that random forests are better than other decision trees. Summary of the reasons, there may be the following aspects. The decision tree ignores the correlation between the attributes in the data set, and the random forest can detect the interaction between the features and the importance of the features during the training process. The decision tree will be confronted with great difficulty in dealing with missing data, and the random forest has strong fault tolerance for training data. When the data is missing it can still maintain the same accuracy.

5   Conclusion and Future Work

In this study, we addressed the problem of automatically difficulty grading for a Chinese question. We propose an Automatically Difficulty Grading Model (ADGM) based on random forest machine learning algorithm. Evaluation results demonstrate that the summaries produced by our proposed model have high accuracy and recall. In the meanwhile, a question bank and a knowledge tree were contributed. In this study, we do not consider the logicality of a question, because it is really very challenging to get the logicality. In the future, besides studying the logicality, we will explore to make use of all of the chapters to improve the knowledge tree, so that the difficulty factors can be automatically applied for generating a test paper.

Acknowledgments. The authors would like to thank Master Weilai Liu for providing the help of the *ADGM* model. The authors would also like to thank the anonymous referees for their valuable comments and helpful suggestions. The work is supported by the Specialized Research Fund for the Doctoral Program of Higher Education of China award (No. 20130031120028), Research Plan in Application Foundation and Advanced Technologies in Tianjin award (No. 14JCQNJC00700), Open Project of the State Key Laboratory of Computer Architecture, Institute of Computing Technology, Chinese Academy of Sciences award under Grant No. CARCH201604.

References

1. Woods, A.M.: Exploiting linguistic features for sentence completion. In: The 54th Annual Meeting of the Association for Computational Linguistics, p. 438 (2016)
2. Sahu, S.K., Anand, A.: Recurrent neural network models for disease name recognition using domain invariant features. arXiv preprint arXiv:1606.09371 (2016)
3. Sapkota, U., Solorio, T., Gomez, M.M., Bethard, S.: Domain adaptation for authorship attribution: improved structural correspondence learning. In: Proceedings of the 54th Annual Meeting of the Association for Computational Linguistics, Berlin, Germany, pp. 2226–2235 (2016)
4. Chang, T.-H., Hsu, Y.-C., Chang, C.-W., Hsu, Y.-C., Chang, J.-I.: Kc99: a prediction system for chinese textual entailment relation using decision tree. In: NTCIR, Citeseer (2013)
5. Gkatzia, D., Lemon, O., Rieser, V.: Natural language generation enhances human decision-making with uncertain information. arXiv preprint arXiv:1606.03254 (2016)
6. Benzschawel, E.: Identifying potential adverse drug events in tweets using bootstrapped lexicons. YEARACL 2016, p. 15 (2016)
7. Varjokallio, M., Klakow, D.: Unsupervised morph segmentation and statistical language models for vocabulary expansion. In: Proceedings of the 54th Annual Meeting of the Association for Computational Linguistics, vol. 2, pp. 175–180 (2016)
8. Qian, P., Qiu, X., Huang, X.: A new psychometric-inspired evaluation metric for chinese word segmentation. In: Proceedings of the 54th Annual Meeting of the Association for Computational Linguistics, pp. 2185–2194 (2016)
9. Feifei, Z., Chengqing, Z.: Research on recognition of fillers in spoken dialogs. J. Chin. Inf. Process. **3**, 017 (2011)
10. Mehta, P.: From extractive to abstractive summarization: a journey. ACL 2016, p. 100 (2016)
11. Wan, X., Wang, T.: Automatic labeling of topic models using text summaries. In: Proceedings of the 54th Annual Meeting of the Association for Computational Linguistics, vol. 1, pp. 2297–2305 (2016)
12. Wei12, Z., Liu, Y., Li, Y.: Is this post persuasive? Ranking argumenta-tive comments in the online forum. In: The 54th Annual Meeting of the Association for Computational Linguistics, p. 195 (2016)
13. Logacheva, V., Lukasik, M., Specia, L.: Metrics for evaluation of word-level machine translation quality estimation. In: Proceedings of the 54th Annual Meeting of the Association for Computational Linguistics, Berlin, Germany (2016)
14. Poursabzi-Sangdeh, F., Boyd-Graber, J., Findlater, L., Seppi, K.: Alto: active learning with topic overviews for speeding label induction and document labeling. In: Proceedings of the 54th Annual Meeting of the Association for Computational Linguistics, vol. 1, pp. 1158–1169 (2016)

15. Kazemian, S., Zhao, S., Penn, G.: Evaluating sentiment analysis in the context of securities trading. In: Proceedings of the 54th Annual Meeting of the Association for Computational Linguistics, vol. 1, pp. 2094–2103 (2016)
16. Chang, T.-H., Lee, C.-H., Chang, Y.-M.: Enhancing automatic chinese essay scoring system from figures-of-speech. In: PACLIC (2006)
17. Jia, H.E., Lin, C.: Optimization of chinese word segmentation based on neural network. J. Chengdu Univ. Inf. Technol. 6, 008 (2006)
18. Mitra, A., Baral, C.: Learning to use formulas to solve simple arithmetic problems. ACL (2016)
19. Pasupat, P., Liang, P.: Inferring logical forms from denotations. arXiv preprint arXiv:1606.06900 (2016)
20. Persing, I., Ng, V.: Modeling stance in student essays. In: Proceedings of the 54th Annual Meeting of the Association for Computational Linguistics, Berlin, Germany, pp. 2174–2184. Association for Computational Linguistics (2016)
21. Yih, W.-T., Richardson, M., Meek, C., Chang, M.-W., Suh, J., Richardson, M., Meek, C., Yih, S.W.-T.: The value of semantic parse labeling for knowledge base question answering. In: Proceedings of ACL (2016)
22. Jia, R., Liang, P.: Data recombination for neural semantic parsing. arXiv preprint arXiv:1606.03622 (2016)
23. Khani, F., Rinard, M., Liang, P.: Unanimous prediction for 100% precision with application to learning semantic mappings. arXiv preprint arXiv:1606.06368 (2016)
24. Yu, D., Ji, H.: Unsupervised person slot filling based on graph mining. In: Proceedings of the 54th Annual Meeting of the Association for Computational Linguistics, Berlin, Germany, pp. 44–53 (2016)
25. Moosavi, N.S., Strube, M.: Which coreference evaluation metric do you trust? A proposal for a link-based entity aware metric. In: Proceedings of the 54th Annual Meeting of the Association for Computational Linguistics, vol. 1, pp. 632–642 (2016)

A Weighted Non-monotonic Averaging Image Reduction Algorithm

Jiaxin Han and Haiyang Xia[✉]

School of Computer Science, Xi'an ShiYou University, Shaanxi 710065, China
jiaxinhan@xsyu.edu.cn, haiyangxia15@gmail.com

Abstract. Image reduction is commonly used as a data pre-processing method in many image processing field, an efficient image reduction operator can underpinning many practical applications. Traditional monotonic averaging image reduction operator may lost some detail features during reduction. However, In certain task those small features have very important significance. Therefore, some scholars proposed a non-monotonic averaging image reduction algorithm, recent works focus on integrate the pixel cluster's space structure information into image representative pixel selection progress, it has certain practical significance but this method is only suitable for specific background pictures. To fill this gap, We propose an novel sigmoid function based weighted image reduction algorithm, which can be used to image reduction under different background colours. Experiments show that the proposed method has better image reduction effect on images with different background colors.

1 Introduction

In image processing, often need to reduce the image to cut down the computational complexity and storage space, so an efficient and accurate image reduction [1] operator is essential to many applications.

The most important task of image reduction is to preserve the image characteristics while minimizing the noise, making reduced image have the ability to express the original features. Furthermore the efficiency of image reduction is also required to match the speed of image acquisition, such as linear run time and parallel calculation. Traditional image reduction technique [14] is based on the image attributes or dimensions, such as rough set [10] theory based method. It takes the entire image as the raw input, making the computational complexity is very high and is not conducive to parallelization. And they don't focus the image size after reduction, however in some specific applications the size of the archived image has strict required.

In recent years, researchers have proposed block-based image reduction techniques [13], this method used a small image block as the smallest unit of the image operation, it characteristics of high efficient and effectiveness. Block-based image reduction focus on reducing image size and preserving image features, it's reduction methods is shown in Fig. 1.

© Springer International Publishing AG 2017
G. Li et al. (Eds.): KSEM 2017, LNAI 10412, pp. 458–465, 2017.
DOI: 10.1007/978-3-319-63558-3_39

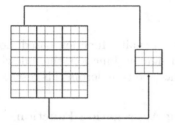

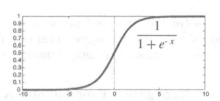

Fig. 1. Block-based image reduction methods

Fig. 2. Sigmoid function

If an image of size M*N, it can be divided into non-overlapping blocks of size m*n. Each blocks using aggregation function to drive an appropriate represent value. Accordingly, the image can be reduced to the size A*B in which A = $\lfloor \frac{M}{m} \rfloor$ and B = $\lfloor \frac{N}{n} \rfloor$. The most commonly used aggregation function in image reduction is averaging aggregation function [11]. Traditional aggregation function is monotonic aggregation function, [5] which means any increase in the input data the aggregation value must not decrease. When the data is noiseless this technique is useful, however in many application domains the raw data is often polluted by certain noise, those corrupted pixel must be disregard before aggregation. In order to better reduce these noisy data, non-monotone averaging aggregation function has been proposed in [2]. Non-monotone means that if the input of the function has slightly increased the corresponding output will also increased, however, when the input value has dramatic increased the output value will be reduced. For a sudden increase value, it's more likely be an outlier and have a negative contribution to the aggregation value. Recent advance in non-monotonic averaging aggregation function have proposed use penalty-based non-monotonic averaging function as aggregation technique and incorporates other relevant information. Such as Tim et al. incorporate the location information of pixels to the problem to identify the significant cluster in the raw image and reduce the image noise. [15] Gleb et al. proposed integration the geometrically compactness of cluster to drive the represent pixel value [4]. Which has a very good image reduction capability, but it has a certain background color limit, To fill this gap, we propose an sigmoid function based weighted image reduction operator, which can obtain a same weight under different background color, and this weight also integrate the spatial organization information of pixel cluster. The experiment result shows, this operator can preserve of fine, pixel-scale details under different background color, that is to say, this technique can provide an unified image reduction framework.

The rest of this paper is structured as follow. We introduce some theoretical basis about the method used in this paper at Sect. 2. In Sect. 3, present the main mechanism of the sigmoid function based background weighted non-monotone averaging image reduction operator. Section 4 describes the experiments and the corresponding analysis. Finally, we present the conclusion on Sect. 5.

2   Preliminaries

We solve the proposed problem by using the penalty function-based non-monotonic averaging aggregation technique and sigmoid function, the following content describe the principal mathematical conceptions underpin this research.

2.1   Aggregation Function and Averaging Aggregation Function

Aggregation function [12] is a function that can aggregate a series of input variables into a output scalar, it has a wide range of applications in many fields such as decision support system and expert systems. The definition of aggregation function as following:

Definition 1. *Aggregation function: A function $f : [a,b]^n \to [a,b]$ is an aggregation function if and only if it satisfies the following conditions:*

(1) bounds preservation: $f(a) = a$ and $f(b) = b$
(2) monotonicity: for all input x_1, $x_2 \in [a,b]^n$, if $x \leq y$ then $f(x_1) \leq f(x_2)$.
    The details and other excellent aggregate functions can find in [8].

But in image reduction background aggregation function must have two properties: averaging and idempotency. Averaging means aggregation function in image reduction filed must have averaging behaviour [6]. This type of aggregate function called averaging aggregation function, it's mathematical definition as follow:

Definition 2. *Averaging aggregation Function: An aggregation function f is averaging aggregation function, if for every input value x the corresponding output of the function is be bound to the interval $[min(x), max(x)]$.*

Idempotency denote that if all the input value are equivalent then the output must equal to the input.

2.2   Non-monotone Averaging Aggregation Function and Penalty Function

As mentioned in the introduction, non-monotone averaging aggregation function [2] has a wider application in the filed of image processing, precious work focus design the non-monotone averaging aggregation function in penalty-based method [7], it has good scalability, and has a certain application in the image reduction field. So this paper continues pursue design such penalty based technique. The definition of non-monotone averaging aggregation function and penalty function [7] as follow:

Definition 3. *Non-monotone Averaging Aggregation: If an averaging aggregation function f satisfy for all input $x_1, x_2 \in [a,n]^n$, and $x_2 = x_1 + \epsilon$, where ϵ is a non-negative constant, if ϵ less than a user specified threshold σ then $f(x_2) > f(x_1)$.*

Penalty function transforms the averaging aggregation problem into find an solution to a minimisation problem of the follow form:

$$f(x) = \underset{y}{\operatorname{argmin}} \; P(x, y) \tag{1}$$

For to maintain the necessary attributes of f, the selected function P must satisfy the following definition.

Definition 4. *Penalty-function: If and only if a function f satisfies the following condition then can be called penalty-function:*

(1) $P(x, y) \geqslant c \; \forall x \in [a, b]$
(2) $P(x, y) = 0$ if all $x_i = y$

The first condition ensure that P has a solution of the minimise problem, and the second condition providing idempotence of P.

3   Sigmoid Function Based Weighted Image Reduction Algorithm

In this section a novel sigmoid function based image reduction algorithm is proposed. Which can be weighted according to the area of different background colors to adapt to different background colors image reduction.

In [15], Wilkin et al. give an penalty function-based image reduction technique, which can favours compact clusters by given the small penalty to those pixels are close to the represent pixel. It was given as follow:

$$P(x, y) = \sum_{i=1}^{n} w_i(y) \rho(x_i, y) \tag{2}$$

where

$$\rho(x_i, y) = \begin{cases} r_{(k)} & r_{(k)} < \tau \\ r_{(k)} & r_{(k)} \geq \tau \end{cases}$$

$\tau = \alpha \, max(\epsilon, r_{(t)})$ and $\alpha > 0$, $0 \leq \beta \leq 1$, $2 \leq t \leq n$, and r_k represent the k th smallest difference of the set of ordered value, y is the given represent value. And $w_i(y)$ is normalised distance between pixel:

$$w_i(y) = \frac{d(x_i, y)}{\sum_{i=1}^{n} d(x_i, y)} \; \forall y = x_j \in x_1, ..., x_p \tag{3}$$

This operator can reduce the image according their pixel's normalised distance, but don't take into account the spatial organization of pixel cluster. So Gleb et al., used fuzzy measure constructs a new weight which can measure the pixel's spatial arrangement [3], it can characterizing compact and scattered groups of same pixel cluster, the closer the spatial arrangement of the pixel cluster the bigger the weight. Figure 3 is an example of different pixel cluster, Fig. 3a and b depict above mentioned spatial arrangement difference and their corresponding weight based on the proposed MST method.

This method can preserve fine pixel-scale details after reduction, but this method has a great drawback is that can only measure the pixel cluster compactness of image of black background. For example, if we want reduce an image of white background and it have the same pixel spatial organization with Fig. 3b, as show in Fig. 4a, we will get a wrong weight 0.95. For the proposed method in that paper, specifies that the vertex is black pixel when conducting the minimal spanning tree clustering [16]. And they assume that $S(\varnothing) = 0$, this means if a pixel block contains only the background color, then the weight is 0. So if we want reduce an image with a black feature on a white background, and if there have a block with no black pixel (as show in Fig. 4b) the reduced image will be black because the weight of this block is 0. It is unreasonable for a white background block reduced to black.

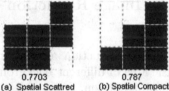

(a) 0.7703   (b) 0.787
Spatial Scattred Spatial Compact

Fig. 3. Different pixel cluster

(a) 0.945   (b)

Fig. 4. White background block with pixel spatial organization like Fig. 3b and White background block

For solving this problem, we using the sigmoid function [9] to automatically select the background and according to the proportion of different background color to give a certain weight. If the background of the image to be reduced is black, then the weight of the black background is 1, the weight of the white background is 0, vice versa. If the background of the image to be reduced is both black and white, the corresponding weight will be given according to the difference in the proportion of the different background colors, the greater the proportion of the color, the greater of the corresponding weight. The formula of the proposed method shows below:

$$f(x) = Sigmoid(-10 * \frac{N_a - N_b}{N})W_A + (1 - Sigmoid(-10 * \frac{N_a - N_b}{N})) * W_B \quad (4)$$

Where N_a represent the number of black pixel and N_b represent the number of white pixel, N represent the total number of all pixels in the image, W_A denote the weight of black background area, W_B denote the weight of white background area, and we set that $W_A(\emptyset) = 0$, $W_B(\emptyset) = 1$, for each color the W is calculated as follows:

$$W = N - \frac{\frac{MST(U)}{TM} + 1}{|U|} \quad (5)$$

where $|U|$ is the number of feature pixel in pixel cluster, T is the cardinality of the pixel cluster, M is the largest distance between the pixel of a cluster, and $N = 1 + \frac{T(M+1)-1}{T^2 M}$. For example, for an 3*3 block, the $T = 3 * 3, M = 2\sqrt{2}$, the MST(U) denote the weight of the minimum spanning trees of U.

Figure 5 shows the different aggregation value in image block of different background colour, whether it is black on a white background, or a white feature on a black background, get the same weight. And this weight can also describe the spatial structure of the pixel cluster.

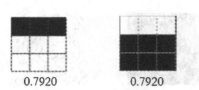

0.7920   0.7920

Fig. 5. Different aggregation value in image block of different background colours

Fig. 6. Test image

4   Experiments and Evaluation

For analyze the performance of designed operator, we apply it to the actual image reduction task. Figure 6 is an often used test image in image reduction task, it depicts a series of concentric circles, which can represent the fine details feature.

We construct our operator based on Eq. 3, by replacing the individual W(i) with sigmoid function based weight f(x) (Eq. 5). Such that

$$P(x,y) = f(x) \sum_{i=1}^{p} \rho(x_i, y) \tag{6}$$

Here x is the feature pixel of each 3*3 block, the 3*3 block can reduce the raw image to $\frac{1}{9}$ scale. First we compare the image reduction performance of proposed method on black background image to the traditional monotone averaging reduction operator and non-monotone averaging reduction operator, the result images are shown in Fig. 7, from this we can find,first, non-monotone averaging reduction operator have a better performance on preserve fine detail image features than monotonic averaging reduction operator.

In the case of arithmetic mean based reduction operator, while the circles are preserved but the radial gradients have been corrupted, so the image becomes blurred and spatially broadened. In the case of median based image reduction operator, we can clearly find it can't preserve the curves after reduction, because within a 3*3 block the curves must appears as an outlier but the background would be as the represent pixel. Second the sigmoid function based reduction operator and cmode operator (Gleb et al. [3]) have similar reduction performance on a black background image, both can preserve fine pixel-scale details after reduction. Then we compare the performance of the Sigmoid function based image reduction operator on the black background image and the white background image, the result is show as Fig. 8.

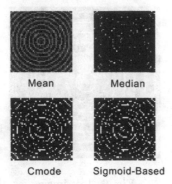

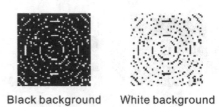

Fig. 7. Reduced test image using different operators

Fig. 8. Reduced different background color image using proposed operator

It is apparent from this result whether in black background image or white background image sigmoid function based image reduction operator has the same good performance. Therefore, the reduction operator based on the sigmoid function just compensates the shortcomings that the cmode operator can only measure the compactness of pixel cluster of black background image.

5   Conclusion

This work has proposed an sigmoid function based weighted image reduction operator, which can weights the pixel cluster's compactness of different background colours. Compare with the traditional monotonic averaging image reduction operator, it has better details feature preserve ability, can preserving more fine pixel details after image reduction. Compare with the latest non-monotonic image reduction operator, it makes up the shortcomings that this method can only measure the compactness of pixel cluster of specific background. In general, this method has better robustness than exiting image reduction operators, which can provide a unified non-monotonic averaging image reduction scheme.

Acknowledgments. The authors would like to thank the anonymous reviewers for their constructive comments. This research is supported by the International Cooperation and Exchanges in Science and Technology Plan Project in Shannxi under the Grant No. 2016kw-047.

References

1. Image reduction. In: Creating and Enhancing Digital Astro Images, pp. 37–73. Springer, London (2007)
2. Beliakov, G., Yu, S., Paternain, D.: Non-monotone averaging aggregation. In: 2011 IEEE International Conference on Fuzzy Systems (FUZZ), pp. 2905–2908. IEEE (2011)
3. Beliakov, G., Li, G., Vu, H.Q., Wilkin, T.: Fuzzy measures of pixel cluster compactness. In: IEEE International Conference on Fuzzy Systems, pp. 1104–1111 (2014)
4. Beliakov, G., Li, G., Vu, H.Q., Wilkin, T.: Characterizing compactness of geometrical clusters using fuzzy measures. IEEE Trans. Fuzzy Syst. **23**, 1030–1043 (2015)
5. Beliakov, G., Pradera, A., Calvo, T. (eds.): Aggregation Functions: A Guide for Practitioners, pp. 139–141. Springer, Heidelberg (2007). doi:10.1007/978-3-540-73721-6
6. Beliakov, G., Pradera, A., Calvo, T. (eds.): Averaging Functions. STUDFUZZ, vol. 221, pp. 39–122 (2007). doi:10.1007/978-3-540-73721-6_2
7. Calvo, T., Beliakov, G.: Aggregation functions based on penalties. Fuzzy Sets Syst. **161**(10), 1420–1436 (2010)
8. Grabisch, M., Marichal, J.L., Mesiar, R., Pap, E.: Aggregation Functions (Encyclopedia of Mathematics and Its Applications). Cambridge University Press, Cambridge (2009)
9. Ito, Y.: Approximation of functions on a compact set by finite sums of a sigmoid function without scaling. Neural Netw. **4**(6), 817–826 (1991)
10. Li, H., Zhang, W., Xu, P., Wang, H.: Rough set attribute reduction in decision systems. In: Wang, G.-Y., Peters, J.F., Skowron, A., Yao, Y. (eds.) RSKT 2006. LNCS, vol. 4062, pp. 135–140. Springer, Heidelberg (2006). doi:10.1007/11795131_20
11. Paternain, D., Bustince, H., Fernandez, J., Beliakov, G., Mesiar, R.: Image reduction with local reduction operators, vol. 23(3), pp. 1–8 (2010)
12. Paternain, D., Bustince, H., Fernandez, J., Beliakov, G., Mesiar, R.: Some averaging functions in image reduction. In: García-Pedrajas, N., Herrera, F., Fyfe, C., Benítez, J.M., Ali, M. (eds.) IEA/AIE 2010. LNCS, vol. 6098, pp. 399–408. Springer, Heidelberg (2010). doi:10.1007/978-3-642-13033-5_41
13. Paternain, D., Lopez-Molina, C., Bustince, H., Mesiar, R., Beliakov, G.: Image reduction using fuzzy quantifiers. In: Melo-Pinto, P., Couto, P., Serôdio, C., Fodor, J., De Baets, B. (eds.) Eurofuse 2011. Advances in Intelligent and Soft Computing, vol. 107. Springer, Berlin (2011)
14. Thangavel, K., Pethalakshmi, A.: Dimensionality reduction based on rough set theory: a review. Appl. Soft Comput. **9**(1), 1–12 (2009)
15. Wilkin, T.: Image reduction operators based on non-monotonic averaging functions. In: IEEE International Conference on Fuzzy Systems, pp. 1–8 (2013)
16. Zahn, C.T.: Graph-theoretical methods for detecting and describing gestalt clusters. In: IEEE Transactions on Computers, pp. 68–86 (1971)

Knowledge Representation and Reasoning

Learning Deep and Shallow Features for Human Activity Recognition

Sadiq Sani[1], Stewart Massie[1(✉)], Nirmalie Wiratunga[1], and Kay Cooper[2]

[1] School of Computing Science and Digital Media, Robert Gordon University,
Aberdeen AB25 1HG, Scotland, UK
{s.sani,s.massie,n.wiratunga}@rgu.ac.uk
[2] School of Health Sciences, Robert Gordon University, Aberdeen, Scotland, UK

Abstract. selfBACK is an mHealth decision support system used by patients for the self-management of Lower Back Pain. It uses Human Activity Recognition from wearable sensors to monitor user activity in order to measure their adherence to prescribed physical activity plans. Different feature representation approaches have been proposed for Human Activity Recognition, including shallow, such as with hand-crafted time domain features and frequency transformation features; or, more recently, deep with Convolutional Neural Net approaches. The different approaches have produced mixed results in previous work and a clear winner has not been identified. This is especially the case for wrist mounted accelerometer sensors which are more susceptible to random noise compared to data from sensors mounted at other body locations e.g. thigh, waist or lower back. In this paper, we compare 7 different feature representation approaches on accelerometer data collected from both the wrist and the thigh. In particular, we evaluate a Convolutional Neural Net hybrid approach that has been shown to be effective on image retrieval but not previously applied to Human Activity Recognition. Results show the hybrid approach is effective, producing the best results compared to both hand-crafted and frequency domain feature representations by a margin of over 1.4% on the wrist.

Keywords: Human activity recognition · Feature representation · Deep learning

1 Introduction

Human activity recognition (HAR) is the computational discovery of human activity from sensor data. It is receiving increasing interest in the areas of health care and fitness [9], largely motivated by the need to find creative ways to encourage physical activity. HAR is generally considered as a Machine Learning classification problem. However, there is little consistency in the research on which feature representation and classification models should be applied; to some extent this may reflect the differences that can exist between the detailed

© Springer International Publishing AG 2017
G. Li et al. (Eds.): KSEM 2017, LNAI 10412, pp. 469–482, 2017.
DOI: 10.1007/978-3-319-63558-3_40

characteristics of different HAR problems. In particular, the sensor type and location may vary between applications, as can the activity types that need to be recognised.

The motivation for this work is to develop an effective HAR component for use in selfBACK[1], an EU funded project that is developing a self-management system for patients with Lower Back Pain (LBP). Recent published guidelines for the management of non-specific LBP advises patients to remain active and avoid excessive bed rest or extended periods of inactivity [1]. The HAR component monitors the patient activity using sensor data that is continuously polled from a wearable device. User activities are recognised in real time and if selfBACK detects continuous periods of sedentary behaviour, a notification is given to alert the user. In addition, a daily activity profile is generated and compared with a prescribed activity plan. The information in this daily profile includes the durations of activities and, for ambulation activities (e.g. walking, running, stair climbing), the intensity of activity, and the counts of steps taken.

Given the importance of accurate recognition of user activity for selfBACK, an effective HAR component becomes crucial. Different feature representation approaches for HAR have been proposed, from shallow hand-crafted features to frequency transformation features e.g. Fast Fourier Transform (FFT) and Discrete Cosine Transform (DCT) coefficients, and more recently, deep learning approaches [9,17,18]. All these approaches have had some degree of success and setbacks in performance [15]. However, previous work has presented mixed results and does not provide a clear answer on which feature representation approach to adopt. Given the intended real-world use case of selfBACK, it is important to determine the appropriate feature extraction approach to use for our preferred sensor configuration (single wrist-mounted accelerometer) and our selected activity classes (walking, jogging, up-stairs, downstairs, standing, sitting). Hence in this paper, we provide an empirical evaluation of 7 different feature representation approaches across the three different classes of features i.e. shallow hand-crafted, shallow frequency transformed, and deep CNN derived.

Wrist data presents additional challenges as it has different characteristics to data collected from other body locations (e.g. thigh). It is more prone to random noise due to increased variations in movement and posture possible with the hand during activities. As a result, the performance of HAR with a wrist-mounted sensor may suffer. However, long-term usage of an application may be improved with a less obtrusive wrist-worn sensor. It is therefore important to compare both locations, in order to assess the trade-off between performance and usability. Hence, we provide an empirical evaluation of wrist-mounted and thigh-mounted accelerometer data.

The main contributions of this work are:

- the evaluation of deep and shallow features for HAR for the specific activities required for the management of LBP, with sensors located on both the wrist and thigh;

---

[1] http://www.selfback.net/.

- a CNN hybrid approach has been introduced for HAR in which deep features are generated from the raw accelerometer data and then fed to a traditional classifiers; and
- a new public dataset has been made available to the research community to support future work in this area.

The rest of this paper is organised as follows: in Sect. 2, we highlight important related work on HAR. Section 3 discusses the different feature representation approaches. Evaluation, describing our dataset and experiments, is presented in Sect. 4 with our conclusions in Sect. 5.

2   Related Work on HAR

In this work, we focus on sensor input from a single tri-axial accelerometer mounted on a person's wrist or thigh. A tri-axial accelerometer sensor measures changes in acceleration in 3 dimensional space [9]. Other types of wearable sensors have also been proposed e.g. gyroscope. A recent study compared the use of accelerometer, gyroscope and magnetometer for activity recognition [19]. The study found the gyroscope alone was effective for activity recognition while the magnetometer alone was less useful. However, the accelerometer still produced the best activity recognition accuracy. Other sensors that have been used include heart rate monitor [20], light and temperature sensors [13]. These sensors are however typically used in combination with the accelerometer rather than independently.

Some studies have proposed the use of a multiplicity of accelerometers [4,12] or combination of accelerometer and other sensor types placed at different locations on the body. These configurations however have limited practical use outside of a laboratory setting. In addition, limited improvements have been reported from using multiple sensors for recognising every day activities [6] which may not justify the inconvenience, especially as this may hinder the real-world adoption of the activity recognition system. For these reasons, some studies e.g. [11] have limited themselves to using single accelerometers which is also the case for selfBACK [3].

Another important consideration is the placement location of the sensor. Several body locations have been proposed e.g. thigh, hip, back, wrist and ankle. Many comparative studies exist that compare activity recognition performance at these different locations [4]. The wrist is considered the least intrusive location and has been shown to produce high accuracy especially for ambulation and upper-body activities [11]. Hence, this is the chosen sensor location for our system.

Many different feature extraction approaches have been proposed for accelerometer data for the purpose of activity recognition [9]. Most of these approaches involve extracting statistics e.g. mean, standard deviation, percentiles etc. on the raw accelerometer data (time domain features). FFTs applied to the raw data have been shown to be beneficial. Typically this requires a further preprocessing step applied to the resulting FFT coefficients in order to extract features that measure characteristics such as spectral energy, spectral entropy and dominant frequency [5]. Another approach is to use coefficients obtained from applying DCT on the raw accelerometer data as features [17]. This is particularly attractive as it avoids the need for further preprocessing of the data to extract features to generate instances for the classifiers.

Recently, deep learning approaches have been applied to the task of HAR due to their ability to extract features in an unsupervised manner. In particular, Convolutional Neural Networks (CNNs), Recurrent Neural Networks (RNNs) and Deep Belief Networks (DBNs), have all been applied with promising results [7]. One of the early approaches to apply deep belief networks for feature learning is [14]. Here the authors use stacked Restricted Boltzmann Machines (RBMs) to learn higher level features from raw accelerometer data. RBMs are fully connected bipartite graphs that generatively model input data by training a set of stochastic binary hidden input units which function as low-level feature detectors. A comparative evaluation with other representation approaches, e.g. FFT and Principal Component Analysis (PCA), did not present RBM as a clear winner. Four different datasets were used in the evaluation, of which only one (Skoda) used a single wrist mounted accelerometer. However, this dataset was collected from a single person and the activity classes in the dataset are car assembly activities which are very different from the usage scenario with selfBACK.

More recently, CNNs have been popularly applied to the task of HAR. One of the main advantages of CNNs is their ability to model local dependencies that may exist between adjacent data points in the accelerometer data [21]. In [16], CNNs are used for representation of both accelerometer and gyroscope data from the Samsung mobile phone dataset [2]. The dataset also provides a set of 561 hand-crafted features. The evaluation compares a CNN classifier architecture (with 3 hidden layers) with four other classifiers, including SVM and Naive Bayes that use the hand-crafted features. Results showed deep features to be on par with hand-crafted features using SVM.

Zeng et al. used CNNs with a modified weight sharing technique called partial weight sharing [21]. Evaluation compared performance of standard CNNs, partial weight sharing CNNs, time-domain hand crafted features, PCA and RBMS on three datasets using accuracy. Results show the CNN with partial weight sharing to perform best. However, only the Skoda datasets used a wrist mounted accelerometer and performance was reported as accuracy which can be misleading given that the datasets contained unbalanced class distributions. Furthermore the comparative study considered only kNN on just the raw features and did not consider state-of-the-classifiers such as SVM.

Huang et al. found that while CNNs are good at learning invariant features, SVMs may be better at producing decision surfaces for classification [8]. For image classification they show that a hybrid approach, with a SVM trained on the features learned by the convolutional network, outperforms standard CNNs. Our work applies a similar approach but instead of image classification our focus is on studying comparative performance of deep hybrid features on HAR.

3   Feature Representation

The feature representations considered in this work can be divided into three categories: hand-crafted features, frequency transformation features and deep features.

3.1   Hand-Crafted Features

Shallow hand-crafted features are the most common used in HAR and involve the computation of a number of defined measures on either the raw accelerometer data (time-domain) or the frequency transformation of the data (frequency domain). These measures are designed to capture the characteristics of the signal that are useful for distinguishing different classes of activities by bridging the semantic gap from low level sensor features to richer representations. Hand-crafted features can be extracted both from the original time series accelerometer values (time-domain features) or from a frequency transformation of the accelerometer data (frequency-domain features). For time and frequency domains, the input is a vector of real values $\vec{v} = v_1, v_2, \ldots .v_n$ for each axis x, y and z. A function θ_i is then applied to each vector to compute a single feature value. The time-domain and frequency domain features used in this work are presented in Table 1, see [17] for more details on these features.

While hand-crafted features have worked well for HAR, a significant disadvantage is that they are sensor-type specific. A different set of features need to be defined for each different type of input data i.e. accelerometer, gyroscope, time-domain and frequency domain. Hence, some understanding of the characteristics of the data is required. In addition, it is not always clear which features are likely to work best. The choice of features is usually made through empirical evaluation of different combinations of features or with the aid of feature selection algorithms.

3.2   Frequency Transform Features

Feature extraction from the frequency domain involves first applying a single function f on the raw accelerometer data to map into an alternative representation space (i.e. the frequency domain), where it is expected that distinctions between different activities will be more emphasised. The intuition is that some activities have particular characteristics, such as repetitive actions, which will become more evident. The main difference between frequency transform and hand-crafted features is that the coefficients of transformation are directly used for feature representation without taking further measurements. Common transformations include FFTs and DCTs.

FFT is an efficient algorithm optimised for computing the discrete Fourier transform of a digital input. Fourier transforms decompose an input signal into its constituent sine waves. In contrast, DCT, a similar algorithm to FFT, decomposes a given signal into it's constituent cosine waves. Also, DCT returns an ordered sequence of coefficients such that the most significant information is

Table 1. Hand-crafted features for both time and frequency domains.

Time domain features
Mean
Standard deviation
Inter-quartile range
Lag-one-autocorrelation
Percentiles (10, 25, 50, 75, 90)
Peak-to-peak amplitude
Power
Skewness
Kurtosis
Log-energy
Zero crossings
Root squared mean
Frequency domain features
Dominant frequency
Spectral centroid
Maximum
Mean
Median
Standard deviation

concentrated at the lower indices of the sequence. This means that higher DCT coefficients can be discarded without losing information, making DCT better for compression.

For frequency transform features, a transformation function (DCT or FFT) ϕ is applied to time-series accelerometer vector $\vec{v}$ of each axis. The output of ϕ is a vector of coefficients which describe the sinusoidal wave forms that constitute the original signal. Accordingly the transformed vector representations, $\mathbf{x}' = \phi(\mathbf{x})$, $\mathbf{y}' = \phi(\mathbf{y})$ and $\mathbf{z}' = \phi(\mathbf{z})$, are obtained for each axis of a given instance. Additionally we derive a further magnitude vector, $\mathbf{m} = \{m_{i1}, \ldots, m_{il}\}$ of the accelerometer data for each instance as a separate axis, where m_{ij} is defined in Eq. 1.

$$m_{ij} = \sqrt{x_{ij}^2 + y_{ij}^2 + z_{ij}^2} \tag{1}$$

As with $\mathbf{x}'$, $\mathbf{y}'$ and $\mathbf{z}'$, we also apply ϕ to $\mathbf{m}$ to obtain $\mathbf{m}' = \phi(\mathbf{m})$. This means that our representation of a training instance consists of the pair $(\{\mathbf{x}', \mathbf{y}', \mathbf{z}', \mathbf{m}'\}, c)$, where c is the corresponding activity class label. Note that the coefficients returned after applying ϕ are combinations of negative and positive real values for DCT, and a combination of real and complex values for FFT.

For the purpose of feature representation, we are only interested in the magnitude of the coefficients, irrespective of (positive or negative) sign. Accordingly for each coefficient e.g. x'_{ij}, we maintain its absolute value $|x'_{ij}|$ and for complex coefficients, we convert them to real numbers. In our approach we retain a subset of the l coefficients. The final feature representation is obtained by concatenating the absolute values of the first l coefficients of $\mathbf{x}'$, $\mathbf{y}'$, $\mathbf{z}'$ and $\mathbf{m}'$ to produce a single feature vector of length $4xl$. An illustration of this feature selection and concatenation appears in Fig. 1.

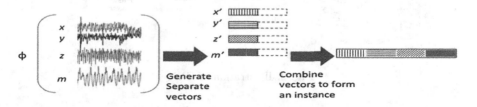

Fig. 1. Feature extraction and vector generation using frequency transforms.

3.3 Deep Features

Deep learning approaches have been applied to the task of HAR due to their ability to extract features in an unsupervised manner. They are able to stack multiple layers of differentiable functions to create a hierarchy of increasingly more abstract features [10]. Recent applications have used more of CNNs due to their ability to model local dependencies that may exist between adjacent data points in the accelerometer data [21]. CNNs are a type of Deep Neural Network that is able to extract increasingly more abstract feature representations by passing the input data through a stack of multiple convolutional layers [10], where each layer in the stack takes as input, the output of the previous layer of convolutional operators. An example of a CNN is shown in Fig. 2.

The input into the CNN in Fig. 2 is a 3-dimensional matrix representation with dimensions $1 \times 28 \times 3$ representing the width, length and depth respectively. Tri-axial acceleromter data typically have a width of 1, a length l and a depth of 3 representing the x, y and z axes. A convolution operation is then applied by passing a convolution filter over the input which exploits local relationships between adjacent data points. This operation is defined by two parameters, D representing the number of convolution filters to apply and C, the dimensions of each filter. For this example, $D = 6$ and $C = 1 \times 5$. The output of the convolution operation is a matrix with dimensions $1 \times 24 \times 6$, these dimensions being determined by the dimension of the input and the parameters of the convolution operation applied. This output is then passed through a Pooling operation which basically performs dimensionality reduction. The parameter P determines the dimensions of the pooling operator which in this example is 1×2, which results in a reduction of the width of its input by half. The output of the pooling layer

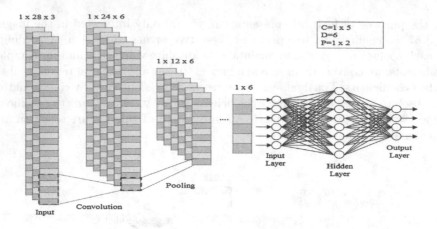

Fig. 2. Illustration of CNN

can be passed through additional Convolution and Pooling layers. The output of the final Pooling layer is then flattened into a 1-dimensional representation and then fed into a fully connected neural network. The entire network (including convolution layers) is trained through back propagation over a number of generations until some convergence criteria is reached. Detailed description of CNNs can be obtained in [10].

Note that once the CNN is fully trained, it can be used to provide feature representations for use with other types of classifiers e.g. SVM, or kNN. This is achieved by cutting off the trained network after the final pooling layer and just before the fully-connected neural network. Each training example is then passed through the convolutional network in order to obtain an abstract representation which is used to train the SVM or kNN classifier. A similar operation is performed for each test example to obtain an abstract representation which is passed to the SVM or kNN for classification. We refer to this as a CNN hybrid approach.

4   Evaluation

A custom dataset was created to reflect the requirements of selfBACK. A group of 34 volunteer participants were used for data collection. The age range of participants is 18 to 54 years and the gender distribution is 52% Female and 48% Male. Data collection captured data for the six activities shown in Table 2.

This set of activities was selected by the health partners of selfBACK as the set of activities of normal daily living that they would like to monitor for the management of LBP. Data was collected using the Axivity Ax3 tri-axial accelerometer[2], at a sampling rate of 100 Hz with a range of ±4g. Accelerometers were mounted on the right-hand wrist and thigh of the participants using

---

[2] http://axivity.com/product/ax3.

Table 2. Details of activity classes in our dataset.

Activity	Description
Walking	Walking at normal pace
Jogging	Jogging on a treadmill at moderate speed
Up stairs	Walking up 4–6 flights of stairs
Down stairs	Walking down 4–6 a flights of stairs
Standing	Standing relatively still
Sitting	Sitting still with hands on desk or thighs

specially designed wristbands provided by Axivity for the wrist, and tape for the thigh. The activities have a balanced distribution between classes, as the participants were asked to carry out each activity for the same period of time (3 min). Each dataset consists of a set of accelerometer readings for each user on each activity, with each reading containing a timestamp and the x, y, z-axis readings for a specific poll of the accelerometer sensor. The wrist and thigh dataset are available on GitHub[3] as an open source dataset.

4.1 Experiment Design

Experiments are reported using our dataset of 34 users. Evaluations are conducted using a leave-one-person-out methodology in which each user is selected in turn for testing and the remaining 33 are used for training. In this way, we are testing the general applicability of the system to users whose data is not included in the trained model. Performance is reported using micro-averaged F1. We compare 7 different feature representations for both wrist and thigh mounted sensors. The 7 representations we use consist of 4 shallow feature representations and 3 deep feature representations:

Shallow Features use a Support Vector Machine (SVM) for classification as it provides state-of-the-art performance and was found to outperform kNN and Naive Bayes in experimentation. The number of coefficients for the frequency domain representations was also selected by initial experimentation. The handcrafted and frequency domain representations reported are:

- *Time*: hand-crafted time domain features, detailed in Table 1;
- *Freq*: hand-crafted frequency domain features, detailed in Table 1;
- *FFT*: applies Fast Fourier Transform and feeds the first 80 coefficients from the x, y, z axis along with the magnitude to an SVM classifier;
- *DCT*: applies Discreet Cosine Transform and feeds the first 80 coefficients from the x, y, z axis along with the magnitude to an SVM classifier;

Deep Features have a vector of accelerometer values as an input into the CNN. After experimenting with different parameter settings, the final configuration

---

[3] https://github.com/rgu-selfback/activity-recognition.

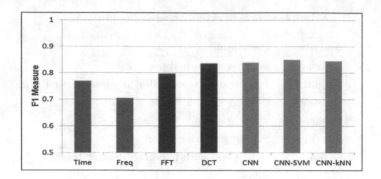

Fig. 3. F1 results for different feature representations for the wrist

used for the wrist data is 5 convolution layers with 150, 100, 80, 60, and 40 convolution filters respectively, created with a filter of size 10 and pooling size of 2. The last pooling layer is connected to a fully connected network with 2 hidden layers, with 900 units in the first layer and 300 units in the second layer and dropout probability of 0.5, and a final output layer with 6 units representing the 6 activity classes in our dataset and soft-max regression. A shallower configuration is used for the thigh data with 3 convolution layers with 150, 100, and 80 filters respectively. For the deep features, we use both a neural network for classification as well as the CNN hybrid approach with traditional classifiers. For the hybrid approach, we cut off the network, after it has been trained, at the last pooling layer and feed the output of this as vector to the SVM or kNN classifier. The purpose of this is to compare the performance of the representation learned by the CNN using the same classifier as was used for the other representations. The deep features reported include a standard CNN and two hybrid approaches:

- *CNN*: uses CNN features with a Neural Network classifier;
- *CNN-SVM*: a CNN hybrid approach using SVM as the classifier with features extracted from the CNN; and
- *CNN-kNN*: a further CNN hybrid approach using kNN as the classifier with features extracted from the CNN.

4.2   Results

Wrist: The first set of results we present is a comparison of the F1 results for the different representations with the accelerometer positioned on the wrist, as shown in Fig. 3. The best results are achieved using the deep features, second best using frequency transform features, with the hand-crafted features performing poorly. Within the deep features, the hybrid approaches of *CNN-SVM* and *CNN-kNN* deliver the highest F1 scores at 0.850 and 0.845 respectively, slightly outperforming *CNN* which has a score of 0.839 although the difference is not significant[4]. *CNN-SVM* significantly outperform both *FFT* and the

---

[4] Significance is tested with a two-tailed student's t-test at p = 0.05.

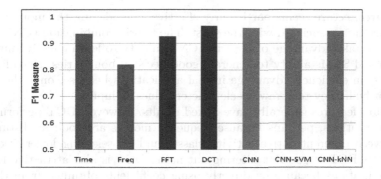

Fig. 4. F1 results for different feature representations for the thigh

hand-crafted features. Among the remaining features *DCT* performs well with an F1 score of 0.836 which significantly outperforms *FFT* (0.797), *Time* (0.769) and *Freq* (0.705). Generally data collected from the wrist tends to be more noisy than data collected from the thigh; the deep features with the 5-layer architecture appear to be more effective at building noise-resistant models. Similarly *DCT*, which disregard higher frequency coefficients, may be more effective than other shallow features at eliminating noise.

Thigh: The second set of results compares the F1 scores for the accelerometer positioned on the thigh (see Fig. 4). The best result is achieved by *DCT* with an F1 score of 0.967, which is slightly better than achieved by the deep learning features of *CNN*, *CNN-SVM* and *CNN-kNN* with scores of 0.959, 0.957 and 0.949 respectively. The differences are not significant, however, *DCT* is significantly better than the other shallow features. Among the deep features CNN and CNN-SVM are on a par, both slightly outperforming CNN-kNN, and significantly better than FFT (0.927), Time (0.936), and Freq (0.820). It is interesting that Time outperforms FFT for this location and may explain hand-crafted time domain features are a popular choice for HAR with data collected from the thigh.

Discussion: HAR with data collected from the thigh for the 6 classes evaluated here is excellent (0.967 for DCT) and superior to collection from the wrist by over 11%. This difference reflects that HAR from the thigh is an easier problem because there is less variation in the data; typically less noise; and data from the thigh is better able to identify leg orientation to distinguish between standing and sitting. However, the 0.850 obtained with deep features is still good performance, particularly if distinguishing between standing/sitting is less important than between active/inactive classes for selfBACK. Wrist movement alone appears insufficient to differentiate between standing/sitting. There is still a clear trade-off between the accuracy of classification from the thigh and the ease of wear on the wrist.

Deep feature representations, particularly using the hybrid approach, perform well on both locations. They provide the best results with data from the wrist; the deeper architecture seems particularly effective at modelling the more complex and noisy data associated with the wrist. The hybrid combination (CNN-SVM) appears to take advantage of both the CNNs ability to learn effective invariant features and SVMs ability to generate good decision boundaries. Deep features also have an efficiency advantage in real applications because, once the model has been built they take raw accelerometer values as inputs.

Shallow features generally have mixed results. However, DCT performs well on both locations, perhaps because frequency models are good at distinguishing between different active activity classes and by selecting lower frequency coefficients DCT is good at filtering out noise. DCT is also attractive for real time applications because by directly using coefficients obtained from the raw accelerometer data as features extensive pre-processing is not required. With hand-crafted features, *Time* performs adequately on the thigh data but is not competitive for wrist data.

5   Conclusion

In this paper, we have presented an analysis of deep and shallow feature representations for accelerometer data on HAR with data collected from the wrist and thigh. Three types of representations are considered: hand-crafted, frequency transform and deep features which includes two hybrid approaches. For the purpose of selfBACK, it is important to determine which representation approach is best suited for HAR using a single wrist-mounted accelerometer on the required activity classes. However, wrist data is more prone to random noise due to the variation in movement and posture of the hand. We evaluate the difference in classification performance with data collected form both locations and find data from the thigh to outperform data from the wrist by 11%. However, classification with data from the wrist still achieves good performance with an F1 score reaching 85%. Comparative evaluation of the seven representation approaches shows deep features to work best on data form the wrist. In particular CNN-SVM, a hybrid approach, that uses CNN to learn features for an SVM classifier outperformed both hand-crafted and frequency transform by a margin of over 1.4%.

References

1. Airaksinen, O., Brox, J., Cedraschi, C.O., Hildebrandt, J., Klaber-Moffett, J., Kovacs, F., Mannion, A., Reis, S., Staal, J., Ursin, H., et al.: Chapter 4 European guidelines for the management of chronic nonspecific low back pain. Eur. Spine J. **15**, s192–s300 (2006)
2. Anguita, D., Ghio, A., Oneto, L., Parra, X., Reyes-Ortiz, J.L.: A public domain dataset for human activity recognition using smartphones. In: ESANN (2013)
3. Bach, K., Szczepanski, T., Aamodt, A., Gundersen, O.E., Mork, P.J.: Case representation and similarity assessment in the SELFBACK decision support system. In: Goel, A., Díaz-Agudo, M.B., Roth-Berghofer, T. (eds.) ICCBR 2016. LNCS, vol. 9969, pp. 32–46. Springer, Cham (2016). doi:10.1007/978-3-319-47096-2_3

4. Bao, L., Intille, S.S.: Activity recognition from user-annotated acceleration data. In: Ferscha, A., Mattern, F. (eds.) Pervasive 2004. LNCS, vol. 3001, pp. 1–17. Springer, Heidelberg (2004). doi:10.1007/978-3-540-24646-6_1

5. Figo, D., Diniz, P.C., Ferreira, D.R., Cardoso, J.M.: Preprocessing techniques for context recognition from accelerometer data. Pers. Ubiquit. Comput. 14(7), 645–662 (2010)

6. Gao, L., Bourke, A., Nelson, J.: Evaluation of accelerometer based multi-sensor versus single-sensor activity recognition systems. Med. Eng. Phys. 36(6), 779–785 (2014)

7. Hammerla, N.Y., Halloran, S., Ploetz, T.: Deep, convolutional, and recurrent models for human activity recognition using wearables. In: Proceedings of the 25th International Joint Conference on AI (2016)

8. Huang, F.J., Lecun, Y.: Large-scale learning with SVM and convolutional for generic object categorization. In: Proceedings of the IEEE Computer Society Conference on Computer Vision and Pattern Recognition, pp. 284–291 (2016)

9. Lara, O.D., Labrador, M.A.: A survey on human activity recognition using wearable sensors. IEEE Commun. Surv. Tutorials 15(3), 1192–1209 (2013)

10. LeCun, Y., Bengio, Y.: Convolutional networks for images, speech, and time series. In: Arbib, M.A. (ed.) The Handbook of Brain Theory and Neural Networks, pp. 255–258. MIT Press, Cambridge (1998)

11. Mannini, A., Intille, S.S., Rosenberger, M., Sabatini, A.M., Haskell, W.: Activity recognition using a single accelerometer placed at the wrist or ankle. Med. Sci. Sports Exerc. 45(11), 2193 (2013)

12. Mäntyjärvi, J., Himberg, J., Seppänen, T.: Recognizing human motion with multiple acceleration sensors. In: 2001 IEEE International Conference on Systems, Man, and Cybernetics, vol. 2, pp. 747–752. IEEE (2001)

13. Maurer, U., Smailagic, A., Siewiorek, D.P., Deisher, M.: Activity recognition and monitoring using multiple sensors on different body positions. In: BSN International Workshop on Wearable and Implantable Body Sensor Networks, 2006. IEEE (2006)

14. Plötz, T., Hammerla, N.Y., Olivier, P.: Feature learning for activity recognition in ubiquitous computing. In: Proceedings of the 22nd International Joint Conference on Artificial Intelligence, pp. 1729–1734. AAAI Press (2011)

15. Ravi, D., Wong, C., Lo, B., Yang, G.Z.: A deep learning approach to on-node sensor data analytics for mobile or wearable devices. IEEE J. Biomed. Health Inform. 21(1), 56–64 (2017)

16. Ronao, C.A., Cho, S.-B.: Deep convolutional neural networks for human activity recognition with smartphone sensors. In: Arik, S., Huang, T., Lai, W.K., Liu, Q. (eds.) ICONIP 2015. LNCS, vol. 9492, pp. 46–53. Springer, Cham (2015). doi:10. 1007/978-3-319-26561-2_6

17. Sani, S., Wiratunga, N., Massie, S., Cooper, K.: SELFBACK—activity recognition for self-management of low back pain. In: Bramer, M., Petridis, M. (eds.) Research and Development in Intelligent Systems XXXIII, pp. 281–294. Springer, Cham (2016). doi:10.1007/978-3-319-47175-4_21

18. Sani, S., Wiratunga, N., Massie, S., Cooper, K.: kNN sampling for personalised human activity recognition. In: Aha, D., Lieber, J. (eds.) Case-Based Reasoning Research and Development. ICCBR 2017. LNCS, vol. 10339, pp. 330–344. Springer, Cham (2017). doi:10.1007/978-3-319-61030-6_23

19. Shoaib, M., Bosch, S., Incel, O.D., Scholten, H., Havinga, P.J.: Fusion of smartphone motion sensors for physical activity recognition. Sensors 14(6), 10146–10176 (2014)

20. Tapia, E.M., Intille, S.S., Haskell, W., Larson, K., Wright, J., King, A., Friedman, R.: Real-time recognition of physical activities and their intensities using wireless accelerometers and a heart rate monitor. In: Proceedings of 11th IEEE International Symposium on Wearable Computers, pp. 37–40 (2007)
21. Zeng, M., Nguyen, L.T., Yu, B., Mengshoel, O.J., Zhu, J., Wu, P., Zhang, J.: Convolutional neural networks for human activity recognition using mobile sensors. In: Proceedings of 6th International Conference on Mobile Computing, Applications and Services, pp. 197–205 (2014)

Transfer Learning with Manifold Regularized Convolutional Neural Network

Fuzhen Zhuang[1,2], Lang Huang[1(✉)], Jia He[1,2], Jixin Ma[3], and Qing He[1,2]

[1] Key Lab of Intelligent Information Processing of Chinese Academy of Sciences (CAS), Institute of Computing Technology, CAS, Beijing 100190, China
{zhuangfz,hej,heq}@ics.ict.ac.cn, laynehuang@outlook.com
[2] University of Chinese Academy of Sciences, Beijing 100049, China
[3] University of Greenwich, London, UK

Abstract. Deep learning has been recently proposed to learn robust representation for various tasks and deliver state-of-the-art performance in the past few years. Most researchers attribute such success to the substantially increased depth of deep learning models. However, training a deep model is time-consuming and need huge amount of data. Though techniques like fine-tuning can ease those pains, the generalization performance drops significantly in transfer learning setting with little or without target domain data. Since the representation in higher layers must transition from general to specific eventually, generalization performance degrades without integrating sufficient label information of target domain. To address such problem, we propose a transfer learning framework called manifold regularized convolutional neural networks (MRCNN). Specifically, MRCNN fine-tunes a very deep convolutional neural network on source domain, and simultaneously tries to preserve the manifold structure of target domain. Extensive experiments demonstrate the effectiveness of MRCNN compared to several state-of-the-art baselines.

Keywords: Transfer learning · Convolutional neural network · Manifold learning

1 Introduction

Recently, deep learning shows great success for learning robust representation and outperforms conventional state-of-the-art methods in computer vision applications. Convolutional neural networks (CNNs), win the ImageNet challenge which is a contest based on a large scale data sets with over 1 million images since 2012 [15,23]. And the key of this success is that the substantially increased depth enlarge the capacity of CNNs and then enable CNNs to fit the data sets

Lang Huang—This work is finished when Lang Huang is an intern (under the supervision of Fuzhen Zhuang) in Institute of Computing Technology, Chinese Academy of Sciences.

© Springer International Publishing AG 2017
G. Li et al. (Eds.): KSEM 2017, LNAI 10412, pp. 483–494, 2017.
DOI: 10.1007/978-3-319-63558-3_41

well. The works [17,24] reinforced this result by showing significant improvement over shallow models when applying a very deep neural network architecture.

On the other hand, as the power of CNNs keeps growing, the complexity of models increases, which further requires more data to avoid overfitting during training process. The problem is that most of data sets are not large enough for training, thus the performance degrades. To take advantage of both huger capacity of deep models and less needed data of shallow ones, a training technique called fine-tuning is proposed. Fine-tuning adopts pretrained models, which are trained on large scale data sets, e.g. ImageNet, on new tasks with only sightly modifying the parameters of the pretrained models. Several studies have reported that fine-tuning obtains outstanding performance and reduces training time from 2 or 3 weeks to few days [18,19].

Although fine-tuning can learn effective representation in various fields, the performance drops significantly when directly applied to transfer learning with insufficient target domain data. Transfer learning aims to improve learning performance in target domain with little or without any label information by leveraging knowledge from auxiliary source domain. Yosinski et al. [20] pointed out that deep feature must transition from general to specific by higher layers of the networks for transfer learning. Hence, only integrating with source domain data will lead to the learned presentation go too specific to source domain.

To address such problem, we propose a manifold regularized convolutional neural network (MRCNN) framework for transfer learning, which aims to use manifold learning approach to regularize fine-tuning progress. Manifold learning approaches are widely adopted in semi-supervised or unsupervised learning, which assumes that data points within a same local structure are likely to have the same label [2]. Therefore, the unlabeled data in target domain can be utilized to preserve such structure in higher layer or output layer by imposing manifold based constraints. By coupling manifold regularization and fine-tuning, we expect that the learned representation in higher layers go more general or more specific to target domain, and thus the knowledge from auxiliary source domain is successfully transferred.

The contributions of this paper are summarized as follow:

1. We propose a unsupervised learning framework that collaborates fine-tuning technique and manifold regularization within deep convolutional neural network for transfer learning.
2. We conduct extensive experiments on several data sets and statistical evidence shows the effectiveness of our framework.
3. Furthermore, we investigate the impact of fine-tuning and manifold regularization on knowledge transfer.

2 Preliminary Knowledge

In this section, we first review convolutional neural networks architecture, fine-tuning technique and manifold regularization, which serve as preliminary knowledge of this paper.

2.1   Convolutional Neural Network

Deep learning approaches have been widely adopted in the last decade [3]. Particularly, convolutional neural network (CNN) is proposed to learn robust representation and achieve satisfying results in computer vision [15,17].

A typical CNN is a feed-forward neural network stacked with multiple convolutional (*conv*) layers, pooling layers (max-pool or average-pool), fully connected (*fc*) layers and a classifier on top of them. Both *conv* and *fc* layers learn non-linear mapping h^l in the lth layer with slight difference. The mapping of *conv* and *fc* layers can be formalized by

$$h^{l+1} = \sigma(w^l * h^l + b^l) \tag{1}$$

$$h^{l+1} = \sigma(w^l h^l + b^l) \tag{2}$$

respectively, where w^l is the weight matrix (kernel in *conv* layers) and b^l is the bias of lth hidden layer, '*' denotes convolution operation and $\sigma(\cdot)$ is an non-linear activation function, e.g., Rectified Linear Unit (ReLU) $\sigma(x) = max(0, x)$ [8] for hidden layers or softmax function $\sigma(x) = \frac{e^x}{\sum_{i=1}^{n} e^{x_i}}$ for output layer. Pooling layers perform a downsampling operation along the spacial dimension. It is useful for reducing computational cost and providing robustness for learning representation [26].

Given data set X with label y, the objective to minimize in CNN is

$$\mathcal{L} = \frac{1}{n} \sum_{i=1}^{n} \mathcal{C}(h_i^l, y_i) \tag{3}$$

where n is the size of data set X and l is the depth of model, h^l is the learned representation of lth hidden layer formulized by Eq. (1) or (2) and $\mathcal{C}(\cdot, \cdot)$ is the cross entropy loss function.

2.2   Fine-Tuning

According to [20], the representation in earlier layers of deep CNNs which are trained on large scale data sets are general to different tasks. Hence, it would be beneficial for both performance boost and time-saving to use those weights as either initialization or feature extractor. Fine-tuning is a training procedure that aims to adopt pretrained models on new tasks.

A standard fine-tuning procedure usually re-initializes the top *fc* layers to match the dimension, since the sizes of input data sets differ between tasks. And for the *conv* layers, there are two major strategies: (1) Fix some earlier layers and only fine-tune the higher layers when data is very limited [18]; (2) Back propagate through all the layers when we have enough data [19].

Note that it's common to use smaller learning rate to avoid overfitting. In this paper, we mainly follow the first fine-tuning strategy and the details will be presented in later sections.

2.3   Manifold Regularization

Manifold learning is a graph based semi-supervised or unsupervised method. It attempts to preserve the manifold structure in a data set. In this paper, we incorporate manifold learning as a regularization term by enforcing neighbors located in the same local structure on embedding space.

Given a data set X, let M be the adjacent matrix of the instances of X and

$$M_{[i,j]} = \begin{cases} 1, \ x_i \in NN(k, x_j) \ or \ x_j \in NN(k, x_i) \\ 0, \ Otherwise \end{cases} \tag{4}$$

where k is a hyper-parameter and $NN(k, x)$ denotes the k nearest neighbors of x. The similarity between x_i and x_j can be measured by cosine distance

$$\cos(x_i, x_j) = \frac{x_i^\top x_j}{\sqrt{x_i^\top x_i} \cdot \sqrt{x_j^\top x_j}} \tag{5}$$

where x_i and x_j are column vectors and $\top$ denotes matrix transpose. Let $D = diag(\sum_j M_{[i,j]})$ and $L = D - M$ be the Laplacian matrix, then the manifold regularized term $\Gamma(X)$ can be written as

$$\begin{aligned} \Gamma(X) &= \sum_{i,j} M_{[i,j]} \|f(x_i) - f(x_j)\|^2 \\ &= trace(F^\top L F) \end{aligned} \tag{6}$$

where $f(\cdot)$ is a map function, $F_{[i,\cdot]}$ denotes ith row of F and $F_{[i,\cdot]} = f(x_i)$.

3   MRCNN Model

In this paper, we focus on unsupervised transfer learning, i.e., only labeled data are available in source domain and unlabeled data in target domain. Thus, we denote source domain as $D_s = \{x_i^{(s)}, y_i^{(s)}\}_{i=1}^{n_s}$ with n_s labeled instances, target domain as $D_t = \{x_i^{(t)}\}_{i=1}^{n_t}$ with n_t unlabeled instances.

Now we are ready to present details of our framework. MRCNN is based on VGG19 Network architecture [17], which is composed of 16 *conv*, 5 max-pooling, 3 *fc* and 1 softmax layers. Following the notation in [17], the 16 *conv* layers are divided into 5 blocks by max-pooling layers, i.e., *conv*1 block consists of *conv* layers before first max-pooling layer, *conv*2 block consists of *conv* layers between first and second max-pooling layer, and so on.

We adopt the weights of VGG19 model pretrained on ImageNet in all *conv* layers and randomly initialize the *fc* layers, which are shared for both source and target domains. Then we further extend VGG Net by integrating manifold regularization, which is imposed on target domain to enforce similar instances to have the same labels, and cross entropy loss on source domain data to incorporate label information. By preserving manifold structure in this manner, we hope the learned representation in higher layers can be well generalized. Figure 1 intuitively illustrates MRCNN framework. Note that *conv*1−*conv*5 each denotes

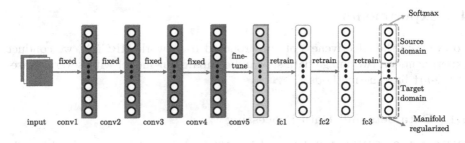

Fig. 1. The framework of MRCNN.

a convolutional block, not a single layer. Due to the limitation of data and the learned representations transition from general to specific along the network, we adopt following three different strategies: (1) randomly initialize 3 *fc* layers because *fc* layers require strict dimensional matching while the sizes of input data sets are different. (2) *conv*5 block, containing 4 *conv* layers, is carefully fine-tuned since the representation in this block is more transferable and needs only sightly tuning. In other word, we apply a smaller learning rate on this block. (3) the *conv*1-*conv*4 blocks consisting of 12 *conv* layers are fixed and used as feature extractor since the representation of these blocks are general.

Let $\boldsymbol{w}$ and $\boldsymbol{b}$ denote collections of weights and biases of the MRCNN, $\boldsymbol{L}$ denotes Laplacian matrix of target domain, the overall objective is written as below:

$$\mathcal{J} = \mathcal{L}(\boldsymbol{w}, \boldsymbol{b}, \boldsymbol{x}^{(s)}, \boldsymbol{y}^{(s)}) + \alpha\varGamma(k, \boldsymbol{L}, \boldsymbol{w}, \boldsymbol{b}, \boldsymbol{x}^{(t)}) + \beta\varOmega(\boldsymbol{w}, \boldsymbol{b}). \qquad (7)$$

The first term of Eq. (7) is the cross entropy loss between the output log-its and labels of source domain as presented in Eq. (3). The second term $\varGamma(k, \boldsymbol{L}, \boldsymbol{w}, \boldsymbol{b}, \boldsymbol{x}^{(t)})$ is the manifold regularization as described in Eq. (6) imposed on target domain and k is the number of nearest neighbors. The last term of Eq. (7) is L2 norm of weight and bias matrices, which controls the complexity of the network structure and is defined as

$$\varOmega(\boldsymbol{w}, \boldsymbol{b}) = \sum_{i=1}^{l}(||\boldsymbol{w}^i||^2 + ||\boldsymbol{b}^i||^2), \qquad (8)$$

where l is the depth of the neural network, i.e., 19 in this paper. α, β are hyper-parameters that balance the importance of manifold regularization and model complexity in the entire framework.

The objective Eq. (7) can be minimized by performing Gradient Descent. Since we implement MRCNN by Deep Learning Library TensorFlow [25], which can automatically compute the gradients and derive the solution, the update rules for parameters will be omitted here.

4   Experiments

To evaluate the effectiveness of our proposed framework MRCNN, we conduct experiments on two image data sets and compare our model with several state-of-the-art baseline methods.

4.1   Data Sets and Data Processing

CIFAR-100. CIFAR-100 data set[1] has 100 classes, which are grouped into 20 superclasses [5], and each contains 600 images. Among these 20 superclasses, we randomly choose two of them '*fruit and vegetables*' and '*household electrical devices*' and take '*fruit and vegetables*' as positive examples and '*household electrical devices*' as negative one. Each superclass of '*fruit and vegetables*' and '*household electrical devices*' has 5 classes. To construct transfer learning classification problems, we randomly choose one class from '*fruit and vegetables*' and one from '*household electrical devices*' as source domains, and then choose another one class of '*fruit and vegetables*' and another one of '*household electrical devices*' from the remaining classes to construct target domain. In this way, we can obtain 400 ($P_5^2 \cdot P_5^2$) classification problems.

Corel. Corel data set[2] consists of two different top categories, '*flower*' and '*traffic*' [9]. Each top category further includes four subcategories. We take '*flower*' as positive class and '*traffic*' as negative one. Then by following the same processing procedure, we can construct 144 ($P_4^2 \cdot P_4^2$) transfer learning classification problems.

Note that we do not perform any data argumentation on these data sets except subtracting mean for CNN based methods and normalizing features to [0, 1] for the other compared competitors.

4.2   Baseline Methods

We compare MRCNN with a variety of baselines,

- Logistic regression (LR), one of the most widely applied supervised learning algorithm without transfer learning technique.
- Transductive Support Vector Machine (TSVM) [1], a transductive learning algorithm to incorporate unlabeled target domain data. However, TSVM assumes the labeled source domain data and unlabeled target domain data follow the same distribution.
- Transfer Component Analysis (TCA) [12], which aims at learning a low-dimensional representation for transfer learning. We use Support Vector Machine (SVM) as the basic classifier for it in this paper.

---

[1] https://www.cs.toronto.edu/~kriz/cifar.html.
[2] http://archive.ics.uci.edu/ml/datasets/Corel+Image+Features.

- Transfer Learning with Deep Autoconders (TLDA) [22], which uses deep autoencoders to find a proper embedding space for both source and target domains, while their distribution are explicitly enforced to be similar.
- Standard VGG Net [17], which finetunes on source domain but without manifold regularization. We denote it as VGG.

4.3 Implementation Details

For LR, we perform grid search for L2 regularization term, and for TSVM, we use SVM$_{lin}$[3] and sample the hyper-parameter in $[10^{-5}, 10^1]$. For TCA, the number of latent dimension is carefully tuned, e.g., for Corel data set, the number varies between 10 and 100 with the step size 10. For TLDA, we adopt author's source code and use the default parameters.

VGG and MRCNN are implemented in Deep Learning Library TensorFlow[4] [25]. And for VGG, the learning rate r and weight decay strength β are set as $10^{-2.5}$ and 10^{-2} for CIFAR-100, 10^{-4} and 10^{-2} for Corel. For MRCNN, r and β are set as $10^{-2.5}$ and 10^{-2} for CIFAR-100, $10^{-2.5}$ and 10^{-1} for Corel. Moreover, the trade-off parameter α is set as 10^{-5} for CIFAR-100, 10^{-3} for Corel, and the number of nearest neighbors k is set as 3 for both CIFAR-100 and Corel. Note that, we use Adam [21] as an optimizer to minimize the objective function for both CNN based models.

4.4 Results

In total, we construct 400 classification tasks for CIFAR-100 and 144 classification tasks for Corel. To make comprehensive comparison, we further divide the classification tasks into two groups for each data set according to the accuracy of LR. Specifically, we first conduct LR model on all classification tasks, and then group them into two groups, i.e., the first group of classification tasks with accuracies from LR lower than 70%, while the other one with accuracies from LR higher than 70%. Finally, we report the average results of two groups, and all the results are shown in Table 1. *Left* and *Right* respectively denote the average performance of classification tasks whose accuracies lower and higher than 70%, and *Total* means the average results over all tasks. Note that lower accuracy from LR indicates more difficult to make transfer, and vice versa.

From these results, we have the following observations:

- TSVM is better than LR, which indicates the importance to consider unlabeled data. However, TSVM can not achieve satisfying results since it assumes the labeled and unlabeled data should follow the same distribution. TCA performs even worse than LR on Corel data set, which may show the difficulty to make transfer on the constructed classification tasks. TLDA delivers best results in most cases compared to above methods, which reveals the power of deep models.

---

[3] http://vikas.sindhwani.org/svmlin.html.
[4] The code is available at https://github.com/LayneH/MRCNN.

- CNN based methods, i.e., VGG and MRCNN, significantly outperform all other conventional methods by a large margin. We attribute it to the fine-tuning procedure, which fixes the earlier 4 *conv* clocks and only back-propagates through the last layers to preserve the generality. This also indicates the essential to adopt deep learning models for classification.
- Among the deep learning methods, MRCNN achieves considerable improvement over standard VGG Net with fine-tuning. This validates that manifold regularization successfully guides the training process to obtain better representation.
- Overall, the incorporation of manifold regularization leads to the success of MRCNN. In other word, MRCNN performs the best on all groups.

4.5   Analysis and Discussion

Although we show the average results in Table 1, you maybe more interest in more detailed results. Here we try our best to present more detailed results, although it is not easy to show all detailed ones of totally 544 (400+144) classification tasks. Taking CIFAR-100 as an example, we average the accuracy of problems with the same target domain ($P_5^1 \cdot P_5^1 = 25$ instances). The accuracies of LR, VGG and MRCNN are presented in Fig. 2(a) in an increasing order of LR's accuracy.

From the results in Fig. 2(a), MRCNN outperforms the baselines most of time. However, MRCNN can not achieve satisfying results at some points. We conjecture that the proposed model heavily depends on the local relationship of target domain data, so the bad quality of nearest neighbors would lead to the poor performance. Motivated by [6], we use ϵ to measure the confusion of nearest neighbors, and it can be formalized as

$$\epsilon = \frac{\sum_{i=1}^{n_t} \sum_{j=1}^{\#nn_i} \mathbb{1}\{label(nn_{i,j}) \neq label(x_i^{(t)})\}}{\sum_{i=1}^{n_t} \#nn_i} \tag{9}$$

Table 1. Average accuracy (%) of different methods.

	LR	TSVM	TCA	TLDA	VGG	MRCNN
CIFAR-100						
Left	64.87	66.90	74.24	68.15	72.50	**76.98**
Right	79.11	80.33	78.38	81.58	82.33	**86.91**
Total	74.91	76.37	77.16	77.62	79.43	**83.98**
Corel						
Left	61.64	78.27	71.04	79.12	86.26	**87.25**
Right	80.62	80.81	75.40	81.04	86.63	**87.54**
Total	74.03	79.93	73.89	80.38	86.50	**87.44**

where $\mathbb{1}\{\cdot\}$ is the indicator function, $\#nn_i$ is the number of nearest neighbors of $x_i^{(t)}$, $nn_{i,j}$ is the jth nearest neighbor of $x_i^{(t)}$ and $label(x)$ is the label of x. Intuitively, the higher ϵ is, the more confusing knowledge are introduced by imposing manifold regularization. We sort the classification problems according to the values of ϵ on target domains and show how ϵ influence the classification accuracy in Fig. 2(b). Moreover, we group the tasks into 3 groups according to the values of ϵ: the first group consists of tasks with $\epsilon < 0.1$, the second one consists of those with $0.1 \leq \epsilon \leq 0.15$ and the rest form the third one. The average accuracy of each group is presented in Table 2.

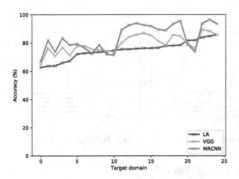

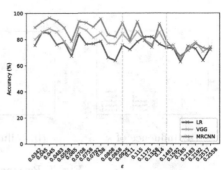

(a) Accuracy in ascending order of LR's.   (b) Influence of ϵ. The yellow dashed line divide target domains into 3 classes.

Fig. 2. Average accuracy of target domains. (Color figure online)

Table 2. Mean accuracy (%) of different groups.

	$\epsilon < 0.1$	$0.1 \leq \epsilon \leq 0.15$	$\epsilon > 0.15$
LR	76.00	77.30	70.18
VGG	82.26	79.88	**72.84**
MRCNN	**90.00**	**82.67**	72.24

From Fig. 2(b) and Table 2, we can find that the classification accuracy substantially drops as ϵ grows for all methods. The reason may be that the positive and negative instances are similar for these classification tasks, and they are not easy to be separated. The above results also reveal that the confusion of nearest neighbors is the key factor that influences the performance of MRCNN. Hence, to further generalize MRCNN on transfer learning tasks, one crucial problem to be enhanced is how to obtain correct nearest neighbors.

4.6   Parameter Sensitivity

In this section, we will discuss how the parameters r, α, β and k in Eq. (7) impact on the experiments. To tune the hyper-parameters, we randomly select 10 problems from CIFAR-100 as validation problems. We sample the learning rate r between $\{10^{-5}, 10^{-4}, 10^{-3}, 10^{-2}, 10^{-1}, 10^{0}, 10^{1}\}$, the trade-off parameter α and β from $\{10^{-6}, 10^{-5}, 10^{-4}, 10^{-3}, 10^{-2}, 10^{-1}, 10^{0}, 10^{1}\}$, and the number of nearest neighbors is selected from $[1, 3, 5, 10, 20, 40]$. From these results from Fig. 3, we then set r, α, β and k to $10^{-2.5}$, 10^{-5}, 10^{-1} and 3 in CIFAR-100 experiment. For the Corel data set, we randomly select 6 problems, and the parameters are similarly tuned.

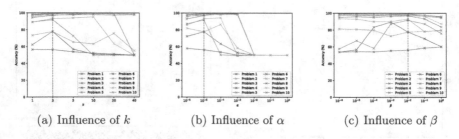

(a) Influence of k (b) Influence of α (c) Influence of β

Fig. 3. Parameter sensitivity

5   Related Work

Transfer learning is the improvement of learning in a new domain by transferring the knowledge from auxiliary source domain [7,11]. It has drawn much attention in past decades for its potential to ease the pain of manual labeling. Feature based approaches are one of the most widely proposed, which aim to learn a good feature representation for both source domain and target domain by reduce the difference between domains or integrating regularization [10,12,14,16]. Among feature based transfer learning methods, several methods have been proposed to reduce the domain discrepancy explicitly. For example, transfer component analysis (TCA) [12] aims to minimize the difference of distributions between domains in a kernel Hilbert space, [10] is trying to find a subspace where training and testing samples are approximately i.i.d. by integrating Bregman divergence-based regularization between distributions of domains. One crucial problem of these methods is that most of them only adopt shallow representation models to reduce the domain discrepancy, which limits their ability to generalize for various tasks.

Deep learning methods show its potential to learn effective and robust representation in recent years. To enjoy such benefit, several frameworks have been introduced. Stacked Denoising Autoencoders (SDAEs) [13] aims to improve the effectiveness of learned representation in Denoising Autoencoders (DAEs) [4] by extending the depth of DAEs, i.e. stack multiple DAEs within the framework. [22] further couples SDAEs and feature based transfer learning approach together where they explicitly minimize the KL divergence between distributions of source and target domains in embedding space.

However, we argue that such methods do not take advantage of the generality brought by large scale data sets. Researches like [18,19] directly adopt the pretrained models on new tasks and deliver state-of-the-art results. [20] investigates the generality of deep neural network and reveals that features must eventually transition from general to specific by the last layer of the network. Motivated by these works, we proposed a manifold regularized convolutional neural networks (MRCNN) framework to enjoy the generality of deep learning models for transfer learning.

6   Conclusion

In this paper, we propose a novel manifold regularized convolutional neural network (MRCNN) framework for transfer learning. This framework adopts a very deep CNN architecture and incorporates manifold regularization component. By imposing manifold based constrains, we enforce the manifold structure of target domain to be preserved while the cross entropy loss of source domain is minimized. We confirm that this manifold regularization can help improve generalization performance and thus successfully leverage knowledge cross domains. Moreover, in order to ease the pain of training such deep CNN, we apply fine-tuning technique to our framework. Finally, we conduct a series of experiments against several competitors to demonstrate the effectiveness of our framework.

Acknowledgment. This work is supported by the National Natural Science Foundation of China (Nos. 61473273, 91546122, 61573335, 61602438), Guangdong provincial science and technology plan projects (No. 2015 B010109005), the Youth Innovation Promotion Association CAS 2017146.

References

1. Joachims, T.: Transductive inference for text classification using support vector machines. ICML **99**, 200–209 (1999)
2. Belkin, M., Niyogi, P., Sindhwani, V.: Manifold regularization: a geometric framework for learning from labeled and unlabeled examples. J. Mach. Learn. Res. **7**, 2399–2434 (2006)
3. Hinton, G.E., Salakhutdinov, R.R.: Reducing the dimensionality of data with neural networks. Science **313**(5786), 504–507 (2006)
4. Vincent, P., Larochelle, H., Bengio, Y., et al.: Extracting and composing robust features with denoising autoencoders. In: Proceedings of the 25th International Conference on Machine Learning, pp. 1096–1103. ACM (2008)
5. Krizhevsky, A., Hinton, G.: Learning multiple layers of features from tiny images (2009)
6. Wu, J., Xiong, H., Chen, J.: Adapting the right measures for k-means clustering. In: Proceedings of the 15th ACM SIGKDD International Conference on Knowledge Discovery and Data Mining, pp. 877–886. ACM (2009)

7. Torrey, L., Shavlik, J.: Transfer learning. Handb. Res. Mach. Learn. Appl. Trends: Algorithms Methods Tech. **1**, 242 (2009)

8. Nair, V., Hinton, G.E.: Rectified linear units improve restricted boltzmann machines. In: Proceedings of the 27th International Conference on Machine Learning (ICML 2010), pp. 807–814 (2010)

9. Zhuang, F., Luo, P., Xiong, H., et al.: Cross-domain learning from multiple sources: a consensus regularization perspective. IEEE Trans. Knowl. Data Eng. **22**(12), 1664–1678 (2010)

10. Si, S., Tao, D., Geng, B.: Bregman divergence-based regularization for transfer subspace learning. IEEE Trans. Knowl. Data Eng. **22**(7), 929–942 (2010)

11. Pan, S.J., Yang, Q.: A survey on transfer learning. IEEE Trans. Knowl. Data Eng. **22**(10), 1345–1359 (2010)

12. Pan, S.J., Tsang, I.W., Kwok, J.T., et al.: Domain adaptation via transfer component analysis. IEEE Trans. Neural Netw. **22**(2), 199–210 (2011)

13. Vincent, P., Larochelle, H., Lajoie, I., et al.: Stacked denoising autoencoders: learning useful representations in a deep network with a local denoising criterion. J. Mach. Learn. Res. **11**, 3371–3408 (2010)

14. Glorot, X., Bordes, A., Bengio, Y.: Domain adaptation for large-scale sentiment classification: a deep learning approach. In: Proceedings of the 28th International Conference on Machine Learning (ICML 2011), pp. 513–520 (2011)

15. Krizhevsky, A., Sutskever, I., Hinton, G.E.: Imagenet classification with deep convolutional neural networks. In: Advances in Neural Information Processing Systems, pp. 1097–1105 (2012)

16. Chen, M., Xu, Z., Weinberger, K., et al.: Marginalized denoising autoencoders for domain adaptation. arXiv preprint arXiv:1206.4683 (2012)

17. Simonyan, K., Zisserman, A.: Very deep convolutional networks for large-scale image recognition. arXiv preprint arXiv:1409.1556 (2014)

18. Hoffman, J., Guadarrama, S., Tzeng, E.S., et al.: LSDA: large scale detection through adaptation. In: Advances in Neural Information Processing Systems, pp. 3536–3544 (2014)

19. Sharif Razavian, A., Azizpour, H., Sullivan, J., et al.: CNN features off-the-shelf: an astounding baseline for recognition. In: Proceedings of the IEEE Conference on Computer Vision and Pattern Recognition Workshops, pp. 806–813 (2014)

20. Yosinski, J., Clune, J., Bengio, Y., et al.: How transferable are features in deep neural networks?. In: Advances in Neural Information Processing Systems, pp. 3320–3328 (2014)

21. Kingma, D., Ba, J.: Adam: a method for stochastic optimization. arXiv preprint arXiv:1412.6980 (2014)

22. Zhuang, F., Cheng, X., Luo, P., et al.: Supervised representation learning: transfer learning with deep autoencoders. In: IJCAI, pp. 4119–4125 (2015)

23. Russakovsky, O., Deng, J., Su, H., et al.: Imagenet large scale visual recognition challenge. Int. J. Comput. Vis. **115**(3), 211–252 (2015)

24. He, K., Zhang, X., Ren, S., et al.: Deep residual learning for image recognition. In: Proceedings of the IEEE Conference on Computer Vision and Pattern Recognition, pp. 770–778 (2016)

25. Abadi, M., Agarwal, A., Barham, P., et al.: Tensorflow: large-scale machine learning on heterogeneous distributed systems. arXiv preprint arXiv:1603.04467 (2016)

26. Goodfellow, I., Bengio, Y., Courville, A.: Deep Learning. MIT Press, Cambridge (2016)

Learning Path Generation Method Based on Migration Between Concepts

Dan Liu[1,3], Libo Zhang[1,2](✉), Tiejian Luo[1], and Yanjun Wu[2]

[1] University of Chinese Academy of Sciences, Beijing 101408, China
zsmj@hotmail.com
[2] Institute of Software, Chinese Academy of Sciences, Beijing 100190, China
icode@iscas.ac.cn
[3] Nankai University, Tianjin 300350, China

Abstract. The learning strategies often have a direct impact on learning effects. Often, the learning guidance is provided by teachers or experts. With the speed of knowledge renewal going faster and faster, it has been completely unable to meet the needs of the learner due to the limitation of individual time and energy. In order to solve this problem, we propose a learning strategy generation method based on migration between concepts, in which the semantic similarity is creatively applied to measure the relevance of concepts. Moreover, the concept of jump steps is introduced in Wikipedia to measure the difficulty of different learning orders. Based on the hyperlinks in Wikipedia, we build a graph model for the target concepts, and achieve multi-target learning path generation based on the minimum spanning tree algorithm. The test datasets include the books about Computer Science in Wiley database and test sets provided by volunteers. Evaluated by expert scoring and path matching, experimental results show that more than 59% of the 860 single-target learning paths generated by our algorithm are highly recognized by teachers and students. More than 60% of the 500 multi-targets learning paths can match the standard path with 0.7 and above.

Keywords: Learning path · Wikipedia · Semantic similarity · Graph model

1 Introduction

In the era of knowledge explosion, finding an efficient way to the target knowledge, called as learning path in this paper, in these too many learning materials is a problem needed to be solved. It is necessary for learners to find a proper way grasping the knowledge they need, as it is universal that people reach their destinations efficiently relying on navigators. Traditionally, most of the learning strategies are made by teachers or experts. As a result, the learning strategies are personally and subjective. In the process of learning, learners will also have problems such as cognitive overload or cognitive impairment due to the inappropriate learning order of concepts which leads to the inefficiency of learning. Therefore, in the information age with knowledge expanding and updating rapidly, it has been far from being able to meet the needs of learners relying solely on the individual or an educational group.

The data generated by human activities are enriching the knowledge space constantly. The development of Internet has greatly accelerated the growth of information,

© Springer International Publishing AG 2017
G. Li et al. (Eds.): KSEM 2017, LNAI 10412, pp. 495–506, 2017.
DOI: 10.1007/978-3-319-63558-3_42

which makes people begin to explore the way to organize the useful information, so there is a knowledge organization form named knowledge base which is a knowledge cluster after the artificial systematization and standardization, such as Wikipedia.

In recent years, the rapid development of the World Wide Web has led to the study of knowledge space. Lexical relations diagram, semantic relevance, knowledge ontology, semantic network, etc. have been applied to the text classification, word disambiguation, machine translation and so on. These studies also provide a theoretical basis for the learning path generation.

The knowledge base absorbs new knowledge and establishes the connection between concepts quickly while the book is authoritative and academic. Therefore, in this paper, we put forward a learning path generation method making full use of the knowledge base and traditional books to help the learners learn the target knowledge better and faster.

Our main contributions are as follows:

- We build a basic concept set in the computer domain and propose a graph model for concepts to express the concepts and their relationships;
- We propose a learning path generation method based on migration and similarity between concepts;
- We build two test data sets and evaluate the generated learning path by scoring and the consistency degree between paths.

In this paper, we creatively apply semantic similarity to the generation of learning path, and propose a new graph model for concepts to express their relationships. We use Wikipedia to construct the graph model for the target concepts automatically, without manual assistance.

Our study takes Computer Science discipline as an example, but our method's scope of application is not limited to Computer Science. Our method is based on Wikipedia and the books on the subject, which makes it possible to migrate to other subject areas at low cost, thus benefiting more learners in various fields.

The rest of this paper is organized as follows. In Sect. 2, we review related works. In Sect. 3, we propose our method of the learning path generation. Section 4 describes the validation of our results based on scoring and path matching. Finally, in Sect. 5, we give the conclusion.

2 Related Work

2.1 Approaches to Learning Path Generation

Nowadays, the speed of knowledge update is going faster and faster, and there is a high demand for learning ability and learning efficiency. The rapid development of the Internet brings people into an era of e-learning, more and more scholars turn to the automatic generation of learning strategies. The existing methods can be divided into three categories: individual based methods, knowledge level based methods, learning difficulty based methods.

Individual based methods often use learners' gender, personality, major, social background and other attributes to recommend learning paths for learners. Lin et al. [1] developed a personalized creative learning system (PCLS) in 2013 based on decision tree in data mining technology. PCLS includes a series of creative tasks and a questionnaire that involves several key volumes. It determines the level of creativity according to individual attributes and tasks. There are some one-sidedness since such methods solely rely on individual attributes, and it is not effective when learners are unwilling to provide such information.

Knowledge level based methods get the knowledge not grasped by learners according to the wrong answered questions in the pre-test. Then the learning path is recommended accordingly. Lendyuk et al. [2] proposed an individual learning path generation method using the learning object sequence to navigate the learning content. They adjusted the difficulty of problem next stage to solve according to the number of right answered questions in previous stage which allowed learners to improve their knowledge level gradually. Such methods depend on the question bank, which requires a lot of manpower to construct. Moreover, the knowledge area covered by the question bank is limited.

Learning difficulty based methods constantly adjust the learning difficulty, which is given by human initially, according to learning effect. Zhao and Wan [3] built a graph model for knowledge units. Each node represents a knowledge unit, and the directed edges represent the relationships between knowledge units: precedence, succession or parallel. The weight of the edge represents the difficulty of learning from the precedence concept to the succession concept. The initial value was given by the teacher, and then updated according to the learning of the previous students. Bonifati et al. [4] proposed a learning path generation method that satisfies the path query conditions of the user on the graph database. The input of the algorithm includes a graph database in which the user marks the node as a positive or negative case based on whether he wants the node to be part of the query result. Such methods not only consume a lot of manpower, but also have a strong subjectivity due to the artificial annotation. Furthermore, it is more and more difficult to keep up with the speed of knowledge updating for an individual even a group of experts.

In view of these problems, the method proposed in this paper combines the Wikipedia and online library resources with the consideration of the update speed and authority of the knowledge. Moreover, the migration between concepts in Wikipedia is used to measure the learning difficulty rather than artificial annotation.

2.2 Word Similarity

The word similarity discussed in this paper is at the semantic level. The methods of calculating word similarity are divided into four categories according to the basic method adopted: word similarity calculating based on vector space model, word similarity calculating based on text attributes, word similarity calculating based on sequence alignment and word similarity calculating based on semantic analysis.

In the vector space model, the text is regarded as a collection of basic language units (word, phrase, etc.), and it is assumed that the text feature is only relevant to the frequency of some basic language units, with its position or order in the text not considered completely. Word similarity calculating based on text attributes is driven by the goal of the task. It often extracts the text attributes that contribute to the certain task. At present, the text attributes used in the study include the new word occurrence rate [5], the overlap rate of words [6], the text graph model [7], etc. The text is treated as a sequence of characters when calculating word similarity based on the sequence alignment, so that the text similarity calculation turn into the calculation of sequence similarity. Commonly used methods include Hamming Distance, Edit Distance, the Longest Common Subsequence, etc. Word similarity calculating based on semantic analysis uses synonymy, antisense, upper and lower relations to calculate the similarity [8]. The advantages and disadvantages of the four kinds of methods are shown in Table 1.

Table 1. Advantages and disadvantages of the four kinds of methods.

Calculation methods	Advantages	Disadvantages
Based on the vector space model	Computational overhead is small	The effect is not good when applied to the text containing less words; Ignore the order and ambiguity of words
Based on text attributes	Focus on a certain task, more flexible and more feasible	Poor portability due to strong targeted
Based on sequence alignment	Consider the text order	Ignore the semantic relations between words
Based on semantic analysis	Discover the deep semantic relations between features	Computational overhead, high resource requirements

We used the vector space model to calculate the similarity between words via the word2vec, a tool provided by Google. We trained a CBOW (Continuous Bag-of-Words Model) model of word2vec using the Wikipedia corpus described in Sect. 3.2.

3   Generating a Learning Path

3.1   Definition of Learning Path

Learning path is a list of relevant concepts. The relevant concepts refers the concepts have semantic connection, which can be measured by word similarity. In this paper, the learning paths are divided into two kinds:

- **Single-Target Learning Path.** An orderly sequence of concepts generated for one target concept that needs to be learned. The relevant concepts are ranked in descending order of the concept's relevance to the target concept. For example, the learning path of the target concept *jQuery* can be the list: *JavaScript, jQuery UI, plugin, jQuery Ajax, HTML, jQuery.*

- **Multi-target Learning Path.** Learners have several concepts that need to be learned. For example, when they preview a new unit, they want to learn some emerging concepts of the unit, which are often relevant. These concepts that need to be learned constitute the target concept set, and the multi-target learning path is a directed graph generated for the target concept set, in which the node represents the target concept, and the direction of the edge represents the learning order. For example, the learning path of the target concept set {*array, tree, stack, list, queue*} can be the learning order shown in Fig. 1.

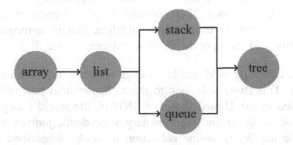

Fig. 1. A multi-target learning path.

3.2 Data Preprocessing

We propose a learning path generation method based on Wikipedia, the authoritative course guidelines and professional books.

- **Wikipedia.** The Internet Encyclopedia is the largest and most complete human knowledge base that is currently available for random access. Wikipedia is the largest Internet multilingual encyclopedia compiled by humans. As of April 8, 2017, Wikipedia has a total of 5,378,718 articles, 41,883,956 pages, which cover most of the human knowledge areas. The original data set used to build the graph model and train the similarity model is a compressed file for all Wikipedia pages (2016.06.01), a 53.4 GB XML file after decompression. In the experiment to determine the relevant concepts of the target, the online Wikipedia is used because of the consideration of web page redirection. We extracted the title of each page as an entry from the original XML file, and extract the text with a hyperlink in the page (hereinafter referred to as hyperlink text) to build data set called as wiki_R. Each line in wiki_R.txt represents the contents of an entry in the following format:

$$title\#outlinks_1\#outlinks_2\#\ldots\#outlinks_n\#$$

Where *title* is the page's title, and $outlinks_i (i = 1, 2, \ldots, n)$ is a hyperlink text in the page corresponding to the entry. "#" is the delimiter.
- **Course Guidelines.** Computer Science is a fast-growing discipline with much knowledge. Learners often feel confused when they study. Therefore, we choose the Computer Science as an example. In the field of computer, ACM and IEEE are the authoritative institutions, and the two agencies jointly issued 13 guidelines to lead

the direction of the development of computer courses: CC1991, CC2001, IS2002, CE2004, SE2004, CC2005, IS2006, IT2008, CS2008, SE2009, IS2010, CS2013 and SE2014, all of which are incorporated into the original data set in this study. The basic concept set volume_R can be obtained by analyzing the distinction of computer-related basic concepts in the document, where $D_i(i = 1, 2, ..., 13)$ represents the original guidelines:

a. Delete the URL and special characters in D_i, and then extract the fields around "()" and "/", capitalized phrases, phrases with hyphens to document d1, d2, d3, d4;
b. Filter the document d1, d2, d3, d4 artificially. Removing those words that not belong to Computer Science. Then construct R_i after duplication eliminating;
c. Repeat the above steps for each course guideline. Finally summarize R_i and eliminate duplication and then get the result document volume_R.

- **Books.** In the computer field, the knowledge update is fast. As a result, the books are too many. This study will look to the computer field part in Wiley database. Wiley, founded in the United States in 1807, is the world's largest independent academic book publisher and the third largest academic journal publisher. In this paper, we use the Wiley online collection of books subscribed by the Chinese Academy of Sciences collecting books from Wiley-Blackwell, Wiley-VCH, Jossey-Bass and so on. We download 181 books in computer field to construct a book library CSLibrary. The 181 books are divided into three categories: 75 of Computer Science, 67 of General Computing, 39 of Information Science and Technology. In order to facilitate the follow-up experiment, we use PDFMiner to turn pdf document into txt document.

3.3   Extraction of Relevant Concepts

How to find the relevant concepts for the target concept is the key to constructing our model. In this section, we extracted relevant concepts from Wikipedia and books based on word similarity discussed in Sect. 2.2.

- **Extract Relevant Concepts from Wikipedia.** In Wikipedia, the concept exists in the form of an entry, and its structure is reflected by the hyperlinks between pages. We find these links have a large or small correlation with the concept tc. Therefore, we use the hyperlinks to extract relevant concepts from Wikipedia. The specific steps are as follows:

a. Extract all the hyperlink texts from the Wikipedia page corresponding to the target concept tc. Name the hyperlink text set as HLinkSet;
b. Calculate the similarity between the target concept tc and each hyperlink text (concept) rc in the HLinkSet and rank the concepts in descending order of the rc's relevance to tc, recorded as HLinkList;
c. Take the first k concepts in HLinkList to form the relevant concept set named as RCsFromWiki of the target concept tc.

- **Extract Relevant Concepts from Books.** A concept is closely relevant to its context. Thus, it can be assumed that the concepts appearing adjacently in the text are related to each other stronger. We try to use the computer course guidelines published by ACM and IEEE to establish a basic concept set in computer domain. This process has been described in detail in Sect. 3.2. It should be noted that the basic concept set volume_R can not completely cover the basic concepts in the book, for which we make full use of the index of the book. We extract relevant concepts from books as follows:

a. For a book b, extract the basic concepts that appear in the book according to volume_R and record the page number it appears in the book. Add these concepts to the set BasicSet;
b. Add the search terms that appear in the index of b and the page number that it appears in the book to BasicSet;
c. Rank the concepts in BasicSet according to the page number they first appear. Then we get BasicList;
d. For a target concept tc in the book b that needs to be learned, search for its index in the BasicList, denoted as $index0$. Construct the candidate relevant concept set CandidateSet with the concept in [$index0$-m, $index$-1] and [$index$+1, $index$+m] scope of BasicList. The size of the set is identified as $2m$ ($2m > k$);
e. For each concept rc in CandidateSet, calculate the similarity between the target concept tc and rc, and rank it according to the similarity degree. The result is denoted as CandidateList;
f. Take the first k relevant concepts in CandidateList to constitute the relevant concept set RCsFromBook, where k is in order to keep consistent with the number of relevant concepts extracted from the Wikipedia.

3.4   The Generation Method

For the two cases, we propose different generation methods respectively.

- **Single-Target Learning Path Generation**

The single-target learning path is to help learners have a better understand of the target concept. The simple-target learning path generation method combines the relevant concepts extracted from Wikipedia and books. The specific steps are as follows:

a. For the single target concept tc given by the learner, extract the candidate relevant concept set from the corresponding Wikipedia page (which may be redirected pages) as RCsFromWiki (size: k); extract the candidate relevant concept set as RCsFromBook (size: k);
b. Take the union of RCsFromWiki and RCsFromBook as RCs (the size of the set is not greater than k). For each concept rc in RCs, calculate the similarity between the target concept tc and rc, and rank concepts according to the similarity to get RCsList;
c. Take the first k concepts in RCsList as the single-target learning path of tc.

- **Multi-target Learning Path Generation**

Given a concept set, how to determine the order of these concepts in order to minimize the difficulty of learning? In other words, how to reduce time and effort spent as far as possible? To solve this problem, the concept of *jumps* is introduced in Wikipedia to measure the learning difficulty between two concepts. For the two pages in Wikipedia *page1* and *page2*, the *jumps* of *page1* to *page2* are defined as:

The minimum number of clicks required clicking on the link in *page1* to finally jump to *page2*. If the number is greater than MAX_CLICK, we said that the distance of *page1* to *page2* is infinity (∞).

After introducing the concept of *jumps*, we can build a weighted directed graph model for a concept set, in which the node represents the target concept, and the direction of the edge represents the jump direction between pages corresponding to the concepts, the weight of the edge is the *jumps*. It can be seen in Fig. 2.

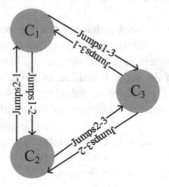

Fig. 2. Graph model based on jumps between Wikipedia pages.

With the definition of *jumps* and the graph model constructed successfully, we can get the multi-target learning path generation method:

a. Calculate the *jumps* for any of the two concepts tc_i and tc_j in the given target concept set TCs and construct the graph model accordingly;
b. Find the "path" to learn the target concepts with the least number of jumps, that is, the minimum support tree for the connected graph built in a., which is the multi-target learning path.

4   Test

We set up a series of experiments to detect the effectiveness of the proposed method.

4.1   Test Data

- **Test Data for Single-Target Learning Path Generation Method**

For the single-target learning path, we selected the book containing more than 150 basic concepts as the original test data set based on CSLibrary described in Sect. 3.2, for a total of 86 books: 60 of Computer Science, 23 of General Computing, 3 of Information Science and Technology. We randomly selected 10 basic concepts from each book to constitute the test set test_860 which contains a total of 860 basic concepts obviously.

- **Test Data for Multi-target Learning Path Generation Method**

For the multi-target learning path, there is some correlation between concepts in the targets set, that is, the learner conducts a study with a certain goal, and the relevant concepts need to be studied under the navigation of the goal. For example, When the learner wants to learn file operation while he learns C++ programing language, his target concept may be *read*, *write*, *buffer*, *seek* etc., but it seems not likely to be the *stack*, *cache*, *router* such concepts that are not consistent with the goal. Therefore, in order to ensure the test is practical enough, we use artificial data sets rather than randomly generated. We collected 500 target concept sets from the student volunteers. The target concept set size varying from 6 to 10. The ratio is 1:1:1:1:1, respectively, denoted as TCSet_6_100, TCSet_7_100, TCSet_8_100, TCSet_9_100, TCSet_10_100.

4.2   Evaluation

- **Evaluation Method for Single-Target Learning Path**

For the evaluation of the single-target learning path, we adopted the teacher and student scoring strategy:

a. Invite 50 teachers, covering lecturers, associate professors, professors, the ratio of which is 2:1:1. Invited 100 students as volunteers, whose study stage cover the undergraduate third grade, fourth grade and graduate first grade, graduate sophomore, graduate third grade, doctor first grade, the proportion of which is 1:2:2:2:2.
b. Each teacher student give a score in 0–10 for the generated learning path, and finally take the average score of all teachers and students respectively, denoted as *teacher_acore* and *student_score*.
c. Take the average of teacher_score and student_score as the comprehensive score of the learning path, denoted as *final_score*.

- **Evaluation Method for Multi-target Learning Path**

For the evaluation of the multi-target learning path, we use two methods:

a. The scoring strategy referred above;
b. Invite 50 teachers to make learning paths for the target concept sets, recorded as standard learning paths. Then map each concept to a character, so that the generated learning path and the standard learning path are mapped to two strings, the result string and the standard string. Finally, calculate the similarity between the two strings based on the Levenshtein Distance:

$$Similarity = (Max(x, y) - Levenshtein)/Max(x, y)$$

Where x is the length of the result string and y is the length of the standard string. The smaller the Levenshtein Distance, the higher the similarity of the two strings, and the more consistent the result learning path and the standard learning path, in other words, the better the results of the proposed method.

4.3   Results

For the target concepts given by the learner, we tested on data set mentioned in Sect. 4.1 and evaluated according to the evaluation methods in Sect. 4.2.

- **Single-Target Learning Path Results**

We experimented on the test set test_860. In the experiment, k was set to 5 while m was set to 8. 860 learning paths were obtained according to the method proposed in Sect. 3.4, and then we calculated *teacher_score*, *student_score* and *final_score* after getting the scores from teachers and students. The scores were divided into five intervals: [0, 2), [2, 4), [4, 6), [6, 8), [8, 10]. The number of learning paths in each interval is shown in Table 2.

Table 2. The number of learning paths in each score interval.

	[0, 2)	[2, 4)	[4, 6)	[6, 8)	[8, 10]
Teachers	24	141	193	256	246
Students	43	102	208	230	277
Final	32	123	201	248	256

We can see that 96% of the 860 learning paths' *final_score* are not less than 2 points and 59% are not less than 6 points. 30% fall in the [8, 10] interval. It can be said that the generated learning paths are highly recognized by the teachers and students. However, there are scores less than 4 points. It seems led by the relativity of the relevance concepts extraction. We extract relevant concepts via hyperlinks in Wikipedia and the context of the target concept in the book. Although we filter out of the concepts with lower similarity with the target concept, the compare is too relative in some cases.

- **Multi-target Learning Path Results**

In the experiment, the MAX_CLICK was set to 6. The experimental results on the 6 test data sets mentioned in Sect. 4.1 were denoted as ResultLP_6_100, ResultLP_7_100, ResultLP_8_100, ResultLP_9_100, ResultLP_10_100.

a. Evaluation Results Using Scoring Strategy

The number of learning paths in each score interval for each of the six result sets is shown in Table 3.

From Table 3 we can see that even if the size of the target concept set is different, the difference of learning paths in each score interval is not more than ±10, and the number of generated learning paths gaining score not less than 6 is more than 62. We can draw the conclusion: The size of the target concept set has less influence on the results. Moreover, more than 62% of the generated learning paths gain a score not less than 6.

Table 3. The number of learning paths in each score interval when the target concept set size varying from 6 to 10.

	6	7	8	9	10
[0, 2)	2	3	2	1	2
[2, 4)	12	10	12	11	9
[4, 6)	23	25	24	23	21
[6, 8)	30	28	25	27	26
[8, 10]	33	34	37	38	42

When building the graph model for the target concept set, the value of *jumps* between two pages depends on the links in the page totally. Due to indeterminacy of the page links and the editors' subjectivity, there will be the case that the value of *jumps* is greater than MAX_CLICK. More generally, there may be the case that the value of *jumps* from concept c_1 to c_2 is equal to the value of *jumps* from concept c_2 to c_1, and the generated learning path will have a random sequence in c_1, c_2 which will lead to the deviation.

b. Evaluation Results of Standard Path Comparison

We invited 50 teachers to make learning paths for TCSet_6_100, TCSet_7_100, TCSet_8_100, TCSet_9_100, TCSet_10_100 as standard learning paths. In accordance with Sect. 4.2, the generated learning path and the standard learning path on the same target concept set were mapped to two strings, and we calculated their similarity to measure the consistency between them. The path consistency with the target concept set size varying from 6 to 10 is shown in Table 4.

Table 4. Path consistency with the target concept set varying from 6 to 10.

	6	7	8	9	10
[0, 0.5)	3	5	4	6	5
[0.5, 0.7)	31	31	36	29	24
[0.7, 1.0)	66	64	60	65	71

It is shown that when the size of the target concept set is 6, 7, 8, 9 or 10, the difference of the number of learning paths falling in each similarity interval is less than ±5, and the number of learning paths with the consistency not less than 0.7 compared to the standard learning path was more than 60. We can draw the conclusion: The size of the target concept set has less influence on the results. Moreover, more than 60% of the generated learning paths can match the standard path with 0.7 and above.

5   Conclusions

In this paper, we focus on the problem how to help learners to learn the knowledge they need better and faster. In view of the limitations of the existing learning path generation methods, we propose a learning path generation method with the combination of the traditional learning and modern network learning aiming at all learners.

We creatively apply semantic similarity to the generation of learning strategies to measure the relevance of concepts and introduce jump steps in Wikipedia to measure the difficulty of different learning orders. Based on the hyperlinks in Wikipedia, the graph model can be built successfully for the target concepts, which is a key step in multi-target learning path generation. We test the proposed method on the books about Computer Science in Wiley database and test sets provided by volunteers. The expert scoring results show that more than 59% of the 860 single-target learning paths generated by our method are highly recognized by teachers and students and more than 62% of the 500 multi-target learning paths gain a score not less than 6. By path matching, it can be seen that more than 60% of the 500 multi-target learning paths can match the standard path given by experts with 0.7 and above.

Considering the truth that the graph model in this paper is built automatically, without any manual work, our method is manpower saving and the results are more objective. In future research, a broader library of books and a basic concept set covering more fields will be built to serve more learners.

References

1. Lin, C.F., Yeh, Y.C., Hung, Y.H., Chang, R.I.: Data mining for providing a personalized learning path in creativity: an application of decision trees. Comput. Educ. **68**, 199–210 (2013)
2. Lendyuk, T., Melnyk, A., Rippa, S., Golyash, I., Shandruk, S.: Individual learning path building on knowledge-based approach. In: 8th International Conference on Intelligent Data Acquisition and Advanced Computing Systems: Technology and Applications (IDAACS), pp. 949–954. IEEE, Warsaw (2015)
3. Zhao, C., Wan, L.: A shortest learning path selection algorithm in e-learning. In: 6th International Conference on Advanced Learning Technologies, pp. 94–95. IEEE, Kerkrade (2006)
4. Bonifati, A., Ciucanu, R., Lemay, A.: Learning path queries on graph databases. In: 18th International Conference on Extending Database Technology (EDBT), Brussels (2015)
5. Youmans, G.: A new tool for discourse analysis: the vocabulary-management profile. Language, 763–789 (1991)
6. Fragkou, P., Petridis, V., Kehagias, A.: A dynamic programming algorithm for linear text segmentation. J. Intell. Inf. Syst. **23**(2), 179–197 (2004)
7. Malioutov, I., Barzilay, R.: Minimum cut model for spoken lecture segmentation. In: 21st International Conference on Computational Linguistics, pp. 25–32. Association for Computational Linguistics, Sydney (2006)
8. Luo, T., Zhang, L., Yang, L., Chen, X.: TACE: a toolkit for analyzing concept evolution in computing curricula. In: 28th International Conference on Software Engineering and Knowledge Engineering, SEKE, Redwood City (2016)

Representation Learning of Multiword Expressions with Compositionality Constraint

Minglei Li$^{(\boxtimes)}$, Qin Lu, and Yunfei Long

Department of Computing, The Hong Kong Polytechnic University,
Hung Hom, Hong Kong
{csmli,csluqin,csylong}@comp.polyu.edu.hk

Abstract. Representations of multiword expressions (MWE) are currently learned either from context external to MWEs based on the distributional hypothesis or from the representations of component words based on some composition functions using the compositional hypothesis. However, a distributional method treats MWEs as a non-divisible unit without consideration of component words. Distributional methods also have the data sparseness problem, especially for MWEs. On the other hand, a compositional method can fail if a MWE is non-compositional. In this paper, we propose a hybrid method to learn the representation of MWEs from their external context and component words with a compositionality constraint. This method can make use of both the external context and component words. Instead of simply combining the two kinds of information, we use compositionality measure from lexical semantics to serve as the constraint. The main idea is to learn MWE representations based on a weighted linear combination of both external context and component words, where the weight is based on the compositionality of MWEs. Evaluation on three datasets shows that the performance of this hybrid method is more robust and can improve the representation.

1 Introduction

Multiword expressions (MWEs) are expressions consisting of two or more words, which can be noun phrases such as *gun dog*, verb phrases such as *break up*, or idioms such as *kick the bucket* [8]. Multiword expressions are important semantic units to express certain specific meanings, especially the non-compositional MWEs. MWEs play important roles in diverse applications such as machine translation and sentiment analysis [15]. According to [19], 57% of sentences from web pages contain at least one MWE. Generally speaking, MWEs can be categorized as either **compositional** such as *traffic light, fresh air*, whose semantics are composed from the semantics of its component words, or **non-compositional** such as idiomaticity such as *couch potato* and *kick the bucket*, whose semantics are not directly composed from its component words. In this paper, we use the term **external context** to refer to the context words surrounding a MWE. Recently, distributed representation of words as dense and low-dimensional vectors, referred to as **word embedding** or **word vector**, has

© Springer International Publishing AG 2017
G. Li et al. (Eds.): KSEM 2017, LNAI 10412, pp. 507–519, 2017.
DOI: 10.1007/978-3-319-63558-3_43

achieved better results on various NLP tasks [4] than symbolic representations which have limited power to directly encode semantic information. Different models are also proposed to learn distributed representation of longer language units, such as phrases [12,25], sentences and documents. Studies on learning MWE representation, which falls into the phrase category, are mainly divided into two approaches.

The first approach, referred to as the **distributional approach**, is based on the distributional hypothesis that words appearing in similar context tend to have similar meanings [6]. This is also the same principle used for word embedding. To apply this principle for MWEs, a MWE is simply treated as a single term and its representation is inferred from its external context in the same way as learning word representation [12,24]. Since this approach treats a MWE as a non-divisible unit, its component words are completely ignored even though this information may be useful, especially for compositional MWEs. For example, the MWE *close interaction* is semantically similar to *contact*, which can be reflected through the word embedding similarities between the component word *interaction* and *contact*. But, by treating *close interaction* as one non-divisible unit, its representation has to be learned independently which is more likely to suffer from data sparseness problem. In the case of an infrequently used MWE which would have insufficient context, this approach can fail to learn their representations. Compared to single words, the sparseness problem of MWEs is indeed more severe given the same corpus.

The second approach, referred to as the **compositional approach**, is based on the principle of compositionality to determine the meaning of a MWE using the combined meanings of its constituents. Based on this principle, this approach employs certain composition function to obtain the representation of a MWE from the representations of its component words [14,25]. The representation of the component words is obtained using distributional models. This approach only uses information of component words and the information of its external context is not directly considered. One key problem of the compositional approach is that it can fail if a MWE is non-compositional because the semantics of non-compositional MWE cannot and should not be derived from its component words. For example, the MWE *monkey business* cannot be composed from *monkey* and *business*. In such a situation, the information of the component words can lead to an erroneous result. The unique meaning of non-compositional MWEs can be lost using current word embedding based compositional models.

We argue that both the external context and the component words provide helpful information to the representation of a MWE. Furthermore, the usefulness of the component words depends on the compositionality of the MWE. If there is a way to measure the compositionality of a MWE, the compositionality can then be used to measure the usefulness of the component words of a MWE. Based on the above analysis, we propose a hybrid method to learn the representation of MWEs from their context with the compositionality constraint. This method can make use of both the external context and component words. Instead of simply combining the two kinds of information, we use the compositionality measures from lexical semantics to serve as a constraint.

In our proposed method, the representation of a MWE is based on a linear combination of external context with a weighted composition of the component words where the weight is based on automatically predicted compositionality. Compared to previous works, our model has the advantages of both previous methods while overcomes their drawbacks. Evaluation based on 3 MWE tasks first shows that compositional method does work better for compositional MWEs while the distributional method works better for non-compositional MWEs. Most importantly, the evaluation shows that our hybrid model gives the overall best performance and is more robust for both compositional and non-compositional MWEs.

2 Related Works

Learning of distributed representation of words can be categorized into counting-based methods and prediction-based methods [1]. Both methods are based on the distributional hypothesis [6]. Counting-based methods perform matrix factorization on the word-context co-occurrence matrix. For example, Latent Semantic Analysis (LSA) performs Singular Value Decomposition (SVD) on the word-context co-occurrence matrix to obtain a low dimensional word vector representation. Glove [16] uses another kind of matrix factorization to directly factorize the co-occurrence matrix into two low-dimensional matrices so that the multiplication of the two matrices is close to the original entry in the co-occurrence matrix. On the other hand, prediction-based methods use neural network models to learn latent representations of words under certain objective functions. For example, the Skip-Gram model tries to maximize the conditional probability of context given a target word and this probability is represented as a vector function of the target word and its context [12]. It is proven that counting-based methods and prediction-based methods are essentially equivalent, differing only in a constant shift [11].

To learn the representation of MWEs, previous works mostly focus on learning phrase representations. One approach treats a phrase as a single unit and learns phrase representations from its external context using the same word representation learning model [12,24]. As for compositional models, there are a number of different methods including addition and multiplication of the component vectors [13]. Baroni and Zamparelli [2] proposes to represent a noun as a vector and an adjective as a matrix and use matrix-vector multiplication to obtain the representation of adjective-noun phrases. Some proposed task specific and supervised composition models include recursive neural networks by matrix multiplication on concatenated component vectors [20]. However, all the above composition based methods assume that phrases and MWEs are compositional. The study by Sun [21] on MWE representation actually makes use of both external context and component words by constraining a WME's vector to be close to the vectors of both its two component words. However, the semantics of many

MWEs are not necessarily similar to both of its component words. Some non-compositional MWEs have no relation to its component words. The work from [7] also considers both the external context and component words with compositionality constraint, which is quite similar to our idea. However, this work can only handle verb-noun phrases. Note that all the above composition models focus on two-word MWEs.

Previous research on MWEs mainly focuses on MWE detection tasks and the compositionality prediction tasks. MWE detection aims to identify MWEs in the text, normally a sequence tagging task [18]. Compositionality prediction aims to identify if a given MWE is compositional or not. Yazdani [23] proposes a semantic composition based method for compositionality detection. Noun compounds that cannot be well modeled by the composition model are considered non-compositional. Another method uses word embedding to predict the compositionality value of MWEs based on the cosine similarity between the representation of the MWE and its component words [17].

3   Our Proposed Model

Our proposed MWE representation model includes two parts. The first part is based on a distributional model to learn word and MWE representation using external context. The second part is the composition model to learn MWE representation from component words. Similar to previous works, we also focus on MWEs that consists of two component words.

Let us first introduce some notations. Given a large corpus S with a set of words $w \in V_W$ and their context $c \in V_C$ where V_W and V_C are the word and context vocabularies. Note that the vocabularies of V_w and V_C may be identical. The distinction is more for conceptual convenience. The context of word w_i is defined as the words surrounding w_i in a window of size L $w_{i-L}, \cdots, w_{i-1}, (w_i)$, $w_{i+1}, \cdots, w_{i+L}$. Let $\#(w)$ denote the frequency of word w, and $\#(c)$ denote the occurrence frequency of context c. Now, let $\#(w, c)$ denote the frequency of a word-context pair (w, c). For MWEs, let V_M denote the set of given MWEs where each MWE $m \in V_M$ consists of two words. We use t_m to denote the compositionality of m and the larger t_m is, the more compositional is the WME. Let D denote the set of (w, c) and (m, c) pairs.

The objective is to learn a vector representation $\boldsymbol{w} \in \mathbb{R}^d$ for each $w \in V_W$, a vector representation $\boldsymbol{c} \in \mathbb{R}^d$ for each context $c \in V_C$, and a vector representation $\boldsymbol{m} \in \mathbb{R}^d$ for each $m \in V_M$. d is the vector dimension.

3.1   Skip-Gram with Negative Sampling (SGNS)

Different models have been proposed to learn the dense word representation as already discussed in Sect. 2. Here we present the widely used SGNS model - the Skip-Gram model trained using negative-sampling [12]. Consider a word-context pair (w, c). Let $p(D = 1|w, c)$ be the probability that (w, c) comes from D and let $p(D = 0|w, c)$ be the probability that (w, c) does not. The basic assumption of SGNS is that the conditional probability of $p(D = 1|w, c)$ should be high if c is the context of word w in corpus D and $p(D = 0|w, c)$ should be high otherwise. $p(c|w)$ is computed as:

$$p(D = 1|w, c) = \sigma(\boldsymbol{w} \cdot \boldsymbol{c}) = \frac{1}{1 + e^{-\boldsymbol{w} \cdot \boldsymbol{c}}}$$

The basic idea behind this is that if word w and context c co-occur, their vectors should have close correlation, modeled by the element-wise multiplication $\boldsymbol{w} \cdot \boldsymbol{c}$. The objective of negative sampling is to maximize the conditional probability $p(D = 0|w, c_N) = \sigma(-\boldsymbol{w} \cdot \boldsymbol{c}_N)$ by randomly samples negative context c_N of w from V_C. This can be translated to maximizing $\sigma(-\boldsymbol{w} \cdot \boldsymbol{c})$. So the objective for a single (w, c) pair is:

$$log(\sigma(\boldsymbol{w} \cdot \boldsymbol{c}) + k \cdot \mathbb{E}_{c_N \sim P_D}[log(-\boldsymbol{w} \cdot \boldsymbol{c}_N)])$$

where k is the number of negative samples and P_D is the empirical unigram distribution $P_D(c) = \frac{\#(c)}{|D|}$.

The final objective function for the whole corpus is:

$$J_S = \sum_{w \in V_W} \sum_{c \in V_C} \#(w, c)\Big(log(\sigma(\boldsymbol{w} \cdot \boldsymbol{c})) + k \cdot \mathbb{E}_{c_N \sim P_D}[log(-\boldsymbol{w} \cdot \boldsymbol{c}_N)]\Big) \quad (1)$$

All the entries in the vectors are treated as model parameters and stochastic gradient is used over the whole set of word-context pairs D in the corpus S. The obtained vector $\boldsymbol{w}$ is the learned representation of w and $\boldsymbol{c}$ is a by-product. Note that the SGNS may have data sparseness problem if the frequency of w is low.

When applying this model to representation learning of MWEs, $m_i \in V_M$, is treated as a single term and representation learning is the same as learning the word representation. However, data sparseness is an even bigger issue for MWEs, especially for MWEs. Also, for compositional MWEs, SGNS completely ignore the component words.

3.2   Composition Model

In a compositional model, the representation of a MWE is inferred from that of its component words. Given a MWE m with two component words w_m^1 and w_m^2 and their respective vector representations $\boldsymbol{w}_m^1$ and $\boldsymbol{w}_m^2$, the representation of m, denoted by $\boldsymbol{m}$, can be computed by a function f:

$$\boldsymbol{m} = f(\boldsymbol{w}_m^1, \boldsymbol{w}_m^2) \quad (2)$$

Different composition models are proposed for f [13]. The weighted additive composition model with weights α and β is defined as a linear composition:

$$\boldsymbol{m} = \alpha \boldsymbol{w}_m^1 + \beta \boldsymbol{w}_m^2 \quad (3)$$

The multiplicative composition model is defined by:

$$m = w_m^1 \cdot w_m^2 \tag{4}$$

Compared to SGNS, the compositional model can make use of component words information. However, this model can produce erroneous representation for non-compositional MWEs, such as *couch potato* whose meaning cannot be composed from its component words *couch* and *potato*.

3.3 The Hybrid Model

The distributional model using SGNS suffers from data sparseness problem for MWEs and cannot make use of component words information. The compositional model alone does not make full use of external context and is not appropriate for non-compositional MWEs. In this work, we propose a hybrid model which makes use of the judgment on compositionality of MWEs and propose a model that can make proper use of the combined compositional model and the distributional model, denoted as the **C&S** model.

Given a corpus S and a MWE set V_M, we aim to learn the vector representation $m \in \mathbb{R}^d$ for every $m \in V_M$, and to learn the vector representation $w \in \mathbb{R}^d$ for every $w \in V_W$. Since learning include both words and MWEs, we first construct the candidate term set $V_T = V_W \cup V_M$ and then build the corresponding context set V_C based on the window size L. For a word w, its representation can be learned according to Formula 1. For a MWE m, the proposed C&S model can be modeled as:

$$J_S = \sum_{m \in V_M} \sum_{c \in V_C} \#(m, c) \Big(log\sigma(m \cdot c) + k \cdot \mathbb{E}_{c_N \sim P_D}[log(-m \cdot c_N)] \\ + \lambda h(t_m, m, w_m^1, w_m^2) \Big) \tag{5}$$

In Formula 5, the first two parts are identical to the SGNS model. The third part is for component words composition with a constant weight λ to balance the overall contributions of the two models. Function h is defined as

$$h(t_m, m, w_m^1, w_m^2) = t_m log\sigma \left(m \cdot f(w_m^1, w_m^2) \right) \tag{6}$$

$f(w_m^1, w_m^2)$ can be any compositional model defined by Formula 2. $\sigma(m \cdot f(w_m^1, w_m^2))$ defines the correlation between the learned MWE representation m and the composed WME representation. The more they are correlated, the larger contribution the third part is to J_S. t_m is the compositionality of m, which can be computed using the method proposed by Salehi et al. [17] based on word embedding as follows:

$$t_m = g \left(m, w_m^1 + w_m^2 \right), \tag{7}$$

g is the cosine similarity according to Salehi. m, w_m are obtained using SGNS. Theoretically speaking, Formula 6 has the following properties:

1. If the compositionality t_m is low (m being more non-compositional), the weight of the correlation between the MWE representation m and the composed representation $f(w_m^1, w_m^2)$ from its component words should be low. It means m should be based mainly on SGNS, namely its external context.
2. If the compositionality t_m is high (m being more compositional), the weight of the correlation between m and $f(w_m^1, w_m^2)$ should be high and the objective function will force m to be similar to the composed $f(w_m^1, w_m^2)$. It means m should consider both the external context and component words.

In principle, the compositionality t_m under our framework can be computed by any compositionality prediction model and the composition part $f(w_m^1, w_m^2)$ can use any proposed composition model. By setting λ to zero, our model degrades to the SGNS model. By setting t_m to a constant, our model changes to a fixed weight model.

Our model can be trained through stochastic gradient descent (SGD) suggested by [12]. The gradient can be directly calculated for each training sample. Both the word vectors and MWE vectors are randomly initialized as what was used by Mikolov.

4   Experiment and Analysis

Wikipedia August 2016 dump[1] is used as our training corpus. In pre-processing, pure digits and punctuations are removed, and all English words are converted to lowercase. The final corpus consists of about 3.2 billion words. During training, only words that occur more than 100 times are kept, resulting in a vocabulary of 204,981 words. The list of MWEs used in the evaluation are from 5 sources: (1) the set of 2,180 MWEs in the Noun-Modifier Composition dataset [22], (2) the DISCo set of 349 MWEs for the 2011 shared task in Distributional Semantics and Compositionality [3], (3) the set of 8,105 MWEs from the SemEval 2013 Task 5A [9], (4) the set of 1,042 MWEs from [5], and (5) the set of 56,850 phrases from [24]. The consolidated MWE list contained has a total of 60,315 after removal of duplication. Their representations are learned using the word2vecf source code [10].

4.1   Evaluation Tasks

We evaluate the representation of our MWEs on three evaluation tasks. The first task is called the **SemEval 2013 Task 5**. The dataset for this task, denoted as **SemEval**, is prepared to judge whether a given bigram-unigram pair is semantically related or not [9]. For example, the bigram *newborn infant* is semantically related to the unigram *neonate*. So, the gold answer for this pair is *(newborn infant, neonate, 1), where the label 1 indicates their relatedness*. On the other hand, the bigram *stable condition* is not related to the unigram *interview*, So, in the gold answer, the entry is *(stable condition, interview, 0)*. The officially released data contains 7,814 test samples and 11,722 training samples.[2] Since

---

[1] https://dumps.wikimedia.org/enwiki/latest/.
[2] https://www.cs.york.ac.uk/semeval-2013/task5.html.

some of the bigrams/unigrams are not contained in the Wikipedia training corpus, we only use the 15,973 samples contained in Wikipedia in our evaluation. Since SemEval 2013 Task 5 is a binary classification problem, we simply use the cosine similarity between the learned bigram embedding and the unigram embedding as the feature and use an SVM to learn the threshold to perform 5-fold cross-validation classification. Accuracy, precision, recall and F-score are used in the evaluation metrics. The second task is called **Phrase Similarity**. This task provides a phrase pair similarity dataset with 324 samples[3] constructed using manually rated scores from 1 to 7 with 7 being the most similar [14]. For example, the phrase pair *(hot weather, cold air)* has a similarity score 2.22. The dataset contains three types of phrases: adjective-nouns, noun-nouns, and verb-objects with 108 samples for each type. All 324 samples are used in evaluation, denoted as **PS**. Cosine similarity used to compare the different MWE vectors and Spearmans ρ correlation coefficient is used to evaluate the performance. The third task is called the **Turney-5**. This dataset in this task is a 7-choice Noun-Modifier Question dataset built from WordNet [22] with 2,180 question groups. For example, in the sample *(small letter, lowercase, small, letter, little, missive, ploughman, debt)*, the first bigram *small leter* is the question and the latter 7 unigrams are the candidate answers. The task is to select the most similar unigram as the answer, which should be *lowercase* in this sample. To remove the bias towards MWEs by following Yu's suggestion [25], we remove the two component words to construct a 5 choice single word questions to form our evaluation dataset, denoted as **T-5**. Again, by removing samples that are not contained in the Wikipedia training corpus, the final evaluation data contains 669 questions. The cosine similarity is used to measure the semantic closeness of a bigram MWE and the unigrams. The one with the highest similarity score is chosen as the answer. Accuracy is used as the evaluation metric.

4.2 Baselines and Experiment Settings

We compare our model with the following baselines:

1. **SGNS:** the original vector representation model to take a MWE as a non-divisible unit [12,24].
2. **SEING:** a modified SGNS model by treating component words as *morphemes* of the MWE with a constraint that the MWE vector should be similar to both the vectors of its the component word regardless of compositionality of the MWE [21].
3. **Comp-Add:** a simple additive composition model to use the vectors of the component words to obtain the MWE's vector.
4. **Comp-Mul:** a simple multiplicative composition model to use the multiplication of the two components vectors to obtain the MWE's vector.
5. **Comp-W1:** a composition model to use the vector of the first component word directly as the vector for the MWE.

---

[3] http://homepages.inf.ed.ac.uk/mlap/index.php?page=resources.

6. **Comp-W2:** a composition model to use the vector of the second component word directly as the vector for the MWE.

Our proposed model has two settings for the compositionality t_p. The first one directly sets t_p as a constant, $t_p = 1$. This means the influence of compositionality is fixed rather than using Formula 7, denoted as **C&S-C**. The second one uses the automatically computed t_p which is MWE dependent, denoted as **C&S-t**.

The size of the context window for all the models is set to 5, negative samples size is 5, and the word vector dimension is 300. We empirically set λ to 8. For the composition model, we empirically evaluate several kinds of combinations such as the additive model with α and β as 1, or the multiplicative model. Experiments show that the additive composition model achieves the best, so we only report the results using the additive composition model. To obtain the compositionality t_p, we first train the representation of MWEs using SGNS and compute its compositionality based on Formula 7.

Table 1. Performance of different MWE representation models on 3 evaluation tasks.

Model	SemEval				PS	T-5
	Acc	Pre	Rec	F	ρ	Acc
SGNS	.645	.654	.616	.634	.155	.535
SEING	.582	.578	.608	.593	.055	.575
Comp-Add	**.795**	**.826**	**.748**	**.785**	**.622**	.603
Comp-Mul	.506	.506	.483	.494	410	.227
Comp-W1	.737	.771	.672	.718	.450	.499
Comp-W2	.759	.796	.697	.743	.500	.463
C& S-C	**.779**	**.808**	**.731**	**.767**	**.595**	**.683**
C& S-t	.764	.794	.711	.750	.580	**.681**

4.3 Performance Evaluation and Analysis

Table 1 shows the performance using the three datasets of evaluation tasks for a total of six baseline systems and two variants of our proposed model.

General Analysis. Note that the first two models, SGNS and SEING are distributional models. Compared to all the other models which have the compositional elements, the distributional models have no advantage. In other words, the semantics of MWE expressions are not fully recognized by using external context. Treating them as a non-divisible unit obviously loses some semantic information. Among the four baseline composition models, Comp-Mul performs the worst, which means that element-wise multiplication can introduce more noise than information. Comp-W1 and Comp-W2 have similar performance with Comp-W2 performing slightly better. A reasonable explanation is that the second component word is more likely the head word. For compositional MWEs, head words have a more defining role semantically. Among all the models, Comp-Add performs the best on the two datasets SemEval and PS. Theoretically speaking, however, our model considers information in both the external and the component words, which should lead to better performance. Our model performs the best on the T-5 data, and outperforms COMP-Add with a large margin (13.3%).

We argue that the performances of different models are dataset dependent, especially dependent on the proportion of non-compositional MWEs. We randomly select 30 MWEs from the three datasets to manually annotate their compositionality and the result shows that the proportions of non-compositional MWEs are about 2.5%, 2.5% and 10% in SemEval, PS and T-5, respectively. Because the compositional model is more suitable on compositional MWEs, the Comp-Add model performs much better than SGNS on SemEval and the gap decreases on the T-5 as the proportion of non-compositional MWEs increases. Comp-Add indicates that the combined use of the vectors of the two component words is more comprehensive than using the external context of the compositional MWEs. On the T-5 dataset, the proportion of the non-compositional MWEs is larger than in the other two sets. So, there are more MWEs which would not work using the compositional method, which means their semantic cannot be composed from the component words. That is why the performance of SGNS increases and our model outperforms Comp-Add.

Note that C&S-t with predicted compositionality performs slightly worse than C&S-C with fixed compositionality. This is mainly because the estimation of compositionality t_p is not sufficiently accurate. We evaluate the predicted compositionality on the MWEs dataset from Farahmand's list [5] which has 1,042 MWES manually annotated with compositionality by calculating the Spearman's ρ correlation between the annotated compositionality and the predicted compositionality. The result shows that ρ only achieves 0.227, which means the current method to predict compositionality still has a lot of room for improvement. Figure 1 shows the effects of the weight λ on the **T-5** dataset, which indicates that C&S-C achieves the best performance when λ equals 8.

Compositionality Analysis. To further explore the effects of compositionality on different models, we further analyze the effect of the proportion of non-compositional MWEs on the SemEval semantic relation task. We manually select 20 non-compositional MWEs from Farahmand's list [5]. Based on the 20 MWEs, we make 20 positive (semantically related) bigram-unigram pairs and 20 negative (not semantically related) bigram-unigram pairs to form a balanced non-compositional sample set for the SemEval task, denoted as **N-Sem**. We also randomly take 60 samples from the original SemEval dataset to form a compositional sample set, denoted as **C-Sem**. In the evaluation, we add the non-compositional MWEs from N-Sem to C-Sem to increase the proportion of non-compositional WMEs until all the non-compositional WMEs are used up (total of 100 samples). We then reduce the compositional portion so that the

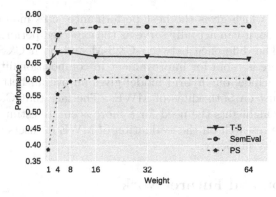

Fig. 1. Performance of C&S-C with different λ values.

non-compositional portion reaches about 70% of the total set (57 samples). We add a **C&S-M** model which takes the compositionality t_p using manually annotated compositionality (1 or 0). F-score is used as the evaluation metric. Because of the limited data size, each model is run 10 times and the average is used.

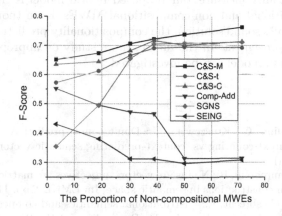

Fig. 2. Performance of increasing the proportion of non-compositional MWEs.

Figure 2 shows that when the proportion of non-compositional MWEs is small, all the compositional models perform better, consistent with the result in Table 1. As the non-compositional portion increases, Comp-Add degrades gradually whereas SGNS increases gradually. This indicates that external context is indeed useful for non-compositional MWEs and the composition model is ill-suited for non-compositional MWEs. The performance of SEING indicates that the constraint to force a MWE's vector to be similar to both of its components can actually bring adverse effect. Over the whole spectrum, our models give much more stable performance and are the overall top performers in all the automatic methods. C&S-M, which gives the best performance, uses manually annotated compositionality. So, it can actually serve as the expected performance for automated systems. The difference between C&S-M and both C&S-t and C&S-C indicates that there is room for improvement in compositionality prediction in the future. To conclude, our hybrid model gives an overall better performance as it is more robust on both kinds of MWEs. The fact that C&S-M gives the best performance highlights the need for a more accurate estimation of compositionality. Future research on the evaluation of MWE datasets should consider the compositionality property.

5   Conclusion and Future Work

In this paper, we propose a hybrid method to learn the representation of MWEs from their context with the compositionality constraint. This method can make use of both the external context and component words of MWEs. Instead of simple combination of the two kinds of information, we use the compositionality measures from lexical semantics to serve as a constraint. Evaluations show that the composition model works better for compositional MWEs whereas the distributional model performs better on non-compositional MWEs. As we introduce compositionality measure, our proposed hybrid model is the most robust on both compositional and non-compositional MWEs. Even though our model gives a theoretically sound solution, the compositionality prediction method still has room for improvement. In the future, more study on appropriate compositionality prediction model can be investigated.

References

1. Baroni, M., Dinu, G., Kruszewski, G.: Don't count, predict! A systematic comparison of context-counting vs. context-predicting semantic vectors. In: ACL, pp. 238–247 (2014)
2. Baroni, M., Zamparelli, R.: Nouns are vectors, adjectives are matrices: representing adjective-noun constructions in semantic space. In: EMNLP, pp. 1183–1193 (2010)
3. Biemann, C., Giesbrecht, E.: Distributional semantics and compositionality 2011: shared task description and results. In: Proceedings of the Workshop on Distributional Semantics and Compositionality, DiSCo 2011, Stroudsburg, PA, USA, pp. 21–28 (2011)
4. Collobert, R., Weston, J., Bottou, L., Karlen, M., Kavukcuoglu, K., Kuksa, P.: Natural language processing (almost) from scratch. JMLR **12**, 2493–2537 (2011)
5. Farahmand, M., Smith, A., Nivre, J.: A multiword expression data set: annotating non-compositionality and conventionalization for English noun compounds. In: Proceedings of the 11th Workshop on Multiword Expressions, NAACL, pp. 29–33 (2015)
6. Harris, Z.S.: Distributional structure. Word (1954)
7. Hashimoto, K., Tsuruoka, Y.: Adaptive joint learning of compositional and non-compositional phrase embeddings. arXiv preprint arXiv:1603.06067 (2016)
8. Korkontzelos, I.: Unsupervised learning of multiword expressions. Ph.D. thesis, University of York, UK (2010)

9. Korkontzelos, I., Zesch, T., Zanzotto, F.M., Biemann, C.: Semeval-2013 task 5: evaluating phrasal semantics. In: Second Joint Conference on Lexical and Computational Semantics (*SEM), vol. 2, pp. 39–47 (2013)
10. Levy, O., Goldberg, Y.: Dependency-based word embeddings. In: Proceedings of ACL, vol. 2, pp. 302–308 (2014). 00054
11. Levy, O., Goldberg, Y.: Neural word embedding as implicit matrix factorization. In: Proceedings of NIPS, pp. 2177–2185 (2014)
12. Mikolov, T., Sutskever, I., Chen, K., Corrado, G.S., Dean, J.: Distributed representations of words and phrases and their compositionality. In: Proceedings of NIPS, pp. 3111–3119 (2013)
13. Mitchell, J., Lapata, M.: Vector-based models of semantic composition. In. In Proceedings of ACL, pp. 236–244 (2008). 00288
14. Mitchell, J., Lapata, M.: Composition in distributional models of semantics. Cogn. Sci. 34(8), 1388–1429 (2010)
15. Moreno-Ortiz, A., Prez-Hernndez, C., Del-Olmo, M., et al.: Managing multiword expressions in a lexicon-based sentiment analysis system for Spanish. In: NAACL HLT 2013, vol. 1 (2013)
16. Pennington, J., Socher, R., Manning, C.D.: Glove: global vectors for word representation. In: Proceedings of EMNLP, pp. 1532–1543 (2014)
17. Salehi, B., Cook, P., Baldwin, T.: A word embedding approach to predicting the compositionality of multiword expressions. In: Proceedings of NAACL-HLT, pp. 977–983 (2015)
18. Schneider, N., Danchik, E., Dyer, C., Smith, N.A.: Discriminative lexical semantic segmentation with gaps: running the MWE gamut. TACL 2, 193–206 (2014)
19. Schneider, N., Onuffer, S., Kazour, N., Danchik, E., Mordowanec, M.T., Conrad, H., Smith, N.A.: Comprehensive annotation of multiword expressions in a social web corpus. In: Proceedings of LREC, Reykjavik, Iceland, pp. 455–461 (2014)
20. Socher, R., Lin, C.C., Manning, C., Ng, A.Y.: Parsing natural scenes and natural language with recursive neural networks. In: Proceedings of ICML, pp. 129–136 (2011)
21. Sun, F., Guo, J., Lan, Y., Xu, J., Cheng, X.: Inside out: two jointly predictive models for word representations and phrase representations. In: Proceedings of AAAI (2016)
22. Turney, P.D.: Domain and function: a dual-space model of semantic relations and compositions. JAIR 44, 533–585 (2012)
23. Yazdani, M., Farahmand, M., Henderson, J.: Learning semantic composition to detect non-compositionality of multiword expressions. In: Proceedings of EMNLP, pp. 1733–1742 (2015)
24. Yin, W., Schtze, H.: An exploration of embeddings for generalized phrases. In: Proceedings of the ACL, Student Research Workshop, pp. 41–47 (2014)
25. Yu, M., Dredze, M.: Learning composition models for phrase embeddings. TACL 3, 227–242 (2015). 00001

Linear Algebraic Characterization
of Logic Programs

Chiaki Sakama[1]([⊠]), Katsumi Inoue[2], and Taisuke Sato[3]

[1] Wakayama University, Wakayama, Japan
sakama@sys.wakayama-u.ac.jp
[2] National Institute of Informatics, Tokyo, Japan
inoue@nii.ac.jp
[3] AI Research Center AIST, Tokyo, Japan
satou.taisuke@aist.go.jp

Abstract. This paper introduces a novel approach for computing logic programming semantics based on multilinear algebra. First, a propositional Herbrand base is represented in a vector space and if-then rules in a program are encoded in a matrix. Then we provide methods of computing the least model of a Horn logic program, minimal models of a disjunctive logic program, and stable models of a normal logic program by algebraic manipulation of higher-order tensors. The result of this paper exploits a new connection between linear algebraic computation and symbolic computation, which has potential to realize logical inference in huge scale of knowledge bases.

1 Introduction

Logic programming (LP) provides languages for declarative problem solving and symbolic reasoning, while proof-theoretic computation like Prolog turns inefficient in real-world applications. Recent studies have developed efficient solvers for *answer set programming* (ASP)—LP under the stable model semantics [1]. In this paper, we take a different approach and introduce a new method of computing LP semantics in vector spaces. There are several reasons for considering linear algebraic computation of LP. First, linear algebra is at the core of many applications of scientific computation, and integrating linear algebraic computation and symbolic computation is considered a challenging topic in AI [13]. Second, linear algebraic computation has potential to cope with Web scale symbolic data, and several studies develop scalable techniques to process huge relational knowledge bases [10,11,18]. Since relational KBs consist of ground atoms, the next challenge is applying linear algebraic techniques to LP and deductive DBs. Third, it would enable us to use efficient (parallel) algorithms of numerical linear algebra for computing LP. Moreover, matrix/tensor factorization techniques would be useful for approximation and optimization in LP.

This work is supported by NII Collaborative Research Program.

G. Li et al. (Eds.): KSEM 2017, LNAI 10412, pp. 520–533, 2017.
DOI: 10.1007/978-3-319-63558-3_44

Several studies attempt to realize logical reasoning using linear algebra. Grefenstette [5] introduces tensor-based predicate calculus in which elements of tensors represent truth values of domain objects and logical operations are realized by third-order tensor contractions. Yang *et al.* [18] introduce a method of mining Horn clauses from relational facts represented in a vector space. Serafini and d'Avila Garcez [16] introduce logic tensor networks that integrate logical deductive reasoning and data-driven relational learning. Sato [14] formalizes Tarskian semantics of first-order logic in vector spaces, and shows how tensorization realizes efficient computation of Datalog [15]. These studies realize linear algebraic computation of predicate calculus over unary/binary relations on a finite domain, while they do not target computing LP semantics. There are studies that compute LP semantics in other computational paradigms (e.g. [8]). To the best of our knowledge, however, there is no study that realizes LP using linear algebra.

In this paper, we develop a theory of LP based on multilinear algebra. First, a propositional Herbrand base is represented in a vector space and if-then rules in a program are encoded in a matrix. Then the least model of a (propositional) Horn logic program is computed using matrix products. Next disjunctive logic programs are represented in third-order tensors and their minimal models are computed by algebraic manipulation of tensors. Normal logic programs are also represented by third-order tensors in terms of disjunctive logic programs, and their stable models are computed using tensor products. The rest of this paper is organized as follows. Section 2 reviews notions in multilinear algebra. Section 3 formulates LP semantics in vector spaces. Section 4 discusses related issues and Sect. 5 concludes the paper.

2   Preliminaries

This section reviews basic notions of tensors used in this paper. The following definitions are from [6]. An *N-th order tensor* is an element of the tensor product of N vector spaces. A first-order tensor is a vector $v \in \mathbb{R}^I$, a second-order tensor is a matrix $M \in \mathbb{R}^{I \times J}$, and a third-order tensor is a three-way array $T \in \mathbb{R}^{I \times J \times K}$. The i-th element of a vector is denoted by a_i, an element (i, j) of a matrix is denoted by a_{ij}, and an element (i, j, k) of a third-order tensor is denoted by a_{ijk}. A tensor or matrix is often written as $T = (a_{ijk})$ or $M = (a_{ij})$, respectively. A *column vector* is an $m \times 1$ matrix which is represented as $(a_1, \ldots, a_m)^\mathsf{T}$ $(m \geq 1)$ where T is the transpose operation. In this paper, a vector means a column vector unless stated otherwise. *Slices* are two-dimensional sections of a tensor, defined by fixing all but two indices. A third-order tensor $T \in \mathbb{R}^{I \times J \times K}$ is decomposed into (frontal) slices by fixing an index k $(1 \leq k \leq K)$. The k-th slice of a third-order tensor T is a matrix and is written as $T_{::k}$. Let $T_{::1}, \ldots, T_{::K}$ be the slices of a third-order tensor $T \in \mathbb{R}^{I \times J \times K}$ such that

$$\begin{pmatrix} a_{111} & \cdots & a_{1J1} \\ \vdots & \ddots & \vdots \\ a_{I11} & \cdots & a_{IJ1} \end{pmatrix} \cdots \begin{pmatrix} a_{11K} & \cdots & a_{1JK} \\ \vdots & \ddots & \vdots \\ a_{I1K} & \cdots & a_{IJK} \end{pmatrix}$$

Then T is *flattened* into the $I \times (J \times K)$ matrix as

$$
\begin{pmatrix}
a_{111} & \cdots & a_{1J1} & & a_{11K} & \cdots & a_{1JK} \\
\vdots & \ddots & \vdots & \cdots & \vdots & \ddots & \vdots \\
a_{I11} & \cdots & a_{IJ1} & & a_{I1K} & \cdots & a_{IJK}
\end{pmatrix}
$$

To distinguish slices in a matrix, we often introduce a vertical line between blocks representing slices $T_{::k}$ and $T_{::k+1}$.

Example 2.1. Consider a third-order tensor $T \in \mathbb{R}^{2 \times 3 \times 2}$ which is decomposed into two slices

$$
\begin{pmatrix} a_1 \ b_1 \ c_1 \\ d_1 \ e_1 \ f_1 \end{pmatrix} \quad \text{and} \quad \begin{pmatrix} a_2 \ b_2 \ c_2 \\ d_2 \ e_2 \ f_2 \end{pmatrix}
$$

Then T is flattened into the 2×6 matrix as

$$
\left(\begin{array}{ccc|ccc} a_1 & b_1 & c_1 & a_2 & b_2 & c_2 \\ d_1 & e_1 & f_1 & d_2 & e_2 & f_2 \end{array} \right)
$$

The (2-*mode*) *product* of a tensor $T \in \mathbb{R}^{I \times J \times K}$ with a vector $v \in \mathbb{R}^J$ is denoted by $T \bullet v$. The result is the $I \times K$ matrix that has the element $(T \bullet v)_{ik} = \sum_{j=1}^{J} x_{ijk} v_j$ where $T = (x_{ijk})$ and $v = (v_1, \ldots, v_J)^{\mathsf{T}}$. For example, let $T \in \mathbb{R}^{2 \times 3 \times 2}$ be the tensor of Example 2.1. Given the vector $v = (\alpha, \beta, \gamma)^{\mathsf{T}}$, $T \bullet v$ becomes

$$
\begin{pmatrix} a_1\alpha + b_1\beta + c_1\gamma & a_2\alpha + b_2\beta + c_2\gamma \\ d_1\alpha + e_1\beta + f_1\gamma & d_2\alpha + e_2\beta + f_2\gamma \end{pmatrix}
$$

When $M \in \mathbb{R}^{I \times J}$ is a matrix and $v \in \mathbb{R}^J$, the product $M \bullet v$ is the standard matrix multiplication that becomes a vector in $\mathbb{R}^I$.

3   Tensor Logic Programming

We consider a language $\mathcal{L}$ that contains a finite set of propositional variables and the logical connectives $\neg$, $\wedge$, $\vee$ and $\leftarrow$. $\mathcal{L}$ contains $\top$ and $\perp$ representing *true* and *false*, respectively. Given a logic program P, the set of all propositional variables appearing in P is the *Herbrand base* of P (written B_P). We assume $\{\top, \perp\} \subseteq B_P$.

3.1   Horn Logic Programs

A *Horn (logic) program* is a finite set of *rules* of the form:

$$
h \leftarrow b_1 \wedge \cdots \wedge b_m \quad (m \geq 0) \tag{1}
$$

where h and b_i are propositional variables (also called atoms) in $\mathcal{L}$. In particular, the rule (1) is a *fact* if "$h \leftarrow \top$" and (1) is a *constraint* if "$\perp \leftarrow b_1 \wedge \cdots \wedge b_m$". The fact and the constraint are simply written as "$h \leftarrow$" and "$\leftarrow b_1 \wedge \cdots \wedge b_m$", respectively. For each rule r of the form (1), the left-hand side of $\leftarrow$ is the *head* and the right-hand side is the *body*. We write $head(r) = h$ and $body(r) = \{b_1, \ldots, b_m\}$. We assume that every program (implicitly) contains the rule $\top \leftarrow \top$. A *definite program* is a Horn program that contains no constraints. For a Horn program P, a set I satisfying $\{\top\} \subseteq I \subseteq B_P$ is an *interpretation* of P. An interpretation I is a *model* of P if $\{b_1, \ldots, b_m\} \subseteq I$ implies $h \in I$ for every rule (1) in P and $\bot \notin I$. We often omit $\top$ in a model, i.e., any model I is semantically identified with $I \setminus \{\top\}$. A model I is the *least model* of P if $I \subseteq J$ for any model J of P. P is *inconsistent* if it has no model. A mapping $T_P : 2^{B_P} \to 2^{B_P}$ is defined as $T_P(I) = \{h \mid h \leftarrow b_1 \wedge \cdots \wedge b_m \in P$ and $\{b_1, \ldots, b_m\} \subseteq I\}$ if $\bot \notin I$. Otherwise, $T_P(I) = B_P$.[1] The powers of T_P are defined as: $T_P^{k+1}(I) = T_P(T_P^k(I))\,(k \geq 0)$ and $T_P^0(I) = I$. Given $\{\top\} \subseteq I \subseteq B_P$, there is a *fixpoint* $T_P^{n+1}(I) = T_P^n(I)\,(n \geq 0)$. For a definite program P, the fixpoint $T_P^n(\{\top\})$ coincides with the least model of P [17].[2]

In this paper, we consider a Horn program P such that for any two rules r_1 and r_2 in P, $head(r_1) = head(r_2)$ implies $|body(r_1)| \leq 1$ and $|body(r_2)| \leq 1$ (called the *multiple definitions (MD) condition*). That is, if two different rules have the same head, those rules contain at most one atom in their bodies. Every Horn program is converted to this class of programs by a simple program transformation. Given a Horn program P, define a set $Q_p \subseteq P$ such that $Q_p = \{r \in P \mid head(r) = p$ and $|body(r)| > 1\}$. Let $Q_p = \{p \leftarrow \Gamma_1, \ldots, p \leftarrow \Gamma_k\}$ where $\Gamma_i\,(1 \leq i \leq k)$ is the conjunction in the body of each rule. Then Q_p is transformed to $Q_p' = \{p_1 \leftarrow \Gamma_1, \ldots, p_k \leftarrow \Gamma_k\} \cup \{p \leftarrow p_i \mid 1 \leq i \leq k\}$ where $p_i\,(\neq p)$ are new propositional variables associated with p and $p_i \neq p_j$ if $i \neq j$. Let P' be the program obtained from P by replacing $Q_p \subseteq P$ with Q_p' for every $p \in B_P$ satisfying the condition: $|\{r \in P \mid head(r) = p\}| > 1$ and $Q_p \neq \emptyset$. Then P has the least model M iff P' has the least model M' such that $M' \cap B_P = M$.

Example 3.1. Consider $P = \{p \leftarrow q \wedge r,\ p \leftarrow r \wedge s,\ p \leftarrow t,\ r \leftarrow t,\ s \leftarrow,\ t \leftarrow\}$ where $B_P = \{p, q, r, s, t\}$. Then $P' = \{p_1 \leftarrow q \wedge r,\ p_2 \leftarrow r \wedge s,\ p \leftarrow t,\ r \leftarrow t,\ s \leftarrow,\ t \leftarrow,\ p \leftarrow p_1,\ p \leftarrow p_2\}$ has the least model $M' = \{p, p_2, r, s, t\}$ and $M' \cap B_P = \{p, r, s, t\}$ is the least model of P.

As such, every Horn program P is transformed to a semantically equivalent program P' that satisfies the MD condition. We then consider programs satisfying the MD condition without loss of generality. Next we represent interpretations/programs in vector spaces.

Definition 3.1 (vector representation of interpretations). Let P be a Horn program and $B_P = \{p_1, \ldots, p_n\}$. Then an interpretation $I \subseteq B_P$ of P is represented by a vector $\boldsymbol{v} = (a_1, \ldots, a_n)^\top$ where each element a_i represents the truth value of the proposition p_i such that $a_i = 1$ if $p_i \in I\,(1 \leq i \leq n)$; otherwise, $a_i = 0$. The vector representing $I = \{\top\}$ is written by $\boldsymbol{v}_0$. We write $\mathrm{row}_i(\boldsymbol{v}) = p_i$.

---

[1] The operator T_P of [17] is applied to Horn programs by viewing each constraint as a rule with the head $\bot$. In this setting, every atom in B_P is derived in $T_P(I)$ if I is inconsistent, i.e., $\bot \in I$. Note that $\top \in T_P(I)$ by $(\top \leftarrow \top) \in P$.

[2] $I = \{\top\}$ is semantically identified with $I = \{\}$.

Definition 3.2 ($\leq$). Let P be a Horn program and $B_P = \{p_1, \ldots, p_n\}$. Suppose two vectors $\boldsymbol{v} = (a_1, \ldots, a_n)^\mathsf{T}$ and $\boldsymbol{w} = (b_1, \ldots, b_n)^\mathsf{T}$ representing interpretations $I \subseteq B_P$ and $J \subseteq B_P$, respectively. Then $\boldsymbol{v} \leq \boldsymbol{w}$ if $a_i \leq b_i$ for every $1 \leq i \leq n$.

A vector $\boldsymbol{v} = (a_1, \ldots, a_n)^\mathsf{T}$ representing an interpretation I is *minimal* if $\boldsymbol{w} \leq \boldsymbol{v}$ implies $\boldsymbol{w} = \boldsymbol{v}$ for any $\boldsymbol{w}$ representing an interpretation J.

Proposition 3.1. *For* $\boldsymbol{v} = (a_1, \ldots, a_n)^\mathsf{T}$ *and* $\boldsymbol{w} = (b_1, \ldots, b_n)^\mathsf{T}$, $\boldsymbol{v} \leq \boldsymbol{w}$ *iff* $a_i = a_i * b_i$ *for* $i = 1, \ldots, n$ *where* $*$ *means algebraic multiplication.*

Definition 3.3 (matrix representation of Horn programs). Let P be a Horn program and $B_P = \{p_1, \ldots, p_n\}$. Then P is represented by a matrix $\boldsymbol{M}_P \in \mathbb{R}^{n \times n}$ such that for each element a_{ij} ($1 \leq i, j \leq n$) in $\boldsymbol{M}_P$, (i) $a_{ij} = 1$ if $p_i = \top$ or $p_j = \bot$; (ii) $a_{ij_k} = \frac{1}{m}$ ($1 \leq k \leq m$; $1 \leq i, j_k \leq n$) if $p_i \leftarrow p_{j_1} \wedge \cdots \wedge p_{j_m}$ is in P; (iii) otherwise, $a_{ij} = 0$. We write $\mathrm{row}_i(\boldsymbol{M}_P) = p_i$ and $\mathrm{col}_j(\boldsymbol{M}_P) = p_j$.

In $\boldsymbol{M}_P$ the i-th row corresponds to the atom p_i appearing in the head of a rule, and the j-th column corresponds to the atom p_j appearing in the body of a rule. The first condition (i) says that every element in the i-th row is 1 if $\mathrm{row}_i(\boldsymbol{M}_P) = \top$ and every element in the j-th column is 1 if $\mathrm{col}_i(\boldsymbol{M}_P) = \bot$. The second condition (ii) says if there is a rule $p_i \leftarrow p_{j_1} \wedge \cdots \wedge p_{j_m}$ in P, then each a_{ij_k} in the i-th row and the j_k-th column has the value $\frac{1}{m}$. The remaining elements are set to $a_{ij} = 0$ in (iii).

Example 3.2. Consider $P = \{p \leftarrow q, \ p \leftarrow r, \ q \leftarrow r \wedge s, \ r \leftarrow, \ \leftarrow q\}$ with $B_P = \{p, q, r, s, \top, \bot\}$. Then $\boldsymbol{M}_P \in \mathbb{R}^{6 \times 6}$ is the matrix (right). The 1st row "011001" represents $p \leftarrow q$, $p \leftarrow r$ and $p \leftarrow \bot (\equiv \top)$. The 2nd row "00 $^1/_2$ $^1/_2$ 01" represents $q \leftarrow r \wedge s$ and $q \leftarrow \bot (\equiv \top)$. The 3rd row "000011" represents $r \leftarrow \top (\equiv r \leftarrow)$ and $r \leftarrow \bot (\equiv \top)$. The 4th row "000001" represents $s \leftarrow \bot (\equiv \top)$. The 5th row "111111" represents $\top \leftarrow p$, $\top \leftarrow q$, $\top \leftarrow r$, $\top \leftarrow s$, $\top \leftarrow \top$, and $\top \leftarrow \bot$, which are all equivalent to $\top$. The 6th row "010001" represents $\bot \leftarrow q (\equiv \leftarrow q)$ and $\bot \leftarrow \bot (\equiv \top)$.

$$
\begin{array}{c@{}c}
 & \begin{array}{cccccc} p & q & r & s & \top & \bot \end{array} \\
\begin{array}{c} p \\ q \\ r \\ s \\ \top \\ \bot \end{array} &
\left(\begin{array}{cccccc}
0 & 1 & 1 & 0 & 0 & 1 \\
0 & 0 & ^1/_2 & ^1/_2 & 0 & 1 \\
0 & 0 & 0 & 0 & 1 & 1 \\
0 & 0 & 0 & 0 & 0 & 1 \\
1 & 1 & 1 & 1 & 1 & 1 \\
0 & 1 & 0 & 0 & 0 & 1
\end{array} \right)
\end{array}
$$

Suppose a matrix $\boldsymbol{M}_P \in \mathbb{R}^{n \times n}$ representing a Horn program. Given a vector $\boldsymbol{v} \in \mathbb{R}^n$ representing an interpretation $I \subseteq B_P$, let $\boldsymbol{M}_P \bullet \boldsymbol{v} = (a_1, \ldots, a_n)^\mathsf{T}$. We transform $\boldsymbol{M}_P \bullet \boldsymbol{v}$ to a vector $\boldsymbol{w} = (a'_1, \ldots, a'_n)^\mathsf{T}$ where $a'_i = 1 \, (1 \leq i \leq n)$ if $a_i \geq 1$; otherwise, $a'_i = 0$. We write $\boldsymbol{w} = \boldsymbol{M}_P \underline{\bullet} \boldsymbol{v}$.

Example 3.3. Consider $\boldsymbol{M}_P$ of Example 3.2. Given $\boldsymbol{v} = (0, 1, 1, 0, 1, 0)^\mathsf{T}$ representing the interpretation $I = \{q, r, \top\}$, it becomes $\boldsymbol{M}_P \bullet \boldsymbol{v} = (2, ^1/_2, 1, 0, 3, 1)^\mathsf{T}$. Then $\boldsymbol{w} = \boldsymbol{M}_P \underline{\bullet} \boldsymbol{v} = (1, 0, 1, 0, 1, 1)^\mathsf{T}$ represents $J = \{p, r, \top, \bot\}$.

In Example 3.3, we can see that $J = T_P(I)$ is computed by $\boldsymbol{w} = \boldsymbol{M}_P \underline{\bullet} \boldsymbol{v}$. Formally, the next result holds.

Proposition 3.2. *Let P be a Horn program and $M_P \in \mathbb{R}^{n \times n}$ its matrix representation. Let $v \in \mathbb{R}^n$ be a vector representing an interpretation $I \subseteq B_P$. Then $w \in \mathbb{R}^n$ is a vector representing $J = T_P(I)$ iff $w = M_P \bullet v$.*

Proof. Suppose $w = M_P \bullet v$. Then for $w = (x_1', \ldots, x_n')^\mathsf{T}$, $x_k' = 1$ $(1 \le k \le n)$ iff $x_k \ge 1$ in $M_P \bullet v$. Let $M_P = (a_{ij})$ and $v = (y_1, \ldots, y_n)^\mathsf{T}$. Then $a_{k1}y_1 + \cdots + a_{kn}y_n = x_k \ge 1$. If $\mathrm{row}_k(M_P) = \top$ then $a_{kj} = 1$ $(1 \le j \le n)$. Otherwise, let $\{b_1, \ldots, b_m\} \subseteq \{a_{k1}, \ldots, a_{kn}\}$ such that $b_i \ne 0$ $(1 \le i \le m)$ and $b_i \ne a_{kj}$ for $\mathrm{col}_j(M_P) = \bot$. Two cases are considered. (i) $b_i = 1$ $(1 \le i \le m)$ and $b_1 y_{j_1} + \cdots + b_m y_{j_m} \ge 1$ $(1 \le j_l \le n, 1 \le l \le m)$. Then there are rules $p_k \leftarrow p_i$ in P such that $p_i = \mathrm{col}_j(M_P)$ for $b_i = a_{kj}$ $(1 \le i \le m)$ and $p_i \in I$ implies $p_k \in T_P(I)$ where $p_k = \mathrm{row}_k(w)$. (ii) $b_i = 1/m$ and $b_1 y_{j_1} + \cdots + b_m y_{j_m} = 1$ $(1 \le j_l \le n, 1 \le l \le m)$. Then there is a rule $p_k \leftarrow p_1 \wedge \cdots \wedge p_m$ in P such that $p_i = \mathrm{col}_j(M_P)$ for $b_i = a_{kj}$ $(1 \le i \le m)$ and $\{p_1, \ldots, p_m\} \subseteq I$ implies $p_k \in T_P(I)$ where $p_k = \mathrm{row}_k(w)$. By putting $J = \{\, \mathrm{row}_k(w) \mid x_k' = 1 \,\}$, $J = T_P(I)$ holds. Conversely, suppose $J = T_P(I)$. Construct a matrix $M_P = (a_{ij}) \in \mathbb{R}^{n \times n}$ as Definition 3.3. For $v = (y_1, \ldots, y_n)^\mathsf{T}$ representing I, $M_P \bullet v = (x_1, \ldots, x_n)^\mathsf{T}$ is a vector such that $x_k \ge 1$ $(1 \le k \le n)$ iff $\mathrm{row}_k(M_P \bullet v) = \top$ or $\mathrm{row}_k(M_P \bullet v) \in T_P(I)$. Define $w = (x_1', \ldots, x_n')^\mathsf{T}$ such that $x_k' = 1$ $(1 \le k \le n)$ iff $x_k \ge 1$ in $M_P \bullet v$. Then w represents $J = T_P(I)$. Hence, $w = M_P \bullet v$. $\qquad \square$

Proposition 3.3. *Let P be a Horn program and $M_P \in \mathbb{R}^{n \times n}$ its matrix representation. Let $v \in \mathbb{R}^n$ be a vector representing an interpretation I. Then I is a model of P iff (i) $w = M_P \bullet v$ and $w \le v$, and (ii) $a_i = 1$ implies $\mathrm{row}_i(v) \ne \bot$ for any element a_i in v.*

Proof. I is a model of a Horn program P iff $T_P(I) \subseteq I$ and $\bot \notin I$ [9]. Then the result holds by Proposition 3.2. $\qquad \square$

Given a matrix $M_P \in \mathbb{R}^{n \times n}$ and a vector $v \in \mathbb{R}^n$, define

$$M_P \bullet^{k+1} v = M_P \bullet (M_P \bullet^k v) \quad \text{and} \quad M_P \bullet^1 v = M_P \bullet v \ (k \ge 1).$$

When $M_P \bullet^{k+1} v = M_P \bullet^k v$ for some $k \ge 1$, we write $\mathsf{FP}(M_P \bullet v) = M_P \bullet^k v$.

Theorem 3.4. *Let P be a Horn program and $M_P \in \mathbb{R}^{n \times n}$ its matrix representation. Then $m \in \mathbb{R}^n$ is a vector representing the least model of P iff $m = \mathsf{FP}(M_P \bullet v_0)$ and $a_i = 1$ implies $\mathrm{row}_i(m) \ne \bot$ for any element a_i in m.*

Proof. Since the least model of P is computed by the fixpoint of $T_P^k(\{\top\})$ $(k \ge 0)$, the result holds by Proposition 3.2. $\qquad \square$

Corollary 3.5. *Let P be a Horn program and $M_P \in \mathbb{R}^{n \times n}$ its matrix representation. Then P is inconsistent iff a vector $w = M_P \bullet^k v_0$ $(k \ge 1)$ has an element $a_i = 1$ $(1 \le i \le n)$ such that $\mathrm{row}_i(w) = \bot$.*

Corollary 3.5 is used for query-answering by refutation in a definite program.

3.2 Disjunctive Logic Programs

A *disjunctive (logic) program* is a finite set of *rules* of the form:

$$h_1 \vee \cdots \vee h_l \leftarrow b_1 \wedge \cdots \wedge b_m \quad (l, m \geq 0) \tag{2}$$

where h_i and b_j are propositional variables in $\mathcal{L}$. A rule (2) is called a *disjunctive rule* if $l > 1$. In particular, the rule is a *(disjunctive) fact* if the body of (2) is $\top$ and it is a *constraint* if the head of (2) is $\bot$. A disjunctive fact is simply written as "$h_1 \vee \cdots \vee h_l \leftarrow$". For each rule r of the form (2), we write $head(r) = \{h_1, \ldots, h_l\}$ and $body(r) = \{b_1, \ldots, b_m\}$. A disjunctive program reduces to a Horn program if $l \leq 1$ for every rule (2) in a program. We assume that every program (implicitly) contains the rule $\top \leftarrow \top$ as before. An interpretation $\{\top\} \subseteq I \subseteq B_P$ is a *model* of a disjunctive program P if $body(r) \subseteq I$ implies $head(r) \cap I \neq \emptyset$ for every rule r in P and $\bot \notin I$. A model I is a *minimal model* of P if there is no model J of P such that $J \subset I$.

We consider a disjunctive program P satisfying the *MD condition*: for any two rules r_1 and r_2 in P, $head(r_1) \cap head(r_2) \neq \emptyset$ implies $|body(r_1)| \leq 1$ and $|body(r_2)| \leq 1$. That is, if two different rules share the same atom in their heads, those rules contain at most one atom in their bodies. The condition coincides with the MD condition of Horn programs when P contains no disjunctive rules. Every disjunctive program is converted to this class of programs by a simple program transformation as the case of Horn programs. Given a disjunctive program P, define a set $Q_p \subseteq P$ such that $Q_p = \{r \in P \mid p \in head(r)$ and $|body(r)| > 1\}$. Let $Q_p = \{\Sigma_1 \leftarrow \Gamma_1, \ldots, \Sigma_k \leftarrow \Gamma_k\}$ where Σ_i (resp. Γ_i) $(1 \leq i \leq k)$ is the disjunction (resp. conjunction) in the head (resp. body) of each rule. Then Q_p is transformed to $Q'_p = \{\Sigma'_1 \leftarrow \Gamma_1, \ldots, \Sigma'_k \leftarrow \Gamma_k\} \cup \{p \leftarrow p_i \mid 1 \leq i \leq k\}$ where $p_i \,(\neq p)$ are new propositional variables associated with p and $p_i \neq p_j$ if $i \neq j$. Σ'_i is obtained from Σ_i by replacing p in Σ_i by p_i. Let P'_p be the program obtained from P by replacing $Q_p \subseteq P$ with Q'_p for an atom $p \in B_P$ satisfying the condition: $|\{r \in P \mid p \in head(r)\}| > 1$ and $Q_p \neq \emptyset$. Repeat such replacement one by one until there is no atom $p \in B_P$ satisfying the above condition. Let P' be the resulting program. Then P has a minimal model M iff P' has a minimal model M' such that $M' \cap B_P = M$.

Example 3.4. Consider $P = \{p \vee q \leftarrow r \wedge s, \; p \vee r \leftarrow t, \; r \leftarrow s, \; s \leftarrow\}$ where $B_P = \{p, q, r, s, t\}$. Then $P' = \{p_1 \vee q \leftarrow r \wedge s, \; p \vee r \leftarrow t, \; r \leftarrow s, \; s \leftarrow, \; p \leftarrow p_1\}$ has two minimal models $M'_1 = \{p, p_1, r, s\}$ and $M'_2 = \{q, r, s\}$ which correspond to the minimal models $M_1 = \{p, r, s\}$ and $M_2 = \{q, r, s\}$ of P.

As such, every disjunctive program P is transformed to a semantically equivalent program P' that satisfies the MD condition. We then consider disjunctive programs satisfying the MD condition without loss of generality. Given a disjunctive program P, its *split program* is a Horn program obtained from P by replacing each disjunctive rule of the form (2) in P with a Horn rule:

$h_i \leftarrow b_1 \wedge \cdots \wedge b_m \ (1 \leq i \leq l)$. By definition, P has multiple split programs in general. When P has k split programs, it is written as $SP_1, \ldots, SP_k \ (k \geq 1)$.

Example 3.5. $P = \{p \vee r \leftarrow s, \ q \vee r \leftarrow, \ s \leftarrow \}$ has the four split programs: $SP_1 = \{p \leftarrow s, \ q \leftarrow, \ s \leftarrow \}$, $SP_2 = \{p \leftarrow s, \ r \leftarrow, \ s \leftarrow \}$, $SP_3 = \{r \leftarrow s, \ q \leftarrow, \ s \leftarrow \}$, and $SP_4 = \{r \leftarrow s, \ r \leftarrow, \ s \leftarrow \}$.

For a set $\mathcal{I}$ of interpretations, define the set of minimal elements as $min(\mathcal{I}) = \{I \mid I \in \mathcal{I} \text{ and there is no } J \in \mathcal{I} \text{ such that } J \subset I \}$.

Proposition 3.6. *Let P be a disjunctive program and $SP_1, \ldots, SP_k$ its split programs. Also let $\mathcal{LM}$ be the set of least models of $SP_1, \ldots, SP_k$. Then $min(\mathcal{LM})$ coincides with the set $\mathcal{MM}$ of minimal models of P.*

Proof. Let M be the least model of some split program SP_j. Then for each rule $r : h_i \leftarrow b_1 \wedge \cdots \wedge b_m$ in SP_j, there is a rule $r' : h_1 \vee \cdots \vee h_l \leftarrow b_1 \wedge \cdots \wedge b_m$ in P such that $h_i \in head(r')$. Since M satisfies r, M satisfies r'. Thus, M is a model of P. Then a minimal set M in $min(\mathcal{LM})$ is a minimal model of P. Hence, $min(\mathcal{LM}) \subseteq \mathcal{MM}$. Conversely, let $M \in \mathcal{MM}$. Then for each rule $r : h_1 \vee \cdots \vee h_l \leftarrow b_1 \wedge \cdots \wedge b_m$ in P, $\{b_1, \ldots, b_m\} \subseteq M$ implies $h_i \in M$ for some $i \, (1 \leq i \leq l)$. In this case, there is a split program SP_j of P in which r is replaced by $h_i \leftarrow b_1 \wedge \cdots \wedge b_m$. Then M is the least model of SP_j. Since M is minimal among models in $\mathcal{LM}$, $\mathcal{MM} \subseteq min(\mathcal{LM})$. $\qquad\square$

Example 3.6. Consider the program P in Example 3.5. The set of least models of split programs is $\mathcal{LM} = \{\{p, q, s\}, \{p, r, s\}, \{q, r, s\}, \{r, s\}\}$. Then $min(\mathcal{LM}) = \{\{p, q, s\}, \{r, s\}\}$, which is the set of minimal models of P.

By definition, if a disjunctive program satisfies the MD condition, its split program satisfies the MD condition for Horn programs. Next we introduce tensor representation of a disjunctive program in terms of its split programs.

Definition 3.4 (tensor representation of disjunctive programs). Let P be a disjunctive program that is split into $SP_1, \ldots, SP_k$ and $B_P = \{p_1, \ldots, p_n\}$. Then P is represented by a third-order tensor $\boldsymbol{U}_P \in \mathbb{R}^{n \times n \times k}$ as follows.

1. Each slice $U_{::h} \ (1 \leq h \leq k)$ of $\boldsymbol{U}_P$ is a matrix $\boldsymbol{M}_h \in \mathbb{R}^{n \times n}$ representing a split program SP_h.
2. Each matrix $\boldsymbol{M}_h \in \mathbb{R}^{n \times n}$ has an element $a_{ij} \ (1 \leq i, j \leq n)$ such that
   (a) $a_{ij} = 1$ if $p_i = \top$ or $p_j = \bot$.
   (b) $a_{ij_l} = \frac{1}{m} \ (1 \leq l \leq m; \ 1 \leq i, j_l \leq n)$ if $p_i \leftarrow p_{j_1} \wedge \cdots \wedge p_{j_m}$ is in SP_h.
   (c) Otherwise, $a_{ij} = 0$.

$U_{::h}$ represents a slice that is obtained by fixing an index $h \, (1 \leq h \leq k)$. The index h represents the h-th split program SP_h of P. In this way, a disjunctive program is represented by a third-order tensor by introducing an additional dimension to the matrix representation of Horn programs.

Suppose a third-order tensor $U_P \in \mathbb{R}^{n \times n \times k}$ of Definition 3.4. Given a vector $v \in \mathbb{R}^n$ representing an interpretation $I \subseteq B_P$, the (2-mode) product $U_P \bullet v$ produces a matrix $(a_{ij}) \in \mathbb{R}^{n \times k}$. We transform $U_P \bullet v$ to a matrix $W_P = (a'_{ij}) \in \mathbb{R}^{n \times k}$ in a way that $a'_{ij} = 1$ $(1 \le i \le n; 1 \le j \le k)$ if $a_{ij} \ge 1$ in $U_P \bullet v$; otherwise, $a_{ij} = 0$. We write $W_P = U_P \underline{\bullet} v$.

Example 3.7. The program P in Example 3.5 is represented by the 3rd-order tensor $U_P \in \mathbb{R}^{5 \times 5 \times 4}$ such that[3]

$$
\begin{array}{c}
p\,q\,r\,s\,\top \\
U_{::1} = \begin{pmatrix} 0&0&0&1&0 \\ 0&0&0&0&1 \\ 0&0&0&0&0 \\ 0&0&0&0&1 \\ 1&1&1&1&1 \end{pmatrix} \begin{array}{l} p \\ q \\ r \\ s \\ \top \end{array}
\end{array}
\quad
U_{::2} = \begin{pmatrix} 0&0&0&1&0 \\ 0&0&0&0&0 \\ 0&0&0&0&1 \\ 0&0&0&0&1 \\ 1&1&1&1&1 \end{pmatrix}
\quad
U_{::3} = \begin{pmatrix} 0&0&0&0&0 \\ 0&0&0&0&1 \\ 0&0&0&1&0 \\ 0&0&0&0&1 \\ 1&1&1&1&1 \end{pmatrix}
\quad
U_{::4} = \begin{pmatrix} 0&0&0&0&0 \\ 0&0&0&0&0 \\ 0&0&0&1&1 \\ 0&0&0&0&1 \\ 1&1&1&1&1 \end{pmatrix}
$$

where $U_{::1}$, $U_{::2}$, $U_{::3}$ and $U_{::4}$ represent SP_1, SP_2, SP_3, and SP_4, respectively. Given the vector $v_0 = (0, 0, 0, 0, 1)^{\top}$ representing $I = \{\top\}$, the product $U_P \bullet v_0$ becomes the matrix in $\mathbb{R}^{5 \times 4}$ where $W_P = U_P \underline{\bullet} v_0 = U_P \bullet v_0$.

In the matrix W_P (right), the 1st column $(0, 1, 0, 1, 1)^{\top}$ represents the interpretation $T_{SP_1}(\{\top\}) = \{q, s, \top\}$, the 2nd column $(0, 0, 1, 1, 1)^{\top}$ represents $T_{SP_2}(\{\top\}) = \{r, s, \top\}$, the 3rd column $(0, 1, 0, 1, 1)^{\top}$ represents $T_{SP_3}(\{\top\}) = \{q, s, \top\}$, and the 4th column $(0, 0, 1, 1, 1)^{\top}$ represents $T_{SP_4}(\{\top\}) = \{r, s, \top\}$.

$$
\begin{array}{c}
\begin{array}{l} p \\ q \\ r \\ s \\ \top \end{array}
\begin{pmatrix} 0&0&0&0 \\ 1&0&1&0 \\ 0&1&0&1 \\ 1&1&1&1 \\ 1&1&1&1 \end{pmatrix}
\end{array}
$$

Given a matrix $M = (a_{ij})$ $(1 \le i \le m; 1 \le j \le n)$, $(a_{1j}, \dots, a_{mj})^{\top}$ is said a *column vector in M*. Let $U_P \in \mathbb{R}^{n \times n \times m}$ and $M \in \mathbb{R}^{n \times m}$. The product $U_P \bullet M$ is then defined as the matrix in $\mathbb{R}^{n \times m}$ such that each column vector in $U_P \bullet M$ is $U_{::k} \bullet (a_{1k}, \dots, a_{nk})^{\top}$ $(1 \le k \le m)$ where $(a_{1k}, \dots, a_{nk})^{\top}$ is a column vector in M. Then $U_P \underline{\bullet} M$ is the matrix obtained from $U_P \bullet M$ by replacing each element a_{ij} in $U_P \bullet M$ by $a'_{ij} = 1$ if $a_{ij} \ge 1$; otherwise, replacing a_{ij} by $a'_{ij} = 0$. Given a tensor $U_P \in \mathbb{R}^{n \times n \times m}$ and a vector $v \in \mathbb{R}^n$, define

$$U_P \underline{\bullet}^{k+1} v = U_P \underline{\bullet}(U_P \underline{\bullet}^k v) \quad \text{and} \quad U_P \underline{\bullet}^1 v = U_P \underline{\bullet} v \ (k \ge 1).$$

When $U_P \underline{\bullet}^{k+1} v = U_P \underline{\bullet}^k v$ for some $k \ge 1$, we write $\mathsf{FP}(U_P \underline{\bullet} v) = U_P \underline{\bullet}^k v$.

Theorem 3.7. *Let P be a disjunctive program and $U_P \in \mathbb{R}^{n \times n \times k}$ its tensor representation of Definition 3.4. Let $M_P = \mathsf{FP}(U_P \underline{\bullet} v_0)$ be a matrix where v_0 represents $I = \{\top\}$. Then a vector $m \in \mathbb{R}^n$ in M_P represents a minimal model of P iff m is minimal among all column vectors in M_P and $a_i = 1$ implies $\mathrm{row}_i(m) \ne \bot$ for any element a_i in m.*

Proof. The result holds by Theorem 3.4 and Proposition 3.6. □

---

[3] Here we omit the row and the column representing $\bot$. When a program contains no constraints, the row and the column representing $\bot$ can be removed (see Sect. 4).

Example 3.8. In Example 3.7, $M_P = U_P \bullet^k v_0 \ (k \geq 2)$ becomes the matrix (below). The 1st column $(1, 1, 0, 1, 1)^\mathsf{T}$ represents the minimal model $\{p, q, s\}$ of SP_1, the 2nd column $(1, 0, 1, 1, 1)^\mathsf{T}$ represents the minimal model $\{p, r, s\}$ of SP_2, the 3rd column $(0, 1, 1, 1, 1)^\mathsf{T}$ represents the minimal model $\{q, r, s\}$ of SP_3, and the 4th column $(0, 0, 1, 1, 1)^\mathsf{T}$ represents the minimal model $\{r, s\}$ of SP_4.

$$\begin{array}{c} p \\ q \\ r \\ s \\ \top \end{array} \left(\begin{array}{c|c|c|c} 1 & 1 & 0 & 0 \\ 1 & 0 & 1 & 0 \\ 0 & 1 & 1 & 1 \\ 1 & 1 & 1 & 1 \\ 1 & 1 & 1 & 1 \end{array} \right)$$

Then two vectors $(1, 1, 0, 1, 1)^\mathsf{T}$ and $(0, 0, 1, 1, 1)^\mathsf{T}$ are minimal in M_P, which represents two minimal models $\{p, q, s\}$ and $\{r, s\}$ of P.[4]

3.3   Normal Logic Programs

A *normal (logic) program* is a finite set of *rules* of the form:

$$h \leftarrow b_1 \wedge \cdots \wedge b_m \wedge \neg b_{m+1} \wedge \cdots \wedge \neg b_n \quad (n \geq m \geq 0) \tag{3}$$

where h and b_i are propositional variables in $\mathcal{L}$. h and b_i $(1 \leq i \leq m)$ are *positive literals*, and $\neg b_j$ $(m + 1 \leq j \leq n)$ are *negative literals*. As before, a rule is a *fact* if the body of (3) is $\top$ and it is a *constraint* if the head of (3) is $\bot$. A program implicitly contains $\top \leftarrow \top$. For each rule r of the form (3), we write $head(r) = h$, $body^+(r) = \{b_1, \ldots, b_m\}$ and $body^-(r) = \{b_{m+1}, \ldots, b_n\}$. A normal program reduces to a Horn program if it contains no negative literals. An interpretation $\{\top\} \subseteq I \subseteq B_P$ is a *model* of a normal program P if $body^+(r) \subseteq I$ and $body^-(r) \cap I = \emptyset$ imply $head(r) \in I$ for every rule r in P and $\bot \notin I$. Given a normal program P, an interpretation I is a *stable model* of P [4] if it coincides with the least model of the Horn program: $P^I = \{ h \leftarrow b_1 \wedge \cdots \wedge b_m \mid h \leftarrow b_1 \wedge \cdots \wedge b_m \wedge \neg b_{m+1} \wedge \cdots \wedge \neg b_n \in P$ and $\{b_{m+1}, \ldots, b_n\} \cap I = \emptyset \}$. A stable model coincides with the least model if P is a Horn program. A normal program may have no, one, or multiple stable models in general. A program is *consistent* if it has at least one stable model; otherwise, the program is *inconsistent*.

In this paper, we consider a normal program P satisfying the following *MD condition*: for two rules r_1 and r_2 in P, (i) $head(r_1) = head(r_2)$ implies $\mid body^+(r_1) \mid \leq 1$ and $\mid body^+(r_2) \mid \leq 1$, and (ii) $body^-(r_1) \cap body^-(r_2) \neq \emptyset$ implies $\mid body^+(r_1) \mid \leq 1$ and $\mid body^+(r_2) \mid \leq 1$. The condition reduces to the one of Horn programs when P contains no negative literals. Every normal program is converted to this class of programs by a simple program transformation. Given a normal program P, define a set $Q_p^+, Q_q^- \subseteq P$ such that $Q_p^+ = \{ r \in P \mid head(r) = p$ and $\mid body^+(r) \mid > 1 \}$ and $Q_q^- = \{ r \in P \mid q \in body^-(r)$ and $\mid body^+(r) \mid > 1 \}$. Let $Q_p^+ = \{ p \leftarrow \Gamma_1, \ldots, p \leftarrow \Gamma_k \}$ and $Q_q^- = \{ h_1 \leftarrow \Upsilon_1 \wedge not\, q, \ldots, h_l \leftarrow \Upsilon_l \wedge not\, q \}$ where $\Gamma_i \, (1 \leq i \leq k)$ or $\Upsilon_j \, (1 \leq j \leq l)$ is the conjunction in the body of each rule. Then Q_p^+ and Q_q^- are respectively transformed to $R_p^+ = \{ p_1 \leftarrow \Gamma_1, \ldots, p_k \leftarrow \Gamma_k \} \cup \{ p \leftarrow p_i \mid 1 \leq i \leq k \}$ and $R_q^- = \{ h_1 \leftarrow \Upsilon_1 \wedge not\, q_1, \ldots, h_l \leftarrow \Upsilon_l \wedge not\, q_l \} \cup \{ q_j \leftarrow q \mid 1 \leq j \leq l \}$ where

---

[4] Here we omit $\top$ in each model.

p_i ($\neq p$) and q_j ($\neq q$) are new propositional variables respectively associated with p and q, and $p_i \neq p_j$ ($q_i \neq q_j$) if $i \neq j$. Let P' be the program obtained from P by replacing $Q_p^+ \subseteq P$ with R_p^+ for every $p \in B_P$ satisfying the condition: $|\{r \in P \mid head(r) = p\}| > 1$ and $Q_p^+ \neq \emptyset$. Next let P'' be the program obtained from P' by replacing $Q_q^- \subseteq P'$ with R_q^- for an atom $q \in B_P$ satisfying the condition: $|\{r \in P' \mid q \in body^-(r)\}| > 1$ and $Q_q^- \neq \emptyset$. Repeat the replacement one by one until there is no atom $q \in B_P$ satisfying the above condition. Let P'' be the resulting program. Then, P has a stable model M iff P'' has a stable model M' such that $M' \cap B_P = M$.

Example 3.9. Consider $P = \{p \leftarrow q \wedge r \wedge not\, s,\ p \leftarrow r \wedge t \wedge not\, s,\ q \leftarrow t,\ r \leftarrow, t \leftarrow\}$. First, it is converted to $P' = \{p_1 \leftarrow q \wedge r \wedge not\, s,\ p_2 \leftarrow r \wedge t \wedge not\, s,\ q \leftarrow t,\ r \leftarrow,\ t \leftarrow,\ p \leftarrow p_1,\ p \leftarrow p_2\}$. Next, it is converted to $P'' = \{p_1 \leftarrow q \wedge r \wedge not\, s_1,\ p_2 \leftarrow r \wedge t \wedge not\, s_2,\ q \leftarrow t,\ r \leftarrow,\ t \leftarrow,\ p \leftarrow p_1,\ p \leftarrow p_2,\ s_1 \leftarrow s, s_2 \leftarrow s\}$. Then P'' has the single stable model $M' = \{p, p_1, p_2, q, r, t\}$ which corresponds to the stable model $M = \{p, q, r, t\}$ of P.

As such, every normal program P is transformed to a semantically equivalent program P'' that satisfies the MD condition. We then assume normal programs satisfying the MD condition without loss of generality. A normal program P is transformed to a disjunctive program with *integrity constraints* as follows [3].

$$P^\varepsilon = \{\, h \vee \varepsilon b_{m+1} \vee \cdots \vee \varepsilon b_n \leftarrow b_1 \wedge \cdots \wedge b_m \mid$$
$$(h \leftarrow b_1 \wedge \cdots \wedge b_m \wedge \neg b_{m+1} \wedge \cdots \wedge \neg b_n) \in P\,\} \cup \{\, \varepsilon p \leftarrow p \mid p \in B_P \setminus \{\top, \bot\}\,\},$$
$$IC_P = \{\, \varepsilon p \Rightarrow p \mid p \in B_P \setminus \{\top, \bot\}\,\}$$

where εp is a new atom associated with p. Let B_{P^ε} be the Herbrand base of P^ε such that $\{\top, \bot\} \subseteq B_{P^\varepsilon}$. An interpretation $\{\top\} \subseteq I \subseteq B_{P^\varepsilon}$ satisfies IC_P iff $\varepsilon p \in I$ implies $p \in I$ for every $\varepsilon p \Rightarrow p$ in IC_P.

Proposition 3.8. [3]. *Let P be a normal program. M is a stable model of P iff $M \cup \varepsilon M$ is a minimal model of P^ε satisfying IC_P where $\varepsilon M = \{\varepsilon p \mid p \in M\}$.*

Example 3.10. Consider $P = \{p \leftarrow \neg q,\ q \leftarrow \neg p\}$. Then $P^\varepsilon = \{p \vee \varepsilon q \leftarrow,\ q \vee \varepsilon p \leftarrow, \varepsilon p \leftarrow p,\ \varepsilon q \leftarrow q\}$ and $IC_P = \{\varepsilon p \Rightarrow p,\ \varepsilon q \Rightarrow q\}$. The program P^ε has three minimal models: $M_1 = \{p, \varepsilon p\}$, $M_2 = \{q, \varepsilon q\}$, and $M_3 = \{\varepsilon p, \varepsilon q\}$. Of which M_1 and M_2 satisfy IC_P. Then $\{p\}$ and $\{q\}$ are two stable models of P.

By definition, if a normal program P satisfies the MD condition, P^ε satisfies the MD condition for disjunctive programs. Suppose that a disjunctive program P^ε is split into $SP_1^\varepsilon, \ldots, SP_k^\varepsilon$. Then P^ε is represented by a third-order tensor $U_{P^\varepsilon} \in \mathbb{R}^{n \times n \times k}$ as in Definition 3.4, and we can compute stable models of P in terms of minimal models of P^ε.

Definition 3.5 ($v_{B_P}, v_{\varepsilon B_P}$). Let $v \in \mathbb{R}^n$ be a vector representing $\{\top\} \subseteq I \subseteq B_{P^\varepsilon}$. Then v is divided into two vectors $v_{B_P} \in \mathbb{R}^m$ and $v_{\varepsilon B_P} \in \mathbb{R}^m$ ($m \leq n$) where v_{B_P} represents elements of $B_P = \{p_1, \ldots, p_m\}$ ($\top, \bot \in B_P$) and $v_{\varepsilon B_P}$ represents elements of $(B_{P^\varepsilon} \setminus B_P) \cup \{\bot, \top\}$. Two vectors satisfy the condition: $row_i(v_{B_P}) = row_i(v_{\varepsilon B_P})$ iff $row_i(v_{B_P}) = \top$ or $\bot$; otherwise, $p_i = row_i(v_{B_P})$ iff $\varepsilon p_i = row_i(v_{\varepsilon B_P})$ ($1 \leq i \leq m$).

Proposition 3.9. *Let P be a normal program and $\mathbf{v} \in \mathbb{R}^n$ a vector representing $\{\top\} \subseteq I \subseteq B_{P^\varepsilon}$. Then I satisfies IC_P iff $\mathbf{v}_{\varepsilon B_P} \leq \mathbf{v}_{B_P}$.*

Proof. Let $\mathbf{v}_{B_P} = (a_1, \ldots, a_m)^\mathsf{T}$ and $\mathbf{v}_{\varepsilon B_P} = (b_1, \ldots, b_m)^\mathsf{T}$ $(m \leq n)$. Then, $\mathbf{v}_{\varepsilon B_P} \leq \mathbf{v}_{B_P}$ iff $b_i = 1$ implies $a_i = 1$ $(1 \leq i \leq m)$ for any $\mathrm{row}_i(\mathbf{v}_{\varepsilon B_P}) = \varepsilon p$ and $\mathrm{row}_i(\mathbf{v}_{B_P}) = p$ iff $\varepsilon p \in I$ implies $p \in I$. Hence, the result holds.   □

Theorem 3.10. *Let P be a normal program and $B_{P^\varepsilon} = \{p_1, \ldots, p_n\}$. Also let $U_{P^\varepsilon} \in \mathbb{R}^{n \times n \times k}$ be a tensor representation of a disjunctive program P^ε where k is the number of split programs of P^ε. Then*

1. *$\mathbf{m} = \mathbf{v}_{B_P}$ represents a stable model of P iff $\mathbf{v} \in \mathbb{R}^n$ represents a minimal model of P^ε and $\mathbf{v}_{\varepsilon B_P} \leq \mathbf{v}_{B_P}$.*
2. *P is inconsistent iff $\mathbf{v}_{\varepsilon B_P} \not\leq \mathbf{v}_{B_P}$ for any $\mathbf{v} \in \mathbb{R}^n$ representing a minimal model of P^ε.*

Proof. Since minimal models of P^ε are computed by Theorem 3.7, the result holds by Propositions 3.8 and 3.9.   □

Example 3.11. The program P^ε of Example 3.10 has four split programs $SP_1^\varepsilon = \{p \leftarrow, q \leftarrow\} \cup R$, $SP_2^\varepsilon = \{p \leftarrow, \varepsilon p \leftarrow\} \cup R$, $SP_3^\varepsilon = \{\varepsilon q \leftarrow, q \leftarrow\} \cup R$, and $SP_4^\varepsilon = \{\varepsilon q \leftarrow, \varepsilon p \leftarrow\} \cup R$ where $R = \{\varepsilon p \leftarrow p, \varepsilon q \leftarrow q\}$. Then P^ε is represented by the tensor $U_{P^\varepsilon} \in \mathbb{R}^{5 \times 5 \times 4}$. $FP(U_{P_\varepsilon} \bullet \mathbf{v}_0)$ is the matrix where $\mathbf{v} = (1,0,1,0,1)^\mathsf{T}$, $\mathbf{v}' = (0,1,0,1,1)^\mathsf{T}$ and $\mathbf{v}'' = (0,0,1,1,1)^\mathsf{T}$ are minimal (right).

$$
\begin{array}{c}
p \\
q \\
\varepsilon p \\
\varepsilon q \\
\top
\end{array}
\left(
\begin{array}{c|c|c|c}
1 & 1 & 0 & 0 \\
1 & 0 & 1 & 0 \\
1 & 1 & 0 & 1 \\
1 & 0 & 1 & 1 \\
1 & 1 & 1 & 1
\end{array}
\right)
$$

Then $\mathbf{v}_{B_P} = (1,0,1)^\mathsf{T}$ (representing p, q, $\top$) and). $\mathbf{v}_{\varepsilon B_P} = (1,0,1)^\mathsf{T}$ (representing εp, εq, $\top$) satisfy $\mathbf{v}_{\varepsilon B_P} \leq \mathbf{v}_{B_P}$. Also, $\mathbf{v}'_{B_P} = (0,1,1)^\mathsf{T}$ (representing p, q, $\top$) and $\mathbf{v}'_{\varepsilon B_P} = (0,1,1)^\mathsf{T}$ (representing εp, εq, $\top$) satisfy $\mathbf{v}'_{\varepsilon B_P} \leq \mathbf{v}'_{B_P}$. By contrast, $\mathbf{v}''$ does not satisfy $\mathbf{v}''_{\varepsilon B_P} \leq \mathbf{v}''_{B_P}$. Then $\mathbf{v}_{B_P} = (1,0,1)^\mathsf{T}$ and $\mathbf{v}'_{B_P} = (0,1,1)^\mathsf{T}$ represent two stable models $\{p\}$ and $\{q\}$ of P.

The result of this section is extended to *normal disjunctive programs* (i.e., programs containing both disjunction and negation) by combining techniques of Sects. 3.2 and 3.3.

4   Discussion

It is known that the least model of a Horn program is computed in $O(N)$ time and space where N is the size of the program, i.e., the total number of occurrences of literals in the program [2]. The proposed method requires $O(n^2)$ space and $O(n^4)$ time (i.e., $O(n^3)$ for matrix multiplication and at most n-iteration of powers) in the worst case where n is the number of propositional variables in B_P. Since the size of a matrix is independent of the size of a program, the proposed method would be advantageous in a large knowledge base on a fixed language. Several optimization techniques are considered for efficient computation in practice. When every element in the i-th row of a matrix M_P is zero except the element $a_{ij} = 1$ for $\mathrm{col}_j(\boldsymbol{M_P}) = \bot$, there is no rule deriving p_i in P. In this case, the i-th row and the i-th column can be removed from $\boldsymbol{M_P}$. For instance, let $\boldsymbol{M_P'}$ be the matrix obtained by removing the 4th row and the 4th column from $\boldsymbol{M_P}$ in Example 3.2. Then $\mathsf{FP}(\boldsymbol{M_P'} \bullet \boldsymbol{v_0})$ produces the least model of P. In particular, when a program P contains no facts (resp. constraints), the row and the column representing $\top$ (resp. $\bot$) could be removed from $\boldsymbol{M_P}$. In this way, we can reduce the size of vector spaces when a program is represented in a sparse matrix/tensor. A technique of dividing a program into subprograms [7] also helps to reduce the size of matrices/tensors in large scale of programs. Tensor LP can also be extended to representing *weight constraints* [12] for ASP. For instance, a weight constraint rule $p_i \leftarrow l \leq \{p_{j_1} = w_{j_1}, \ldots, p_{j_m} = w_{j_m}\}$ $(w_{j_k} \geq 0;$ $1 \leq k \leq m)$ is represented by a matrix $\boldsymbol{M_P} \in \mathbb{R}^{n \times n}$ such that $\mathrm{row}_i(\boldsymbol{M_P}) = p_i,$ $\mathrm{col}_j(\boldsymbol{M_P}) = p_j$ $(1 \leq i, j \leq n)$, and $a_{ij_k} = w_{j_k}/l$. Given a vector $\boldsymbol{v} = (b_1, \ldots, b_n)^{\mathsf{T}}$ representing an interpretation, p_i becomes true if the i-th element of $\boldsymbol{M_P} \bullet \boldsymbol{v}$ has a value $b_{j_1} w_{j_1}/l + \cdots + b_{j_m} w_{j_m}/l \geq 1$, thereby computed by $\boldsymbol{M_P} \underline{\bullet} \boldsymbol{v}$. In this way, weight constraints are effectively translated into linear algebraic calculation.

5   Conclusion

This paper introduced linear algebraic characterization of logic programs. The result of this paper bridges logic programming and linear algebraic approaches, which would contribute to a step for realizing logical inference in huge scale of knowledge bases. This paper focuses on theoretical aspects of LP in vector spaces. Future research includes implementation and evaluation of the proposed approach as well as development of techniques for robust and scalable inference in a huge scale of programs.

References

1. Brewka, G., Eiter, T., Truszczynski, M. (eds.): Special issue on answer set programming. AI Mag. **37**(3), 5–80 (2016)
2. Dowling, W.F., Gallier, J.H.: Linear-time algorithms for testing the satisfiability of propositional Horn formulae. J. Logic Program. **1**(3), 267–284 (1984)
3. Fernandez, J.A., Lobo, J., Minker, J., Subrahmanian, V.S.: Disjunctive LP + integrity constraints = stable model semantic. Ann. Math. AI **8**(3–4), 449–474 (1993)
4. Gelfond, M., Lifschitz, V.: The stable model semantics for logic programming. In: Proceedings of the 5th International Conference and Symposium on Logic Programming, pp. 1070–1080. MIT Press (1988)
5. Grefenstette, E.: Towards a formal distributional semantics: simulating logical calculi with tensors. In: Proceedings of the 2nd Joint Conference on Lexical and Computational Semantics, pp. 1–10 (2013)
6. Kolda, T.G., Bader, B.W.: Tensor decompositions and applications. SIAM Rev. **51**(3), 455–500 (2009)
7. Lifschitz, V., Turner, H.: Splitting a logic program. In: Proceedings of the 5th International Conference on Logic Programming, pp. 23–37. MIT Press (1994)

8. Liu, G., Janhunen, T., Niemelä, I.: Answer set programming via mixed integer programming. In: Proceedings of the 13th International Conference on Knowledge Representation and Reasoning, pp. 32–42 (2012)
9. Lloyd, J.W.: Foundations of Logic Programming, 2nd edn. Springer, Heidelberg (1987)
10. Nickel, M., Murphy, K., Tresp, V., Gabrilovich, E.: A review of relational machine learning for knowledge graphs. Proc. IEEE **104**(1), 11–33 (2016)
11. Rocktaschel, T., Bosnjak, M., Singh, S., Riedel, S.: Low-dimensional embeddings of logic. In: Proceedings of the ACL 2014 Workshop on Semantic Parsing, pp. 45–49 (2014)
12. Simons, P., Niemela, I., Soininen, T.: Extending and implementing the stable model semantics. Artif. Intell. **138**, 181–234 (2002)
13. Saraswat, V.: Reasoning 2.0 or machine learning and logic-the beginnings of a new computer science. Data Science Day, Kista, Sweden (2016)
14. Sato, T.: Embedding Tarskian semantics in vector spaces. In: Proceedings of the AAAI-2017 Workshop on Symbolic Inference and Optimization (2017)
15. Sato, T.: A linear algebraic approach to Datalog evaluation. TPLP **17**(3), 244–265 (2017)
16. Serafini, L., d'Avila Garcez, A.S.: Learning and reasoning with logic tensor networks. In: Adorni, G., Cagnoni, S., Gori, M., Maratea, M. (eds.) AI*IA 2016. LNCS, vol. 10037, pp. 334–348. Springer, Cham (2016). doi:10.1007/978-3-319-49130-1_25
17. van Emden, M.H., Kowalski, R.A.: The semantics of predicate logic as a programming language. JACM **23**(4), 733–742 (1976)
18. Yang, B., Yih, W.-T., He, X., Gao, J., Deng, L.: Embedding entities and relations for learning and inference in knowledge bases. In: Proceedings of the International Conference on Learning Representations (2015)

Representation Learning with Entity Topics for Knowledge Graphs

Xin Ouyang[1]($\boxtimes$), Yan Yang[1]($\boxtimes$), Liang He[1,2], Qin Chen[1],
and Jiacheng Zhang[1]

[1] Institute of Computer Applications, East China Normal University,
Shanghai, China
{xouyang,yyang,lhe9191,qchen,jchzhang}@ica.stc.sh.cn
[2] Shanghai Engineering Research Center of Intelligent Service Robot,
Shanghai, China

Abstract. Knowledge representation learning which represents triples as semantic embeddings has achieved tremendous success these years. Recent work aims at integrating the information of triples with texts, which has shown great advantages in alleviating the data sparsity problem. However, most of these methods are based on word-level information such as co-occurrence in texts, while ignoring the latent semantics of entities. In this paper, we propose an entity topic based representation learning (ETRL) method, which enhances the triple representations with the entity topics learned by the topic model. We evaluate our proposed method knowledge graph completion task. The experimental results show that our method outperforms most state-of-the-art methods. Specifically, we achieve a maximum improvement of 7.9% in terms of hits@10.

Keywords: Knowledge representation · Entity topics · Topic model · Knowledge graph completion

1 Introduction

Knowledge Graphs (KGs) aim at storing the facts of the real world with (h, r, t) triples, where h (or t) denotes the head (or tail) entity and r represents the relation between entities. Large-scale KGs such as Freebase [1] have been widely used in NLP tasks, including Question Answering (QA) [7]. However, with the increasing of triple amounts, KGs often suffer from data sparseness and incompleteness problems [11].

To alleviate these problems, Knowledge Representation Learning (KRL) which learns low-dimensional semantic embeddings of entities and relations has attracted extensive attention recently. Most of existing KRL methods learn the semantic representations from triples and show good performance in knowledge graph completion task. However, these methods cannot represent very well triples that suffer from sparsity. For example, more than 20% entities in FB15k [2]

G. Li et al. (Eds.): KSEM 2017, LNAI 10412, pp. 534–542, 2017.
DOI: 10.1007/978-3-319-63558-3_45

are associated with less than 20 triples, which leads to poor entity representations and diminishes the KRL performance. Therefore, recent researches turn to utilize the text information to enhance the performance of KRL [10,11]. In particular, Xie et al. [11] proposed DKRL model, which explored convolutional neural networks (CNN) for encoding entity descriptions. In [10], a TEKE model was proposed to learn the co-occurrence word networks for entities based on the Wikipedia pages. However, these methods only utilize the word-level information, and cannot well capture the latent topics of entities which are assumed to be useful to improve the KRL performance in this paper. For instance, "Agatha" is a playwright as demonstrated in the triple and has the descriptions as shown in Fig. 1. If a new entity "Joseph" has a description, namely "Joseph was an American satirical **novelist, short story writer** and **dramatist**", it is reasonable to infer that "Joseph" is also a playwright since he has the common topics like "writer" with "Agatha" and "dramatist" with "Playwright".

(**Agatha Christie**, /people/person/profession, **Playwright**)

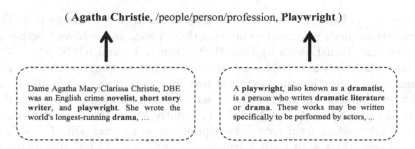

Dame Agatha Mary Clarissa Christie, DBE was an English crime **novelist, short story writer**, and **playwright**. She wrote the world's longest-running **drama**, ...

A **playwright**, also known as a **dramatist**, is a person who writes **dramatic literature** or **drama**. These works may be written specifically to be performed by actors, ...

Fig. 1. Example of triple and entity descriptions.

Motivated by the potential of the latent topics for KRL, we propose an entity topic based representation learning (ETRL) method in this paper. Specifically, we first use entity descriptions from KGs for topic modeling, which shows strong semantic correlations with entities. The Nonnegative Matrix Factorization (NMF) model is applied to learn the topic representations of entities. We propose our basic model (ETRL(basic)) that treats the obtained topic representations as constants, and integrates them with the triple representations, and then propose the advanced model (ETRL(adv)) that learns both topic and triple representations simultaneously. The contributions of our works are as follows:

1. We propose two variants of ETRL models, which integrates the topic information based on the entity descriptions with the triple structure for knowledge representation learning.
2. We utilize projection matrices in ETRL to map topic and triple representations into a united semantic space.
3. We evaluate our models on knowledge graph completion task. The experimental results show that our models outperforms most of the state-of-the-art models.

The outline of this paper is as follows. Section 2 elaborates the recent works related to our works. Section 3 reveals the details of our method. Section 4 evaluates our models and analyzes the results. Section 5 concludes our works.

2   Related Works

In this section, we will introduce several works aiming at embedding the entities and relations into low-dimensional continuous vector space to address the shortages of large-scale KGs mentioned above.

TransE [2], one of the most famous methods, has shown its advantages among the methods proposed before. For each correct (h, r, t) triple, TransE regards that $h + r \approx t$, which means that the entity h is translated to the entity t through the relation r. TransE defines the score function as shown in (1).

$$e(\mathbf{h}, \mathbf{r}, \mathbf{t}) = \|\mathbf{h} + \mathbf{r} - \mathbf{t}\|_2^2 \tag{1}$$

However, it has issues in modeling the 1-N, N-1 and N-N relations [10]. There are various methods proposed to address these issues and achieved better performance than TransE, such as TransH [9], TransR [6], and KB2E [4].

The methods based on triples are suffering from data sparsity problem. There are increasing methods fusing external text information and triple information to alleviate this problem and enrich the semantic information of representations. The model proposed by Zhong et al. [12] jointly learns the triple information and text information from entity descriptions in word-level. DKRL [11] learns text representations with deep learning method CNN. TEKE [10] memorizes the words co-occurring in contexts of entities from wikipedia pages and then constructs co-occurrence networks to bridge the KG and text-corpus. However, most of these methods disregard the entity topics which contain strong semantic relevance between entities. Although the state-of-the-art method TEKE [10] has achieved some success in discovering entity topics for regarding co-occurrence words as topics of entities, it is still limited by the performance of entity linking tools and the completeness of external text-corpus.

3   Methodology

In this section, we will present our ETRL method. The structure of our method is shown in Fig. 2. Firstly, we use entity descriptions in KGs as our text-corpus and learn entity topic representations with NMF [5]. Then, we construct ETRL(basic) model to incorporate topic embeddings into triple embeddings. In addition, we build ETRL(adv) model to jointly learn two embeddings simultaneously. Finally, we train our models with Adagrad method.

3.1   Notations and Definitions

For a given KG, We denote E as the set of entities and R as the set of relations. Meanwhile, $\mathbf{E}$, $\mathbf{R}$ is the set of embeddings of E, R respectively. Then, we denote

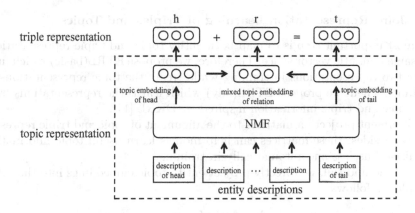

Fig. 2. The structure of ETRL(basic and adv).

T as the set of $(\mathbf{h}, \mathbf{r}, \mathbf{t})$ triples in KG where $\mathbf{h}, \mathbf{t} \in \mathbf{E}$ and $\mathbf{r} \in \mathbf{R}$. T is also the train set for our models. For a given text-corpus D, we denote W as the set of words arising in D and denote $\mathbf{v_e}$ as the topic embedding of entity $e \in E$. It is noteworthy that all embeddings $\mathbf{h}, \mathbf{r}, \mathbf{t}, \mathbf{v_h}, \mathbf{v_t} \in \mathbb{R}^k$, where k is the embedding vector size.

3.2 Topic Representation of Entities

We use entity descriptions as our text-corpus for it is much easier for us to obtain entity descriptions, and most entities in KGs are associated with their own descriptions which can be regarded as semantic supplements.

Considering about the performance, efficiency, and complexity of topic models [8], we adopt NMF [5] to learn topic representations of entities. Regarding each entity description as input document, NMF is defined as follows:

$$M = VS \tag{2}$$

where M is the $n \times m$ word-frequency matrix of entities preprocessed from entity descriptions, V is the matrix of entity topic representations while S is the matrix of word topic representations.

We select Euclidean distance defined in (3) as the convergence criterion of NMF to factorize matrix M into entity and word topic embeddings.

$$L_{nmf} = \sum_{i=1}^{n} \sum_{j=1}^{m} \|M_{i,j} - \mathbf{v_{e_i}} \mathbf{s_{w_j}}^\top\|_2^2 \tag{3}$$

where $M_{i,j}$ is the frequency of word w_j appearing in the description of entity e_i, $\mathbf{v_{e_i}}$ and $\mathbf{s_{w_j}}$ are the topic embeddings of entity e_i and word w_j respectively. Note that in NMF, for all $e_i \in E$ and $w_j \in W$ we have $\mathbf{v_{e_i}} \geq 0$ and $\mathbf{s_{w_j}} \geq 0$. We pre-train the NMF and obtain the entity topic representations in store for triple representation learning.

3.3 Joint Representation Learning of Triples and Topics

The most important step is to map both entity topic and triple representations into same semantic vector space. Therefore, we propose ETRL(basic) which integrates two representations in training while treats the topic representations as constants. Then, we propose ETRL(adv) which learns the representations from both topic and triple information inspired by DKRL [11].

We present projection matrices to the alignment of topic and triple representations. Besides, these matrices can help models learn useful topic information and discard useless one automatically in training.

For head and tail, we directly project entity topic embeddings into the space of triples as follows:

$$\mathbf{h}_\perp = \mathbf{v_h} M_e + \mathbf{h} \tag{4}$$

$$\mathbf{t}_\perp = \mathbf{v_t} M_e + \mathbf{t} \tag{5}$$

where $\mathbf{v_h}$ (or $\mathbf{v_t}$) is the entity topic embedding of head (or tail), $\mathbf{h}$ and $\mathbf{t}$ are the original entity representations in (1), and M_e is the $k \times k$ projection matrix.

To strengthen the capability of representing complex relations, we average the head and tail topic representations as relation topic representations $\mathbf{v_r}$. Then we define $\mathbf{r}_\perp$ as follows:

$$\mathbf{r}_\perp = \mathbf{v_r} M_r + \mathbf{r} \tag{6}$$

where M_r is the $k \times k$ projection matrix. It is worth noting that, for complex relations, different head-tail pairs may have different $\mathbf{v_r}$ to affect $\mathbf{r}_\perp$, thus improving the variousness and robustness of relation representations.

Then we replace $\mathbf{h}, \mathbf{r}, \mathbf{t}$ in (1) with $\mathbf{h}_\perp, \mathbf{r}_\perp, \mathbf{t}_\perp$ and redefine the score function for both two models as follows:

$$e_\perp(\mathbf{h}, \mathbf{r}, \mathbf{t}) = \|\mathbf{h}_\perp + \mathbf{r}_\perp - \mathbf{t}_\perp\|_2^2 \tag{7}$$

3.4 Loss Optimization and Training

ETRL(basic). To force the scores of correct triples as close as possible to zero and make the wrong's as far as possible, we define the loss function in (8).

$$L_s = \sum_{(h,r,t) \in T} \sum_{(h',r',t') \in T'} max(\gamma + e_\perp(\mathbf{h}, \mathbf{r}, \mathbf{t}) - e_\perp(\mathbf{h}', \mathbf{r}', \mathbf{t}'), 0) \tag{8}$$

where $max(\cdot, 0)$ is hinge loss, T is the train set consisting correct triples while T' is the set of negative samples denoted in (9).

$$T' = \{(h', r, t) | h' \in E\} \cup \{(h, r', t) | r' \in R\} \cup \{(h, r, t') | t' \in E\} \tag{9}$$

where h', r' and t' are selected by a certain probability to satisfy $\forall (\mathbf{h}, \mathbf{r}, \mathbf{t}) \in T'$, $(\mathbf{h}, \mathbf{r}, \mathbf{t}) \notin T$. We follow the'bern' method mentioned in TransH [9] to select the negative samples.

ETRL(adv). Inspired by the works of DKRL [11], we propose an advanced model to jointly learn both the topic and triple representations. The loss function is defined in (10).

$$L_a = L_s + L_{nmf} \tag{10}$$

where we optimize triple loss (8) and topic loss (3) simultaneously.

Training. We adopt Adagrad in the optimization for a better performance than SGD. In order to accelerate the speed of convergence, we initialize the entities and relations with the vectors produced by TransE and initialize the projection matrices as identify matrices.

4   Experiments

In this section, we will evaluate our models on knowledge graph completion task. Then we analyze results comparing with other state-of-the-art methods.

4.1   Datasets and Experiment Settings

Datasets. In this paper, we adopt FB15k dataset extracted from Freebase [1] as experimental dataset. FB15k is first proposed by the authors of TransE [2] and is widely used in KRL methods. The statistics of FB15k are listed in Table 1.

Table 1. Statistics of dataset.

Dataset	Entity	Relation	Train	Valid	Test
FB15k	14,951	1,345	483,142	50,000	59,071

Text-Corpus. We extract the English entity descriptions from the latest Freebase Dump [3] in FB15k as our text-corpus. The average number of words in each entity description is 124. Then we filter out the stop words, connect phrases which contain the entity names and remove the suffixes of the words in text-corpus.

Experiment Settings. We train our two models and set the best hyperparameters as follows: the number of iterations $iter = 1000$, embeddings size and topic numbers $k = 100$, margin $\gamma = 2$ and the learning rate scaling factor $\epsilon = 1 \times 10^{-7}$ by experience. We select TransE and two state-of-the-art methods, TEKE and DKRL as our baseline. We also compare our models to other methods such as TransH and TransR.

4.2   Knowledge Graph Completion

The task of knowledge graph completion (also called link prediction) aims at completing the triples which have missing components. For instance, given a triple (h, r, t) where head entity h is missing, the model firstly fill all entities $e \in E$ as candidate entities in the triple and then compute the score of triple (e, r, t) for each entity. Finally, it ranks the scores of candidate entities and completes the triple with the entity e which has the lowest score.

Two evaluation metrics [2] are adopted in this task: mean rank (MR) and hits@10. The former means the average rank of the correct one in the list of candidates while the latter means the rate of correct one ranked in the top 10. We also follow the two settings as "Raw" and "Filter" mentioned in TransE [2].

Entity Prediction Results. We remove the head and tail entities respectively of triples in test set and compute MR and hits@10. Then we take the average MR of heads and tails as the final results. The results listed in Table 2 show that our models especially ETRL(adv) outperform all the baseline except the mean rank(filter). By comparing with other experimental results listed in Table 2 we find that:

1. All evaluation metrics of our models are better than baselines which shows that the rich semantic information in entity topics can improve the performance. Furthermore, our models could be treated as the topic supplement for TransE and achieve 28.6% improvement than TransE in terms of hits@10.
2. Because TransR has more projection matrices [6], mean rank(filter) of TransR is better than our models. This shows that it is necessary to adopt more projection matrices for the learning of entities and relations.
3. The performance of ETRL(adv) is better than ETRL(basic). We can infer that joint models are more suitable for incorporating topic and triple embeddings with more smoothy influence.

Table 2. Entity prediction results on FB15k.

Metrics	Mean Rank		Hits@10	
	Raw	Filter	Raw	Filter
TransE	243	125	34.9	47.1
TransH	212	87	45.7	64.4
TransR	198	**77**	48.2	68.7
DKRL(CNN)	200	113	44.3	57.6
DKRL(CNN) + TransE	181	91	49.6	67.4
TEKE_TransE	233	79	43.5	67.6
ETRL(basic)	174	90	54.8	74.8
ETRL(adv)	**170**	83	**60.3**	**75.5**

Complex Relation Modeling Problem. Complex relation modeling problem, which leads to low performance in modeling 1-N, N-1 and N-N relations, is one of the most important problems in KRL. According to the unbalance of the ratio of head to tail, researchers divide relations into 1-1, 1-N, N-1 and N-N relations where the latter three are called complex relations. We test the ability of modeling complex relations of our models. The results listed in Table 3 show our models outperform others in modeling 1-N, N-1 and N-N relations. Based on the results, we believe that with the addition of projection matrices both on entities and relations, our models can have advantages in handling the complex relation modeling problem.

Table 3. Entity prediction for complex relation problem on FB15k.

Metrics	Hits@10			
	1-1	1-N	N-1	N-N
TransE	43.7	42.7	42.5	48.6
TransH	66.2	63.7	56.0	65.9
TransR	**79.0**	63.3	62.3	70.7
TEKE_TransE	47.6	61.2	63.8	76.5
ETRL(basic)	63.1	**66.5**	63.1	77.4
ETRL(adv)	62.5	65.0	**67.1**	**78.8**

5   Conclusion and Future Work

In this paper, to make full use of entity topic information in entity descriptions, we propose ETRL model using NMF to learn entity topic embeddings and mapping both topic and triple embeddings into same semantic space with projection matrices. Experimental results show that our models outperform most state-of-the-art methods which enhancing KRL with external text information. However, there also are some deficiencies to be solved in our future works. To lower the error brought by entity topic information, we will extend our models with more projection matrices to overcome the complex relations problem inspired by TransR [6]. We will extend our models to other state-of-the-art triple based representation models and try to improve the performance of them.

Acknowledgments. This work was supported by the National Key Technology Support Program (No. 2015BAH01F02), Shanghai Municipal Commission of Economy and Information Under Grant Project (No. 201602024), the Natural Science Foundation of Shanghai (No. 17ZR1444900)and the Science and Technology Commission of Shanghai Municipality (No. 15PJ1401700).

References

1. Bollacker, K., Evans, C., Paritosh, P., Sturge, T., Taylor, J.: Freebase: a collaboratively created graph database for structuring human knowledge. In: ACM SIGMOD International Conference on Management of Data, SIGMOD 2008, Vancouver, BC, Canada, June, pp. 1247–1250 (2008)

2. Bordes, A., Usunier, N., Garcia-Duran, A., Weston, J., Yakhnenko, O.: Translating embeddings for modeling multi-relational data. In: NIPS, pp. 2787–2795 (2013)
3. Google: Freebase data dumps. https://developers.google.com/freebase/data
4. He, S., Liu, K., Ji, G., Zhao, J.: Learning to represent knowledge graphs with gaussian embedding. In: CIKM, pp. 623–632. ACM (2015)
5. Lee, D.D., Seung, H.S.: Learning the parts of objects by non-negative matrix factorization. Nature **401**(6755), 788–791 (1999)
6. Lin, Y., Liu, Z., Sun, M., Liu, Y., Zhu, X.: Learning entity and relation embeddings for knowledge graph completion. In: AAAI, pp. 2181–2187 (2015)
7. Miller, A.H., Fisch, A., Dodge, J., Karimi, A., Bordes, A., Weston, J.: Key-value memory networks for directly reading documents. In: EMNLP (2016)
8. Stevens, K., Kegelmeyer, P., Andrzejewski, D., Buttler, D.: Exploring topic coherence over many models and many topics. In: EMNLP-CoNLL, pp. 952–961 (2012)
9. Wang, Z., Zhang, J., Feng, J., Chen, Z.: Knowledge graph embedding by translating on hyperplanes. In: AAAI, pp. 1112–1119. Citeseer (2014)
10. Wang, Z., Li, J.: Text-enhanced representation learning for knowledge graph. In: IJCAI, pp. 1293–1299. AAAI Press (2016)
11. Xie, R., Liu, Z., Jia, J., Luan, H., Sun, M.: Representation learning of knowledge graphs with entity descriptions. In: AAAI, pp. 2659–2665 (2016)
12. Zhong, H., Zhang, J., Wang, Z., Wan, H., Chen, Z.: Aligning knowledge and text embeddings by entity descriptions. In: EMNLP, pp. 267–272 (2015)

Robust Mapping Learning for Multi-view Multi-label Classification with Missing Labels

Weijieying Ren[1], Lei Zhang[2], Bo Jiang[2], Zhefeng Wang[1],
Guangming Guo[1], and Guiquan Liu[1(✉)]

[1] School of Computer Science and Technology,
University of Science and Technology of China, Hefei, China
{wjyren,zhefwang,guogg}@mail.ustc.edu.cn, gqliu@ustc.edu.cn
[2] School of Computer Science and Technology, Anhui University, Hefei, China
{zl,jiangbo}@ahu.edu.cn

Abstract. The multi-label classification problem has generated significant interest in recent years. Typical scenarios assume each instance can be assigned to a set of labels. Most of previous works regard the original labels as authentic label assignments which ignore missing labels in realistic applications. Meanwhile, few studies handle the data coming from multiple sources (multiple views) to enhance label correlations. In this paper, we propose a new robust method for multi-label classification problem. The proposed method incorporates multiple views into a mixed feature matrix, and augments the initial label matrix with label correlation matrix to estimate authentic label assignments. In addition, a low-rank structure and a manifold regularization are used to further exploit global label correlations and local smoothness. An alternating algorithm is designed to slove the optimization problem. Experiments on three authoritative datasets demonstrate the effectiveness and robustness of our method.

Keywords: Multi-label · Multi-view · Label correlations · Missing labels

1 Introduction

Multi-label learning occupies a decisive position in many domains such as automatic multi-media annotation [1] and image annotation [8]. In multi-label learning, each instance is associated with a set of labels which share similar semantic spaces. The major issue in multi-label classification is how to solve label correlations to improve the generalization performance [10].

In most of the existing studies on multi-label learning, a basic assumption is that all the proper and actual labels of each instance are given [1]. However, in practical applications, training labels are obtained via crowd-sourcing and we may not have access to true labels of each training sample [12]. To tackle this problem, various methods have been further proposed along different directions, including metric learning algorithm [10], a semi-supervised low-rank mapping

© Springer International Publishing AG 2017
G. Li et al. (Eds.): KSEM 2017, LNAI 10412, pp. 543–551, 2017.
DOI: 10.1007/978-3-319-63558-3_46

method [8], etc. However, as stated in [12], these methods based on modeling the original label matrix may not accurately capture the relations between labels and features due to the missing labels.

Recently, more works considered data coming from multiple views to solve missing label problems. A classic methodology pursued in multi-view learning considers a shared structure, which can generate multiple types of views. Existing works included a feature selection method [11] that enforced a sparsity-inducing norm on the concentrated features from multiple views; a multi-view matrix recovery method [9] which sought a shared low-rank feature representation, etc. The nuclear norm was often adopted to address the rank minimization problem. These algorithms emphasized the significance of a shared feature representation of multiple views, while overlooking label relevance. Meanwhile, a low-rank constraint was often restricted to obtain a robust subspace, and the relationships among nuclear norm and other norms (e.g. Frobenius norm) need to be further studied.

To tackle above challenges, we propose a Robust Label Mapping method to address the Multi-view Multi-label Classification problem with Missing Labels (abbreviated as RLM-MCML framework). In this paper, we integrate input data from multiple views into a mixed feature matrix and augment the initial label matrix with correlation matrix to obtain a more authentic label assignments. A group-structured sparsity norm is adopted on the weight matrix for the ease of feature selection. Then, a nuclear norm is imposed to exploit global label correlations, and the manifold regularization is introduced to enhance local smoothness. The major contributions of this paper are summarized as follows:

- Introducing a principle way of exploiting label correlations in the presence of noisy and incomplete labels.
- Adopting Frobenius norm as a substitution for the general low-rank constraint with theoretic analysis.
- Conducting experiments on three authoritative datasets with six metrics to demonstrate the robustness and effectiveness of the proposed RLM-MCML method.

2   The RLM-MCML Approach

Given a set of N training instances coming from K views where the training set for the i-th view is $D^i = \{X^i, Y\}$. $X^i \in \Re^{d_i}$ is the input matrix for the i-th view, $Y \in \Re^{N \times m}$ is the corresponding label matrix, and m is the total label size. To strengthen label correlations among multiple views, we adopt the method in [3] which integrates input data X^i of each view into a mixed feature matrix $X = [X^1; \ldots; X^k] \in \Re^{N \times d}$, with $d = \sum_{i=1}^{k} d_i$. If we apply the generally used least square loss function, the labeling approximation error of N instances given in the weight matrix $W \in \Re^{m \times d}$ can be written as:

$$\mathcal{L}(W) = \sum_{i=1}^{N} ||y_i - W x_i||_2^2, \tag{1}$$

Unfortunately, incomplete labels exist in realistic scenarios and only a partial set of labels are available. We can translate this problem into another two tangible guidelines: (1) authentic label assignments can be estimated via label correlations; (2) the points close to each other are more likely to share a label, and multi-view data can be leveraged to preserve the label structure.

To achieve this, we introduce label correlation matrix S and decompose this into two components to formulize the two assumptions:

$$\Psi(S) = \Psi_g(S) + \Psi_l(S)$$

where $\Psi_g(S)$ is to approximate the authentic labels from global way and $\Psi_l(S)$ utilizes multi-view data and label correlations to preserve the local geometry structure of labels. Inspired by the idea of label propagating [12], we also supply the initial label matrix Y with correlation matrix S to approximate the authentic label assignments $\tilde{Y}$, i.e., $\tilde{Y} \approx YS$. Notably, since some types of features are not useful to categorize certain specific classes, we imposed a group-sparsity norm on the weight matrix to select important features. Given a group structure $\phi = \{\Omega_1, \Omega_1, \ldots, \Omega_q\}$, x_{Ω_i} indicates a sub-vector of the features in the i-th group, d_i expresses weight scalar for the group Ω_i, and $||X||_{2,1}^{\Omega} = \sum_{i=1}^{q} d_i ||x_{\Omega_i}||_2$. A general classification model can be trained by solving the following problem:

$$\min_{S,W,E} \quad ||XW - YS||_2^2 + \lambda_1 ||W||_{2,1}^{\Omega} + \lambda_2 ||E||_F + \Psi_g(S) + \Psi_l(S)$$
$$\text{s.t.} \quad Y = YS + E, \tag{2}$$

where the Frobenius norm is used to control the difference matrix E between the initial label assignments and the targets. In the next subsections, the two basic stages of RLM-MCML, i.e., global structure of label correlations and local smoothness of label structure will be scrutinized respectively.

2.1 Global Structure of Label Correlations

The motivation of this part comes from the observation that similar labels exhibits strong label co-occurrence dependencies, hence the label set demonstrates a strong block structure [12]. A rank regularization is leveraged to encode global label correlations, which means relative labels are strongly correlated with each other while the label correlations should be shared by all the instances. The rank minimization problem is NP-hard, many works adopted its convex envelope, the nuclear norm to substitute the original problem:

$$\Psi_g(S) = ||S||_* \tag{3}$$

2.2 The Local Smoothness of Label Structure

Building off of the exploration of label correlations from global aspect, we can go one step further and preserve the label structure from local way. A smoothness assumption is usually adopted that the points close to each other are more likely to share a label. We can naturally induce from this assumption that the distance of two instances in their feature space can measure the similarities of their corresponding labels. Furthermore, if the initial two label vectors y_i and y_j are close in the intrinsic geometry of the label distribution, then the authentic label vector $\hat{y}_i$ and $\hat{y}_j$ are also close to each other. The local invariance assumption can be formulated via the manifold regularization:

$$\Psi_l(S) = \sum_{i,j}^{N} z_{i,j} \|\hat{y}_i - \hat{y}_j\|^2 = tr((YS)L(YS)^T), \tag{4}$$

where $L = D - Z$ is the Laplacian matrix, and D is a diagonal matrix whose main diagonal entries are column sums of Z, i.e., $d(i,i) = \sum_{j}^{N} z_{ij}$. Note that the heat kernel weight with self-tuning technique [8] is often adopted to measure the edge weight $z_{i,j} = exp(\frac{\|x_i - x_j\|^2}{\sigma})$, while this method is incapable of dealing with points near the intersection of two subspaces and also sensitive to the neighbor size [4]. In the following, we construct a low-rank graph to obtain the affinity matrix Z, which contains more comprehensive and complementary information than the neighborhood graph.

To guarantee the affinity matrix achieve good performance, we expect it to satisfy the following two assumptions: (1) each instances can be represented by its nearest neighbors; (2) similar instances can be clustered together. Inspired by the success of self-representation which has been widely used in subspace learning [4], we adopt this technique to encode the relationship between each instance and its neighbors. To further explore the comprehensiveness of features derived from multiple views, we seek a low-rank structure in the feature space to approximate the affinity matrix Z:

$$\min_{A,Z} \quad \|Z\|_* + \frac{\alpha}{2} \|X - A\|_F^2 \tag{5}$$
$$\text{s.t.} \quad A = AZ,$$

where A denotes the dictionary matrix which can be regarded as an approximation of the input matrix X. Based on the certification in [6], the minimal cost of Eq. (5) can be denoted as: $\mathcal{F} = r + \frac{\alpha}{2} \sum_{j>r} \sigma_j^2$. Where r, σ is the rank and singular value of the input matrix X, respectively. Here we also restrict the affinity matrix Z with Frobenius norm to substitute the nuclear norm.

$$\min_{A,Z} \quad \|Z\|_F^2 + \frac{\alpha}{2} \|X - A\|_F^2 \tag{6}$$
$$\text{s.t.} \quad A = AZ,$$

The minimal cost of problem (6) can be essentially roughly equal to the result in (5), while there is a closed-form solution in the former formulation. And the proof is provided in a longer version of this paper.

2.3 Problem Formulation

Combining global label correlation term and local label smoothness term, the final optimization formulation can be written as:

$$\min_{W,S,E} \quad ||XW - YS||_F^2 + \lambda_1||W||_{2,1}^{\Omega} + \lambda_2||E||_2$$
$$+ \lambda_3||S||_* + \lambda_4 tr((YS)L(YS)^T) \tag{7}$$
$$\text{s.t.} \quad Y = YS + E.$$

where $\lambda_1, \lambda_2, \lambda_3, \lambda_4$ are the non-negative model parameters. The proposed model is able to learn a label mapping function to approximate the authentic label assignments, and select significant features from multiple views to boost the classification results.

3 Experiments

3.1 Datasets

Pascal VOC dataset [5] comprises 9963 images which can be classified into 20 categorizations. In this paper, we chose three representative feature views: the global GIST [15], the local SIFT [14] and the tag information. The dimensions of GIST, SIFT, and tags are 512, 1000 and 804, respectively.

NUS-WIDE dataset [2] comprises 30000 images with 31 classes, which contains 225-dimension block-wise color moments (CM), 64-D color histogram (CH), 144-D color correlation (CoRR), 128-D wavelet texture (WT), 73-D edge distribution (EDH), and 500-D SIFT-based BoW histograms.

Mirflickr dataset [7] comprises 25000 instances with 38 classes, which contains 512-D global GIST, 1000-D local SIFT and 804-D tag information. We randomly sampled a subset of 12200 instances with 12 labels.

3.2 Evaluation Criteria and Algorithms

We apply 8 metrics [14] which are generally used in multilabel classification: *mAUC, Rank Loss (RL), Average Precision (AP), Hamming Loss (HL), Macro-F1, Micro-F1, Accuracy* and *Coverage*. Lower values indicate a better performance in terms of RL, HL and Coverage, while higher values sign better performance for the rest of criteria.

To demonstrate the performance of the proposed method, we adopt six state-of-the-art algorithms as our baselines: (1) ML-KNN [13]: the classic extension of KNN algorithm which generates a set of independent classifiers. (2) SLRM [8]: constructing a neighborhood graph, and learning a lowrank label mapping. (3) ML-LRC [12]: adopting a low-rank constraint to capture the label correlations globally. (4) Best Single View (B-SV): using the single view feature which achieved the best classification performance on the corresponding dataset.

(5) lrMMC [9]: proposing a low-rank multi-view matrix completion method.
(6) SMML [11]: integrating heterogeneous features by using the joint structured sparsity regularization. Since the former three methods are designed for single-view multi-label classification, we take the concatenation of all features as their input data. There are two variants of our proposed method: (1) **RLM-MCML**^{N} which constructs a shared low-rank affinity matrix for Eq. (5); (2) **RLM-MCML**^{F} which takes Frobenius norm as a surrogate for the nuclear norm. We run ML-KNN and lrMMC with the code provided by authors. All the parameters are set as what papers suggested or tuned by 5-fold cross-validation. Meanwhile, we set all the parameters from $[10^{-3}, 10^{-2}, \ldots, 10^2, 10^3]$ to get the best trade-off performance.

3.3 Classification Results

Experimental results are demonstrated in Table 1. Overall, our proposed method shows its significant performance on all the datasets under most criteria. Taking the advantage of comprehensive information derived from multiple views, RLM-MCML outperforms B-SV. Besides, lrMMC which seeks a low-dimensional subspace shared by individual of views, is inferior to other methods. There are two

Table 1. The comparison results (%) of six algorithms on three datasets with respect to 8 metrics. The best result is marked in bold. $\downarrow(\uparrow)$ implies the smaller (larger), the better.

Dataset	Algorithm	mAUC↑	RL↓	AP↑	HL↓	Macro-F1↑	Micro-F1↑	Accuracy↑	Coverage↓
NUS	MLKNN	0.7968	0.0827	0.8929	**0.1222**	0.5229	**0.8032**	**0.8778**	5.3269
	SLRM	0.8171	0.0823	0.8888	0.1520	0.4266	0.7199	0.8480	5.1820
	ML-LRC	0.7855	0.0874	0.8854	0.1448	0.4143	0.7482	0.8552	5.3015
	B-SV	0.7397	0.1170	0.8539	0.1645	0.4292	0.7238	0.8355	5.8507
	lrMMC	0.7748	0.1163	0.853	0.5048	0.4724	0.5589	0.4952	5.9018
	SMML	0.8179	0.0820	0.8897	0.1354	0.4986	0.7684	0.8646	5.1894
	RLM-MCML^N	0.8192	0.0819	0.8901	0.1506	0.4360	0.7238	0.8494	5.1885
	RLM-MCML^F	**0.8267**	**0.0791**	**0.8933**	0.1294	**0.5675**	0.7952	0.8706	**5.1583**
VOC	MLKNN	0.7890	0.1523	0.5971	0.0638	0.1724	0.3522	0.9362	4.0626
	SLRM	0.8588	0.1123	0.7244	0.0501	0.4727	0.5336	0.9499	3.3361
	ML-LRC	0.8296	0.1320	0.6751	0.0731	0.1678	0.3784	0.9076	3.7129
	B-SV	0.7893	0.1673	0.5685	0.0672	0.4443	0.4487	0.9456	4.4620
	lrMMC	0.7831	0.1997	0.4764	0.3263	0.1280	0.2366	0.4737	5.2408
	SMML	0.8639	0.1095	0.7270	0.0487	**0.5091**	**0.5822**	0.9513	3.2620
	RLM-MCML^N	0.8646	0.1082	0.7301	0.0497	0.4735	0.5387	0.9503	**3.2192**
	RLM-MCML^F	**0.8655**	**0.1067**	0.7411	**0.0476**	0.5048	0.5609	**0.9524**	3.2400
Mir	MLKNN	0.7738	0.1978	0.8034	0.2706	0.6175	0.6581	0.7294	6.8030
	SLRM	0.8166	0.1681	0.8305	0.2386	0.6845	0.7116	0.7614	6.6005
	ML-LRC	0.8117	0.1689	0.8328	0.2508	0.7055	0.7291	0.7492	6.3856
	B-SV	0.7857	0.1921	0.8085	0.2768	0.6825	0.7027	0.7232	6.8355
	SMML	0.8161	0.1694	0.8296	**0.2381**	0.6855	0.7117	**0.7619**	6.6252
	lrMMC	0.7398	0.2560	0.7296	0.5769	0.5836	0.5946	0.4231	7.5630
	RLM-MCML^N	0.8254	0.1699	0.8290	0.2470	0.6927	0.7111	0.7530	6.3323
	RLM-MCML^F	**0.8275**	**0.1647**	**0.8340**	0.2508	**0.7134**	0.7419	0.7492	**6.3305**

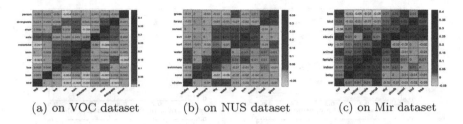

(a) on VOC dataset	(b) on NUS dataset	(c) on Mir dataset

Fig. 1. The visualization of low-rank property on three datasets.

possible reasons: (1) the shared subspace may drop the discriminative yet critical components contained in each view. (2) the ignorance of label correlations makes the prediction inaccurate. We also observe that ML-KNN gains advantages over our methods on NUS dataset, w.r.t. HL, Micro-F1, and Accuracy. Because the top 13 selected labels lead to various instances containing the same label, and in this case instances will share similar labels with its neighbors. It indicates that SMML do improve the predictive performance on VOC dataset. This observation stresses the importance of feature selection.

Besides, we select 10 representative classes from three datasets individually, and visualize the low-rank structure of label correlations in Fig. 1. Instance points sampled from three datasets all exhibit a strong block structure which confined with our assumption. Specially, as shown in Fig. 1(c), the label correlations in Mir dataset are more compact than the other two datasets.

3.4   Impact of Missing Labels

In realistic scenarios, its barely possible to obtain complete label assignments. An important question is that whether comparison algorithms, especially for our proposed method, can handle the incomplete labels problem effectively? To answer this question, we randomly dropped out different ratios of missing labels, from 10% to 50%, with 10% as an interval. For the space limitation, we only present the experiments on the Mir dataset under four assessment criteria (i.e., Macro-F1, Micro-F1, Hamming Loss and Accuracy). In order to make sure the figure is displayed more clearly, we employ RLM-MCML^N method as our representative algorithm. All the experiments are run with 5-fold crossvalidation, and the average results are shown in Fig. 2.

It is obvious that proposed method achieves comparable or better performances across all the datasets. Its noteworthy that when the ratios of missing labels are increased, the prominent performance of our method gradually reveals. For example, when we mask 10% of labels, the performance of **RLM-MCML**^{N} is inferior to SMML. Eventhough **RLM-MCML**^{N} shows relative improvements of 2.2%, 3.3%, and 5.3% w.r.t 30%, 40% and 50% of missing labels.

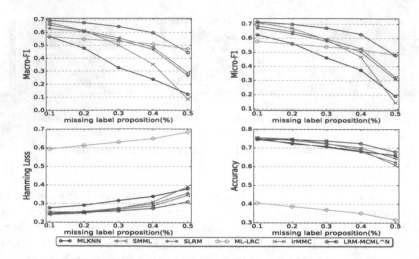

Fig. 2. Performance for missing labels on the Mir dataset.

4   Conclusions

This work proposed a novel model to handle the problem of multi-view multi-label learning with missing labels. The proposed method leveraged label correlations to estimate authentic targets and exploit label correlations from both global and local perspective. The affinity matrix appeared in the local smoothness term was also built with the Frobenius norm to substitute nuclear norm.

Acknowledgement. This work was partially supported by the Natural Science Foundation of China (Grant No. 61502001) and by the Academic and Technology Leader Imported Project of Anhui University (No. J01006057).

References

1. Boutell, M.R., Luo, J., Shen, X., Brown, C.M.: Learning multi-label scene classification. Pattern Recogn. **37**(9), 1757–1771 (2004)
2. Chua, T.-S., Tang, J., Hong, R., Li, H., Luo, Z., Zheng, Y.: NUS-WIDE: a real-world web image database from National University of Singapore. In: CIVR, p. 48. ACM (2009)
3. Deng, C.: Lv, Z., Liu, W., Huang, J., Tao, D., Gao, X.: Multi-view matrix decomposition: a new scheme for exploring discriminative information
4. Elhamifar, E., Vidal, R.: Sparse subspace clustering: algorithm, theory, and applications. TPAMI **35**(11), 2765–2781 (2013)
5. Everingham, M., Van Gool, L., Williams, C.K., Winn, J., Zisserman, A.: The Pascal visual object classes (VOC) challenge. IJCV **88**(2), 303–338 (2010)
6. Favaro, P., Vidal, R., Ravichandran, A.: A closed form solution to robust subspace estimation and clustering. In: CVPR, pp. 1801–1807. IEEE (2011)
7. Huiskes, M.J., Lew, M.S.: The MIR flickr retrieval evaluation. In: ACM MM, pp. 39–43. ACM (2008)

8. Jing, L., Yang, L., Yu, J., Ng, M.K.: Semi-supervised low-rank mapping learning for multi-label classification. In: CVPR, pp. 1483–1491 (2015)

9. Liu, M., Luo, Y., Tao, D., Xu, C., Wen, Y.: Low-rank multi-view learning in matrix completion for multi-label image classification. In: AAAI, pp. 2778–2784 (2015)

10. Verma, Y., Jawahar, C.V.: Image annotation using metric learning in semantic neighbourhoods. In: Fitzgibbon, A., Lazebnik, S., Perona, P., Sato, Y., Schmid, C. (eds.) ECCV 2012. LNCS, vol. 7574, pp. 836–849. Springer, Heidelberg (2012). doi:10.1007/978-3-642-33712-3_60

11. Wang, H., Nie, F., Huang, H., Ding, C.: Heterogeneous visual features fusion via sparse multimodal machine. In: CVPR, pp. 3097–3102 (2013)

12. Xu, L., Wang, Z., Shen, Z., Wang, Y., Chen, E.: Learning low-rank label correlations for multi-label classification with missing labels. In: ICDM, pp. 1067–1072. IEEE (2014)

13. Zhang, M.-L., Zhou, Z.-H.: ML-KNN: a lazy learning approach to multi-label learning. Pattern Recogn. **40**(7), 2038–2048 (2007)

14. Zhang, M.-L., Zhou, Z.-H.: A review on multi-label learning algorithms. IEEE TKDE **26**(8), 1819–1837 (2014)

Fast Subsumption Between Rooted Labeled Trees

Olivier Carloni[(✉)] [iD]

Sem Spirit, Montpellier, France
semspirit.contact@gmail.com
http://www.semspirit.com/

Abstract. This paper presents two data structures designed to efficiently query a set of rooted labeled trees (forest) defined in a language based on a relational vocabulary Σ and provided with a set-theoretic semantics and a subsumption relation matching the existential conjunctive fragment of the description logic $\mathcal{ALC}$. Given a tree query with q nodes and a forest with n nodes, after showing the equivalence between subsumption and homomorphism, an $O(q \cdot n)$ algorithm is proposed to compute all homomorphisms/subsumptions from the query to the forest. Then, are presented the two search data structures for faster homomorphism/subsumption retrieval. The first one provides a query time of $O(q)$ for a structure size of $O(2^n)$; and the second one provides a trade-off between the query time of $O(k^2 \cdot q)$ and the structure size of $O(k^2 \cdot |\Sigma| \cdot 2^{\lceil n/k \rceil})$, for a fixed integer k.

1 Introduction

The multiplication of data sources and the raise of data volumes have led to the emergence of many efficient search data structures. The first approaches formerly introduced in [10, 11] and still inspiring researchers, are the *trie* (or digital tree) and the binary search tree, each of them providing query time in $O(q)$ and $O(q \cdot \log n)$ for a storage size of $O(n)$ and $O(n \cdot \log n)$ where the query is a sequence of q bits and the data set consists in n bit sequences of fixed size. Those two data structures were the starting point for multidimensional range search data structures. Progressively, it emerged that to provide faster query time, the size of the data structure needs to increase exponentially with the tuples dimension, leading to the so-called *curse of dimensionality*. Thus, proposed search data structures were efficient for a small dimension d: as kD-trees [2] ($O(n^{1-1/d})$ time for $O(n)$ size), range-trees [3] ($O(\log^d n)$ time for $O(n \log^{d-1} n)$ size), and quadtrees [7]. More recently, skip lists and quadtrees were combined to propose an innovative data structure [6] that guarantees with some probability the correctness of the query result set. Since tuples are closed to trees, similar issues have been raised in the fields of hierarchical data management and graph theory. In particular, an efficient method was introduced to query XML documents [4] or graphs [9] with queries defined in the *twig* formalism that captures a small useful fragment of the XML query language. Automata theory has also led to some innovative

© Springer International Publishing AG 2017
G. Li et al. (Eds.): KSEM 2017, LNAI 10412, pp. 552–559, 2017.
DOI: 10.1007/978-3-319-63558-3_47

methods, as the one proposed in [8] that uses pushdown-automata to search a rooted tree query pattern within a rooted data tree in a query time linear in the given tree pattern.

The work presented here is also based on automata theory, with the difference that our automata are *word* finite states automata. The main contribution of the paper is the definition of two data structures, both used to store a forest of n nodes whose trees are defined in a knowledge representation (KR) language relying on a relational vocabulary Σ and equipped with a set-theoretic semantics and a subsumption relation matching the existential conjunctive fragment of the description logic $\mathcal{ALC}$. The two data structures are supplied with an efficient interrogation mechanism that complies with the semantics. Given a query of size q, the first one provides $O(q)$ query time for a size of $O(2^n)$; and the second one makes possible a trade-off depending on k between a query time of $O(k^2 \cdot q)$ and a size of $O(k^2 \cdot |\Sigma| \cdot 2^{\lceil n/k \rceil})$. The paper is organized in three parts: the first section presents the tree KR language, its concrete and abstract syntaxes with the set-theoretic semantics and subsumption relation. Then, is introduced the labeled tree homomorphism (shown equivalent to the subsumption) with a linear time algorithm. The second section defines the notion of tour language of a tree, and establishes the equivalence between tour language containment and homomorphism/subsumption. Finally, the third section makes use of the automata theory machinery to build the aforementioned data structures.

2   Trees, Subsumption and Homomorphism

2.1   Concrete Trees

Given a relational vocabulary Σ, the concrete representation of a labeled rooted tree T (in short *concrete tree* or *tree*) defined on Σ is a tuple $T = (N, r, s)$ where N is the non-empty set of nodes, $r \in N$ is the root and $s : N \times \Sigma \to 2^N$ is the successor function between nodes of N. A node $n \in N$ is a λ-*successor* of a node $p \in N$ iff $n \in s(p, \lambda)$. The successor function s is defined such that the node r is the root and s does not contain any cycle. $|T|$ denotes the size in nodes of the tree T and identity between trees is defined by means of tree isomorphism.

2.2   Abstract Syntax

A string a is an abstract representation of a tree (in short *abstract tree*) defined on the relational vocabulary Σ iff either a is the empty string, $a = \lambda X \varnothing$ or $a = XY$ where $\lambda \in \Sigma$, $\varnothing$ is a special symbol not in Σ and X, Y are abstract trees defined on Σ. Let $T = (N_T, r_T, s_T)$ be a concrete tree, an *abstract representation* α_T of T is either the empty string if N_T is a singleton; or else an abstract tree in the form $\alpha_T = \lambda \alpha_u \varnothing \alpha_V$ such that (1) u is one λ-successor subtree of r_T; and (2) V is the concrete tree resulting from the removal of u among the λ-successor subtrees of r_T in T. Let a be an abstract tree defined on Σ, the *concrete representation* $\gamma(a)$ of a is the concrete tree $\gamma(a) = (N_a, r_a, s_a)$ such that: (1) if $a = \lambda x \varnothing Y$

where x and Y are (possibly empty) abstract trees then $\gamma(a)$ is the copy of $\gamma(Y) = (N_Y, r_Y, s_Y)$ such that $r_a = r_Y$ and $\gamma(x)$ belongs to the λ-successor subtrees of r_a; else (2) if a is the empty string then $N_a = \{r_a\}$ is a singleton and s_a is undefined for the root r_a which is the unique node of $\gamma(a)$.

Semantics. An interpretation structure I of a relational vocabulary Σ is a tuple $I = \langle D, i \rangle$ where D is a non-empty set, called the domain and i is an interpretation function that maps every (relational) symbol from Σ to a subset of $D \times D$; and every abstract tree a defined according to Σ to a subset of D such that (1) if a is empty $i(a) = D$, else (2) if $a = \lambda x \varnothing Y$ then $i(a) = i(Y) \bigcap \{o \in D | (o, o') \in i(\lambda) \, and \, o' \in i(x)\}$ where Y and x are abstract trees defined on Σ. This interpretation is equivalent to the one given to the existential conjunctive fragment of the description logic $\mathcal{ALC}$ [1] (i.e. the fragment limited to existential restriction and intersection operators). The equivalence is proven by considering Σ as role names and by defining a bijective function f that recursively translates an abstract tree $a = \lambda x \varnothing Y$ into its equivalent $\mathcal{ALC}$ concept definition $f(a) = (\exists \lambda f(x)) \sqcap f(Y)$. Let a and b be abstract trees, we say that a *subsumes* b (b is subsumed by a), written $b \sqsupseteq a$ ($a \sqsubseteq b$), iff $i(b) \subseteq i(a)$ for all interpretation structures $I = \langle D, i \rangle$ of Σ.

2.3   Rooted Labeled Tree Homomorphism

A rooted labeled tree homomorphism h from a tree A into a tree B is a function $h : N_A \to N_B$ from the nodes of A into the nodes of B that maps the roots $h(r_A) = r_B$ of the trees and preserves the successor function such that for all $\lambda \in \Sigma$ and $n, p \in N_A$ if n is a λ-successor of p in A then $h(n)$ is a λ-successor of $h(p)$ in B. $h(A)$ denotes the subtree in B that is the image of A by h.

Algorithm. As shown in [12], given a labeled graph with n nodes and a labeled tree with m nodes, there is an algorithm that computes all homomorphisms from the tree into the graph in $O(mn)$ time. Restricting this problem to two trees A and B, the set $\mathcal{H} = \{h | h = A \to B\}$ of all homomorphisms from A to B can be computed in time $O(|A| \cdot |B|)$ with the two following steps.

   The first step consists in running the following procedure $homs(a, A, X, B)$ that returns the set $\{x \in X$ such that there is an homomorphism $h = A \to B$ and $h(a) = x\}$:

1. For all λ-successors a_i of the node a in A, for any arbitrary symbol $\lambda \in \Sigma$
2. $C_{x,a_i} \leftarrow \{x_j$ of λ-successors of $x\}$, for every $x \in X$
3. $X_i \leftarrow \bigcup_{x \in E} C_{x,a_i}$
4.   let R_i be the result of $homs(a_i, A, X_i, B)$
5. $D_{x,a_i} \leftarrow C_{x,a_i} \cap R_i$, for every $x \in X$
6. $X \leftarrow \{x \in X$ such that $D_{x,a_i} \neq \emptyset\}$
7. return X

The second step for enumerating all the homomorphisms h of $\mathcal{H}$, consists in keeping the D_{x,a_i} sets outside the function *homs*. Doing so, one can build each homomorphism h by browsing recursively the structure D_{x,a_i} in order to select an image $h(a_i')$ for each node a_i' of A fitting the one selected for its parent a'; following a recursive traversal of A from its root to its leaves. More precisely, $h \in \mathcal{H}$ is exhibited by choosing $h(a) \in homs(a, A, X, B)$, then recursively: if $h(a') = x$ and a_i' is a λ-successor of a' then choose $h(a_i') \in D_{x,a_i'}$ such that $h(a_i')$ is a λ-successor of x.

Subsumption and Homomorphism Equivalence. *Let a and b be two abstract trees, a subsumes b iff there exists an homomorphism $h = \gamma(a) \to \gamma(b)$ between their respective concrete representations $\gamma(a)$ and $\gamma(b)$.* This can be shown by recurrence. When a is empty, it is obvious. Then by using the recursive definitions of subsumption/homomorphism we prove that: given a λ symbol of a and its corresponding node n (which is a λ-successor of some other) in $\gamma(a)$, there is a symbol in b equal to λ establishing the subsumption (so far) iff its corresponding node n' in $\gamma(b)$ (which is a λ-successor of some other) is a valid image for n, establishing the homomorphism (so far).

3 Tour Language of a Tree

3.1 Automata Induced by a Tree

INFA *(resp IDFA)*. Given an alphabet Σ, an incomplete NFA or INFA *(resp. incomplete DFA or IDFA)* is a tuple $A = (Q, \Sigma, \delta, q_0, F)$ where Q is the set of states, δ the transition function $\delta : Q \times \Sigma \to 2^Q$ (which may not be total) *(resp. $\delta : Q \times \Sigma \to Q$)*, q_0 the initial state and F the final states set. Let δ^* be the function such that given a word $w = \lambda w'$ with $\lambda \in \Sigma$ $\delta^*(x, w) = \delta(x, \lambda)$ when $w' = \epsilon$ or otherwise $\delta^*(x, w) = \bigcup\{\delta^*(x', w') | x' \in \delta(x, \lambda)\}$ *(resp. $\delta^*(x, w) = \delta^*(\delta(x, \lambda), w')$)*. An INFA *(resp. IDFA)* A accepts a word w if $\delta^*(x, w) \cap F \neq \emptyset$ *(resp. $\delta^*(x, w) \in F$)* in a time $O(|w|.|A|^2)$ (resp. $O(|w|)$) and the language recognized by A is the set L_A of all words accepted by A. The notation of δ^* is extended to allow δ^* to take a set $X \subseteq Q$ as parameter such that $\delta^*(X, w) = \bigcup\{\delta^*(x', w) | x' \in X\}$.

Tours, Tour Language and Tour Automaton. A word t is a tour of a tree $T = (N, r, s)$ if t is a finite string and (1) either t is empty, (2) either $t = \lambda t' \emptyset t''$ such that t'' is a tour of T and t' is a tour of a subtree of T starting at one of the λ-successors of the root r of T. The tour language L_T of a tree T is the set of all tours of T. A tour t of a tree T is said to be *eulerian* iff $\gamma(t)$ is isomorphic to T. An abstract representation α_T of a tree T is an eulerian tour. Given a tree $T = (N, r, s)$, let $infa(T) = (Q, \Sigma', \delta, q_0, F)$ be the INFA induced by T as the INFA defined on the alphabet $\Sigma' = \Sigma \cup \{\emptyset\}$ such that $Q = N$, $q_0 = r$, $F = \{r\}$ and $\delta(x, \lambda) = X'$ and $\delta(x', \emptyset) = \{x\}$ for all $x' \in X'$ iff $s(x, \lambda) = X'$. *Tour language recognition:* Given a tree $T = (N, r, s)$, the INFA $infa(T)$ recognizes the tour language L_T of T.

3.2   Tour Languages, Containment and Homomorphism

By showing that a tour language L_T of a concrete tree T is equal to the set of every abstract tree t for which an homomorphism exists from $\gamma(t)$ to T, it can be established for two trees T_1 and T_2 that **(1)** *there exists an homomorphism* $h = T_1 \rightarrow T_2$ *iff the language containment* $L_{T_1} \subseteq L_{T_2}$ *holds; and* **(2)** *given an eulerian tour* e *of* T_1, $e \in L_{T_2}$ *iff* $L_{T_1} \subseteq L_{T_2}$.

Forest and Tour Language Union. Given a tree T and a forest B, h is an homomorphism $h = T \rightarrow B$ from T into B iff there exists a tree $U \in B$ such that h is an homomorphism $h = T \rightarrow U$ from T to U. The tour language L_B of a forest B is the union of the tour languages L_T of all the trees $T \in B$ in the forest. We extend the definition of $infa$ to forests: given a forest B, the INFA $infa(B)$ recognizing the tour language L_B of the forest B contains an initial state q_0 and the INFAs $infa(T_i)$ with initial states q_0^i of the trees $T_i \in B$ such that there is in $infa(B)$ a λ-transition from q_0 to a state x iff there is in $infa(T_i)$ a λ-transition from q_0^i to this state x. Given a forest B and a tree T, there exists an homomorphism $h = T \rightarrow B$ iff $L_T \subseteq L_B$.

4   Search Data Structures

Given a forest B and a tree T with abstract syntax t, we have: $t \in L_B$ iff $L_T \subseteq L_B$ iff there exists an homomorphism $h = T \rightarrow B$. Thus, the automaton $infa(B)$ accepts t iff there is an homomorphism $h = T \rightarrow B$. However checking for an homomorphism by using $infa(B)$ as well as the algorithm given in Sect. 2.3 is done in a time that strongly depends in the size $|B|$ of the forest. The main benefit of the two following search data structures is to decrease this dependency by reducing the non-determinism when testing membership in $infa(B)$.

4.1   Complete Determinization

The INFA $infa(B)$ is determinized into an IDFA $idfa(B)$ with the so-called *powerset construction* algorithm presented in [13]. Let $infa(B) = (Q_N, \Sigma, \delta_N, q_0, F_N)$ be the INFA for the forest B, the IDFA $idfa(B) = (Q_D, \Sigma, \delta_D, X_0, F_D)$ equivalent to $infa(B)$ is such that **(1)** $X_0 = \{q_0\}$ and $X_0 \in Q_D$, **(2)** if $\lambda \in \Sigma$ and $X \in Q_D$ then $\delta_D(X, \lambda) = \{y | x \in X \, and \, y \in \delta_N(x, \lambda)\}$ and $\delta_D(X, \lambda) \in Q_D$; and finally **(3)** for all $X \in Q_D$, X is an accepting state $(X \in F_D)$ in $idfa(B)$ iff at least one of its member x is an accepting state $(x \in F_N)$ in $infa(B)$. The size of $idfa(B)$ is $|idfa(B)| = O(2^{|B|})$.

Homomorphism Test. Given a forest B and a tree $A = (N, r, s)$ with abstract syntax (or eulerian tour) a, there is an homomorphism $h = A \rightarrow B$ iff a is accepted by $idfa(B)$. Checking if such an homomorphism h exists is done in time $O(|A|)$.

Homomorphisms Reconstruction. Given an abstract representation $a = \alpha_A$ of a tree A, let f_A be the function that maps an index $0 \leq i \leq |a| - 1$ of a symbol $a[i]$ in a with its corresponding node in A. f_A is not injective, thus one node x from A may have several antecedents indices and $f_A^{-1}(x)$ is a set. Let g_a be the bijective function that maps an index $0 \leq i \leq |a| - 1$ of a symbol in a with the state $\delta^*(q_0, a[0, i])$ of the run reached by the word $a[0, i]$. Now, μ_A is the function mapping each node x of A with a state $\mu_A(x)$ of the run such that $\mu_A(x) = g_a(i_x)$ where i_x is the largest index in $f_A^{-1}(x)$. Thus, $\mu_A(x)$ is the state of the run that recognizes in a the symbol at the farthest position i_x (from the start of the string) and that corresponds to the node $x = f_A(i_x)$. Each state s of the $idfa(B)$ reached by a string w represents a set of nodes $n(s)$ from B in bijection with the set S of states reached in $infa(B)$ by w. Let ω_s be the function partitionning the nodes of $n(s)$ according to their parents such that for a state s in $idfa(B)$ and a node x in B, $\omega_s(x) = n(s) \cap S_x$ where S_x is the set of successors of x. Let π be the function returning the parent node $p = \pi(x)$ in B of any given successor node x of p in B. The set of homomorphisms $\mathcal{H}$ from A into B can be reconstructed with the following procedure.

Initialisation Step. Let $x = f_A(|a| - 1)$ be the node in A corresponding to the last symbol in a leading to the accept state in $\mathcal{A}_B$. Every node x' of the domain $dom(\omega_s)$ of ω_s where $s = \mu_A(x)$ is an image of x according to an homomorphism from A to B. Thus we add in $\mathcal{H}$ as many partial homomorphisms h as there exist nodes $x' \in dom(\omega_s)$ such that $h(x) = x'$. These partial homomorphisms will be completed by the following loop until they become full homomorphisms from A to B. Let $\mathcal{H}'$ be an empty set. At each step, the following loop generates a new set $\mathcal{H}'$ containing further completed copies of the homomorphisms in $\mathcal{H}$; and at the end of the step: $\mathcal{H}'$ replaces $\mathcal{H}$.

Reconstruction Loop. Loop until $\mathcal{H}$ and $\mathcal{H}'$ become identical such that, for every homomorphism $h \in \mathcal{H}$ and every node $x \in A$ where $h(x) = x'$: (1) if x is a successor of p in A then we add in $\mathcal{H}'$ a copy h' of h such that $h'(p)$ is set to the unique parent node $\pi(x')$ of x' in B; (2) if y is a λ-successor of x in A then for all node $y' \in \omega_s(x')$ where $s = \mu_A(y)$ we add in $\mathcal{H}$ a copy h' of h such that $h'(y) = y'$.

Reconstruction Time. If π and ω are precomputed maps using a dichotomic method (as BST) to search the unique parent $\pi(x)$ or the children set $\omega(x)$ of a given node x in B, then the search in the maps is performed in $O(\log |B|)$. Since, the $idfa(B)$ accepts $a = \alpha_A$ in time $O(|A|)$ and at most $O(|A|)$ lookups are done in π and ω maps in order to build each homomorphism of $\mathcal{H}$; then supposing that there are K homomorphisms to be returned, the reconstruction time is $O(K.|A|.\log |B|)$.

4.2   Trade Off Between Space and Time

As shown in [5], given an integer k, an INFA of size n and an input word of size w, there exists a data-structure that provides a trade-off between its size of

$O(k^2 \cdot |\Sigma| \cdot 2^{\lceil n/k \rceil})$ and the time $O(w \cdot k^2)$ needed to simulate the INFA on the input word. This data-structure is an array storing k equal sized arbitrary partitions of the INFA such that inside each partition the simulation is deterministic. The remaining non-determinism occurs when multiple partitions have to be combined to compute the next step during an INFA simulation. Let X_i be k arbitrary subsets of size at most $\lceil n/k \rceil$ partitionning the n states of the INFA $infa(B)$ of the forest B. Each subsets $Y_i \subseteq X_i$, within each subset X_i, are indexed by integers from 0 to $2^{\lceil n/k \rceil}$. Let $\mathbb{D}$ be a function such that $\mathbb{D} : [0, k - 1] \times [0, k - 1] \times \Sigma \times [0, 2^{\lceil n/k \rceil}] \to [0, 2^{\lceil n/k \rceil}]$. An argument (i, j, λ, x_i) of $\mathbb{D}$ is composed of two values $0 \leq i \leq k - 1$ and $0 \leq j \leq k - 1$ identifying two subsets X_i and X_j of the partition, a symbol $\lambda \in \Sigma$ and the integer $0 \leq x_i \leq 2^{\lceil n/k \rceil} - 1$ indexing the subset Y_i of X_i. The value $\mathbb{D}(i, j, \lambda, x_i) \in [0, 2^{\lceil n/k \rceil}]$ is an integer indexing a subset Y_j of X_j such that a state s belongs to Y_j iff s belongs to X_j and there is a state in Y_i that transitions to s on input symbol λ. Let a be an abstract tree and b a sequence of k bits b_i. For the initialisation, X_{i_0} is the partition subset containing the initial state of $infa(B)$ and all bits of the sequence b are set to zero excepted the bit whose index corresponds to this initial state. Let $p = 0$ be the position of the symbol $a[p]$ currently read in a. Each integer value $\mathbb{D}(i, j, a[p], b)$ is the index of a subset of the states of the set X_j that can be reached by a transition on $a[p]$ from a state in the set Y_i identified by the sequence b as a subset of the set X_i. Let Z be the set of all integer values $\mathbb{D}(i, j, a[p], b)$ for every pair $i, j \in [0, k - 1]$ and let z be the result of the OR bitwise binary operation of all the indices in Z. Now, z is the bit sequence indexing the unique subset Y_j of the set of states X_j reachable in $infa(B)$ by the substring $a[0, p]$ of a. If $p < |a| - 1$ then increment p, set the sequence b equal to z and start again this procedure. Otherwise $p = |a| - 1$: in this case, the string a has been read till the end and the simulation is finished. If Y_j contains at least one accepting states from $infa(B)$ then a is accepted, otherwise it is rejected. The space required is the size for storing the function $\mathbb{D}$ which is at most the cardinal of $[0, k - 1] \times [0, k - 1] \times \Sigma \times [0, 2^{\lceil n/k \rceil}]$, and thus $O(k^2 \cdot |\Sigma| \cdot 2^{\lceil n/k \rceil})$. For fixed p, $\lambda = a[p]$ and b, each step of the simulation that reads a symbol of a checks the value $\mathbb{D}(i, j, \lambda, b)$ for all $(i, j) \in [0, k - 1] \times [0, k - 1]$. Thus each step takes time $O(k^2)$, and since there are as many steps as $|a|$ symbols in a then the total simulation time is $O(|a| \cdot k^2)$, which is equal to $O(|A| \cdot k^2)$ where A is the concrete representation of a.

5   Conclusion

The KR language presented in this paper provides concrete and abstract syntaxes for labeled rooted trees defined according to a relational vocabulary Σ. On the one hand, the abstract syntax is provided with a set-theoretic semantics and a subsumption relation matching the existential conjunctive fragment of the description logic $\mathcal{ALC}$. On the other hand, a notion of homomorphism is defined between concrete trees with a linear time algorithm; and shown to be equivalent to subsumption. Moreover, given a tree T with abstract syntax t and a forest

B, there is an homomorphism from T to B: (1) iff T subsumes a tree in B; and (2) iff t belongs to the set L_B of all tours of B. Making use of automata theory, the membership $t \in L_B$ can be checked by running on input string t the automata $infa(B)$ that recognizes the tour language L_B. The first data structure proposed in this paper is obtained by determinizing $infa(B)$ into a deterministic automaton of size at most $O(2^{|B|})$ that reads t in time $O(|t|)$. If t is accepted and K homomorphisms from T to B have to be returned, they can be reconstructed from the run in time $O(K \cdot |T| \cdot \log |B|)$. If the determinization can not be total (e.g. because of space limitations), it is possible to build a second data structure of size $O(k^2 \cdot |\Sigma| \cdot 2^{\lceil |B|/k \rceil})$ that simulates in time $O(k^2 \cdot |t|)$ a run of the automaton $infa(B)$ on input string t; providing a trade-off between time and size depending on the parameter k.

References

1. Baader, F., Horrocks, I., Sattler, U.: Description logics. In: Staab, S., Studer, R. (eds.) Handbook on Ontologies. IHIS, pp. 21–43. Springer, Heidelberg (2009). doi:10.1007/978-3-540-92673-3_1
2. Bentley, J.L.: Multidimensional binary search trees used for associative searching. Commun. ACM **18**(9), 509–517 (1975)
3. Bentley, J.L.: Decomposable searching problems. Inf. Process. Lett. **8**(5), 244–251 (1979)
4. Bruno, N., Koudas, N., Srivastava, D.: Holistic twig joins. In: Proceedings of the 2002 ACM SIGMOD International Conference on Management of Data, SIGMOD 2002. ACM Press (2002)
5. Eppstein, D.: Are there small machines which can efficiently match regular expressions? (Krishnaswami) (2010). https://cstheory.stackexchange.com/questions/11 32/are-there-small-machines-which-can-efficiently-match-regular-expressions/1273 #1273
6. Eppstein, D., Goodrich, M.T., Sun, J.Z.: Skip quadtrees: dynamic data structures for multidimensional point sets. Int. J. Comput. Geom. Appl. **18**(01n02), 131–160 (2008)
7. Finkel, R.A., Bentley, J.L.: Quad trees a data structure for retrieval on composite keys. Acta Inform. **4**(1), 1–9 (1974)
8. Flouri, T., Janoušek, J., Melichar, B., Iliopoulos, C.S., Pissis, S.P.: Tree template matching in ranked ordered trees by pushdown automata. In: Bouchou-Markhoff, B., Caron, P., Champarnaud, J.-M., Maurel, D. (eds.) CIAA 2011. LNCS, vol. 6807, pp. 273–281. Springer, Heidelberg (2011). doi:10.1007/978-3-642-22256-6_25
9. Gou, G., Chirkova, R.: Efficient algorithms for exact ranked twig-pattern matching over graphs. In: Proceedings of the 2008 ACM SIGMOD International Conference on Management of Data, SIGMOD 2008. ACM Press (2008)
10. Knuth, D.E.: Optimum binary search trees. Acta Inform. **1**(1), 14–25 (1971)
11. McClellan, M.T., Minker, J., Knuth, D.E.: The art of computer programming, vol. 3: sorting and searching. Math. Comput. **28**(128), 1175 (1974)
12. Mugnier, M.-L.: On generalization/specialization for conceptual graphs. J. Exp. Theor. Artif. Intell. **7**(3), 325–344 (1995)
13. Rabin, M.O., Scott, D.: Finite automata and their decision problems. IBM J. Res. Dev. **3**(2), 114–125 (1959)

Author Index

Printed in the United States
by Bookmasters